MW00843819

Newton's *Principia* for the Common Reader

NEWTON
Qui genus humanum ingenio superavit

Newton's *Principia* for the Common Reader

S. CHANDRASEKHAR

I rejoice to concur with the common reader; for by the common sense of readers, uncorrupted by literary prejudices, after all the refinements of subtilty and the dogmatism of learning, must be generally decided all claim to poetical honours.
Dr Samuel Johnson

CLARENDON PRESS · OXFORD
1995

Oxford University Press, Walton Street, Oxford OX2 6DP

Oxford New York
Athens Auckland Bangkok Bombay
Calcutta Cape Town Dar es Salaam Delhi
Florence Hong Kong Istanbul Karachi
Kuala Lumpur Madras Madrid Melbourne
Mexico City Nairobi Paris Singapore
Taipei Tokyo Toronto

and associated companies in
Berlin Ibadan

Oxford is a trade mark of Oxford University Press

Published in the United States
by Oxford University Press Inc., New York

First published 1995
Reprinted 1995

A catalogue record for this book is available from the British Library

Library of Congress Cataloging in Publication Data
Chandrasekhar, S. (Subrahmanyan), 1910–
Newton's Principia for the common reader / S. Chandrasekhar.
1. Newton, Isaac, Sir, 1642–1727. Principia. I. Title.
QA803.C48 1995 531—dc20 95-1800
ISBN 0 19 851744 0

Printed in Great Britain by Butler & Tanner Ltd, Frome, Somerset

For Lalitha

Contents

The quotations from Newton's *Principia* are from Florian Cajori's edition of Andrew Motte's English translation of 1729.

Acknowledgements

During the period of my writing this book (April 1992 to June 1994) I have received generous assistance from:

- Tristan Needham, for correspondence relating to the supplement to Chapter 6 (pp. 119–25) and for providing me with copies for the illustrations on pp. 119–21;
- Raghavan Narasimhan, for providing me with English translations of relevant extracts from the treatises of Tisserand, Laplace, and Lagrange;
- Valeria Ferrari and Andrea Malagoli, for the illustrations on pp. 176–9;
- Noel Swerdlow, for reading the first 21 Chapters for the overall perspective.

Of my indebtedness to Sotirios Persides and Åron Grant, I can make no adequate acknowledgements. Persides undertook the massive task of critically reading the entire manuscript and helping to eliminate many errors and obscurities. And Åron Grant, similarly, undertook the onerous task of checking the proofs in all stages for misprints both in the text and in the mathematical developments. But I am of course, responsible for whatever errors that may still remain.

I have been ever grateful to my wife, Lalitha, for her enduring patience and understanding during many hours of discussions and shared enthusiasms.

And finally, I am grateful to the Clarendon Press for their invariable courtesy and cooperation and for bringing to this book their surpassing excellence of craftsmanship and of typography.

Prologue

The present book derives from a study of the *Principia*, off and on for some years, that became earnest only latterly. And the opportunity I had to give two series of ten lectures each in Chicago (in 1990) and in Oxford (in 1991) strengthened the base of my understanding.

The manner of my study of the *Principia* was to read the enunciations of the different propositions, construct proofs for them independently *ab initio*, and then carefully follow Newton's own demonstrations. In the presentation of the propositions, the proofs that I constructed (which cannot substantially differ from what any other serious student can construct) often precede Newton's proofs arranged in a linear sequence of equations and arguments, avoiding the need to unravel the necessarily convoluted style that Newton had to adopt in writing his geometrical relations and mathematical equations in connected prose. With the impediments of language and of syntax thus eliminated, the physical insight and mathematical craftsmanship that invariably illuminate Newton's proofs come sharply into focus. On occasions, I provide supplementary comments and explanations, sometimes quoting from the masters of earlier centuries.

In the course of my study, I made no serious attempt to enlarge my knowledge derived from the *Principia* by any significant collateral reading. The book must, therefore, be assessed—for what it may be worth—as an undertaking by a practising scientist to read and comprehend the intellectual achievement that the *Principia* is.

I should add that on account of diverse constraints, I have not been able to study the entire *Principia* in the manner that I had adopted. I had to content myself, instead, to only those parts of the *Principia* that seemed to me in the direct line leading to Newton's formulation of his universal law of gravitation. However, in the last four Chapters, I consider a few additional matters that may give the reader a flavour of what the *Principia* contains besides.

❖1❖

The beginnings and the writing of the Principia

1. Introduction

Some acquaintance, with the antecedents of the less than two years—from the late autumn of 1684 to the early summer of 1686—in which Newton composed the entire *Principia*, is essential to a proper appreciation of the range and variety of the topics that are treated in it in depth and with rare perception. The literature on this 'pre-*Principia*' period is vast; but it is mostly not relevant to the purposes of this book as stated in the Prologue. And consistent with those purposes, the account that follows is a bare record of events that are not disputed.

In describing the origins of the *Principia* one distinguishes three epochs: the plague years 1665–66, 1679, and 1684.

2. The plague years

There is sufficient interlocking evidence that Newton did attempt a test of the inverse-square law of gravitational attraction during the plague years when he was mostly sojourning in Woolsthorpe. References to the test occur in the following accounts by Newton, Whiston, Pemberton, and Stukeley. (The underlining is by the author.)

Newton (in a memorandum in the Portsmouth collection, written in 1714?):

> And the same year [1666] I began to think of gravity extending to the orb of the Moon, and having found out how to estimate the force with which [a] globe revolving within a sphere presses the surface of the sphere, from Kepler's Rule of the periodical times of the Planets being in a sesquialternate proportion of their distances from the centers of their Orbs I deduced that the forces which keep the Planets in their Orbs must [be] reciprocally as the squares of their

> distances from the centers about which they revolve: and thereby compared the force requisite to keep the Moon in her Orb with the force of gravity at the surface of the earth, and found them answer pretty nearly. All this was in the two plague years of 1665 and 1666, for in those days I was in the prime of my age for invention, and minded Mathematicks and Philosophy more than at any time since.

(There is an alternative version in a letter Newton addressed to Pierre Des Maizeaux in 1718; but it is not substantially enough different from the foregoing to concern us here.)

W. Wiston (in his *Memoirs* published in 1749):

> Upon Sir Isaac's First Trial, when he took a Degree of a great Circle on the Earth's Surface, whence a Degree at the Distance of the Moon was to be determined also, to be 60 measured Miles only, according to the gross Measures then in Use. He was, in some Degree, disappointed, and the Power that restrained the Moon in her Orbit, measured by the versed Sines of that Orbit, appeared not to be quite the same that was to be expected, had it been the Power of Gravity alone, by which the Moon was there influenc'd. Upon this Disappointment, which made Sir Isaac suspect that this Power was partly that of Gravity, and partly that of Cartesius's Vortices, he threw aside the Paper of his Calculation and went to other Studies.

H. Pemberton (in his preface to the third edition of the *Principia*):

> Supposing therefore the power of gravity, when extended to the moon, to decrease in the same manner, he computed whether that force would be sufficient to keep the moon in her orbit. In this computation, being absent from books, he took the common estimate in use among geographers and our seamen, before Norwood had measured the earth, that 60 English miles were contained in one degree of latitude on the surface of the earth. But as this is a very faulty supposition, each degree containing about $69\frac{1}{2}$ of our miles, his computation did not answer expectation; whence he concluded, that some other cause must at least join with the action of the power of gravity on the moon. On this account he laid aside for that time any farther thoughts upon this matter.

W. Stukeley (in his Memoirs of Sir Isaac Newton's Life):

> After dinner, [on 15th April 1726] the weather being warm, we went into the garden and drank thea, under the shade of some appletrees, only he and myself. Amidst other discourse, he told me, he was just in the same situation, as when formerly, the notion of gravitation came into his mind. It was occasion'd by the fall of an apple, as he sat in a contemplative mood. Why should that apple

> always descend perpendicularly to the ground, thought he to himself. Why should it not go sideways or upwards, but constantly to the earths centre? Assuredly, the reason is, that the earth draws it. There must be a drawing power in the matter: and the sum of the drawing power in the matter of the earth must be in the earths center, not in any side of the earth. Therefore dos this apple fall perpendicularly, or towards the center. If matter thus draws matter, it must be in proportion of its quantity. Therefore the apple draws the earth, as well as the earth draws the apple. That there is a power, like that we here call gravity, which extends its self thro' the universe.

The apparent discrepancy between Newton's 'pretty nearly' as opposed to Whiston's 'in some degree disappointed' and Pemberton's 'his computations did not answer expectations' has been discussed at great length in the literature. But the discussion would seem to have become moot in view of the identification of the manuscript described by David Gregory in his account relating to his visit to Newton in Cambridge in 1694:

> I saw a manuscript [written] before the year 1669 (the year when its author Mr Newton was made Lucasian Professor of Mathematics) where all the foundations of his philosophy are laid: namely the gravity of the Moon to the Earth, and of the planets to the Sun. And in fact all these even then are subjected to calculation. I also saw in that manuscript the principle of equal times of a pendulum suspended between cycloids, before the publication of Huygens's *Horologium Oscillatorium.*

A translation of the manuscript in Gregory's account (identified as M.S. Add 3958 (5) folios, 87; 89 (left half)) has been provided by J. Herivel in his book *The background of Newton's Principia* (pp. 195–198). The analysis of this manuscript is presented below without the encumbrances of style and language of the original.

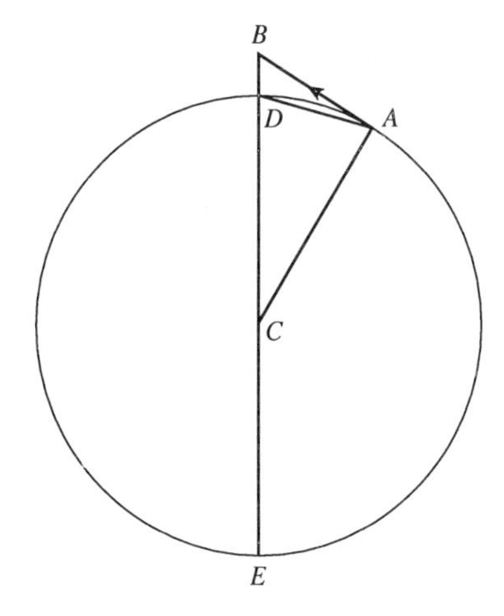

Let $DAED$ represent the assumed circular orbit of a radius R described by the Moon, with a uniform velocity v about the centre of the Earth at C. And let A be the instantaneous position of the Moon displaced by an infinitesimal amount from D and AB the direction of its motion; and finally let B be the point, on the prolongation of CD that the Moon would have arrived at after an interval of time $\mathrm{d}t$ if the radial attraction of the Earth had not acted. If a_{cc} denotes the gravitational attraction to which the Moon is subjected, then it follows from Galileo's law,

$$\tfrac{1}{2}a_{\mathrm{cc}}(\mathrm{d}t)^2 = BD. \qquad (1)$$

If T denotes the period of revolution,

$$\frac{\mathrm{d}t}{T} = \frac{AD}{2\pi R}; \qquad (2)$$

whence

$$\tfrac{1}{2}a_{\mathrm{cc}} = \frac{(2\pi R)^2}{T^2}\frac{BD}{AD^2}. \qquad (3)$$

But, by elementary geometry, $AB^2 = BD\,.\,BE$ or, neglecting quantities of the second order,

$$AD^2 \simeq BD\,.\,BE \sim BD\,.\,DE = BD\,.\,2R; \qquad (4)$$

and it follows from (3):

$$\tfrac{1}{2}a_{\mathrm{cc}} = \frac{(2\pi R)^2}{2R}\frac{1}{T^2}; \qquad (5)$$

that is, 'the required line (namely the third proportional of the circumference to the diameter) through which its [the Moon's] endeavour of receding from the centre would impel the body in the time of a complete revolution when applied constantly in a straight line' as Newton stated the result. Rewriting equation (5) in the form,

$$a_{\mathrm{cc}} = \frac{4\pi^2 R}{T^2}, \qquad (6)$$

and applying it to the Earth–Moon system, with R and T having their current values, we find

$$a_{\mathrm{cc}}(\mathbb{C}) = \frac{39{\cdot}48 \times 3{\cdot}84 \times 10^{10}}{(27{\cdot}32 \times 24 \times 3600)^2} \simeq 0{\cdot}272. \qquad (7)$$

The acceleration of gravity, g, on the surface of the Earth is

$$g = 980 = 3602 \times a_{\mathrm{cc}}(\mathbb{C}). \qquad (8)$$

Newton found for this ratio,

$$\frac{g}{a_{\mathrm{cc}}(\mathbb{C})} \simeq \text{'4000 and more'}. \qquad (9)$$

It is this discrepancy between ~ 3600 and $\sim 4000+$ that 'did not answer expectation' and 'disappointed' Newton. Pemberton and others have explained the discrepancy as a consequence of Newton having assumed the value ('absent from books'):

$$\text{number of miles per degree of latitude on Earth} = 60, \tag{10}$$

instead of the true value,

$$\text{number of miles per degree of latitude on Earth} = 69\tfrac{1}{2}, \tag{11}$$

and Newton's customary usage of 5000 ft (instead of 5280 ft) for a mile. With these two assumptions Newton *should* have found

$$\frac{g}{a_{cc}(\mathbb{C})} \simeq 3600 \times \left(\frac{69{\cdot}5 \times 5000}{60 \times 5280}\right)^2 \simeq 4332, \tag{12}$$

which he apparently 'rounded' to 4000+.

It should be noted that writing

$$\mathrm{d}t = \frac{AD}{v}, \tag{13}$$

instead of (2), Newton doubtless obtained for a_{cc} the value,

$$a_{cc} = 2\frac{BD}{AD^2}v^2 = \frac{v^2}{R} \qquad [\text{by } (4)], \tag{14}$$

a relation which, as he later explained in his Scholium to Proposition IV, could also be derived by considering a sequence of 'reflections from the circle at the several angular points' [of an inscribed polygon] and summing the 'force with which at every reflection it strikes the circle'. This method *may* have been his original derivation (as recorded in the 'Waste book'); but it is hardly to be doubted that Newton did not notice the simpler alternative derivation (14) at the time he wrote the manuscript that is presently being considered. In any event it is with the aid of the formula (14) that Newton deduced, as he stated, 'from Kepler's rule of the periodical times of the planets being in sesquialternate proportion of their distances from the centers of the Orbs that the forces which keep the planets in their Orbs must [be] reciprocally as the squares of their distances from the centers about which they revolve'. More explicitly, from the equation

$$T = \frac{2\pi R}{v}, \tag{15}$$

and 'Kepler's rule',

$$T \propto R^{3/2}, \tag{16}$$

we deduce

$$v^2 = \frac{4\pi^2 R^2}{T^2} \propto \frac{1}{R}; \tag{17}$$

or, by equation (14),

$$a_{cc} \propto R^{-2}. \tag{18}$$

In concluding this section, we may address ourselves to the further question that has often been asked and discussed in the literature: namely, why Newton did not pursue the matter further since after his return to Cambridge from Woolsthorpe he must have become aware of his erroneous assumption (10) and corrected for it. For my part, I accept the view of J. C. Adams quoted by Rouse Ball:

> On the other hand, the late Prof. Adams told me that he believed that Pemberton and Whiston were mistaken as to the insufficiency of the verification. Newton knew that the orbit was not actually circular, and that his numerical data were only approximate; hence he could have expected only a rough verification of the hypothesis, and as he asserted that he found his results agree or "answer pretty nearly," Prof. Adams considered that these calculations were sufficient to convince Newton that it was gravity alone that retained the moon in its orbit, and further he strongly suspected that Newton already believed that gravity was due to the fact that every particle of matter attracts every other particle, and that this attraction varied as the product of the masses and inversely as the square of the distance between them. Any opinion that Prof. Adams expressed on the subject must carry great weight, and the matter is one which may be fairly left to the judgment of the reader. Fortunately, the question whether Newton in 1666 came to the conclusion that gravity is only the chief cause (as Pemberton and Whiston imply), or whether he then came to the conclusion that it was the sole cause by which the moon is retained in its orbit, is comparatively unimportant, because there is no doubt as to what his conclusions ultimately were, and the question of the date when he convinced himself that gravity was sufficient by itself, and that the Cartesian vortices did not exist, is mainly a matter of antiquarian interest.

Besides, it is difficult for me to imagine that Newton with all his insight and perception did not realize that the assumption,

$$\frac{a_{cc}(\mathbb{C})}{g} = \left(\frac{\text{radius of the Earth}}{\text{radius of the Moon's orbit}}\right)^2, \tag{19}$$

implies that the Earth attracts objects on its surface as if its entire mass is concentrated at the centre—an assumption most emphatically against 'common sense' (unless one had known of its truth already). Newton was to prove the theorem in question in 1685 which he had not 'suspected' before the demonstration. It appears to me most likely that the untenability of the basic assumption underlying equation (19)—so Newton must have thought—discouraged him sufficiently to lay the entire matter aside. I shall return to this question in §5.

3. The year 1679

There appears to be little doubt that Newton did not pursue (after the plague years) the subject of gravitational attraction and planetary motions until the year 1679 though his interest was probably revived—albeit very briefly—in 1673 when he received his presentation copy of *De Horologio Oscillatorio* from Huygens and again in 1677, when, as Newton wrote to Halley on 27 May 1686, he had 'discoursed' with Sir Christopher Wren and Dr. Donne on 'this problem of determining the planetary motions upon philosophical principles'.

In 1679 Hooke initiated a correspondence with Newton and the two pairs of letters they exchanged at the time and their aftermath have been the subject of a vast literature. We shall desist from entering that thorny realm. It will suffice for our purposes to note only that, as a result of the correspondence, Newton's interest in dynamics was revived sufficiently for him to realize for the first time the real meaning of Kepler's law of areas. And as he wrote, 'I found now that whatsoever was the law of the force which kept the Planets in their Orbs, the area described by a radius drawn from them to the Sun would be proportional to the times in which they were described'; and he proved the two propositions that

> 'all bodies circulating about a centre sweep out areas proportional to the time'

and that

> 'a body revolving in an ellipse ... the law of attraction directed to a focus of the ellipse ... is inversely as the square of the distance',

as he was to state them in his *De Motu Corporum in Gyrum* written five years later (as Proposition 1 and Problem 3, Proposition 3, respectively). That Newton proved these propositions at this time is confirmed by his statement that he composed 'in December 1679, the 1st and the 11th Propositions' of Book I of the *Principia* (see §5 below).

The resurrection of Kepler's law of areas in 1679 was a triumphant breakthrough from which the *Principia* was later to flow. But meantime, Newton's interest lapsed again.

4. The year 1684

This was to be a fateful year. The salient facts are these: Halley visited Newton in Cambridge in August 1684 (or May 1684 as Herivel believes) to inquire what locus a body would describe under an inverse-square law of attraction. He was clearly unprepared for Newton's ready response that he had proved it to be an ellipse some years earlier. When asked for his demonstration, Newton was unable to find it among his papers; and he promised Halley to rework his proof and send it to him. Newton kept his promise and in November he sent through Dr. Edward Paget his reworked proposition and some additional propositions. Halley seems to have been so struck with the novelty and the originality of the propositions that he visited Newton once again in November and succeeded in persuading him to publish his results. For, as recorded in the minutes of the meeting of the Royal Society on 10 December 1684:

> Mr. Halley gave an account that he had lately seen Mr. Newton at Cambridge who had shewed him a curious treatise *de Motu*; which upon Mr. Halley's desire, was, he said, promised to be sent to the Society to be entered upon their register. Mr. Halley was desired to put Mr. Newton in mind of his promise for the securing of his invention to himself till such time as he could at leisure publish it.

The same story is told in more dramatic terms by Abraham De Moivre recalling, as he said, what Newton had related to him:

> In 1684 Dr Halley came to visit him at Cambridge, after they had been some time together, the D^{r} asked him what he thought the curve would be that would be described by the planets supposing the force of attraction towards the sun to be reciprocal to the square of their distance from it. Sr Isaac replied immediately that it would be an Ellipsis, the Doctor struck with joy & amazement asked him how he knew it, why saith he I have calculated it, whereupon Dr Halley asked him for his calculation without any farther delay, Sr Isaac looked among his papers but could not find it, but he promised him to renew it, & send it.
>
> Sr Isaac in order to make good his promise fell to work again but he could not come to that conclusion wch he thought he had before examined with care, however he attempted a new way which tho^{u} longer than the first, brought him again to his former conclusion, then he examined carefully what might be the reason why the calculation he had undertaken before did not prove right, & . . . he made both his calculations agree together.

The same story, in nearly identical terms, has been recorded by Conduit (see Plate 1 on p. 15).

What seems to have happened, between the two visits of Halley, is that Newton's interest in dynamics was set afire—a description by no means exaggerated as will presently appear—that he not only reworked the demonstrations of the two propositions that he had proved in 1679 but wrote an entire tract *De Motu Corporum in Gyrum* (the motion of revolving bodies) which went far beyond. There seems to be some doubt whether or not *De Motu* represents the substance of the lectures he gave during the Michelmas term of 1684. But there is considerable evidence that a slightly recast 'Version II' of the *De Motu Corporum* was what Paget took to London for transmission to Halley. (Translations of Version I and those parts of Versions II and III that differ from it are now available: J. Herivel, *The background of Newton's Principia.*)

Let us pause to look at the contents of *De Motu Corporum in Gyrum.* It consists of eleven propositions prefaced by three definitions, four hypotheses, and two lemmas. The eleven propositions are listed below; and the corresponding propositions in the *Principia* to which they correspond are noted in parentheses

(i) Kepler's law of areas (Book I, Proposition I).

(ii) The basic relations governing circular orbits described uniformly about the centre (Book I, Proposition IV; the relations in question are those given in equations (6) and (14) in §2).

(iii) The centripetal attraction under which a given locus can be described (Book I, Proposition VI).

(iv) Application to a body revolving in the circumference of a circle under attraction from a point on the circumference (Book I, Proposition VII of the first edition; the crucial Corollaries 1–3 appear for the first time in the second edition).

(v) An ellipse about its centre will be described under a centripetal attraction proportional to the distance (Book I, Proposition X).

(vi) An ellipse about a focus will be described under a centripetal attraction inversely as the square of the distance; and a Scholium on the application of the result to planetary motions (Book I, Proposition XI).

(vii) Kepler's third law and a Scholium on the application of the result of planetary motions (Book I, Proposition XV).

(viii) Given that the centripetal force is inversely as the square of the distance and given also its magnitude, to determine the ellipse which a body will describe when projected from a given point with a given velocity in an assigned direction (Book I, Proposition XVII).

(ix) The motion of a body falling radially towards the centre under a law of attraction inversely as the square of the distance (Book I, Proposition XXXII).

(x) The motion of a particle in a resisting medium, the resistance being proportional to the velocity, under no external force (Book II, Proposition II).

(xi) Same as (x) above but under a constant centripetal force (Book II, Propositions III and IV).

It is staggering—if not bewildering—to realize that all of the foregoing propositions were worked out—almost in the forms that they were later to be included in the *Principia*—in the interval between Halley's first visit in August (or May) and Paget's taking the manuscript to London in November not allowing for the time it must have taken to make the two versions in longhand. (Autograph photographic facsimile copies of these manuscripts are now available: *The preliminary manuscripts of Sir Isaac Newton's 1687 Principia*, 1684–1686: with an introduction by T. Whiteside, Cambridge, 1989.)

No wonder that Halley scurried back to Cambridge on seeing what Paget had brought him!

One additional comment: Newton seems to have corrected his 1666 'Moon test' (§2) before the summer of 1685 at the latest, for, as he writes in Version III of *De Motu* (see Herivel, p. 302),

> My calculations reveal that the centripetal force by which our Moon is held in her monthly motion about the Earth is to the force of gravity at the surface of the Earth very nearly as the reciprocal of the square of the distance [of the Moon] from the centre of the Earth.

To summarize: after Halley's first visit, Newton reconstructed the demonstrations of the two propositions (that he had mislaid) and added several more. The result was Version I of *De Motu Corporum in Gyrum*, composed in the 'Autumn' of 1684. Version II, a fair copy of Version I, was sent to Halley, through Paget, in November. During his second visit in November, Halley was able to persuade Newton to publish his discoveries; and his report to the Royal Society at its meeting on 10 December followed. Finally, as Newton's letter of 23 February 1685 to Aston (one of the then Secretaries of the Royal Society) attests, the *Propositions de Motu* was entered in the Register Book of the Royal Society:

> I thank you for entering in your Register my notions about motion. I designed them for you before now, but the examining several things has taken a greater part of my time than I expected, and a great deal of it to no purpose. And now I am to go into Lincolnshire for a month or six weeks. Afterwards I intend to finish it as soon as I can conveniently.

Newton was, at long last, earnestly embarked on writing his *Principia.*

5. The years 1685–1686: the writing of the *Principia*

The end of the Michelmas term of 1684 found Newton in the grip of what was to become his *Philosophiae Naturalis Principia Mathematica.*

Newton had originally thought of his projected book as no more than an expanded version of *De Motu Corporum in Gyrum.* The extant autograph copy of *De Motu Corporum Liber Primus* (deposited as his 'Lucasian Lectures' for the years beginning October 1684 and October 1685) suggests as much. But by early spring of 1685, the prospect had changed radically. This is confirmed by what Newton wrote to Aston on February 1685 (see §4), and also in his Preface to the first edition of the *Principia*:

> But after I had begun to consider the inequalities of the lunar motions, and had entered upon some other things relating to the laws and measures of gravity

> and other forces; and the figures that would be described by bodies attracted according to given laws; and the motion of several bodies moving among themselves; the motion of bodies in resisting mediums; the forces, densities, and motions, of mediums; the orbits of the comets, and such like, I deferred that publication till I had made a search into those matters, and could put forth the whole together.

The book, however, was finished and ready for press in less than two years. As stated by Brewster, quoting from a memorandum in 'Sir Isaac Newton's own handwriting':

> In the tenth proposition of the second book, there was a mistake in the first edition, by drawing the tangent of the arch GH from the wrong end of the arch, which caused an error in the conclusion; but in the second edition I rectified the mistake. And there may have been some other mistakes occasioned by the shortness of the time in which the book was written, and by its being copied by an amanuensis [Humphrey Newton] who understood not what he copied; besides the press faults, for I wrote it in seventeen or eighteen months, beginning in the end of December 1684, and sending it to the Royal Society in May 1686, excepting that about ten or twelve of the propositions were composed before, viz., the 1st and 11th in December 1679, the 6th, 7th, 8th, 9th, 10th, 12th, 13th, and 17th, Lib. I., and the 1st, 2d, 3d, and 4th, Lib. II., in June and July 1684.

And this is in agreement with what Pemberton wrote in his Preface to the third edition of the *Principia*:

> This treatise full of such a variety of profound inventions, was composed by him from scarce any other materials than the few propositions before mentioned, in the space of one year and an half.

It will be noted that the propositions, listed in the memorandum quoted by Brewster as having been composed in June and July 1684 are, with the exception of the relatively minor Propositions VIII, IX, XII, and XIII, the same as those we have noted in §4 as included in *De Motu Corporum in Gyrum*.

The significant event in the early spring of 1685, whose impact J. C. Adams and J. W. L. Glaisher fully recognized, is Newton's determination of the attraction of a spherical body ('everywhere similar, at every given distance from the centre, on all sides round about', as he was to describe later in the *Principia*) on any external point. What Glaisher said in his address on the occasion of the bicentenary of the publication of the *Principia* is worth quoting:

> No sooner had Newton proved this superb theorem—and we know from his own words that he had no expectation of so beautiful a result till it emerged from his mathematical investigation—than all the mechanism of the universe at once lay spread before him. When he discovered the theorems that form the first three sections of the Book I., when he gave them in his lectures of 1684, he was unaware that the sun and earth exerted their attractions as if they were but points. How different must these propositions have seemed to Newton's eyes when he realised that these results, which he had believed to be only approximately true when applied to the solar system, were really exact! Hitherto they had been true only in so far as he could regard the sun as a point compared to the distance of the planets or the earth as a point compared to the distance of the moon—a distance amounting to only about sixty times the earth's radius—but now they were mathematically true, excepting only for the slight deviation from a perfectly spherical form of the sun, earth, and planets. We can imagine the effect of this sudden transition from approximation to exactitude in stimulating Newton's mind to still greater efforts. It was now in his power to apply mathematical analysis with absolute precision to the actual problem of astronomy.

That Newton attached the greatest significance to the result that he had established is evident from what he says in Book III, Proposition VIII:

> After I had found that the force of gravity towards a whole planet did arise from and was compounded of the forces of gravity towards all its parts, and towards every one part was in the inverse proportion of the squares of the distances from the part, I was yet in doubt whether that proportion inversely as the square of the distance did accurately hold, or but nearly so, in the total force compounded of so many partial ones; for it might be that the proportion which accurately enough took place in greater distances should be wide of the truth near the surface of the planet, where the distances of the particles are unequal, and their situation dissimilar. But by the help of Prop. LXXV and LXXVI, Book I, and their Corollaries, I was at last satisfied of the truth of the Proposition, as it now lies before us.

Equally significant is what Newton wrote to Halley on 20 June 1686:

> I never extended the duplicate proportion lower than to the superficies of the earth, and before a certain demonstration I found last year, have suspected it did not reach accurately enough down so low; and therefore in the doctrines of projectiles never used it nor considered the motions of heavens.

Contrary to what has been commonly asserted, I share a view attributed to Adams and Glaisher that Newton's reluctance (even after 1679) to pursue his dynamical investigations arose from his dissatisfaction in not being able to prove or disprove—he probably did not try hard enough—a proposition on which the exactitude of the entire theory rests. It is scarcely to be thought of that a person of Newton's intellectual perception *and* standards did not realize the principal lacuna in his 'Moon test'. It is equally likely, in my view, that he was persuaded to write *De Motu Corporum in Gyrum* in 1684, after Halley's visit, by the furor that the questions were causing in London.

It is perhaps significant in this context that *De Motu Corporum, Liber Primus* (commonly considered as an early draft of Book I of the *Principia*) ends abruptly after his demonstration of Propositions XXXIX, XL, XLI, and XLIII (which are the key Propositions LXX, LXXI, LXXII, and LXIII of Book I). And equally significant, perhaps, is that no work sheets of the *Principia* beyond this point have been found. Is it so implausible that Newton destroyed them anticipating the cry of latter day historians, 'Whatever happened to the work sheets of the *Principia*?'

To resume the story: the manuscript of Book I went to press before 7 June 1686; for in the records of the meetings of the Royal Society for 2 June and 7 June we read, respectively:

> ... it was ordered, that Mr. Newton's book be printed, and that Mr. Halley undertake the business of looking after it, and printing it as his own charge; which he engaged to do.

and

> Ordered ... that the president be desired to license Mr. Newton's book intitled *Philosophia Naturalis Mathematica* and dedicated to the Society.

The rough manuscript of the second book was finished in the summer of 1685; but was, apparently, not written as a final copy for the press before 20 June 1686.

The preparation of the third book required knowledge of basic astronomical data with which Newton was not previously acquainted. He was fortunate that he could consult John Flamsteed on these matters. The rough manuscript was finished by June 1686. Except for the parts dealing with cometary motions, the manuscript for the third book was ready for press in March 1687.

The *Principia* containing all three books was published in July 1687. As Halley wrote to Newton on 5 July 1687,

> I have at length brought your book to an end and hope that it will please you.

It is fitting that we conclude this historical introduction with Newton's acknowledgment to Halley in his Preface to the first edition of the *Principia*:

> In the publication of this work the most acute and universally learned Mr. *Edmund Halley* not only assisted me in correcting the errors of the press and preparing the geometrical figures, but it was through his solicitations that it came to be published; for when he had obtained of me my demonstrations of the figure of the celestial orbits, he continually pressed me to communicate the same to the *Royal Society*.

To quote Conduit of Halley as

> The Ulysses who produced this Achilles.

*

joy & amazement asked him how he knew it,
why saith he I have calculated it, whereupon
Dr Halley asked him for his calculation without
any farther delay, Sr Isaac looked among his
papers but could not find it, but he promised
him to renew it, & then to send it him, Sr
Isaac in order to make good his promise fell to
work again, but he could not come to that conclusion
wch he thought he had before examined with care,
however he attempted a new way which tho' longer
than the first, brought him again to his former
conclusion, then he examined carefully what
might be the reason why the calculation he had
undertaken before did not prove right, & he
found that having drawn an Ellipsis coursely
with his own hand, he had drawn the two Axes
of the Curve, instead of drawing two Diameters
somewhat inclined to one another, whereby he
might have fixed his imagination to any two
conjugate diameters, which was requisite he
should do, that being perceived, he made both
his calculations agree together.

After this Dr Halley was (I think)
sent down to Cambridge by the Royal
Society to prevail with Sr Isaac to
print his discoveries wch gave rise
to the Principia.

Dr Halley has often valued himself to me
for having been the Ulysses who produced this
Achilles.

Plate 1 Copy of the original De Moivre memorandum, in the Joseph Halle Schaffner Collection of Scientific Manuscripts, Department of Special Collections of the University of Chicago Library; and published here with the permission of the Library.

❖2❖

Basic concepts: Definitions and Axioms

6. Introduction

Two lessons on *Definitions* and *Axioms* (or *Laws of Motion*) precede the formal opening of Book I: *The motion of bodies* of the *Principia.* In these two lessons, Newton formulates and elaborates the basic concepts and laws that are the underpinnings of his Natural Philosophy. The concepts are complex and are interrelated. And Newton's Definitions and Axioms should be read in their *totality* and in their context keeping in mind that they 'Shall be explained at large in the following treatise' (for which end it was composed!).

We shall quote rather more extensively than may be considered necessary; but they draw attention to issues that are commonly ignored; and Newton states his ideas with such clarity that it is a joy to learn them anew.

7. Basic concepts: Definitions

Definition I

The quantity of matter is the measure of the same arising from its density and bulk conjointly.

Thus air of a double density, in a double space, is quadruple in quantity; in a triple space, sextuple in quantity. The same thing is to be understood of snow, and fine dust or powders, that are condensed by compression or liquefaction, and of all bodies that are by any causes whatever differently condensed. I have no regard in this place to a medium, if any such there is, that freely pervades the interstices between the parts of bodies. It is this quantity that I mean hereafter everywhere under the name of body or mass. And the same is known by the weight of each body, for it is proportional to the weight, as I have found by experiments on pendulums, very accurately made, which shall be shown hereafter.

It will be noticed that while Newton is careful in defining the notion of *mass* (as a quantitative measure of *quantity of matter*), he leaves the notion of *weight* unspecified except to say that 'by experiments on pendulums, very accurately made', he has shown that mass is proportional to weight. The reason for this partial explanation is that the precise distinction between 'mass' and 'weight' cannot be made without reference to the Second Law of Motion (yet to be formulated in terms of concepts yet to be introduced). This fact is made clear in Newton's account of 'his experiments made with the greatest accuracy' in Proposition XXIV of Book II (with explicit reference to the Second Law of Motion). We shall presently consider (out of context! in §10) this proposition to emphasize that the Definitions and Laws must be read in their totality and not singly.*

Definition II

The quantity of motion is the measure of the same, arising from the velocity and quantity of matter conjointly.

The motion of the whole is the sum of the motions of all the parts; and therefore in a body double in quantity, with equal velocity, the motion is double; with twice the velocity, it is quadruple.

Quantity of motion (or motion for short) is momentum in our present terminology:

$$\text{quantity of } \mathbf{motion} = \text{mass} \times \mathbf{velocity}. \tag{1}$$

(Quantities in bold face indicate that they are vectors.)

Definition III

The vis insita, *or innate force of matter, is a power of resisting, by which every body, as much as in it lies, continues in its present state, whether it be of rest, or of moving uniformly forwards in a right line.*

This force is always proportional to the body whose force it is and differs nothing from the inactivity of the mass, but in our manner of conceiving it. A body, from the inert nature of matter, is not without difficulty put out of its state of rest or motion. Upon which account, this *vis insita* may, by a most significant name, be called inertia (*vis inertiae*) or force of inactivity. But a body only exerts this force when another force, impressed upon it, endeavours to change its condition; and the exercise of this force may be considered as both resistance and impulse; it is resistance so far as the body, for maintaining its present state, opposes the force impressed; it is impulse so far as the body, by not easily giving way to the impressed force of another, endeavours to change the

* To avoid ambiguity it may be noted explicitly that the distinction that is made here is between *inertial mass* (m_i) and gravitational mass (m_g). By *mass* Newton means the inertial mass m_i and by *weight* he means $g \times m_g$ where m_g denotes the gravitational mass and g the value of *gravity* at the location of the mass. For example, at the surface of the Earth, we should write $g = G \times (\text{mass of Earth})/(\text{radius of Earth})^2$, where G denotes the constant of gravitation, that is, Newton's constant!

state of that other. Resistance is usually ascribed to bodies at rest, and impulse to those in motion; but motion and rest, as commonly conceived, are only relatively distinguished; nor are those bodies always truly at rest, which commonly are taken to be so.

There is hardly anything that one can usefully add to Newton's careful explanation of the concept of *inertia.* But note particularly the statements that are are underlined.

Definition IV

An impressed force is an action exerted upon a body, in order to change its state, either of rest, or of uniform motion in a right line.

This force consists in the action only, and remains no longer in the body when the action is over. For a body maintains every new state it acquires, by its inertia only. But impressed forces are of different origins, as from percussion, from pressure, from centripetal force.

Definition V

A centripetal force is that by which bodies are drawn or impelled, or any way tend, towards a point as to a centre.

Of this sort is gravity, by which bodies tend to the centre of the Earth; magnetism, by which iron tends to the loadstone; and that force, whatever it is, by which the planets are continually drawn aside from the rectilinear motions, which otherwise they would pursue, and made to revolve in curvilinear orbits. A stone, whirled about in a sling, endeavours to recede from the hand that turns it; and by that endeavour, distends the sling, and that with so much the greater force, as it is revolved with the greater velocity, and as soon as it is let go, flies away. That force which opposes itself to this endeavour, and by which the sling continually draws back the stone towards the hand, and retains it in its orbit, because it is directed to the hand as the centre of the orbit, I call the centripetal force. And the same thing is to be understood of all bodies, revolved in any orbits. They all endeavour to recede from the centres of their orbits; and were it not for the opposition of a contrary force which restrains them to, and detains them in their orbits, which I therefore call centripetal, would fly off in right lines, with a uniform motion.

. . .

And after the same manner that a projectile, by the force of gravity, may be made to revolve in an orbit, and go round the whole Earth, the Moon also, either by the force of gravity, if it is endued with gravity, or by any other force, that impels it towards the Earth, may be continually drawn aside towards the

Earth, out of the rectilinear way which by its innate force it would pursue; and would be made to revolve in the orbit which it now describes; nor could the Moon without some such force be retained in its orbit. If this force was too small, it would not sufficiently turn the Moon out of a rectilinear course; if it was too great, it would turn it too much, and draw down the Moon from its orbit towards the Earth. It is necessary that the force be of a just quantity, and it belongs to the mathematicians to find the force that may serve exactly to retain a body in a given orbit with a given velocity; and *vice versa*, to determine the curvilinear way into which a body projected from a given place, with a given velocity, may be made to deviate from its natural rectilinear way, by means of a given force.

In the first of the two extracts quoted, Newton makes the distinction between *his* definition of *centripetal force* and Huygens's definition of *centrifugal force* (without so explicitly stating).

The second extract recalls the *raison d'être* of his 1666 Moon test. And the problem formulated in the last sentence is solved in Proposition XVII of Book I (see §30).

Definition VI

The absolute quantity of a centripetal force is the measure of the same, proportional to the efficacy of the cause that propagates it from the centre, through the spaces round about.

Thus the magnetic force is greater in one loadstone and less in another, according to their sizes and strength of intensity.

Definition VII

The accelerative quantity of a centripetal force is the measure of the same, proportional to the velocity which it generates in a given time.

Thus the force of the same loadstone is greater at a less distance, and less at a greater: also the force of gravity is greater in valleys, less on tops of exceeding high mountains; and yet less (as shall hereafter be shown), at greater distances from the body of the Earth; but at equal distances, it is the same everywhere; because (taking away, or allowing for, the resistance of the air), it equally accelerates all falling bodies, whether heavy or light, great or small.

Definition VIII

The motive quantity of a centripetal force is the measure of the same, proportional to the motion which it generates in a given time.

Thus the weight is greater in a greater body, less in a less body; and, in the same body, it is greater near to the Earth, and less at remoter distances. This sort of quantity is the centripetency, or propension of the whole body towards the centre, or, as I may say, its weight; and it is always known by the quantity of an equal and contrary force just sufficient to hinder the descent of the body.

These quantities of forces, we may, for the sake of brevity, call by the names of motive, accelerative, and absolute forces; and, for the sake of distinction, consider them with respect to the bodies that tend to the centre, to the places of those bodies, and to the centre of force towards which they tend; that is to say, I refer the motive force to the body as an endeavour and propensity of the whole towards a centre, arising from the propensities of the several parts taken together; the accelerative force to the place of the body, as a certain power diffused from the centre to all places around to move the bodies that are in them; and the absolute force to the centre, as endued with some cause, without which those motive forces would not be propagated through the spaces round about; whether that cause be some central body (such as is the magnet in the centre of the magnetic force, or the Earth in the centre of the gravitating force), or anything else that does not yet appear. For I here design only to give a mathematical notion of those forces, without considering their physical causes and seats.

Wherefore the accelerative force will stand in the same relation to the motive, as celerity does to motion. For the quantity of motion arises from the celerity multiplied by the quantity of matter; and the motive force arises from the accelerative force multiplied by the same quantity of matter.* For the sum of the actions of the accelerative force, upon the several particles of the body, is the motive force of the whole. Hence it is, that near the surface of the Earth, where the accelerative gravity, or force productive of gravity, in all bodies is the same, the motive gravity or the weight is as the body; but if we should ascend to higher regions, where the accelerative gravity is less, the weight would be equally diminished, and would always be as the product of the body, by the accelerative gravity. So in those regions, where the accelerative gravity is diminished into one-half, the weight of a body two or three times less, will be four or six times less.

I likewise call attractions and impulses, in the same sense, accelerative, and motive; and use the words attraction, impulse, or propensity of any sort towards a centre, promiscuously, and indifferently, one for another; considering those forces not physically, but mathematically: wherefore the reader is not to imagine that by those words I anywhere take upon me to define the kind, or the manner of any action, the causes or the physical reason thereof, or that I attribute forces,

* These sentences state in words that the ratios of acceleration to motive force and of velocity to motion are the same, being the mass in each case.

> in a true and physical sense, to certain centres (which are only mathematical points); when at any time I happen to speak of centres as attracting, or as endued with attractive powers.

Newton's comments on this Definition are quoted *in extenso*. The parts underlined state, already at this early stage, his view of natural philosophy. His comments in particular: 'I here design only to give a mathematical notion of those forces, without considering their physical causes and seats' and 'the reader is not to imagine that . . . I anywhere take upon me to define the kind, or the manner of any action, the causes of the physical reason thereof' are evocative of the climactic statement at the conclusion of the *Principia* 'I feign no hypothesis',

The long Scholium which concludes this lesson on 'Definitions' is perhaps the part of the *Principia* most commented upon by historians and philosophers. But most of these commentaries (if not all of them) do not heed Newton's concluding statement that further elucidation must be sought in the consistency of their usage in the 'treatise' that is to follow.

Nevertheless concerning the notions of '*absolute time*' and '*absolute space*' on which Newton bases his dynamics, it will suffice to say that, in current terminology, the space-time manifold that is assumed is the Cartesian product,

$$t \otimes \text{Euclidean 3-space}, \tag{2}$$

where t is Newton's 'equable time'.

8. Basic concepts: the Laws of Motion

After the introductory lesson on fundamental notions (§7), Newton proceeds to his second lesson to formulate the basis for his entire dynamics in the form of three Laws of Motion and five corollaries (which are an essential part of the Laws). Again, the Laws and their corollaries must be considered in their totality and not singly. This need was, for example, fully recognized by Maxwell who reformulated Newton's first two Laws of Motion to 'render more precise [their] ennunciation' (see §10 below).

Law I

> *Every body continues in its state of rest, or of uniform motion in a right line, unless it is compelled to change that state by forces impressed upon it.*
>
> Projectiles continue in their motions, so far as they are not retarded by the resistance of the air, or impelled downwards by the force of gravity. A top, whose parts by their cohesion are continually drawn aside from rectilinear motions, does not cease its rotation, otherwise than as it is retarded by the air. The greater bodies of the planets and comets, meeting with less resistance in

> freer spaces, preserve their motions both progressive and circular for a much longer time.

The statement of this Law is not 'precise' (to quote Maxwell), since what we are to understand by 'body' is not made clear. The statement as it stands is valid if a point particle or a 'rigid body' is intended (as the qualification 'cohesion' in the second illustrative example suggests). But the content of the Law is far wider, when considered in the contexts of Laws II and III and Corollaries IV and V, as Maxwell's reformulation of these Laws makes it explicit (Maxwell's version is given in §10).

It will be noticed that in the example with the top, the implied arguments are the same as in Chapter 1, §2 (cf. equation (1)).

Law II

> *The change of motion is proportional to the motive force impressed; and is made in the direction of the right line in which that force is impressed.*

The statement of the Law is self-explanatory. In current terminology it states:

$$\begin{aligned} \textbf{force} &= \text{change in } \textbf{motion} \\ &= \text{change in } [\text{mass} \times \textbf{velocity}] \\ &= \text{mass} \times \text{change in } \textbf{velocity} \\ &= \text{mass} \times \textbf{acceleration}. \end{aligned} \tag{1}$$

Law III

> *To every action there is always opposed an equal reaction: or, the mutual actions of two bodies upon each other are always equal, and directed to contrary parts.*
>
> Whatever draws or presses another is as much drawn or pressed by that other. If you press a stone with your finger, the finger is also pressed by the stone.
>
> . . . .
>
> If a body impinge upon another, and by its force change the motion of the other, that body also (because of the equality of the mutual pressure) will undergo an equal change, in its own motion, towards the contrary part. The changes made by these actions are equal, not in the velocities but in the motions of bodies; that is to say, if the bodies are not hindered by any other impediments. For, because the motions are equally changed, the changes of the velocities made towards contrary parts are inversely proportional to the bodies. This Law takes place also in attractions.

This Law is central to proving the important Corollaries IV and V; and its importance is further emphasized in the Scholium (see §9 where Maxwell and Thomson and Tait are quoted).

Corollary I

A body, acted on by two forces simultaneously, will describe the diagonal of a parallelogram in the same time as it would describe the sides by those forces separately.

If a body in a given time, by the force M impressed apart in the place A, should with a uniform motion be carried from A to B, and by the force N impressed apart in the same place, should be carried from A to C, let the parallelogram $ABCD$ be completed, and, by both forces acting together, it will in the same time be carried in the diagonal from A to D. But it will move in a right line from A to D, by Law I.

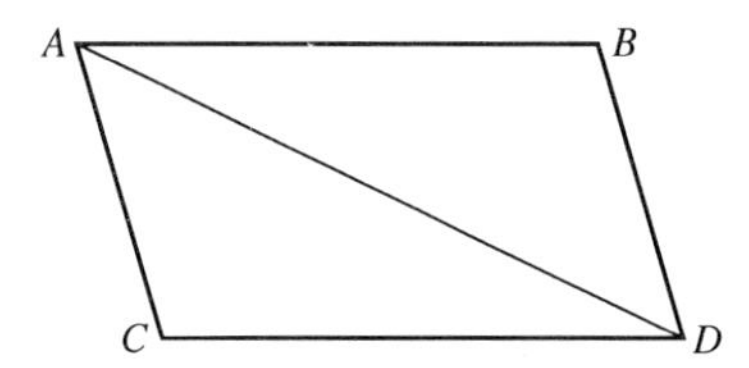

From the manner in which this '*parallelogram law of forces*' is proved, it is clear that the law applies equally to **velocities**, **motions**, and **accelerations**; and Newton does use the law in these other contexts.

Corollary II

And hence is explained the composition of any one direct force AD, out of any two oblique forces AC and CD; and, on the contrary, the resolution of any one direct force AD into two oblique forces AC and CD: which composition and resolution are abundantly confirmed from mechanics.

As if the unequal radii OM and ON drawn from the centre O of any wheel, should sustain the weights A and P by the cords MA and NP; and the forces of those weights to move the wheel were required. Through the centre O draw the right line KOL, meeting the cords perpendicularly in K and L; and from the centre O, with OL the greater of the distances OK and OL, describe a circle, meeting the cord MA in D; and drawing OD, make AC parallel and DC perpendicular thereto. Now, it being indifferent whether the points K, L, D, of the cords be fixed to the plane of the wheel or not, the weights will have the same effect whether they are suspended from the points K and L, or from D and L. Let the whole force of the weight A be represented by the line AD.

With the foregoing construction, it follows from Corollary I that the force,

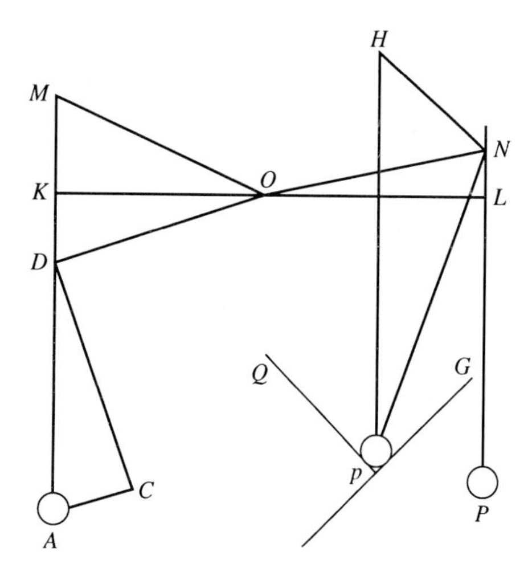

$$\overrightarrow{DA} = \overrightarrow{DC} + \overrightarrow{CA}. \tag{2}$$

But $\overrightarrow{CA}$, being parallel to $\overrightarrow{OD}$, will be ineffective in turning the wheel. And $\overrightarrow{DC}$ 'drawing the radius $\overrightarrow{OD}$ perpendicularly will have the same effect' as P acting on $\overrightarrow{OL}$. Hence

$$\frac{P}{A} = \frac{DC}{DA}. \tag{3}$$

Since $\Delta ADC \equiv \Delta DOK$,

$$\frac{DC}{DA} = \frac{OK}{OD} = \frac{OK}{OL} \quad \text{(since by construction, } OD = OL\text{).} \tag{4}$$

We conclude that

$$\frac{P}{A} = \frac{OK}{OL} \quad \text{or} \quad P \times OL = A \times OK, \tag{5}$$

which is the *law of the lever*. And Newton concludes:

> the use of this Corollary spreads far and wide, and by that diffusive extent the truth thereof is further confirmed. For on what has been said depends the whole doctrine of mechanics variously demonstrated by different authors. For from hence are easily deduced the forces of machines, which are compounded of wheels, pulleys, levers, cords, and weights, ascending directly or obliquely, and other mechanical powers; as also the force of the tendons to move the bones of animals.

Corollary III

The quantity of motion, which is obtained by taking the sum of the motions directed towards the same parts, and the difference of those that are directed to contrary parts, suffers no change from the action of bodies among themselves.

For action and its opposite reaction are equal, by Law III, and therefore, by Law II, they produce in the motions equal changes towards opposite parts. Therefore if the motions are directed towards the same parts, whatever is added to the motion of the preceding body will be subtracted from the motion of that which follows; so that the sum will be the same as before. If the bodies meet, with contrary motions, there will be an equal deduction from the motions of both; and therefore the difference of the motions directed towards opposite parts will remain the same.

It is important to notice that the proof of this corollary (used in the demonstration of Corollaries V and VI) depends explicitly on *both* Laws II and III.

Corollary IV

The common centre of gravity of two or more bodies does not alter its state of motion or rest by the actions of the bodies among themselves; and therefore the common centre of gravity of all bodies acting upon each other (excluding external actions and impediments) is either at rest, or moves uniformly in a right line.

This and the following two corollaries are central to Newtonian dynamics (as we shall further elaborate in §10).

In establishing Corollary IV, Newton first considers the case when the 'bodies' in question are mass points, m_i $(i = 1, \ldots, n)$ moving uniformly with velocities $\vec{v}_i$, that is,

$$\vec{r}_i = \vec{v}_i t + \vec{a}_i \qquad (i = 1, \ldots, n), \tag{6}$$

where $\vec{r}_i$ denotes the position vector of m_i. The corollary states that the centre of mass of the particles,

$$\vec{R} = \frac{1}{M} \sum_{i=1}^{n} m_i \vec{r}_i, \qquad \text{where} \quad M = \sum_{i=1}^{n} m_i, \tag{7}$$

moves uniformly in a straight line. This follows directly from equations (6) and (7); thus,

$$\vec{R} = \frac{1}{M} \sum_{i=1}^{n} m_i(\vec{v}_i t + \vec{a}_i) = \vec{V}t + \vec{A}, \tag{8}$$

where

$$\vec{V} = \frac{1}{M} \sum_{i=1}^{n} m_i \vec{v}_i \qquad \text{and} \qquad \vec{A} = \frac{1}{M} \sum_{i=1}^{n} m_i \vec{a}_i. \tag{9}$$

It is instructive to follow Newton's proof. He makes use of Lemma XXIII established later in Book I:

Lemma XXIII

If two given right lines, as AC, BD, terminating in given points A, B, are in a given ratio one to the other, and the right line CD, by which the indetermined points C, D are joined is cut in K in a given ratio: I say, that the point K will be placed in a given right line.

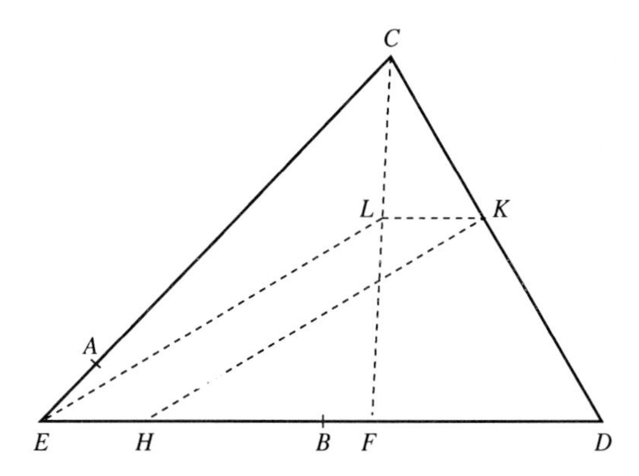

We are required to find the locus of K, given the fixed points E, A, and B and varying points C, D, and K satisfying the requirements

$$\frac{BD}{AC} = \alpha \quad \text{and} \quad \frac{CK}{KD} = \beta, \tag{10}$$

where α and β are assigned constants.

Let F be a point on ED such that

$$\frac{EF}{EC} = \alpha. \tag{11}$$

Draw KL parallel to DF; then

$$\frac{CL}{LF} = \frac{CK}{KD} = \beta. \tag{12}$$

Since the triangles ECL, ELF, and ECF remain similar to themselves as C, L, and F vary in the manner prescribed (by equations (11) and (12)), EL will remain a constant straight line as L varies along with C and F. In other words the locus of L is the constant straight line EL prolonged.

Now draw KH parallel to EL. Then

$$EH = LK = FD\frac{CL}{CF} \qquad \text{(by the similarity of } \Delta\text{s } CLK \text{ and } CFD\text{)}$$

$$= (ED - EF)\frac{CL}{CL + LF}$$

$$= (EB + BD - \alpha EC)\frac{\beta}{\beta + 1} \qquad \text{(by equations (11) and (12))}$$

$$= [(EB + \alpha AC - \alpha(EA + AC)]\frac{\beta}{\beta + 1} \qquad \text{(by equation (10))}$$

$$= (EB - \alpha EA)\frac{\beta}{\beta + 1}; \tag{13}$$

that is, EH is determined by the initially given quantities, and therefore, remains constant as C, D, and K vary as prescribed. Hence the locus of K is the straight line HK prolonged parallel to EL.

Newton's proof (using Lemma XXIII) proceeds as follows:

> For if two points proceed with a uniform motion in right lines, and their distance be divided in a given ratio, the dividing point will be either at rest, or proceed uniformly in a right line. This is demonstrated hereafter in Lem. XXIII and Corollary, when the points are moved in the same plane; and by a like way of arguing, it may be demonstrated when the points are not moved in the same plane. Therefore if any number of bodies move uniformly in right lines, the common centre of gravity of two of them is either at rest, or proceeds uniformly in a right line; because the line which connects the centres of those two bodies so moving is divided at that common centre in a given ratio. In like manner the common centre of those two and that of a third body will be either at rest or moving uniformly in a right line; because at that centre the distance between the common centre of the two bodies, and the centre of this last, is divided in a given ratio.

The proof is now completed by induction (a favourite device of Newton's). The remaining part of the proof, when 'bodies' more general than mass points are considered, is worth quoting *in extenso* to see how Newton skirts ambiguities.

> Moreover, in a system of two bodies acting upon each other, since the distances between their centres and the common centre of gravity of both are reciprocally as the bodies, the relative motions of those bodies, whether of approaching to or of receding from that centre, will be equal among themselves.

Therefore since the changes which happen to motions are equal and directed to contrary parts, the common centre of those bodies, by their mutual action between themselves, is neither accelerated nor retarded, nor suffers any change as to its state of motion or rest. But in a system of several bodies, because the common centre of gravity of any two acting upon each other suffers no change in its state by that action; and much less the common centre of gravity of the others with which that action does not intervene; but the distance between those two centres is divided by the common centre of gravity of all the bodies into parts inversely proportional to the total sums of those bodies whose centres they are; and therefore while those two centres retain their state of motion or rest, the common centre of all does also retain its state: it is manifest that the common centre of all never suffers any change in the state of its motion or rest from the actions of any two bodies between themselves. But in such a system all the actions of the bodies among themselves either happen between two bodies, or are composed of actions interchanged between some two bodies; and therefore they do never produce any alteration in the common centre of all as to its state of motion or rest. Wherefore since that centre, when the bodies do not act one upon another, either is at rest or moves uniformly forwards in some right line, it will, notwithstanding the mutual actions of the bodies among themselves, always continue in its state, either of rest, or of proceeding uniformly in a right line, unless it is forced out of this state by the action of some power impressed from without upon the whole system. And therefore the same law takes place in a system consisting of many bodies as in one single body, with regard to their persevering in their state of motion or of rest. For the progressive motion, whether of one single body, or of a whole system of bodies, is always to be estimated from the motion of the centre of gravity.

I cannot desist from observing that the essential parts of the argument in the three sentences (of ten, four, and six lines, respectively) with many semicolons and a colon is reminiscent of the style of Henry James. (Newton's style is in fact Jamesian throughout.)

Corollary V

The motions of bodies included in a given space are the same among themselves, whether that space is at rest, or moves uniformly forwards in a right line without any circular motion.

For the differences of the motions tending towards the same parts, and the sums of those that tend towards contrary parts, are, at first (by supposition), in both cases the same; and it is from those sums and differences that the collisions and impulses do arise with which the bodies impinge one upon another. Wherefore (by Law II), the effects of those collisions will be equal in both cases;

and therefore the mutual motions of the bodies among themselves in the one case will remain equal to the motions of the bodies among themselves in the other. A clear proof of this we have from the experiment of a ship; where all motions happen after the same manner, whether the ship is at rest, or is carried uniformly forwards in a right line.

Newton returns to this corollary in Section XI of the *Principia* (see §59) where he repeats the same argument more concisely in the context of the '*motions of bodies tending to each other with centripetal forces*'.

This corollary is used later in Proposition LXV (case 2) of Book I in his preliminary considerations towards the problem of many bodies under their mutual gravitational attractions where the argument is repeated in almost identical terms (see §68).

Corollary VI

If bodies, moved in any manner among themselves, are urged in the direction of parallel lines by equal accelerative forces, they will all continue to move among themselves, after the same manner as if they had not been urged by those forces.

For these forces acting equally (with respect to the quantities of the bodies to be moved), and in the direction of parallel lines, will (by Law II) move all the bodies equally (as to velocity), and therefore will never produce any change in the positions or motions of the bodies among themselves.

Notice that this corollary shows that Newton's laws hold even in some accelerated frames.

The proofs of Corollaries V and VI are based on Law II (explicitly quoted) and on Corollary III (which is based on Law III as well). It is important to emphasize the roles of Laws II and III in the context of these corollaries.

9. The Scholium to the Laws of Motion

In a long Scholium (somewhat longer in fact than the Scholium for the first lesson), Newton discusses a variety of matters, historical, analytical, and amplificatory.

The Scholium begins with some historical remarks relating to Galileo's discoveries:

Scholium

Hitherto I have laid down such principles as have been received by mathematicians, and are confirmed by abundance of experiments. By the first two Laws and the first two Corollaries, *Galileo* discovered that the descent of bodies varied as the square of the time (*in duplicata ratione temporis*) and that the motion of projectiles was in the curve of a parabola; experience agreeing with both, unless so far as these motions are a little retarded by the resistance of the air. When a body is falling, the uniform force of its gravity acting equally, impresses, in equal intervals of time, equal forces upon that body, and therefore generates equal velocities; and in the whole time impresses a whole force, and generates a whole velocity proportional to the time. And the spaces described in proportional times are as the product of the velocities and the times; that is, as the squares of the times. And when a body is thrown upwards, its uniform gravity impresses forces and reduces velocities proportional to the times; and the times of ascending to the greatest heights are as the velocities to be taken away, and those heights are as the product of the velocities and the times, or as the squares of the velocities. And if a body be projected in any direction, the motion arising from its projection is compounded with the motion arising from its gravity. Thus, if the body A by its motion of projection alone could describe in a given time the right line AB, and with its motion of falling alone could describe in the same time the altitude AC; complete the parallelogram $ABCD$, and the body by that compounded motion will at the end of the time be found in the place D; and the curved line AED, which that body describes, will be a parabola, to which the right line AB will be a tangent at A; and whose ordinate BD will be as the square of the line AB. On the same Laws and Corollaries depend those things which have been demonstrated concerning the times of the vibration of pendulums, and are confirmed by the daily experiments of pendulum clocks.

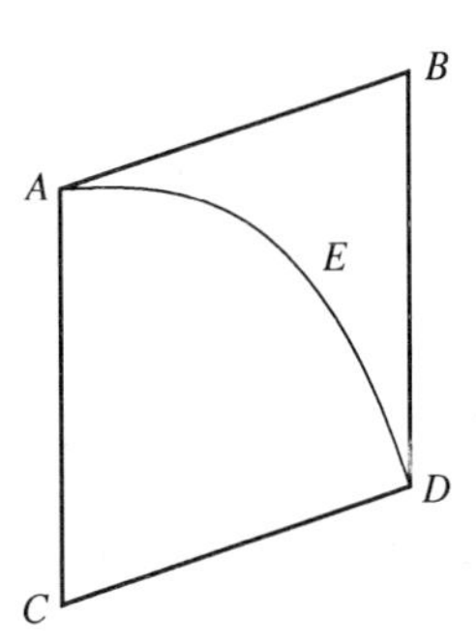

The following remarks by Herivel (*The background to Newton's Principia*, pp. 35–37), quoting the statement underlined, are apposite in this context:

> It is equally certain that Galileo never *enunciated* the principle of inertia, and indeed could not have done so correctly, since 'horizontal' motion was for him always at the surface of the earth, equidistant from its centre, and therefore in reality circular and not rectilinear.

. . . .

> Nevertheless, although Galileo's principle of inertia was thus restricted to a very special terrestrial case, this restriction did not obtrude itself in his vivid physical discussion of inertial motion on a horizontal plane, especially in his discussion of the motion of a projectile. And Newton would have been powerfully impressed and influenced by this discussion.
>
> It is not at all clear how far, if at all, Galileo's understanding of force had progressed along the road leading to the second law of motion. Newton himself, as we have seen, seems to imply that this law was known to Galileo
>
> . . . .
>
> In certain respects this statement is entirely clear and unexceptionable. Galileo *did* discover both the law of falling bodies and the parabolic path of a projectile, basing his derivations of them on an inertial principle and a method of compounding motions identical with that found in the first corollary to Newton's laws of motion. To what extent, however, did Galileo's inertial principle agree with Newton's principle of inertia, and how far if at all did Galileo either recognize or use the second law of motion?

We shall not pursue further Herivel's discussion of Newton's indebtedness (limited or otherwise) to Galileo. But it is relevant to note that neither Herivel, nor anyone else, to the extent I can judge, suggests Newton's indebtedness to Galileo (or to anyone else) for his formulation of the important Corollaries IV and V.

The experiment and theory that Newton refers to in the last underlined sentences in the part of the Scholium quoted are considered in §10.

After a long discussion of how resistance by air affects his experiments in the pendulums, he returns to a further elaboration of his Third Law of Motion.

> And thus the third Law, so far as it regards percussions and reflections, is proved by a theory exactly agreeing with experience.
>
> In attractions, I briefly demonstrate the thing after this manner. Suppose an obstacle is interposed to hinder the meeting of any two bodies *A*, *B*, attracting one the other: then if either body, as *A*, is more attracted towards the other body *B*, than that other body *B* is towards the first body *A*, the obstacle will be more strongly urged by the pressure of the body *A* than by the pressure of the body *B*, and therefore will not remain in equilibrium: but the stronger pressure will prevail, and will make the system of the two bodies, together with the obstacle, to move directly towards the parts on which B lies; and in free spaces, to go forwards *in infinitum* with a motion continually accelerated; which is absurd and contrary to the first Law. For, by the first Law, the system ought to continue in its state of rest, or of moving uniformly forwards in a right line; and therefore the bodies must equally press the obstacle, and be equally attracted

one by the other. I made the experiment on the loadstone and iron. If these, placed apart in proper vessels, are made to float by one another in standing water, neither of them will propel the other; but, by being equally attracted, they will sustain each other's pressure, and rest at last in an equilibrium.

So the gravitation between the Earth and its parts is mutual. Let the Earth *FI* be cut by any plane *EG* into two parts *EGF* and *EGI*, and their weights one towards the other will be mutually equal. For if by another plane *HK*, parallel to the former *EG*, the greater part *EGI* is cut into two parts *EGKH* and *HKI* whereof *HKI* is equal to the part *EFG*, first cut off, it is evident that the middle part *EGKH* will have no propension by its proper weight towards either side, but will hang as it were, and rest in an equilibrium between both. But the one extreme part *HKI* will with its whole weight bear upon and press the middle part towards the other extreme part *EGF*; and therefore the force with which *EGI*, the sum of the parts *HKI* and *EGKH*, tends towards the third part *EGF*, is equal to the weight of the part *HKI*, that is, to the weight of the third part *EGF*. And therefore the weights of the two parts *EGI* and *EGF*, one towards the other, are equal, as I was to prove. And indeed if those weights were not equal, the whole Earth floating in the non-resisting ether would give way to the greater weight, and, retiring from it, would be carried off *in infinitum*.

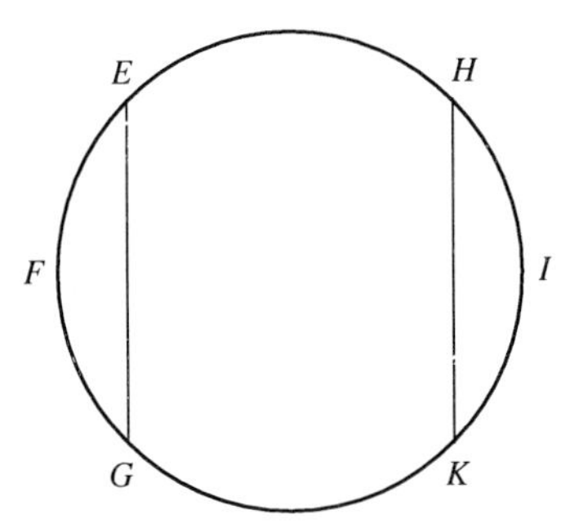

Newton's concluding remarks, again bearing on the Third Law of Motion, are pregnant with meaning as has been noticed by Maxwell and by Thomson and Tait:

The power and use of machines consist only in this, that by diminishing the velocity we may augment the force, and the contrary; from whence, in all sorts of proper machines, we have the solution of this problem: *To move a given weight with a given power*, or with a given force to overcome any other given resistance. For if machines are so contrived that the velocities of the agent and resistant are inversely as their forces, the agent will just sustain the resistant, but with a greater disparity of velocity will overcome it. So that if the disparity of velocities is so great as to overcome all that resistance which commonly arises either from the friction of contiguous bodies as they slide by one another, or from the cohesion of continuous bodies that are to be separated, or from the weights of bodies to be raised, the excess of the force remaining, after all those resistances are overcome, will produce an acceleration of motion proportional thereto, as well in the parts of the machine as in the resisting body. But to treat of mechanics is not my present business. I was aiming only to show by those examples the great extent and certainty of the Third Law of Motion. For if we estimate the action of the agent from the product of its force and velocity, and likewise the reaction of the impediment from the product of the velocities of its several parts, and the forces of resistance arising from the friction, cohesion, weight, and acceleration of those parts, the action and reaction in the use of all sorts of machines will be found always equal to one another.* And so far as the action is propagated by the intervening instruments, and at last impressed upon the resisting body, the ultimate action will be always contrary to the reaction.

Quoting the statement underlined in the foregoing extract, Maxwell writes in his *Theory of heat* (Chapter IV, p. 91):

> Newton, in a Scholium to his Third Law of Motion, has stated the relation between work and kinetic energy in a manner so perfect that it cannot be improved, but at the same time with so little apparent effort or desire to attract attention that no one seems to have been struck with the great importance of the passage till it was pointed out recently (1867) by Thomson and Tait.

and the relevant passage in Thomson and Tait's *Natural philosophy* is:

> In the scholium appended, he makes the following remarkable statement, introducing another description of actions and reactions subject to his third law, the full meaning of which seems to have escaped the notice of commentators.

. . . .

we may read the above statement as follows:—

> *If the Activity of an agent be measured by its amount and its velocity conjointly; and if, similarly, the Counter-activity of the resistance be measured by the velocities of its several parts and their several amounts conjointly, whether these arise from friction, cohesion, weight, or acceleration;—Activity and Counter-activity, in all combinations of machines, will be equal and opposite.*

* Or, in other words: the change in the kinetic energy is equal to the work done by the forces during the motion. This is D'Alembert's principle as Kelvin points out in the quotation that follows.

Farther on we shall give an account of the splendid dynamical theory founded by D'Alembert and Lagrange on this most important remark.

Newton, in the passage just quoted, points out that forces of resistance against acceleration are to be reckoned as reactions equal and opposite to the actions by which the acceleration is produced. Thus, if we consider any one material point of a system, its reaction against acceleration must be equal and opposite to the resultant of the forces which that point experiences, whether by the actions of other parts of the system upon it, or by the influence of matter not belonging to the system. In other words, it must be in equilibrium with these forces. Hence Newton's view amounts to this, that all the forces of the system, with the reactions against acceleration of the material points composing it, form groups of equilibrating systems for these points considered individually. Hence, by the principle of superposition of forces in equilibrium, all the forces acting on points of the system form, with the reactions against acceleration, an equilibrating set of forces on the whole system. This is the celebrated principle first explicitly stated, and very usefully applied, by D'Alembert in 1742, and still known by his name. We have seen, however, that it is very distinctly implied in Newton's own interpretation of his third law of motion.

I have quoted from Maxwell and Thomson and Tait so extensively, since I have found nowhere else comments on the Scholia with the same degree of perception and understanding.

10. Additional amplifications

In this section we shall amplify the account of Newton's Definitions and Axioms given in §§7–9.

(*a*) *The proportionality of mass and weight and the experiments on the pendulums*

As we have remarked in the context of Definition I in §7, the distinction between *mass* and *weight* cannot be made without reference to the Second Law of Motion. Newton makes this abundantly clear in Proposition XXIV of Book II. In view of the key importance of this proposition for an understanding of the basic concepts, we shall here give an account of it. The proposition in question is:

Proposition XXIV. Theorem XIX

The quantities of matter in pendulous bodies, whose centres of oscillation are equally distant from the centre of suspension, are in a ratio compounded of the

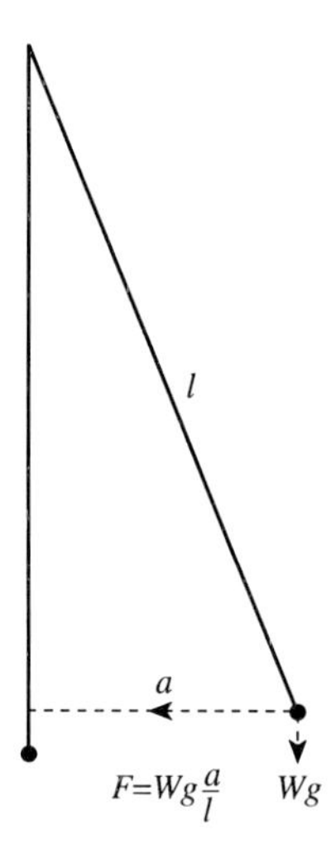

ratio of the weights and the squared ratio of the times of the oscillations in a vacuum.

Newton's proof (paraphrased) is along the following lines.

If a motive force acts on a Mass, M, for a time Δt, then by the Second Law of Motion, the velocity v it will generate is given by

$$M \times \Delta v = \text{motive force} \times \Delta t. \tag{1}$$

On the other hand the force acting vertically *downward* on the pendulum of length l, displaced by a distance a from the vertical, is given by

$$\text{downward force} = Wg \tag{2}$$

where W denotes the gravitational mass and g the value of gravity. The motive force acting *horizontally* is therefore,

$$\text{motive force} = Wg(a/l). \tag{3}$$

By considering two pendulums of equal length, l, displaced by the same amount, a, Newton argues: 'If two [such] bodies by oscillating describe equal arcs [e.g., A] and those arcs [A] are divided into equal parts [Δa], the times [Δt] in which each of the corresponding parts [Δa are described] are as the times of the oscillation [T]'; and he concludes

$$\Delta t : \Delta v = T : a/T. \tag{4}$$

By equations (1) and (4),

$$M \propto \frac{\text{motive force} \times T}{a/T} = \frac{\text{motive force} \times T^2}{a}. \tag{5}$$

Combining equations (3) and (5) we obtain,

$$M \propto WgT^2/l, \tag{6}$$

which is Newton's result (Q.E.D.!)

The seven corollaries which Newton appends to this result are a marvel of clarity:

> COR. I. Therefore if the times are equal, the quantities of matter in each of the bodies are as the weights.
>
> COR. II. If the weights are equal, the quantities of matter will be as the squares of the times.
>
> COR. III. If the quantities of matter are equal, the weights will be inversely as the squares of the times.
>
> COR. IV. Since the squares of the times, other things being equal, are as the lengths of the pendulums, therefore if both the times and the quantities of matter are equal, the weights will be as the lengths of the pendulums.
>
> COR. V. And, in general, the quantity of matter in the pendulous body is directly as the weight and the square of the time, and inversely as the length of the pendulum.
>
> COR. VI. But in a non-resisting medium, the quantity of matter in the pendulous body is directly as the comparative weight and the square of the time, and inversely as the length of the pendulum. For the comparative weight is the motive force of the body in any heavy medium, as was shown above; and therefore does the same thing in such a non-resisting medium as the absolute weight does in a vacuum.
>
> COR. VII. And hence appears a method both of comparing bodies one with another, as to the quantity of matter in each; and of comparing the weights of the same body in different places, to know the variation of its gravity. And by experiments made with the greatest accuracy, I have always found the quantity of matter in bodies to be proportional to their weight.

The experiments themselves are described in greater detail in the Scholium to Section VI, Book II and more briefly in Proposition VI, Book III. This continued and repeated reference to his experiments on the pendulums shows the importance that Newton (justly) attached to them.

(*b*) *Maxwell's reformulation of Newton's Laws of Motion*

James Clerk Maxwell's *Matter and motion* of some 120 pages, first published in 1877 (of which the Dover reprint of Larmor's edition is still available), is a rarely sensitive presentation of the basic concepts of Newtonian dynamics. In particular, Maxwell's reformulation of Newton's Laws of Motion in Chapter IV is so completely in the spirit of the *Principia* and illuminating by itself, that I reproduce the relevant sections (59–65, inclusive) of the chapter, in their entirety.

ON THE PROPERTIES OF THE CENTRE OF MASS OF A MATERIAL SYSTEM

59. Definition of a mass-vector

We have seen that a vector represents the operation of carrying a tracing point from a given origin to a given point.

Let us define a mass-vector as the operation of carrying a given mass from the origin to the given point. The direction of the mass-vector is the same as that of the vector of the mass, but its magnitude is the product of the mass into the vector of the mass.

Thus if $\overline{OA}$ is the vector of the mass A, the mass-vector is $\overline{OA}\,.\,A$.

60. Centre of mass of two particles

If A and B are two masses, and if a point C be taken in the straight line $\overline{AB}$, so that $\overline{BC}$ is to $\overline{CA}$ as A to B, then the mass-vector of a mass $A + B$ placed at C is equal to the sum of the mass-vectors of A and B. For

$$\begin{aligned}\overline{OA}\,.\,A + \overline{OB}\,.\,B &= (\overline{OC} + \overline{CA})A + (\overline{OC} + \overline{CB})B \\ &= \overline{OC}(A + B) + \overline{CA}\,.\,A + \overline{CB}\,.\,B.\end{aligned}$$

Now the mass-vectors $\overline{CA}\,.\,A$ and $\overline{CB}\,.\,B$ are equal and opposite, and so destroy each other, so that

$$\overline{OA}\,.\,A + \overline{OB}\,.\,B = \overline{OC}(A + B)$$

or, C is a point such that if the masses of A and B were concentrated at C, their mass-vector from any origin O would be the same as when A and B are in their actual positions. The point C is called the *Centre of Mass* of A and B.

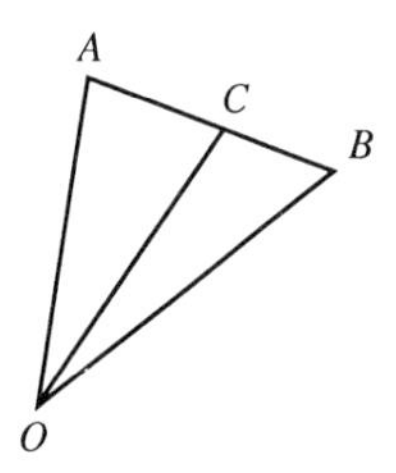

61. Centre of mass of a system

If the system consists of any number of particles, we may begin by finding the centre of mass of any two particles, and substituting for the two particles a particle equal to the sum placed at their centre of mass. We may then find the centre of mass of this particle, together with the third particle of the system, and place the sum of the three particles at this point, and so on till we have found the centre of mass of the whole system.

The mass-vector drawn from any origin to a mass equal to that of the whole system placed at the centre of mass of the system is equal to the sum of the mass-vectors drawn from the same origin to all the particles of the system.

It follows, from the proof in Article 60, that the point found by the construction here given satisfies this condition. It is plain from the condition itself that only one point can satisfy it. Hence the construction must lead to the same result, as to the position of the centre of mass, in whatever order we take the particles of the system.

The centre of mass is therefore a definite point in the diagram of the configuration of the system. By assigning to the different points in the diagrams of displacement, velocity, total acceleration, and rate of acceleration, the masses of the bodies to which they correspond, we may find in each of these diagrams a point which corresponds to the centre of mass, and indicates the displacement, velocity, total acceleration, or rate of acceleration of the centre of mass.

62. *Momentum represented as the rate of change of a mass-vector*

In the diagram of velocities, if the points o, a, b, c, correspond to the velocities of the origin O and the bodies A, B, C, and if p be the centre of mass of A and B placed at a and b respectively, and if q is the centre of mass of $A + B$ placed at p and C at c, then q will be the centre of mass of the system of bodies A, B, C, at a, b, c, respectively.

The velocity of A with respect to O is indicated by the vector $\overline{oa}$, and that of B and C by $\overline{ob}$ and $\overline{oc}$. $\overline{op}$ is the velocity of the centre of mass of A and B, and $\overline{oq}$ that of the centre of mass of A, B, and C, with respect to O.

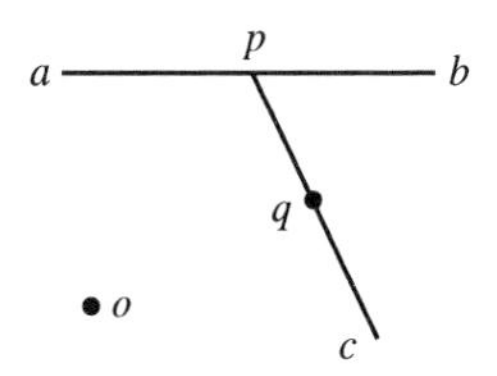

The momentum of A with respect to O is the product of the velocity into the mass, or $\overline{oa}\,.\,A$, or what we have already called the mass-vector, drawn from o to the mass A at a. Similarly the momentum of any other body is the mass-vector drawn from o to the point on the diagram of velocities corresponding to that body, and the momentum of the mass of the system concentrated at the centre of mass is the mass-vector drawn from o to the whole mass at q.

Since, therefore, a mass-vector in the diagram of velocities is what we have already defined as a momentum, we may state the property proved in Article

61 in terms of momenta, thus: The momentum of a mass equal to that of the whole system, moving with the velocity of the centre of mass of the system, is equal in magnitude and parallel in direction to the sum of the momenta of all the particles of the system.

63. Effect of external forces on the motion of the centre of mass

In the same way in the diagram of Total Acceleration the vectors $\overline{\omega\alpha}$, $\overline{\omega\beta}$, etc., drawn from the origin, represent the change of velocity of the bodies A, B, etc., during a certain interval of time. The corresponding mass-vectors, $\overline{\omega\alpha}\,.\,A$, $\overline{\omega\beta}\,.\,B$, etc., represent the corresponding changes of momentum, or, by the second law of motion, the impulses of the forces acting on these bodies during that interval of time. If κ is the centre of mass of the system, $\overline{\omega\kappa}$ is the change of velocity during the interval, and $\overline{\omega\kappa}\ (A + B + C)$ is the momentum generated in the mass concentrated at the centre of gravity. Hence, by Article 61, the change of momentum of the imaginary mass equal to that of the whole system concentrated at the centre of mass is equal to the sum of the changes of momentum of all the different bodies of the system.

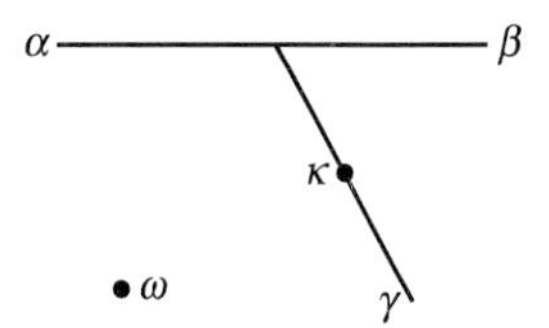

In virtue of the second law of motion we may put this result in the following form:

The effect of the forces acting on the different bodies of the system in altering the motion of the centre of mass of the system is the same as if all these forces had been applied to a mass equal to the whole mass of the system, and coinciding with its centre of mass.

64. The motion of the centre of mass of a system is not affected by the mutual action of the parts of the system

For if there is an action between two parts of the system, say A and B, the action of A and B is always by the third law of motion, equal and opposite to the reaction of B on A. The momentum generated in B by the action of A during any interval is therefore equal and opposite to that generated in A by the reaction of B during the same interval, and the motion of the centre of mass of A and B is therefore not affected by their mutual action.

We may apply the result of the last article to this case and say, that since the forces on A and on B arising from their mutual action are equal and opposite, and since the effect of these forces on the motion of the centre of mass of the system is the same as if they had been applied to a particle whose mass is equal to the whole mass of the system, and since the effect of two forces equal and opposite to each other is zero, the motion of the centre of mass will not be affected.

65. First and second laws of motion

This is a very important result. It enables us to render more precise the enunciation of the first and second laws of motion, by defining that by the velocity of a body is meant the velocity of its centre of mass. The body may be rotating, or it may consist of parts, and be capable of changes of configuration, so that the motions of different parts may be different, but we can still assert the laws of motion in the following form:

Law I.—The centre of mass of the system perseveres in its state of rest, or of uniform motion in a straight line, except in so far as it is made to change that state by forces acting on the system from without.

Law II.—The change of momentum of the system during any interval of time is measured by the sum of the impulses of the external forces during that interval.

(c) *The Newtonian principle of relativity*

By the First Law of Motion, the rectilinear way, which a 'body' may pursue, will, by virtue of the innate force (or inertia) remain unaffected so long as there are no 'external actions or impediments'. Uniform rectilinear motion is thus distinguished in Newtonian dynamics. But to the extent the direction and the magnitude of the rectilinear motion are unspecified, to that extent we can refer the motion equally to another frame of reference obtained by the transformation:

$$\vec{r}' = \vec{R}\,.\,\vec{r} + \vec{v}t + \vec{d} \qquad \text{and} \qquad t' = t + \tau, \tag{7}$$

where $\vec{v}$, $\vec{d}$, and τ are constants and $\vec{R}$ is any orthogonal matrix with constant coefficients. If O and O' denote the unprimed and the primed coordinate systems, then for a stationary observer in O', the coordinate system O will appear as rotated by $\vec{R}$ and, moving with a uniform velocity $\vec{v}$ displaced at $t = 0$ by $\vec{d}$; and further, for the observer in O', the clock in O will be running behind his own by the time τ. One can, on this account, say that *inertial frames are undistinguished: any one frame will serve as equally as any other.* This is the *principle of relativity of Galileo and Newton* as it is sometimes called (see for example Weyl (1922) *Space, time, and matter*, p. 152; Methuen, London). In view of Herivel's comments quoted in §9, the coupling of Galileo's name with Newton's would not appear to have an historical basis.

The preceding formulation of the principle of relativity leaves open the question when rectilinear motion *naturally* occurs. By Corollary IV (to quote it once again) '*the common centre of gravity of all bodies acting upon each other (excluding external actions and impediments) is either at rest or moves uniformly in right lines*'. In other words, the principle of inertia together with the First Law of Motion in Maxwell's reformulation provide the real basis for the Newtonian principle of relativity.*

Another way of formulating the principle of relativity is illustrated by considering the equations of motion governing n mass points $m_i\,(i = 1, \ldots, n)$ attracting each other according to the law:

$$\text{force of attraction between } m_i \text{ and } m_j = -m_i m_j f(|\vec{r}_j - \vec{r}_i|)\,\frac{\vec{r}_j - \vec{r}_i}{|\vec{r}_j - \vec{r}_i|}, \tag{8}$$

where f is some function of the relative distance, $|\vec{r}_j - \vec{r}_i|$, between m_i and m_j. The equations are (anticipating the problems considered in Section XI of the *Principia* and in §59):

$$\frac{\mathrm{d}^2\vec{r}_i}{\mathrm{d}t^2} = -\sum_{j \neq i}^{j=1,\ldots,n} m_j f(|\vec{r}_j - \vec{r}_i|)\,\frac{\vec{r}_j - \vec{r}_i}{|\vec{r}_j - \vec{r}_i|}, \qquad (i = 1, \ldots, n). \tag{9}$$

It can readily be verified that equation (9) retains its form when the coordinate system in which equation (9) is written is subject to the same transformation (7). From equation (9) it also follows that

$$\frac{\mathrm{d}}{\mathrm{d}t}\left[\frac{\mathrm{d}}{\mathrm{d}t}\left(\frac{1}{M}\sum_{i=1}^{n} m_i\vec{r}_i\right)\right] = 0 \qquad \left(M = \sum_{i=1}^{n} m_i\right), \tag{10}$$

proving for this system the uniform rectilinear motion of the centre of mass.

The matter is entirely summed up in Newton's concluding statement in his Definition III:

> 'But motion and rest, as commonly conceived, are only relatively distinguished; nor are those bodies always truly at rest, which commonly are taken to be so.'

* Cf. Eugene Wigner's statement 'It is fitting this principle was enunciated in full clarity, by Newton in his *Principia*' (Symmetries and Reflections: Scientific Essays of Eugene P. Wigner, (page 5) Indiana University Press, Bloomington, London, 1967).

❖3❖

On the notion of limits and the ratios of evanescent quantities

11. Introduction

In Section I, Book I of the *Principia*, Newton describes the elementary notions of differential calculus (without explicitly so stating) that are at the base of his entire treatment. And, as he explains in the Scholium at the end of the section,

> These Lemmas are premised to avoid the tediousness of deducing involved demonstrations *ad absurdum*, according to the method of the ancient geometers. For demonstrations are shorter by the method of indivisibles; but because the hypothesis of indivisibles seems somewhat harsh, and therefore that method is reckoned less geometrical, I chose rather to reduce the demonstrations of the following Propositions to the first and last sums and ratios of nascent and evanescent quantities, that is, to the limits of those sums and ratios, and so to premise, as short as I could, the demonstrations of those limits

12.

Lemma I

Quantities, and the ratios of quantities, which in any finite time converge continually to equality, and before the end of that time approach nearer to each other than by any given difference, become ultimately equal.

More explicitly the lemma states: If two quantities $X(t)$ and $Y(t)$, depending continuously on 'time' t, and neither of which vanishes in the range $t_0 < t < \infty$, and are such that

$$\lim_{t \to t_1} \left[\frac{X(t)}{Y(t)} \right] \to 1, \qquad (1)$$

for some assigned $t = t_1$, then

$$X(t_1) = Y(t_1). \tag{2}$$

Newton's proof of this lemma is to observe that if $X(t_1) \neq Y(t_1)$ then we should have a contradiction with the initial supposition (1).

An alternative version of the lemma which may satisfy the purist is to formally write the equality (1) as

$$X(t) \asymp Y(t) \qquad \text{for } t = t_1, \tag{3}$$

and observe that the basic theorems on limits allows us to conclude that

$$X(t) \asymp Y(t) \quad \text{for } t = t_1 \Leftrightarrow X(t) \text{ and } Y(t) \text{ become equal for } t = t_1. \tag{4}$$

This simple notational device (suggested by Tristan Needham) allows us 'to draw on the intuitive power of infinitesimal geometry while continuing to pay lip service to the tyrannical legacy of Cauchy and Weierstrass'.

If $X(t)$ and $Y(t)$ tend to zero for $t \to t_1$ (i.e. become 'evanescent') then greater care is necessary. As Newton explains later in the Scholium,

> .... by the ultimate ratio of evanescent quantities is to be understood the ratio of the quantities not before they vanish, nor afterwards, but with which they vanish.

And more explicitly:

> For those ultimate ratios with which quantities vanish are not truly the ratios of ultimate quantities, but limits towards which the ratios of quantities decreasing without limit do always converge; and to which they approach nearer than by any given difference, but never go beyond, nor in effect attain to, till the quantities are diminished *in infinitum.*

The underlined statement emphasizes that the limit (A/B) is not always the same as $(\lim A)/(\lim B)$.

13. Lemmas II, III, and IV

Lemma II

If in any figure AacE, terminated by the right lines Aa, AE, and the curve acE, there be inscribed any number of parallelograms Ab, Bc, Cd, etc., comprehended under equal bases AB, BC, CD, etc., and the sides, Bb, Cc, Dd, etc., parallel to

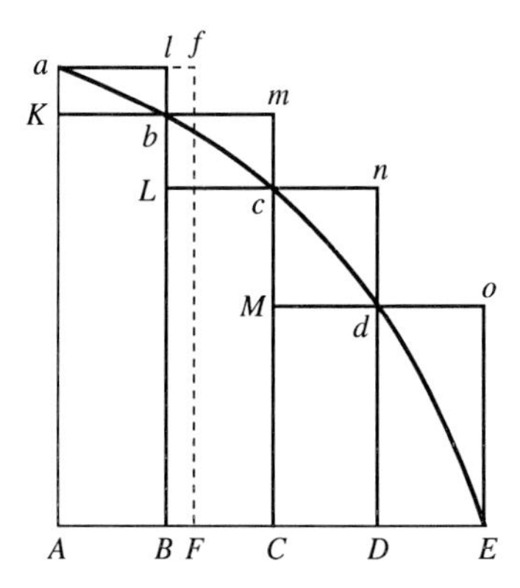

one side Aa of the figure; and the parallelograms aKbl, bLcm, cMdn, etc., are completed: then if the breadth of those parallelograms be supposed to be diminished, and their number to be augmented in infinitum, *I say, that the ultimate ratios which the inscribed figure AKbLcMdD, the circumscribed figure AalbmcndoE, and curvilinear figure AabcdE, will have to one another, are ratios of equality.*

Proof:

$$\begin{aligned}&\text{area of the circumscribed parallelograms } Aalbmcnd\ldots oE\\&\qquad - \text{area of inscribed parallelograms } AKbLcMd\ldots DE\\&= \text{area } AalB = Aa \times AB;\end{aligned}$$

or,

$$\frac{Aalbmcd\ldots oE}{AKbLcMd\ldots DE} = 1 + AB\frac{Aa}{AKbLcMd\ldots DE}.$$

We conclude

$$Aalbmcd\ldots oE \asymp AKbLcMd\ldots DE \qquad (AB \to 0);$$

that is, when the division becomes infinitely fine. Also since each of the parallelograms, *Kalb*, *Lbmc*, *Mcnd*, and so on, are greater in area than the curvilinear figures *Kab*, *Lbc*, *Mcd*, and so on, respectively, it equally follows that

$$Aalbmcd\ldots oE \asymp AKbLcMd\ldots DE \asymp Aabcd\ldots E.$$

In other words, in the limit $AB \to 0$ areas of the circumscribed and the inscribed parallelograms become equal and coincide by definition with that of the curvilinear figure.

Even for a modern reader, Newton's proofs of Lemmas III and IV cannot be improved upon.

Lemma III

The same ultimate ratios are also ratios of equality, when the breadths AB, BC, DC, etc., of the parallelograms are unequal, and are all diminished in infinitum.

For suppose *AF* equal to the greatest breadth, and complete the parallelogram *FAaf*. This parallelogram will be greater than the difference of the inscribed and circumscribed figures; but, because its breadth *AF* is diminshed *in infinitum*, it will become less than any given rectangle. Q.E.D.

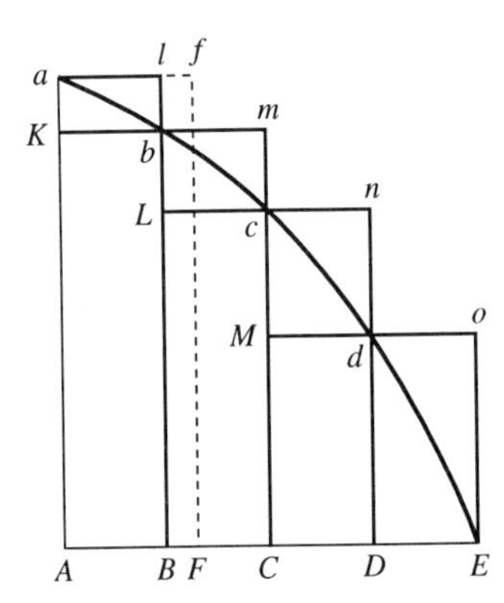

COR. I. Hence the ultimate sum of those evanescent parallelograms will in all parts coincide with the curvilinear figure.

COR. II. Much more will the rectilinear figure comprehended under the chords of the evanescent arcs *ab*, *bc*, *cd*, etc., ultimately coincide with the curvilinear figure.

COR. III. And also the circumscribed rectilinear figure comprehended under the tangents of the same arcs.

COR. IV. And therefore these ultimate figures (as to their perimeters *acE*) are not rectilinear, but curvilinear limits of rectilinear figures.

Lemma IV

If in two figures AacE, PprT, there are inscribed (as before) two series of parallelograms, an equal number in each series, and, their breadths being diminished in infinitum, *if the ultimate ratios of the parallelograms in one figure to those in the other, each to each respectively, are the same: I say, that those two figures, AacE, PprT, are to each other in that same ratio.*

For as the parallelograms in the one are severally to the parallelograms in the other, so (by composition) is the sum of all in the one to the sum of all in the other; and so is the one figure to the other; because (by Lem. III) the former figure to the former sum, and the latter figure to the latter sum, are both in the ratio of equality. Q.E.D.

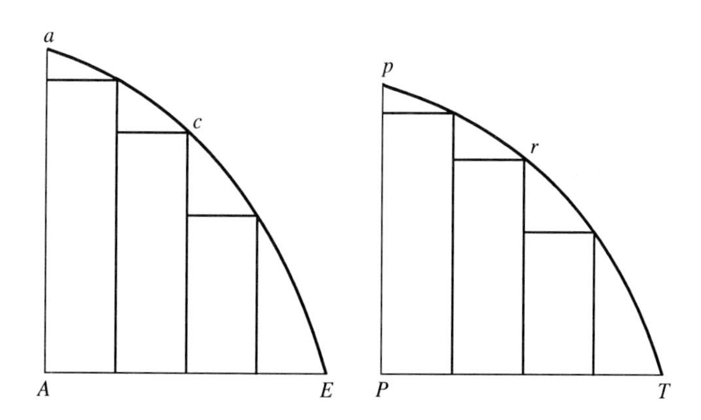

The corollary to Lemma IV states effectively that: if we consider the increasingly finer division of two right lines $A_0, \ldots, A_n$ and $B_0, \ldots, B_n$,

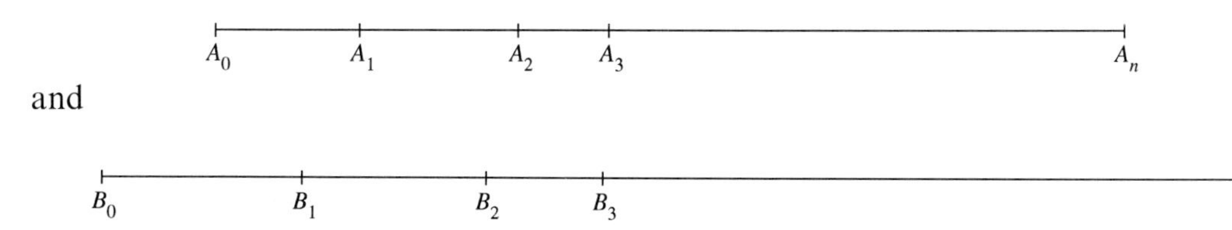

and

such that, at each division, $(A_{m+1} - A_m)/(B_{m+1} - B_m) = \alpha$ $(m = 0, \ldots, n - 1)$ where α is a constant, then in the limit $n \to \infty$ and $A_{m+1} - A_m \to 0$, the lengths of the two right-lines are in the ratio α.

14. Lemmas V–VIII

Lemma V

All homologous sides of similar figures, whether curvilinear or rectilinear, are proportional; and the areas are as the squares of the homologous sides.

Without any comments, Newton passes on to the next lemma.

Lemma VI

If any arc ACB, given in position, is subtended by its chord AB, and in any point A, in the middle of the continued curvature, is touched by a right line AD, produced both ways; then if the points A and B approach one another and meet, I say, the angle BAD, contained between the chord and the tangent, will be diminished in infinitum, *and ultimately will vanish.*

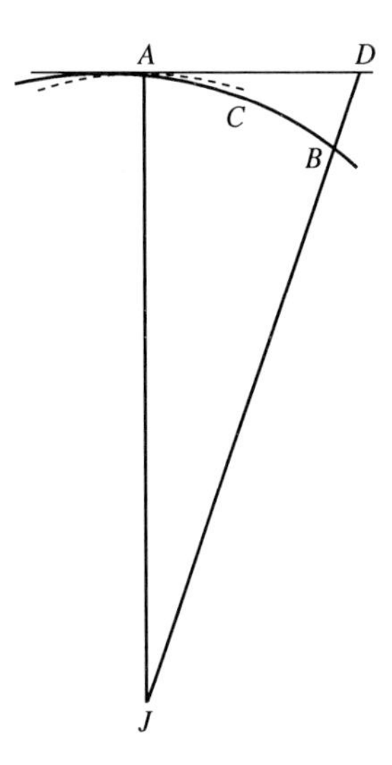

Newton proves this lemma with the observation that if the angle of contact at A does not vanish, then the assumption that the curve continues beyond A continuously will be contradicted. Later in the Scholium he amplifies this brief comment by:

> we have all along supposed the angle of contact to be neither infinitely greater nor infinitely less than the angles of contact made by circles and their tangents; that is, that the curvature at the point A is neither infinitely small nor infinitely great, and that the interval AJ is of a finite magnitude.

Lemma VII

The same things being supposed, I say that the ultimate ratio of the arc, chord, and tangent, any one to any other, is the ratio of equality.

Normally we should prove this lemma as follows. Let AD denote the tangent and AR the normal at A; and R be the centre of curvature. Extend the circle of contact at A to its antipodal point A'. If B is a point on the curve sufficiently near to A and the angle $\angle ARB = \delta\theta$, then by elementary geometry,

$$\begin{aligned}\text{arc } \widehat{ACB} &= R\delta\theta,\\ \text{chord } AB &= 2R \sin \tfrac{1}{2}\delta\theta, \text{ and}\\ \text{tangent } AD &= AB \sec \tfrac{1}{2}\delta\theta,\\ &= 2R \tan \tfrac{1}{2}\delta\theta;\end{aligned}$$

and it follows that

$$\text{arc } ACB : \text{chord } AB : \text{tangent } AD \to 1:1:1 \quad \text{as } \delta\theta \to 0.$$

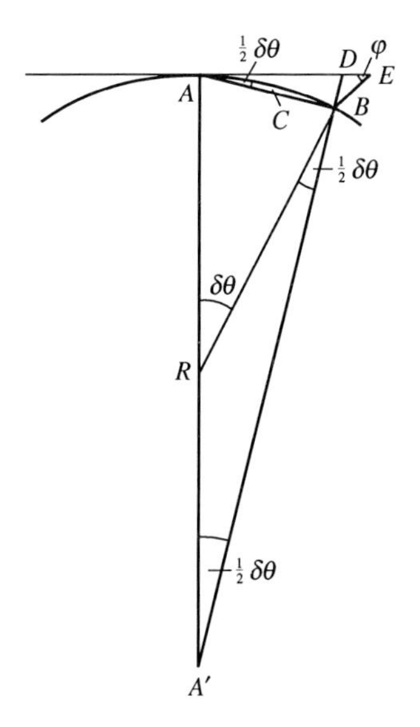

Corollaries I, II, and III

Also if from B we now draw BE inclined at an angle φ with EA and keep $\angle BEA$ constant as $B \to A$, then,

$$\frac{AE}{AB} = \frac{\sin(180^\circ - \varphi - \tfrac{1}{2}\delta\theta)}{\sin\varphi} = \frac{\sin(\varphi + \tfrac{1}{2}\delta\theta)}{\sin\varphi} \to 1 \quad \text{as } \delta\theta \to 0.$$

It follows, that in the limit AD, AE, AB, and arc $\widehat{ACB}$ become mutually equal.

Newton proves the lemma differently.

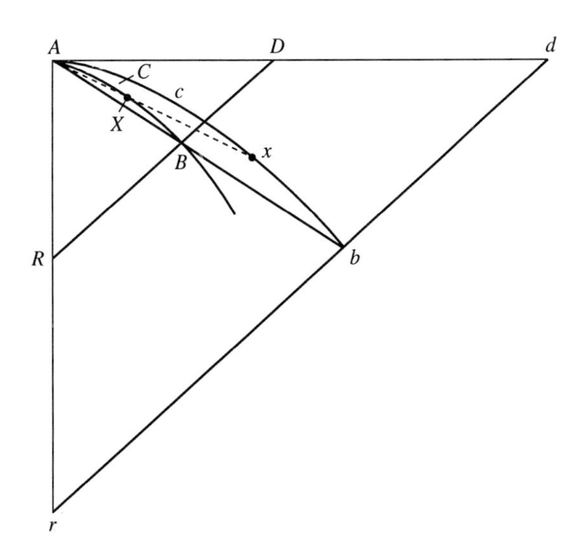

Consider a point B on a smooth curve $\overset{\frown}{ACB}$ approaching the point A. Let AD be the tangent to the curve at A. Prolong the right lines AB and AD to 'remote' points b and d. Draw the arc $\overset{\frown}{Acb}$ similar to $\overset{\frown}{ACB}$, that is, if X is any point on $\overset{\frown}{ACB}$ then AX intersects $\overset{\frown}{Acb}$ at x at a distance Ax such that $AX : Ax = AB : Ab$. As B approaches A, let the secant BD remain parallel to itself and to the 'remote' secant bd. It is clear that by this construction, the ratios of the lengths of the chord AB, the arc $\overset{\frown}{ACB}$, and the tangent AD will remain, mutually among themselves, in the same constant ratios as the remote (and finite) chord Ab, arc $\overset{\frown}{Acb}$ and tangent Ad as B continues to approach A. It follows:

$$\lim \frac{AB}{\overset{\frown}{ACB}} = \lim \frac{Ab}{Acb} = 1,$$

$$\lim \frac{AB}{AD} = \lim \frac{Ab}{Ad} = 1,$$

and

$$\lim \frac{\overset{\frown}{ACB}}{\mathrm{AD}} = \lim \frac{\overset{\frown}{Acb}}{\mathrm{Ad}} = 1. \qquad \text{Q.E.D.}$$

Simple and elegant!

Lemma VIII

If the right lines AR, BR, with the arc ACB, the chord AB, and the tangent AD, constitute three triangles RAB, RACB, RAD, and the points A and B approach and meet: I say, that the ultimate form of these evanescent triangles is that of similitude, and their ultimate ratio that of equality.

By the same construction as in Lemma VII, the $\triangle$s RAB and rAb, $\triangle$s $R\overset{\frown}{ACB}$, and $r\overset{\frown}{Acb}$, and $\triangle$s RAD and rAd remain (respectively) similar as B approaches A and in the limit. Q.E.D.

15. Lemmas IX and X

Lemma IX

If a right line AE, and a curved line ABC, both given by position, cut each other in a given angle, A; and to that right line, in another given angle, BD, CE are ordinately applied, meeting the curve in B, C; and the points B and C together approach towards and meet in the point A: I say, that the areas of the [curvilinear] *triangles ABD, ACE, will ultimately be to each other as the squares of homologous sides.*

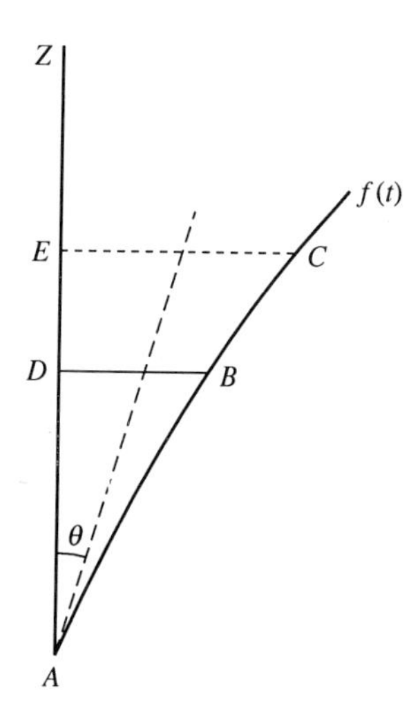

Let the curve AC be the trajectory described by a body; and let the equation of the curve be $f(t)$, t denoting the 'time' measured along AZ. Further let D and E denote varying times along AZ maintaining a constant ratio between AD and AE. Then the space described by B and C will be given by

$$\triangle \widehat{ABD} = \int_0^t DB\,\mathrm{d}t = \int_0^t f(t)\,\mathrm{d}t,$$

and

$$\triangle \widehat{ACE} = \int_0^t EC\,\mathrm{d}t = \alpha \int_0^t f(\alpha t)\,\mathrm{d}t.$$

We assume that

$$f(t) \to f'(0)t \quad \text{as } t \to 0.$$

Then

$$\triangle \widehat{ABD} \to \tfrac{1}{2}f'(0)t^2 \quad \text{and} \quad \triangle \widehat{AEC} \to \tfrac{1}{2}f'(0)\alpha^2 t^2 \qquad \text{as } t \to 0.$$

Therefore

$$\triangle \widehat{ABD} : \triangle \widehat{AEC} \to \alpha^2 \quad \text{for } t \to 0. \qquad \text{Q.E.D.}$$

Lemma X

The spaces which a body describes by any finite force urging it, whether that force is determined and immutable, or is continually augmented or continually diminished, are in the very beginning of the motion to each other as the squares of the times.

This is a restatement of Lemma IX explicitly when the trajectory is that described under the action of a force.

Corollaries I, II, and III are equivalent to the statement

$$x(t) = \int_0^t \mathrm{d}t\, v(t) = \frac{1}{m}\int_0^t \mathrm{d}t_1 \int_0^{t_1} \mathrm{d}t_2\, F(t_2) \quad (m = \text{the inertial mass}).$$

and that, if

$$F(t_1) \to F_0 \quad \text{for } t \to 0,$$

then,

$$x(t) \to \frac{1}{m}\int_0^t \mathrm{d}t_1\, t_1 F_0 = \frac{F_0}{2m}\, t^2.$$

The most important are the Corollaries IV and V:

> COR. IV. And therefore the forces are directly as the spaces described in the very beginning of the motion, and inversely as the squares of the times.
>
> COR. V. And the squares of the times are directly as the spaces described, and inversely as the forces.

16. Lemma XI

The evanescent subtense of the angle of contact, in all curves which at the point of contact have a finite curvature, is ultimately as the square of the subtense of the conterminous arc.

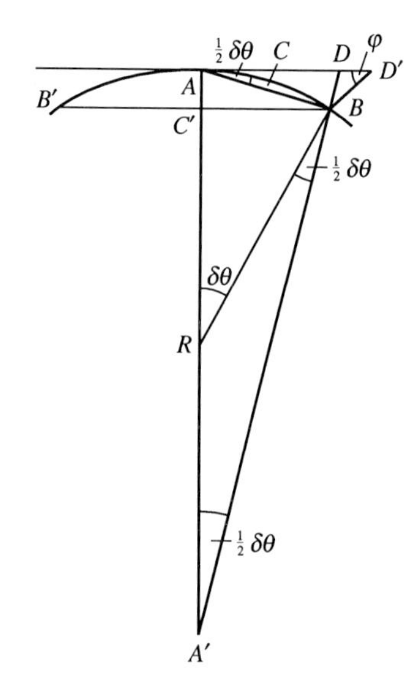

Newton's proof is not different from what one might give currently. Notice, however, that in Newton's terminology:

subtense of the angle of contact at $A = BD$ perpendicular to the tangent AD at A,

and

versed sine of the arc $(\widehat{BAB'})$ is AC' bisecting the chord $(B'B)$.

(Note that this figure is virtually the same as that accompanying Lemma VII.)

Case 1. As in Lemma VII

$$\text{Arc } AB = R\delta\theta, \qquad \text{chord } AB = 2R \sin \tfrac{1}{2}\delta\theta$$

and

$$AD = 2R \tan \tfrac{1}{2}\delta\theta \qquad (\delta\theta \to 0).$$

We therefore have

$$BD = AB \sin \tfrac{1}{2}\delta\theta = 2R \sin^2 \tfrac{1}{2}\delta\theta.$$

We conclude that

$$BD = \frac{1}{2R} AB^2 + O(\delta\theta) \qquad (\text{as } B \to A).$$

(Note that from the similarity of $\triangle$s $AA'B$ and BAD,

$$\frac{AB}{AA'} = \frac{BD}{AB} \qquad \text{or} \qquad AB^2 = AA' \cdot BD = 2R.BD,$$

the relation that Newton uses.)

Case 2. Given that BD' is inclined at some (unspecified but fixed angle φ) to AD, then

$$\frac{AD'}{AB} = \frac{\sin(180^\circ - \varphi - \frac{1}{2}\delta\theta)}{\sin \varphi} = \frac{\sin(\varphi + \frac{1}{2}\delta\theta)}{\sin \varphi} \to 1 \qquad (\delta\theta \to 0)$$

independently of φ. Hence,

$$BD \to \frac{1}{2R} AB^2 \to \frac{1}{2R} (AD')^2.$$

A number of corollaries follow.

Corollary I

Since $AB/AD \to 1$ as $B \to A$, it follows that

$$\frac{BD}{AD^2} \to \frac{1}{2R}.$$

Corollary II

Versed sine $AC = AB$; therefore $2R.BD \to (\text{versed sine})^2$.

Corollary III

If the arc AB is described uniformly with a velocity v (so that $R\delta\theta = v\,\mathrm{d}t$) it follows that

$$BD \to \frac{1}{2R}(AB)^2 = \frac{1}{2R}v^2(\mathrm{d}t)^2.$$

Corollary IV

$$\text{area of } \triangle ABD = \tfrac{1}{2}AD.BD = \frac{1}{4R}(AD)^3.$$

Corollary V

Letting $BD = y$ and $AD = x$, the relation between y and x is, ultimately,

$$y = \frac{x^2}{2R},$$

that is, parabolic. Hence the curvilinear triangle

$$\triangle \widehat{ABD} = \int_0^x y\,\mathrm{d}x = \frac{x^3}{6R} = \frac{(AD)^3}{6R}.$$

Therefore, by Corollary IV,

$$\frac{\triangle \widehat{ABD}}{\triangle ABD} \to \tfrac{2}{3}.$$

Further, since the area of the segment AB bounded by the arc $AB = \triangle ABD - \triangle \widehat{ABD} = AD^3/12R$

$$\frac{\text{area of segment}}{\triangle ABD} \to \tfrac{1}{3}.$$

In the Scholium Newton points out that the relation

$$BD \propto AD^2 \qquad (\text{as } B \to A)$$

is a necessary and sufficient condition for the contact at A to be the same as for circles and their tangents. He considers more generally the case

$$BD \propto (AD)^{\alpha}$$

and remarks that if $\alpha > 2$, the angle of contact is infinitely less than when $\alpha = 2$ and similarly, the angle of contact is infinitely greater than when $\alpha = 2$, if $\alpha < 2$; and concludes with the remark

Nor is Nature confined to any bounds.

4

On the motion of particles under centripetal attraction: an introduction to Newton's treatment

'Since I was aware that there exists an infinite number of points on the orbit and accordingly an infinite number of distances [from the Sun] the idea occurred to me that the sum of these distances is contained in the *area* of the orbit. For I remembered that in the same manner Archimedes too divided the area of a circle into an infinite number of triangles.'

(Kepler on his discovery of the law of areas)

'Why should I mince my words? The truth of Nature, which I had rejected and chased away, returned by stealth through the backdoor, disguising itself to be accepted .... I thought and searched, until I went nearly mad, for a reason, why the planet preferred an elliptical orbit ....'

(Kepler on his discovery of the orbit of Mars as an ellipse)

17. Introduction

Newton's treatment of Kepler's laws of planetary motion is among the most celebrated parts of the *Principia.* To appreciate the novelty and the originality of Newton's treatment, one must have in mind, for comparison, a more conventional treatment to which one is accustomed. Besides, it is important to have a treatment of elliptical motion under the inverse-square law of attraction that is readily adapted to obtaining the equations governing the perturbations of ellipitic motion based on the 'variation of the constants'. For as Tisserand* has stated, Newton undoubtedly had in his possession these variational equations for this treatment of lunar perturbations. We shall consider these matters in due course (Chapter 22); in the meantime, we shall give a brief account of the dynamics of a particle under general centripetal attraction and in particular the

* F. Tisserand, *Mecanique celeste*, Chapter III, p. 27–44. On p. 33 he states, 'I am inclined to think that he [Newton] knew all of the formulae (a) [the variational equations (1)–(6) given in Chapter 13, pp. 233–4]; but that instead of publishing them, he preferred deducing a large number of geometric propositions from them which he obtained by considering in each case one of the components.'

case of the inverse-square law of attraction. We shall follow in main the text of Harry Pollard.*

18. The dynamics of a particle under a general law of centripetal attraction

We shall assume that the force acting on a particle of inertial mass, m, at a distance, r, from the centre of attraction is

$$\vec{F} = -mf(r)\frac{\vec{r}}{r} \qquad (r = |\vec{r}|) \tag{1}$$

where $\vec{r}/r$ is a unit vector in the direction of increasing $\vec{r}$, and m, equal to the gravitational mass, is set equal to the inertial mass m. The equation of motion governing the particle is

$$\frac{\mathrm{d}^2\vec{r}}{\mathrm{d}t^2} = -f(r)\frac{\vec{r}}{r}, \tag{2}$$

where we observe that m does not occur in this equation. Letting

$$\vec{v} = \frac{\mathrm{d}\vec{r}}{\mathrm{d}t}, \tag{3}$$

we can rewrite equation (2) in the alternative form

$$\frac{\mathrm{d}\vec{v}}{\mathrm{d}t} = -f(r)\frac{\vec{r}}{r}. \tag{4}$$

(a) The conservation of angular momentum

From equation (4) it follows:

$$\frac{\mathrm{d}}{\mathrm{d}t}(\vec{r} \times \vec{v}) = \frac{\mathrm{d}\vec{r}}{\mathrm{d}t} \times \vec{v} + \vec{r} \times \frac{\mathrm{d}\vec{v}}{\mathrm{d}t} = \vec{0}, \tag{5}$$

the two terms on the right-hand side vanishing by virtue of equations (3) and (4) respectively. Letting

$$\vec{h} = \vec{r} \times \vec{v}, \tag{6}$$

we conclude from equation (5) that

$$\vec{h} = \text{a constant.} \tag{7}$$

This equation expresses the *conservation of angular momentum*.

* Harry Pollard (1966) '*Mathematical introduction to celestial mechanics.*' Prentice Hall, New Jersey.

Since, by equation (6),

$$\vec{h}\cdot\vec{r} = 0 \tag{8}$$

the *motion of the particle is restricted to the plane orthogonal to* $\vec{h}$ provided $\vec{h} \neq 0$. If

$$\vec{h} = \vec{0}, \tag{9}$$

we must proceed differently. Evaluating the time derivative of $\vec{r}/r$ we find:

$$\frac{d}{\mathrm{d}t}\frac{\vec{r}}{r} = \frac{(\vec{r} \times \vec{v}) \times \vec{r}}{r^3} = \frac{\vec{h} \times \vec{r}}{r^3}. \tag{10}$$

It follows that if $\vec{h} = \vec{0}$, the *motion is radial.*

(*b*) *The law of areas*

If $\vec{h} \neq \vec{0}$, the orbital plane is normal to $\vec{h}$. Choosing the z-axis along the direction of $\vec{h}$, we may write (see the illustration)

$$\vec{h} = (0, 0, h) \qquad \text{and} \qquad \vec{r} = (r \cos \varphi, r \sin \varphi, 0). \tag{11}$$

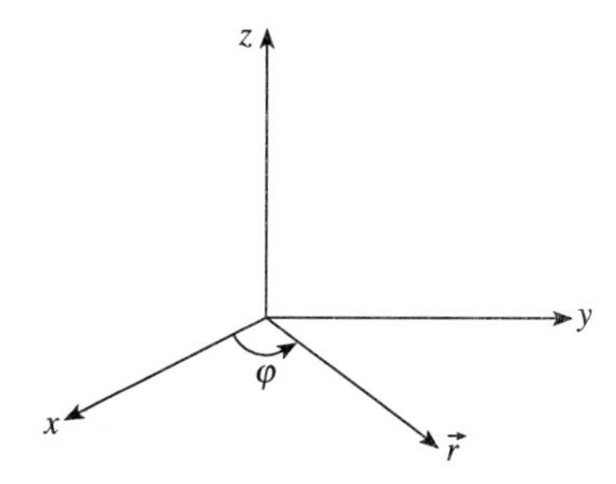

Accordingly

$$\vec{v} = \left(\frac{\mathrm{d}r}{\mathrm{d}t} \cos \varphi - r \sin \varphi \frac{\mathrm{d}\varphi}{\mathrm{d}t}, \frac{\mathrm{d}r}{\mathrm{d}t} \sin \varphi + r \cos \varphi \frac{\mathrm{d}\varphi}{\mathrm{d}t}, 0\right); \tag{12}$$

and we verify that

$$(\vec{r} \times \vec{v}) = r^2 \frac{\mathrm{d}\varphi}{\mathrm{d}t} \frac{\vec{z}}{|z|} = \text{constant vector}; \tag{13}$$

or if A denotes the area swept by the radius vector in the orbital plane

$$\frac{\mathrm{d}A}{\mathrm{d}t} = \tfrac{1}{2}r^2 \frac{\mathrm{d}\varphi}{\mathrm{d}t} = \tfrac{1}{2}h = \text{constant}. \tag{14}$$

This is Kepler's *law of areas.*

(c) The conservation of energy

By equation (4)

$$\vec{v}\cdot\frac{\mathrm{d}\vec{v}}{\mathrm{d}t} = -\frac{f(r)}{r}(\vec{r}\cdot\vec{v}) = -\frac{f(r)}{r}\left(\vec{r}\cdot\frac{\mathrm{d}\vec{r}}{\mathrm{d}t}\right)$$

$$= -f(r)\frac{\mathrm{d}r}{\mathrm{d}t}. \tag{15}$$

Therefore,

$$\tfrac{1}{2}\frac{\mathrm{d}}{\mathrm{d}t}v^2 = -\frac{\mathrm{d}}{\mathrm{d}t}\int^r f(r)\,\mathrm{d}r. \tag{16}$$

Assuming that $f(r) < r^{-1}$, and letting

$$V(r) = \int_r^\infty f(r)\,\mathrm{d}r, \tag{17}$$

we obtain, from equation (16),

$$\tfrac{1}{2}v^2 = V(r) + E. \tag{18}$$

where E is a constant. As defined, $-V$ denotes the *potential energy* and E *the total energy*.

Since

$$(\vec{r}\cdot\vec{v})^2 + |\vec{r}\times\vec{v}|^2 = r^2v^2 \tag{19}$$

we obtain the alternative form of the energy integral:

$$(\vec{r}\cdot\vec{v})^2 + h^2 = 2r^2(V + E). \tag{20}$$

(d) The equation governing r in the orbital plane

From $r^2 = |\vec{r}|^2$, we obtain

$$r\frac{\mathrm{d}r}{\mathrm{d}t} = \vec{r}\cdot\frac{\mathrm{d}\vec{r}}{\mathrm{d}t} = \vec{r}\cdot\vec{v}. \tag{21}$$

Differentiating this equation once again with respect to t, we obtain,

$$r\frac{\mathrm{d}^2r}{\mathrm{d}t^2} + \left(\frac{\mathrm{d}r}{\mathrm{d}t}\right)^2 = \vec{r}\cdot\frac{\mathrm{d}\vec{v}}{\mathrm{d}t} + v^2; \tag{22}$$

or, making use of the identity (19) on the right-hand side, we obtain

$$r\frac{\mathrm{d}^2r}{\mathrm{d}t^2} + \left(\frac{\mathrm{d}r}{\mathrm{d}t}\right)^2 = \vec{r}\cdot\frac{\mathrm{d}\vec{v}}{\mathrm{d}t} + \frac{h^2}{r^2} + \frac{1}{r^2}(\vec{r}\cdot\vec{v})^2. \tag{23}$$

Since, by equations (4) and (21),

$$\vec{r}\cdot\frac{\mathrm{d}\vec{v}}{\mathrm{d}t} = -rf(r) \qquad \text{and} \qquad \frac{1}{r^2}(\vec{r}\cdot\vec{v})^2 = \left(\frac{\mathrm{d}r}{\mathrm{d}t}\right)^2 \tag{24}$$

equation (23) simplifies to give

$$\frac{\mathrm{d}^2 r}{\mathrm{d}t^2} - \frac{h^2}{r^3} = -f(r). \tag{25}$$

This equation, together with the equation

$$r^2\frac{\mathrm{d}\varphi}{\mathrm{d}t} = h \tag{26}$$

enables us to obtain the equation governing the trajectory in the orbital plane. Letting

$$r = u^{-1}, \tag{27}$$

we obtain, by familiar transformations,

$$\frac{\mathrm{d}^2 u}{\mathrm{d}\varphi^2} + u = \frac{f(r)}{h^2 u^2}. \tag{28}$$

19. The dynamics of a particle under the inverse-square law of attraction

When the inverse-square law of attraction obtains,

$$f(r) = \frac{\mu}{r^2} \qquad (\mu\text{, a positive constant}). \tag{1}$$

Specializing the equations of §18, we have:

$$\frac{\mathrm{d}\vec{v}}{\mathrm{d}t} = -\frac{\mu}{r^3}\vec{r} \qquad \text{and} \qquad \vec{h} = \vec{r}\times\vec{v} = \text{a constant vector.} \tag{2}$$

We also have

$$r^2\frac{\mathrm{d}\varphi}{\mathrm{d}t} = h, \qquad V(r) = \int_r^\infty \frac{\mu}{r^2}\,\mathrm{d}r = \frac{\mu}{r}, \tag{3}$$

and

$$\tfrac{1}{2}v^2 = E + \mu/r. \tag{4}$$

(a) *The Lenz vector and the Lenz equation*

An integral, peculiar to the inverse-square law, is the following. By making use of equation (2) we can rewrite equation (10) of §18 in the manner

$$-\mu \frac{d}{dt}\frac{\vec{r}}{r} = -\mu \frac{\vec{h} \times \vec{r}}{r^3} = \left(\vec{h} \times \frac{d\vec{v}}{dt}\right). \tag{5}$$

This equation integrates to give

$$\mu \frac{\vec{r}}{r} = (\vec{v} \times \vec{h}) - \mu\vec{e}, \tag{6}$$

where $\vec{e}$ is a constant vector. It is the *Lenz vector*. We rewrite equation (6) in the form

$$\mu\left(\vec{e} + \frac{\vec{r}}{r}\right) = \vec{v} \times \vec{h}. \tag{7}$$

The scalar product of this equation with $\vec{h}$ gives:

$$\vec{e}\cdot\vec{h} = 0. \tag{8}$$

that is, the Lenz vector lies in the orbital plane.

Multiplying equation (7) scalarly by $\vec{r}$, we obtain the equation

$$\vec{e}\cdot\vec{r} + r = h^2/\mu. \tag{9}$$

If we now choose the origin of φ along $\vec{e}$, equation (9) gives*

$$r(1 + e\cos\varphi) = h^2/\mu, \tag{10}$$

or,

$$r = \frac{h^2}{\mu}\frac{1}{1 + e\cos\varphi}, \tag{11}$$

which is the equation of an ellipse with eccentricity e and semilatus rectum,

$$l = h^2/\mu. \tag{12}$$

Since

$$l = a(1 - e^2), \tag{13}$$

where a denotes the semimajor axis of the ellipse,

$$a = \frac{l}{1 - e^2} = \frac{h^2}{\mu(1 - e^2)}. \tag{14}$$

* If instead, the orbit is a hyperbola, equation (10) must be written in the form,

$$r(e\cos\varphi - 1) = h^2/\mu; \tag{10}$$

and corresponding changes must be made in the subsequent equations: for example, $(e^2 - 1)$ in place of $(1 - e^2)$ in equations (13) and (14).

The principal significance of the *Lenz vector of length e in the orbital plane is that it ensures the fixity in space* (*or, as* Newton *calls 'the quiescence of the aphelion points' in Proposition II, Book III*)* *of the direction of the major axis of the elliptical orbits.*

An identity which follows from the Lenz equation (7) is worth noticing. The square of this equation gives

$$\mu^2\left(e^2 + \frac{2}{r}\,\vec{e}\cdot\vec{r} + 1\right) = v^2h^2. \tag{15}$$

Making use of equation (9) and the energy integral (4), we obtain, on further simplification,

$$\mu^2(e^2 - 1) = 2h^2E; \tag{16}$$

or since

$$h^2/\mu = a(1 - e^2), \tag{17}$$

we have

$$a = -\tfrac{1}{2}\mu/E. \tag{18}$$

(*b*) *Kepler's third law*

Since

$$\text{the area of the elliptical orbit} = \pi ab = \pi a^2\sqrt{(1 - e^2)} \tag{19}$$

and

$$\text{the constant of areas} = A_{,t} = \tfrac{1}{2}h, \tag{20}$$

it follows that the period P of the orbit is given by

$$P = \frac{2\pi}{h}\,a^2\sqrt{(1 - e^2)}; \tag{21}$$

or since (cf. equation (14))

$$h = \sqrt{[\mu a(1 - e^2)]}, \tag{22}$$

we find

$$P = \frac{2\pi}{\sqrt{\mu}}\,a^{3/2}. \tag{23}$$

This is Kepler's third law.

* Or, more explicitly, 'The aphelions are immovable; and so are the planes of the orbit' as in Proposition XIV of Book III, (Chapter (9)).

(c) *An alternative derivation of the elliptical orbit*

Inserting for f its present value given in equation (1), equation (28) of §18 becomes

$$\frac{d^2u}{d\varphi^2} + u = \frac{\mu}{h^2}. \tag{24}$$

The solution of this equation is

$$\frac{1}{r} = u = \frac{\mu}{h^2}(1 + e \cos \varphi), \tag{25}$$

in agreement with the solution (11) obtained earlier with the Lenz equation.

20. The accelerations and velocities along a curved orbit

We conclude this chapter by a summary of what we know about the velocities and the accelerations along a curved orbit.

S: imovable centre of force;
r: radius vector directed outward from S;
s: arc length measured along trajectory;
C: centre of curvature;

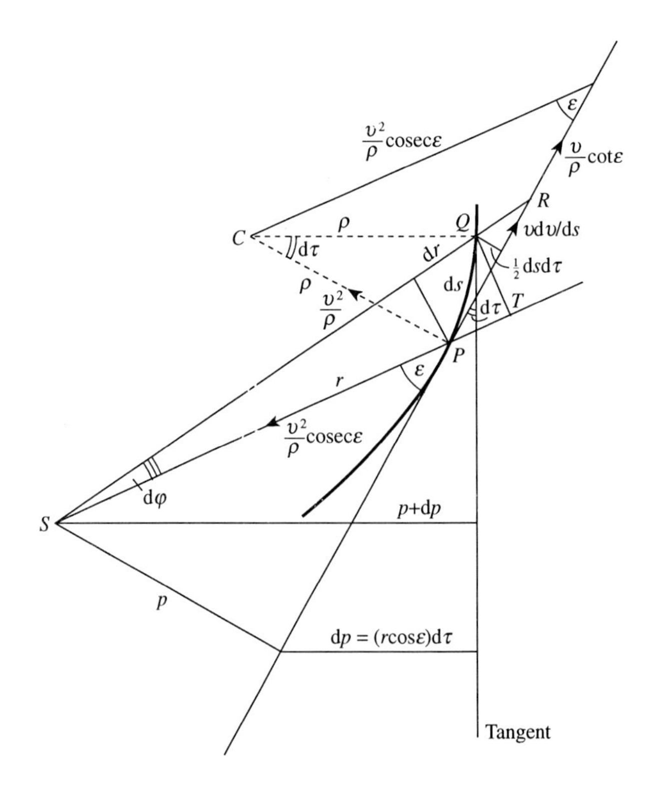

ds: infinitesimal element of arc;
ϵ: angle between the directions of r and tangent at P;
p: perpendicular distance from S on instantaneous tangent: $p = r \sin \epsilon$;
dτ: angle subtended by ds at C: $\mathrm{d}s = \rho \, \mathrm{d}\tau$;
ρ: radius of curvature;
v: velocity along the trajectory: $v = \mathrm{d}s/\mathrm{d}\tau$;
a_s: acceleration along the tangential direction: $a_s = v \, \mathrm{d}v/\mathrm{d}s$;
a_ρ: acceleration along the inward normal: $a_\rho = v^2/\rho$;
a_r: acceleration along the radius vector towards S:

$$a_r = \text{centripetal force} = v^2 \operatorname{cosec} \epsilon/\rho.$$

5

The law of areas and some relations which follow

Since the equable description of areas indicates that there is a centre to which tends that force by which the body is most affected, and by which it is drawn back from its rectilinear motion, and retained in its orbit, why may we not be allowed, in the following discourse, to use the equable description of areas as an indication of a centre, about which all circular [circulatory]* motion is performed in free spaces?

Newton in the Scholium to Propositions I, II, and III

21. Introduction

In this chapter we consider the propositions and lemmas in Section II, Book I of the *Principia*. These propositions and lemmas (as those of Section III) are, in the main, included in *De Motu Corporum in Gyrum* written in the summer of 1684 (cf. Chapter 1, §4) but not with the completeness or incisiveness as in the *Principia*.

22. The area theorem

Proposition I. Theorem I

The areas which revolving bodies describe by radii drawn to an immovable centre of force do lie in the same immovable planes, and are proportional to the times in which they are described.

Let S be the centre of attraction. Consider the orbit as described under a centripetal force of attraction, acting intermittently, as impulses, at equal intervals of time δt apart; and let B, C, D, E, F, etc., be the positions of the particle at times δt, $2\delta t$, $3\delta t$, $4\delta t$, etc. (We shall eventually let $\delta t \to 0$.)

The particle, initially at A and moving with a velocity $\vec{v}$, will, by the operation of inertia (by Law 1), travel a distance $v\delta t$, in the direction of AB and *in the plane containing* SA and $\vec{v}$, and arrive at B. At B the particle will be subject to the centripetal force in the direction BS (acting impulsively, as stated) and be displaced to V (say). In the absence

* In the context, 'circulatory' is more appropriate than 'circular'; and Newton must have meant circulatory: for Propositions I, II, and III are entirely general and are not in any way restricted to circular orbits. Circular orbits are considered in Proposition IV that follows.

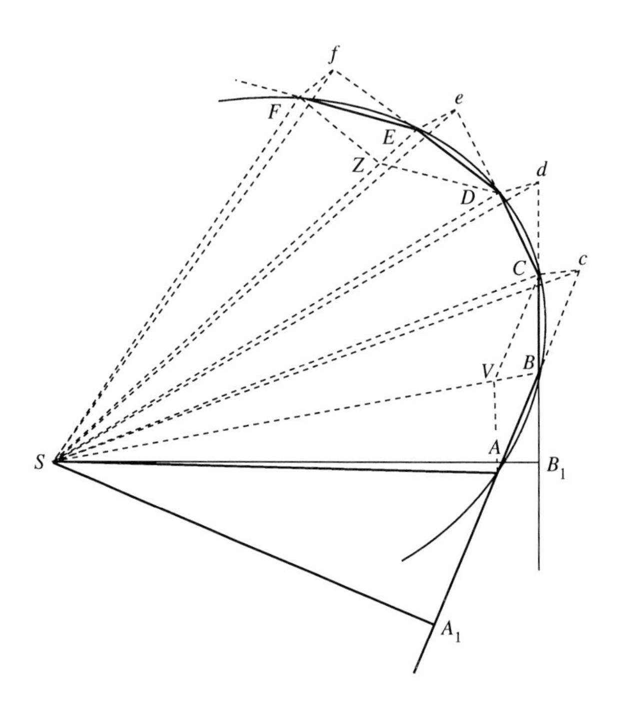

of the impulse it received at *B*, the particle, by inertia, will continue to move on in the direction *AB* and arrive at *c* after having travelled the same distance $v\delta t = AB$ (as in the first interval); and in the *same* plane. By the composition of the displacements *BV* and *Bc*, the particle at the end of the second interval $2\delta t$, will find itself (by Corollary I of the Laws (p. 24)) at *C* where *cC* is equal and parallel to *BV*. By these constructions, the areas of the $\triangle$s *SAB* and *SBc* will be equal (having equal bases *AB* and *Bc*); and the areas of $\triangle$s *SBc* and *SBC* will, likewise, be equal (because *cC* and *BS* are parallel).

Therefore, the

$$\text{area of } \triangle SAB = \text{area of } \triangle SBC.$$

In other words, the areas described in two successive intervals of time, δt, apart, are equal; while the particle continues to move in the same plane *SAB*.

By similar arguments, at the end of the third interval $3\delta t$, the particle will find itself at *D*, in the same plane as *SAB*, and the

$$\text{area of } \triangle SBC = \text{area of } \triangle SCD.$$

Therefore the particle describes the same area during the third interval δt as during each of the first two intervals. By induction, equal areas are described in *all* intervals of time, δt apart. It further follows that the areas described in two sequences of intervals will be proportional to the durations of the two sequences. And as Newton continues:

> Now let the number of those triangles be augmented, and their breadth diminished *in infinitum*; and (by Cor. IV, Lem. III) their ultimate perimeter *ADF*

will be a curved line: and therefore the centripetal force, by which the body is continually drawn back from the tangent of this curve, will act continually; and any described areas *SADS*, *SAFS*, which are always proportional to the times of description, will, in this case also, be proportional to those times. Q.E.D.

Corollary I

We have shown that

$$AB = v_A\delta t, \qquad BC = v_B\delta t, \qquad CD = v_C\delta t, \qquad \text{etc.};$$

and since the areas of $\triangle$s *SAB*, *SCB*, *SCD*, etc. are equal,

$$p_A AB = p_B BC = p_C CD = \text{etc.},$$

where p_A, p_B, p_C, etc., are the lengths of the perpendiculars from S to *AB*, *BC*, *CD*, etc. By combining the foregoing relations, we obtain

$$v_A p_A = v_B p_B = v_C p_C = \text{etc.}$$

Therefore,

> The velocity of a body attracted towards an immovable centre, in spaces void of resistance, is inversely as the perpendicular let fall from that centre on the right line that touches the orbit.

or, in modern terminology: *the angular momentum is conserved.*

Corollary II

The diagonal *BV* of the parallelogram *ABCV* in the limit $\delta t = 0$, passes through S.

Corollaries III and IV

The displacements *BV* and *EZ*, which tend to the versed sines of the arcs *AC* and *DF* as $\delta t \to 0$, are caused by the impulses acting intermittently at B and D. In the limit $\delta t \to 0$, when the centripetal force acts continuously,

> The forces by which bodies, in spaces void of resistance, are drawn back from rectilinear motions, and turned into curvilinear orbits, are to each other as the versed sines of arcs described in equal times.

Corollary V

The preceding corollaries apply directly to the parabolic arcs described by projectiles in uniform gravity acting vertically.

Corollary VI

The foregoing proposition and corollaries all hold good (by Corollary V of the Laws, Chapter 2, p. 29) if *S*, together with the plane of motion, instead of being at rest, move uniformly in a straight line.

Proposition II. Theorem II

> *Every body that moves in any curved line described in a plane, and by a radius drawn to a point either immovable, or moving forwards with a uniform rectilinear motion, describes about that point areas proportional to the times, is urged by a centripetal force directed to that point.*

In other words, given that in a planar orbit equal areas are described in equal times, to show that the orbit is described under the action of a centripetal force. Or, in the manner of considering the problem as in Proposition I, we are given that the areas of △s *SAB*, *SBC*, *SCD*, *SDE*, etc., are equal; and we are to deduce the centripetal character of the force.

If we extend *AB* by an equal length to *c*, then by the action of inertia alone the particle would have arrived at this same point *c*; and the areas of the △s *SAB* and *SBC* are equal. The areas of △s *SBc* and *SBC* are therefore equal; and by Proposition XL, Book I, *Elements of Euclid*, *cC* must be parallel to *SB*, that is, the impulse *cC* that draws the particle from *Bc* to *BC* acts in the radial direction in the limit $C \to B$. Similarly, the impulse, *dD*, that draws the particle from *Cd* to *CD* acts in the radial direction *CS*. (Q.E.D!)

The same arguments apply if *S*, the centre of the system, instead of being at rest, moves uniformly in a straight line.

Corollaries I and II

If under the same assumptions, as in the Proposition, in equal times unequal areas are described, we must conclude that the particle must be acted on by forces in addition to a pure centripetal attraction. If the areas described in equal times continually increase (or decrease), then the additional force acting on the particle must accelerate (or decelerate) it in the direction of the motion. And the same is true even if the orbit should be described in a resisting medium.

In stating the two foregoing corollaries, Newton has already in mind the effect of perturbations on the motions of the Moon and of the planets considered later in Proposition LXVI (see Chapter 14).

In the Scholium that follows Propositions I and II, Newton points out that the motion in the orbital plane will be unaffected by any force that acts normally to the plane.

Proposition III. Theorem III

A body, that by a radius drawn to the centre of another body, howsoever moved, describes areas about that centre proportional to the times, is urged by a force compounded of the centripetal force tending to that other body, and of all the accelerative force by which that other body is impelled.*

Given that L describes, about T, equal areas in equal times in a planar orbit, while T is subject to extraneous forces, then L must be subject to the combined action of a centripetal attraction towards T *and* all the other forces that may be acting on T.

To show this, we subject L and T, simultaneously, to parallel forces in directions opposite to that acting on T. Then T will be reduced to rest, while by Corollary VI of the Laws (Chapter 2, p. 30) the motion of L relative to T will remain unaffected; and it will therefore continue to describe about T equal areas in equal times in a planar orbit. By Proposition II the force acting on L must be a centripetal attraction towards T. Therefore the difference of the forces acting on L and T is a centripetal attraction of L towards T.

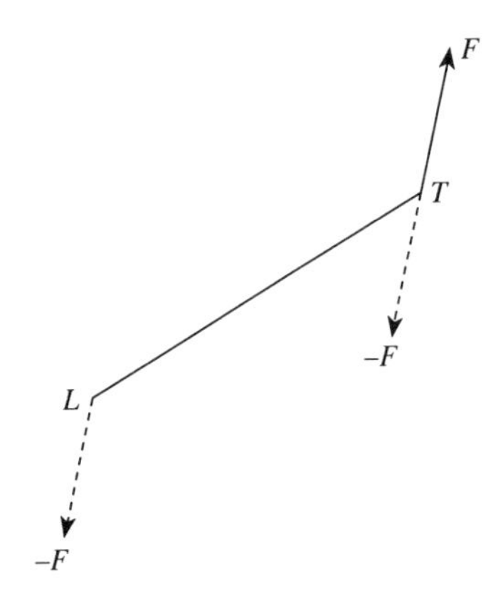

Corollary I

If under the circumstances of the Proposition, we subject L and T simultaneously, in parallel directions, to forces equal and opposite to that acting on L, then T will be subject to a centripetal attraction *towards L*.

Corollaries II and III

If unequal areas are described by L, relative to T, in equal times then the difference of the forces acting on L and T cannot be a pure centripetal attraction of one towards the other.

* 'Every' in Motte's and in Cajori's translations is clearly erroneous: in the French translation (*Principes mathematiques de la philosopie naturelle*, Par feue Madame la Marquise du Chastellet, Tome premier, 1756, Paris, p. 53) it is 'Si'.

Corollary IV

If T moves in a straight line while L moves about it in a planar orbit describing equal areas in equal times, then T must be subject to 'very powerful actions of other forces'; and the same 'powerful actions' must also act on L.

The Scholium which concludes these three propositions is quoted as the motto for this chapter.

Proposition IV. Theorem IV

The centripetal forces of bodies, which by equable motions describe different circles, tend to the centres of the same circles; and are to each other as the squares of the arcs described in equal times divided respectively by the radii of the circles.

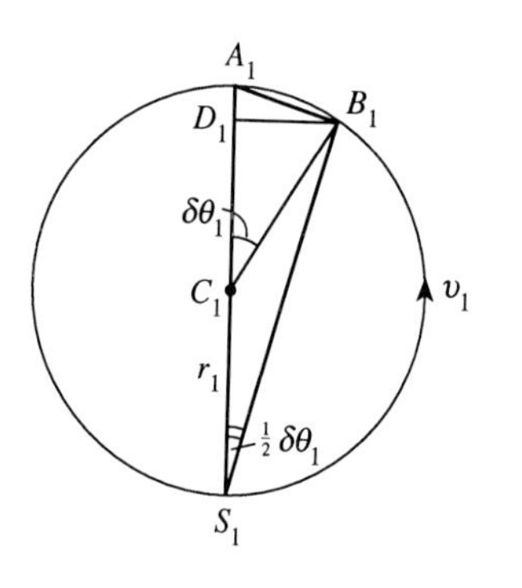

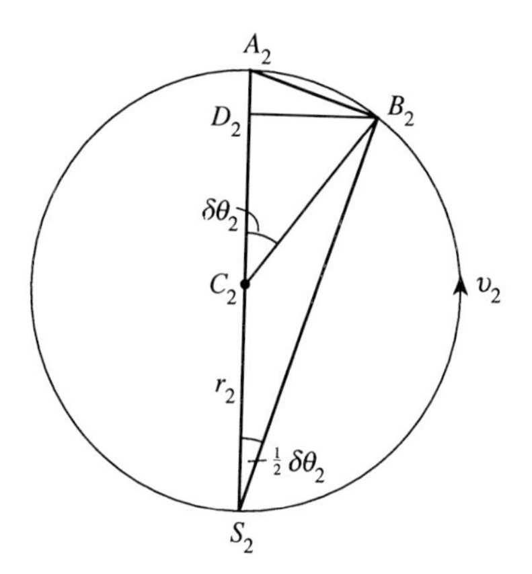

Given two circles of radii r_1 and r_2 described uniformly with velocities v_1 and v_2 and that the arcs A_1B_1 $(=r_1\delta\theta_1)$ and A_2B_2 $(=r_2\delta\theta_2)$ are described in times δt_1 and δt_2 (all, respectively!), we have

$$\text{arc } A_1B_1 = r_1\delta\theta_1 = v_1\delta t_1 \quad \text{and} \quad \text{arc } A_2B_2 = r_2\delta\theta_2 = v_2\delta t_2. \tag{1}$$

Also, by elementary geometry,*

$$A_1S_1 \times A_1D_1 = (A_1B_1)^2 \quad \text{and} \quad A_2S_2 \times A_2D_2 = (A_2B_2)^2. \tag{2}$$

By Lemma I, it follows that†

$$A_1D_1 \asymp \frac{(r_1\delta\theta_1)^2}{2r_1} \quad \text{and} \quad A_2D_2 \asymp \frac{(r_2\delta\theta_2)^2}{2r_2}. \qquad (3)\dagger$$

* Note: $$A_1D_1 = \frac{(r_1\delta\theta_1)^2}{2r_1}\cdot\left(\frac{A_1B_1}{\widehat{A_1B_1}}\right)^2 \to \frac{(r_1\delta\theta_1)^2}{2r_1}.$$

† The symbol $\asymp$ is explained in §12, equations (3) and (4).

Since A_1D_1 and A_2D_2 are the versed sines of twice the arcs A_1B_1 and A_2B_2, they are the distances travelled in times δt_1 and δt_2 by the centripetal forces, (C.F.)$_1$ and (C.F.)$_2$, attracting the two particles towards the respective centres. (Here and elsewhere, the inertial mass is set equal to 1.) Accordingly,

$$\left.\begin{aligned} \tfrac{1}{2}(\text{C.F.})_1 &\asymp \frac{A_1D_1}{(\delta t_1)^2} \asymp \frac{1}{2r_1}\frac{(r_1\delta\theta_1)^2}{(\delta t_1)^2} \\ \tfrac{1}{2}(\text{C.F.})_2 &\asymp \frac{A_2D_2}{(\delta t_2)^2} \asymp \frac{1}{2r_2}\frac{(r_2\delta\theta_2)^2}{(\delta t_2)^2} \end{aligned}\right\} \tag{4}$$

and the result stated follows. Also, if $\delta t_1 = \delta t_2$, then

$$\frac{(\text{C.F.})_1}{(\text{C.F.})_2} = \frac{(r_1\delta\theta_1)^2}{r_1} : \frac{(r_2\delta\theta_2)^2}{r_2}, \tag{4'}$$

Corollary I

By using equation (1), equations (4) give

$$(\text{C.F.})_1 : (\text{C.F.})_2 = v_1^2/r_1 : v_2^2/r_2. \tag{5}$$

Corollary II

If the periodic times in which the circular orbits are described are T_1 and T_2, then,

$$T_1 : T_2 = r_1/v_1 : r_2/v_2 \qquad \text{or} \qquad v_1 : v_2 = r_1/T_1 : r_2/T_2, \tag{6}$$

and from (5) it follows that

$$(\text{C.F.})_1 : (\text{C.F.})_2 = r_1/T_1^2 : r_2/T_2^2. \tag{7}$$

Corollary III

If $T_1 = T_2$ then by (7)

$$(\text{C.F.})_1 : (\text{C.F.})_2 = r_1 : r_2; \tag{8}$$

and conversely.

Corollary IV

If

$$T_1 : T_2 = v_1 : v_2 = \sqrt{r_1} : \sqrt{r_2} \tag{9}$$

then

$$(\text{C.F.})_1 : (\text{C.F.})_2 = 1 : 1 \tag{10}$$

and conversely.

Corollary V

If

$$T_1 : T_2 = r_1 : r_2. \tag{11}$$

then

$$\frac{2\pi r_1}{T_1} = \frac{2\pi r_2}{T_2} \quad \text{or} \quad v_1 = v_2$$

and by (4′)

$$(\text{C.F.})_1 : (\text{C.F.})_2 = r_1^{-1} : r_2^{-1}; \tag{12}$$

and conversely.

Corollary VI

If

$$T_1 : T_2 = r_1^{3/2} : r_2^{3/2}, \tag{13}$$

then by (7)

$$(\text{C.F.})_1 : (\text{C.F.})_2 = r_1^{-2} : r_2^{-2}; \tag{14}$$

and conversely.

Corollary VII

More generally, if

$$T_1 : T_2 = r_1^n : r_2^n, \tag{15}$$

then

$$(\text{C.F.})_1 : (\text{C.F.})_2 = r_1^{-(2n-1)} : r_2^{-(2n-1)}; \tag{16}$$

and conversely.

Corollary VIII

This corollary is best stated in Newton's own words since it represents a sweeping generalization of all the preceding.

> The same things hold concerning the times, the velocities, and the forces by which bodies describe the similar parts of any similar figures that have their centres in a similar position with those figures; as appears by applying the demonstration of the preceding cases to those. And the application is easy, by only substituting the equable description of areas in the place of equable motion, and using the distances of the bodies from the centres instead of the radii.

Corollary IX

The relations (1) and (4) are clearly valid in finite forms. Therefore,

$$\text{arc} = r\theta = vt \qquad \text{and} \qquad \text{C.F.} = v^2/r \tag{17}$$

where θ is the angle described in a time t. Therefore

$$(\text{arc})^2 = (\text{C.F.})rt^2 = 2rs, \tag{18}$$

where s is the linear distance travelled by a particle under a force equal to the centripetal force under which the circular orbit is described.

Alternatively,

$$\text{arc} = \sqrt{(2rs)}. \tag{19}$$

(Note that the *mean proportional* of two quantities a and b is $\sqrt{(ab)}$, i.e. the *geometric mean.*)

Notice the exceptional completeness with which Newton states all the relations relevant for his Moon test of 1662 (see also his renewed applications of these relations in Proposition IV, Book III (Chapter 19)).

We have already considered in Chapter II the Scholium that concludes these four propositions.

Proposition V. Problem I

There being given, in any places, the velocity with which a body describes a given figure, by means of forces directed to some common centre: to find that centre.

Let PT, TQV, and VR be the directions of the velocities at three points, P, Q, and R along the orbit so that these are also the directions of the tangents to the orbit at these points.

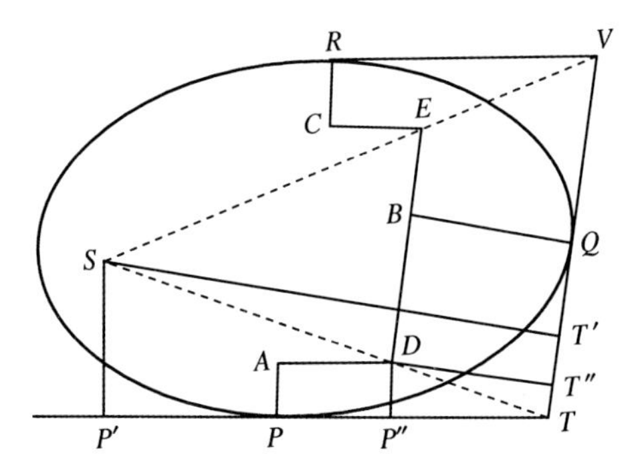

We are required to find the location of the centre of the centripetal attraction, S, under which the orbit is described.

Let PA, QB, and RC be normal to the tangents at P, Q, and R and of lengths inversely proportional to the velocities v_P, v_Q, and v_R at these points; that is,

$$PA:QB:RC = v_P^{-1}:v_Q^{-1}:v_R^{-1}.$$

If S should be the (yet unspecified) centre of attraction, drop the perpendiculars SP' and ST' to the tangents at P and Q. And draw (as illustrated) CE, EBD, and DA parallel to the tangents at R, Q, and P and intersecting at the points E and D. Now drop the perpendiculars DP'' and DT'' to the tangents at P and Q. Then by the constructions and Corollary I, Proposition I,

$$\frac{SP'}{ST'} = \frac{v_Q}{v_P} = \frac{AP}{BQ} = \frac{DP''}{DT''},$$

or

$$\frac{SP'}{DP''} = \frac{ST'}{DT''}.$$

Therefore, S, D, and T are collinear. Similarly, S, E, and V are collinear. Hence S is at the intersection of the lines TD and VE. (Q.E.D!)

23. Newton's relations for determining the law of centripetal attraction from the orbit

In this section we shall present Newton's geometrical relations which are central to his derivation of the law of centripetal attraction when the orbit that is described is given.

Proposition VI. Theorem V

In a space void of resistance, if a body revolves in any orbit about an immovable centre, and in the least time describes any arc just then nascent; and the versed sine of that arc is supposed to be drawn bisecting the chord, and produced passing through the centre of force: the centripetal force in the middle of the arc will be directly as the versed sine and inversely as the square of the time.

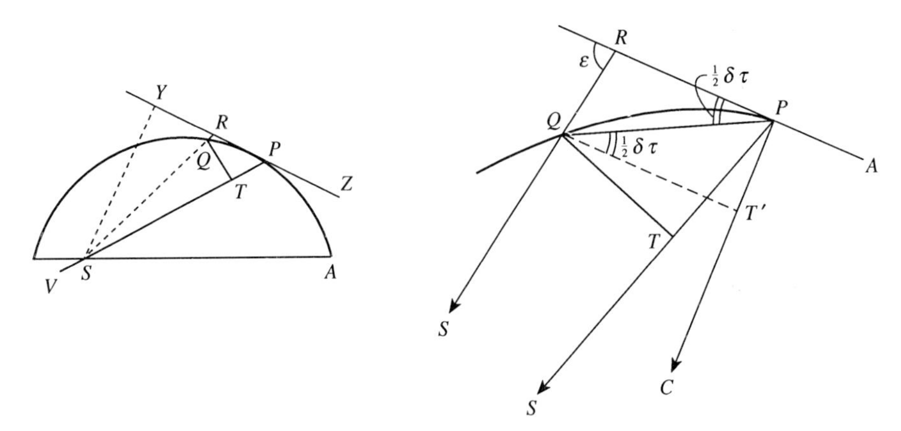

Let APQ be the orbit described by a particle about the centre of attraction S; P and Q neighbouring positions of the particle at an interval δt apart; YPZ the tangent at P specifying the direction of the motion; and QR the continuation of SQ (and is the versed sine of twice the arc PQ). (Note that QR may also be considered as drawn parallel to SP since the difference between them vanishes, to the *second order*, in the limit $\delta t \to 0$—a limit we always consider). Finally let SY be the perpendicular to the tangent at P. (The figure on the right is an enlargement of the area around $PTQR$.)

In the absence of centripetal attraction the particle initially at P will by inertia proceed in the direction of the tangent at P and in a time δt will arrive at R (say). By the centripetal force acting for a time δt, the particle is drawn to Q. Hence by Galileo's theorem, as stated by Newton (cf. Chapter 2, §9 in the Scholium),

$$QR = \tfrac{1}{2}(\text{centripetal force}) \times (\delta t)^2 = \tfrac{1}{2}(\text{C.F.})(\delta t)^2, \tag{1}$$

where, for brevity, we shall write C.F. for centripetal force, (where in accordance with the convention we have adopted, the inertial mass has been set equal to 1).

Corollary I

By Proposition I,

$$\text{The area of the triangle } SPQ = \tfrac{1}{2}SP\,.\,QT \asymp (\mathrm{d}A/\mathrm{d}t)\delta t, \tag{2}$$

where $\mathrm{d}A/\mathrm{d}t$, the constant rate at which the area is described, is the *constant of areas*. Eliminating δt between equations (1) and (2), we obtain the basic relation,

$$\text{C.F.} \asymp 8\left(\frac{\mathrm{d}A}{\mathrm{d}t}\right)^2 \frac{QR}{SP^2 QT^2}. \tag{3}$$

Corollary II

$$\text{C.F.} \propto \asymp \frac{QR}{SY^2 QP^2}, \tag{4}$$

where '$\propto \asymp$' has (here and elsewhere) the meaning that the quantity on the left-hand side is proportional to the limit of the quantity on the right-hand side when $Q \to P$.

This relation follows directly from the limiting similarity of $\triangle$s SYP and QTP as $Q \to P$; thus,

$$\frac{SY}{SP} \asymp \frac{QT}{QP} \quad \text{or} \quad \frac{SY\,.\,QP}{SP\,.\,QT} \asymp 1. \tag{5}$$

Corollary III

It is convenient to redraw the diagrams of this proposition to emphasize that at P the centre of the circle C, passing through P and Q and touching YP at P is on PC

perpendicular to YP; and that in the limit $Q \to P$, the circle becomes the *circle of contact.*

Relations that follow directly from the diagram are:

$$QP^2 = PC\,.\,PT' \quad (=2\rho PT', \text{ where } \rho \text{ is the radius of curvature at } P) \tag{6}$$

and

$$PT' = QR \sin \epsilon. \tag{7}$$

Hence

$$QR = \frac{QP^2}{PC \sin \epsilon} \asymp \frac{(\mathrm{d}s)^2}{2\rho} \operatorname{cosec} \epsilon, \tag{8}$$

where ds denotes the arc length measured along the orbit. Substituting for QR in (4) from equation (8) and noting that

$$PV = PC \sin \epsilon, \tag{9}$$

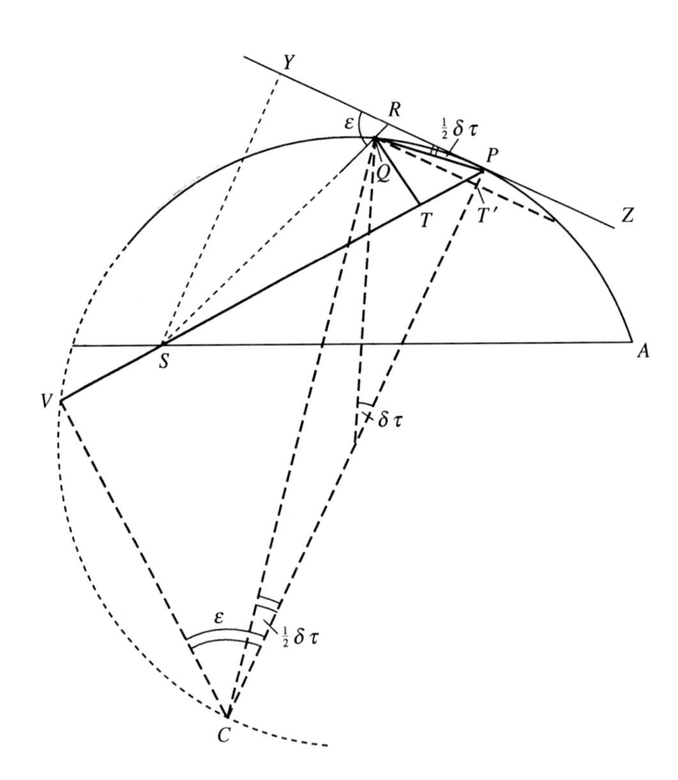

we obtain

$$\text{C.F.} \propto \frac{1}{SY^2 PV}, \tag{10}$$

where it will be noticed that we have dropped the symbol $\asymp$ since the right-hand side is *locally defined* and does not depend on the limiting process, $Q \to P$, implied.

Corollary IV

Since the velocity, v, along the orbit is the reciprocal of SY (by Corollary I, Proposition I),

$$\text{C.F.} \propto \frac{v^2}{PV}, \tag{11}$$

or

$$\text{C.F.} \propto \frac{v^2}{2\,\rho \sin \varepsilon}. \tag{12}$$

Corollary V

Collecting the various relations obtained in Corollaries I–IV, we have

$$(\text{C.F.})^{-1} \propto SY^2 PV \propto \asymp \frac{SP^2 QT^2}{QR} \asymp \frac{SY^2 QP^2}{QR}. \tag{13}$$

Proposition VII. Problem II

If a body revolves in the circumference of a circle, it is proposed to find the law of centripetal force directed to any given point.

The diagram is the same as the one for Corollary III, Proposition VI, except that what was the circle of contact is now the entire orbit. Also QR is now drawn parallel to SP

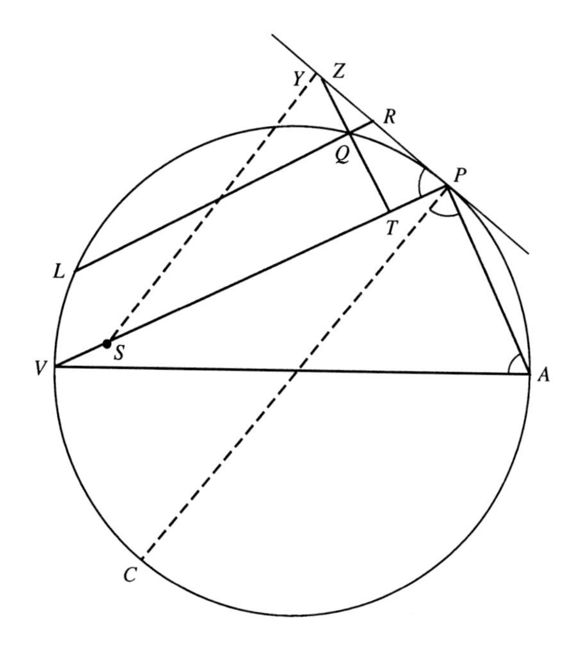

instead of continuing SQ to intersect the tangent at P, since the difference is of the second order as $Q \to P$ and can be ignored.

It is clear from the diagram that the $\triangle$s VPA, ZTP, and ZQR are similar; and it follows that:

$$\frac{AV}{PV} = \frac{PZ}{ZT} = \frac{RZ}{ZQ} = \frac{PZ - RZ}{ZT - ZQ} = \frac{RP}{QT}.$$

Accordingly,

$$QT^2 = RP^2 \frac{PV^2}{AV^2} = (LR.RQ)\frac{PV^2}{AV^2}.$$

Hence

$$\frac{SP^2QT^2}{QR} = LR\frac{SP^2PV^2}{AV^2} \asymp \frac{SP^2PV^3}{AV^2},$$

since $LR \to PV$ when $Q \to P$. Since AV, the diameter of the given circle, is a constant, we conclude from Corollary V of Proposition VI, that

$$\text{C.F.} \propto \frac{1}{SP^2PV^3}.$$

The same otherwise

Since the right-angled $\triangle$s SYP and VPA are similar (the $\angle$s VAP and SPY being equal)

$$SY = \frac{SP.PV}{AV}.$$

Therefore, by Corollary V, Proposition VI,

$$\frac{SP^2PV^3}{AV^2} = SY^2PV \propto (\text{C.F.})^{-1}.$$

Corollary I

If the orbit passes through the centre of attraction $SP = PV$ and remembering that AV is a constant, it follows that

$$\text{C.F.} \propto (SP)^{-5};$$

that is, the inverse fifth-power law. The significance of the inverse fifth-power law in this context will be considered in Chapter 6.

Corollary II

Given, that a body revolves in the same circular orbit, with the same period of revolution, under the centripetal attraction from two *different* centres, to find the ratio of the centripetal attractions towards the two centres.

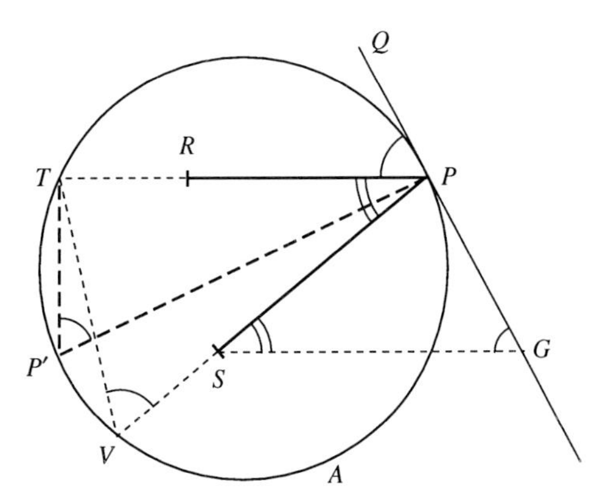

Let R and S be the two centres of attraction and let the circular orbit that the body describes under the centripetal attraction from R *or* S, be $PTP'VA$ (where P' is the antipode of P); and further, let GPQ be the tangent at P. Continue PR and PS to intersect the circle at T and V; and draw SG parallel to RP.

By the proposition just established,

$$\frac{(\text{C.F.})_S}{(\text{C.F.})_R} = \frac{RP^2 \,.\, PT^3}{SP^2 \,.\, PV^3} = SP \,.\, RP^2 \left(\frac{PT}{SP \,.\, PV}\right)^3. \tag{i}$$

Since $\angle PSG = \angle TPV$ and $\angle SGP\,[= \angle RPQ = \angle TP'P] = \angle TVP$, the $\triangle$s SPG and PTV are similar; and therefore,

$$\frac{SP}{SG} = \frac{PT}{PV} \quad \text{or} \quad \frac{1}{SG} = \frac{PT}{SP \,.\, PV}. \tag{ii}$$

Inserting this last relation in (i) we obtain the required ratio:

$$\frac{(\text{C.F.})_S}{(\text{C.F.})_R} = \frac{SP \,.\, RP^2}{(SG)^3}. \tag{iii}$$

The reader may perhaps wish that an explanation be given for the requirement of the equality of the orbital periods in the statement of the corollary. The reason is that in deriving the basic relation (4) in Proposition VI, Corollary I (p. 77) the constant proportionality factor $8(\mathrm{d}A/\mathrm{d}t)^2$ was omitted. The same factor is also omitted in the other relations (13) collected in Corollary V. The omission of this factor is of no consequence so long as one's consideration is restricted to the motion described under a given law of centripetal attraction. But when the same relation is used in a comparison of the motions under two different laws, as in the present instance, equality of the constant of areas in the two cases is implied. And when the orbits are circular the equality of the orbital periods follows from the equality of the constant of areas. All this must, of course, have been obvious to Newton!

Corollary III

Since the relation (iii), established in the preceding Corollary II, is purely *local*, in that it involves no quantity that is not uniquely defined by the point P, it will suffice to require

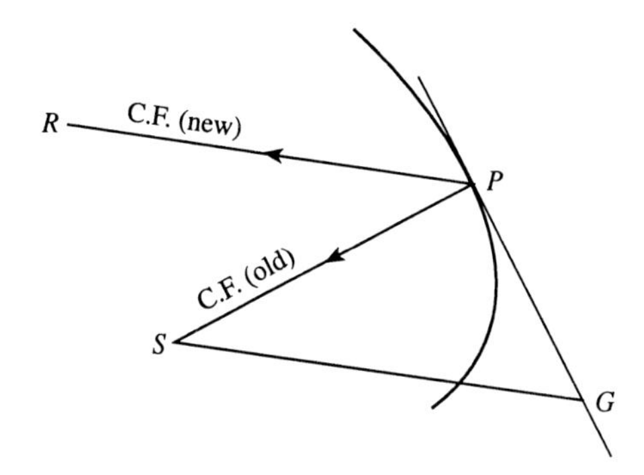

only that the *same orbit* (not necessarily circular) is described under the centripetal attraction of two different laws provided the constant of areas is the *same*. We can state this more general result as follows.

Given that an orbit is described under a known law of centripetal attraction, then the same orbit, with the same constant of areas, will be described under a new law of centripetal attraction related to the old by

$$(\text{C.F.})_{R;\,\text{new}} = (\text{C.F.})_{S;\,\text{old}} \frac{SG^3}{SP\,.\,RP^2}. \tag{iv}$$

The depth of the corollaries of this proposition has not been recognized until recently. It is therefore of some interest to note that these corollaries are not included in the first 1687 edition of the *Principia*; they appear for the first time in the second 1713 edition. We shall return to this matter in some detail in the Supplement to Chapter 6.

Proposition VIII. Problem III

If a body moves in the semicircumference PQA; it is proposed to find the law of the centripetal force tending to a point S, so remote, that all the lines PS, RS drawn thereto, may be taken for parallels.

By the similarity of the $\triangle$s CPM, PZT and RZQ we have the relations:

$$\frac{CP}{PM} = \frac{PZ}{ZT} = \frac{RZ}{ZQ} = \frac{PZ - RZ}{ZT - ZQ} = \frac{RP}{QT}.$$

Therefore,

$$\frac{CP^2}{PM^2} = \frac{RP^2}{QT^2} = \frac{RQ(RN + QN)}{QT^2} \asymp \frac{2PM\,.\,QR}{QT^2};$$

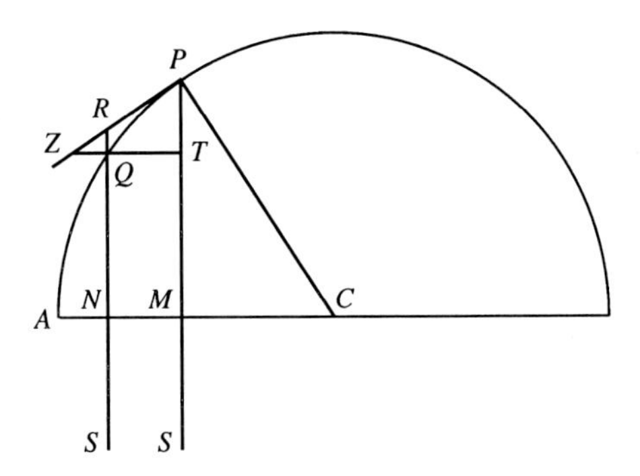

and we obtain

$$\frac{SP^2QT^2}{QR} \asymp \frac{2SP^2PM^3}{CP^2}.$$

Since SP^2/CP^2 is a given quantity, we conclude from relation (3) of Proposition VI, Corollary II, that

$$\text{C.F.} \propto (PM)^{-3}.$$

This result can also be derived directly from Proposition VII by drawing the diameter of the circle (on page 79) normal to PV and RL and intersecting them at M and N, respectively; and by making the correspondences that exist between the diagrams of Propositions VII (p. 79) and VIII (p. 83).

The significance of the 'inverse-cube' law which emerges here will become apparent in the context of Proposition XLV, Example III (Chap. 10, p. 197).

24. Two simple illustrations of the basic relation

As a preliminary to the solution of the Kepler problem in Section III, Newton first considers two simple cases for illustrating the application of the geometric relation of Proposition VI, Corollary I.

Proposition IX. Problem IV

> *If a body revolves in a spiral PQS, cutting all the radii SP, SQ, &c., in a given angle; it is proposed to find the law of the centripetal force tending to the centre of that spiral.*

The requirement that the radius vector, drawn from a fixed point to any point on the curve, makes a constant angle to the tangent at that point, defines an *equiangular spiral.* The defining equation, therefore, is (see the figure),

$$\cot \epsilon = -\frac{\mathrm{d}r}{r\,\mathrm{d}\varphi} = \frac{\mathrm{d}u}{u\,\mathrm{d}\varphi} \qquad (u = r^{-1}). \tag{i}$$

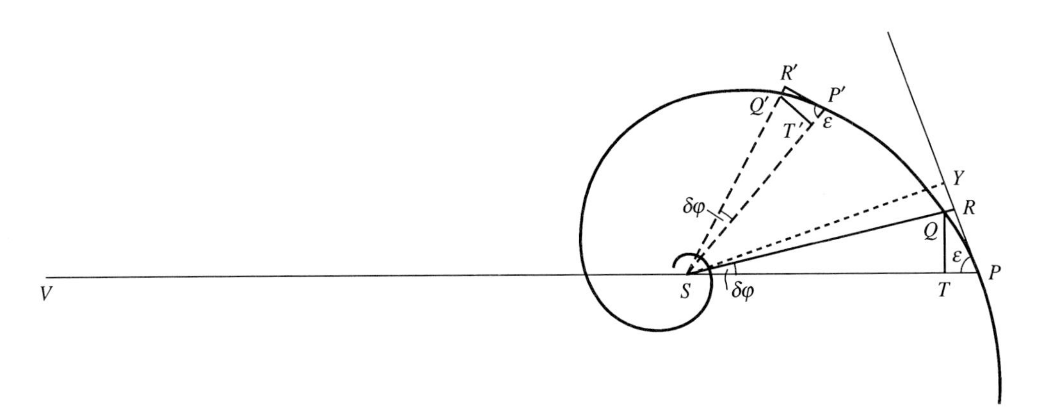

The solution of this equation is

$$u = \exp(\varphi \cot \epsilon), \tag{ii}$$

where a constant of integration has been absorbed in u.

Before presenting Newton's solution of the problem, we shall find it useful to obtain it (as one currently would) with the aid of the general equation,

$$\text{C.F.} = P(u) = h^2u^2\left(u + \frac{d^2u}{d\varphi^2}\right) \tag{iii}$$

derived in Chapter 4, Section 18 (equation (28)). Inserting for u its present solution (ii), we obtain

$$\text{C.F.} = (h^2 \operatorname{cosec}^2 \epsilon)u^3 = \frac{h^2}{r^3 \sin^2 \epsilon}. \tag{iv}$$

In other words, a particle that describes an equiangular spiral, under a centripetal attraction, can do so only under an inverse-cube law of attraction.

In his method of solution, Newton starts with his basic relation (Proposition VI, Corollary I, equation (3)),

$$\text{C.F.} = \frac{QR}{QT^2SP^2}. \tag{v}$$

By equation (8) of Proposition VI, Corollary III,

$$QR = \frac{QP^2}{2\rho \sin \epsilon}, \tag{vi}$$

where ρ is the radius of curvature at P; and an alternative form of equation (v) is

$$\text{C.F.} \propto \;\asymp \frac{1}{2\rho \sin \epsilon} \frac{QP^2}{QT^2} \frac{1}{SP^2}; \tag{vii}$$

or, since $QP/QT = \operatorname{cosec} \epsilon$,

$$\text{C.F.} \propto \frac{1}{2\rho}(\operatorname{cosec}^3 \epsilon) SP^{-2}. \qquad \text{(viii)}$$

But by equation (9) of Proposition VI, Corollary III (see the figure)

$$PV \asymp 2\rho \sin \epsilon. \qquad \text{(ix)}$$

Hence, finally

$$\text{C.F.} \propto \frac{SP}{PV}(\operatorname{cosec}^2 \epsilon) SP^{-3}. \qquad \text{(x)}$$

Equation (x) is entirely general: it applies to *any* orbit described under centripetal attraction.

The arguments leading to equation (x) are essentially Newton's, except that he prefers to argue in terms of the similarity of the figures $QTPR$ and $Q'T'P'R'$ constructed at any two points P and P' with equal $\angle$s PSR and $P'SR'$ (see the figure)—he does not, for example, make use (explicitly) of the relation $QP/QT = \operatorname{cosec} \epsilon$. Having reached the stage of equation (x), Newton states, as self evident (see below), that the ratio SP/PV is independent of P; and he deduces his result:

$$\text{C.F.} \propto (SP)^{-3}. \qquad \text{(xi)}$$

The same otherwise

In his alternative derivation of the result (xi), Newton starts with the relation (10) of Proposition VI, Corollary III,

$$\text{C.F.} \propto \frac{1}{SY^2 PV}. \qquad \text{(xii)}$$

Since $SY = SP \sin \epsilon$, we obtain,

$$\text{C.F.} \propto \frac{\operatorname{cosec}^2 \epsilon}{SP^2 PV} = \operatorname{cosec}^2 \epsilon \frac{SP}{PV}(SP)^{-3}.$$

the same as equation (x). With the statement, 'the chord of the circle concentrically cutting the spiral [at V in the figure] is in a given ratio to the height SP', Newton completes the proof.

An analytical verification of the invariance of SP/PV along an equiangular spiral is of some interest in enabling us to infer what Newton *must* have known at the time.

We start with equation (viii),

$$\text{C.F.} \propto \frac{1}{2\rho}(\operatorname{cosec}^3 \epsilon) u^2, \qquad \text{(xiii)}$$

an alternative form of equation (x). By a known formula, the radius of curvature along a curve, $r(\varphi)$ (given in plane polar coordinates) is

$$\frac{1}{\rho} = \frac{u + \mathrm{d}^2u/\mathrm{d}\varphi^2}{[1 + (\mathrm{d}u/u\,\mathrm{d}\varphi)^2]^{3/2}} \tag{xiv}$$

or by equation (i)

$$\frac{1}{\rho} = \left(u + \frac{\mathrm{d}^2u}{\mathrm{d}\varphi^2}\right)\sin^3\epsilon. \tag{xv}$$

For an equiangular spiral this equation gives

$$\frac{1}{\rho u} = \sin\epsilon \qquad \text{or} \qquad \frac{1}{\rho u \sin\epsilon} = \frac{2SP}{PV} = 1, \tag{xvi}$$

confirming Newton's statement. However, greater interest attaches to the result obtained by combining equations (xiii) and (xv), namely

$$\text{C.F.} \propto \tfrac{1}{2}u^2\left(u + \frac{\mathrm{d}^2u}{\mathrm{d}\varphi^2}\right), \tag{xvii}$$

or, returning to equation (v) from which we started, we have the identity:

$$\frac{QR}{QT^2SP^2} \asymp \propto u^2\left(u + \frac{\mathrm{d}^2u}{\mathrm{d}\varphi^2}\right), \tag{xviii}$$

thus coming full circle. From the fact that in his demonstration of Proposition X Newton had explicitly arrived at equation (x) (which is the same as equation (viii)) it is hard to imagine that he was not fully aware of the identity (xviii).

Finally we may note that in Proposition XLI, Corollary III, Newton presents the complete solution to the problem of motion in an inverse-cube law of centripetal attraction.

Lemma XII

All parallelograms circumscribed about any conjugate diameters of a given ellipse or hyperbola are equal among themselves.

This is demonstrated by the writers on the conic sections.

Proposition X. Problem V

If a body revolves in an ellipse; it is proposed to find the law of the centripetal force tending to the centre of the ellipse.

Considering two infinitely close points P and Q on the ellipse (with centre C and semiaxes CA and CB), we draw the lines QT and QR perpendicular and parallel,

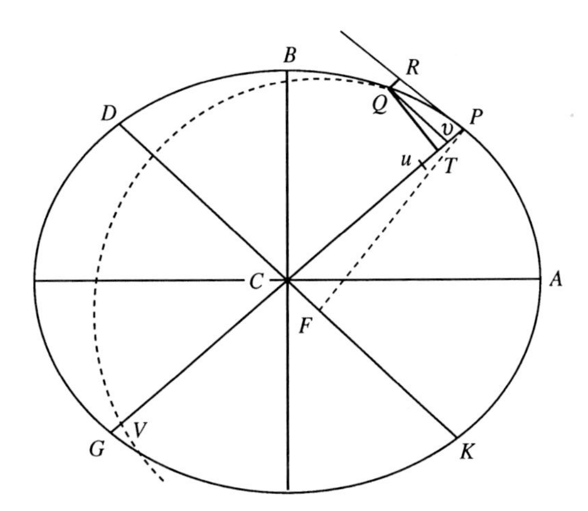

respectively to CP. Let DCK be the diameter conjugate to PCG and parallel to PR, the tangent at P. Finally, draw PF perpendicular to DCK and Qv parallel to RP (and DCK).

By a known property of the ellipse,

$$\frac{Pv\,.\,vG}{Qv^2} = \frac{CP^2}{CD^2}; \tag{i}$$

while by the similarity of the $\triangle$s QvT and PCF,

$$\frac{Qv^2}{QT^2} = \frac{CP^2}{PF^2}; \tag{ii}$$

also

$$\frac{QR}{Pv} = 1, \tag{iii}$$

since $QRPv$ is, by construction, a parallelogram. Now multiplying these three equations, we obtain,

$$\frac{QR\,.\,vG}{QT^2} = \frac{CP^4}{PF^2\,.\,CD^2}. \tag{iv}$$

But by Lemma XII

$$PF^2\,.\,CD^2 = CA^2\,.\,CB^2; \tag{v}$$

and equation (iv) gives,

$$\frac{QR\,.\,vG}{QT^2} = \frac{CP^4}{CA^2\,.\,CB^2}. \tag{vi}$$

But

$$vG \asymp 2CP. \tag{vii}$$

Hence

$$\frac{QR}{QT^2 \,.\, CP^2} \asymp \frac{1}{2CA^2 \,.\, CB^2} CP. \tag{viii}$$

Therefore by the basic relation of Proposition VI, Corollary I,

$$\text{C.F.} \propto CP. \tag{ix}$$

In other words, a particle will describe an ellipse about its centre under a centripetal attraction proportional to the distance. (Q.E.D!)

The directness, the absence of superfluity, and the entirely elementary character of the proof are sparkling; but they are obscured by Newton's (Jamesian) style in writing mathematical derivations in continuous prose.

The same otherwise

Newton's alternative proof is admittedly more involved than his first; but it exhibits his extraordinary virtuosity in devising geometrical constructions that exactly serve the purpose.

The surprising elements in Newton's present construction are *first* to associate v with a point u on the left, equidistant from T so that

$$Tv = Tu; \tag{x}$$

and *second* to define another point V by the condition

$$uV = (Qv)^2/Pv, \tag{xi}$$

or, by equation (i),

$$\frac{uV}{vG} = \frac{(Qv)^2}{Pv \,.\, vG} = \frac{CD^2}{CP^2}. \tag{xii}$$

A surprising identity follows from these definitions.

By equation (xi) we have,

$$(Qv)^2 + Pu \,.\, Pv = Pv(Pu + uV) = Pv \,.\, PV; \tag{xiii}$$

while by an alternative reduction making use of equation (x), we have

$$\begin{aligned}(Qv)^2 + Pu \,.\, Pv &= (Qv)^2 + (PT + Tu)(PT - Tv) \\ &= (Qv)^2 + PT^2 - Tv^2.\end{aligned} \tag{xiv}$$

The $\triangle$s QPT and QTV being right angled,

$$PT^2 = PQ^2 - QT^2 = PQ^2 - (Qv^2 - Tv^2). \tag{xv}$$

By combining equations (xiii), (xiv), and (xv), we obtain the identity,

$$Pu \,.\, PV = PQ^2. \tag{xvi}$$

It follows from this identity that *the circle of contact at P passing through Q intersects PG at V.* (How did Newton know this beforehand?)

By Proposition VI, Corollary III, it now follows that,

$$\text{C.F.} \propto \frac{1}{PF^2 PV} \qquad (PF = \text{`}SY\text{'}), \tag{xvii}$$

or, by equation (v)

$$\text{C.F.} \propto \frac{CD^2}{(CA\,.\,CB)^2 PV}. \tag{xviii}$$

But, according to equation (xii),

$$\frac{CD^2}{CP^2} = \frac{uV}{vG} \asymp \frac{PV}{2PC} \qquad \text{or} \qquad CD^2 = \tfrac{1}{2}PV\,.\,PC. \tag{xix}$$

Hence finally,

$$\text{C.F.} \propto \frac{1}{2(CA\,.\,CB)^2} PC, \tag{xx}$$

which is the required result.

> COR. I. And therefore the force is as the distance of the body from the centre of the ellipse; and, *vice versa*, if the force is as the distance, the body will move in an ellipse whose centre coincides with the centre of force, or perhaps in a circle into which the ellipse may degenerate.

Newton does not provide an explicit demonstration of the '*vice versa*' part of the corollary. But, by his familiarity with the theory of the simple pendulum (see Chapter 2, §10) he must undoubtedly have considered it as obvious (needing no explanation) that the solution of the pair of equations,

$$\frac{d^2x}{dt^2} = -\omega^2 x \qquad \text{and} \qquad \frac{d^2y}{dt^2} = -\omega^2 y \quad (\omega = \text{a constant}),$$

namely,

$$x = a\cos(\omega t - \delta_1) \qquad \text{and} \qquad y = b\sin(\omega t - \delta_2)$$

(where a, b, δ_1, and δ_2 are constants), represents motions in an elliptic orbit; and *vice versa*!

Besides, in Proposition XLII (see Chapter 9) Newton formulates and solves the *initial-value* problem for motion under the action of an entirely general centripetal attraction.

COR II. And the periodic times of the revolutions made in all ellipses whatsoever about the same centre will be equal.

It is clear that it will suffice to consider the two cases: a pair of similar ellipses and a pair of ellipses with equal major (or minor) axes. The general case will follow from combining the results for the two cases.

(a) The case of two similar ellipses is illustrated in the adjoining figure. With the same notation as in the earlier sections, the following sequence of steps is self-explanatory.

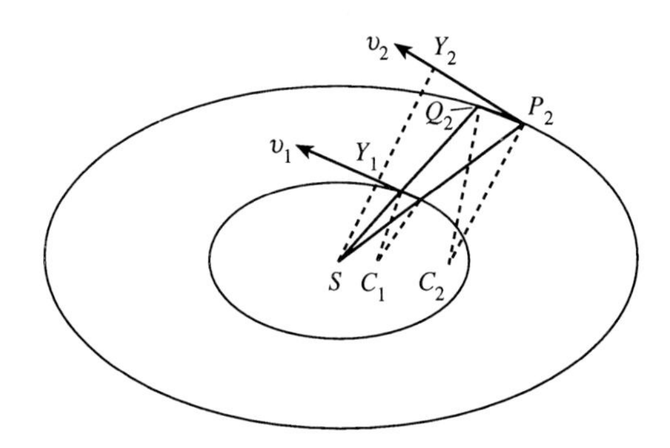

(i) C.F. $= \dfrac{v^2}{\rho}$ cosec ϵ.

(ii) $(\text{C.F.})_1 : (\text{C.F.})_2 = r_1 : r_2 = (v_1^2/\rho_1) : (v_2^2/\rho_2)$ ($\because$ ϵ is the same).

(iii) $v_1^2 : v_2^2 = r_1\rho_1 : r_2\rho_2 = r_1^2 : r_2^2$ (by the similarity of the ellipses).

(iv) $v_1 : v_2 = r_1 : r_2$.

(v) $$T_1 : T_2 = \frac{(\text{area of ellipse})_1}{(\text{constant of areas})_1} : \frac{(\text{area of ellipse})_2}{(\text{constant of areas})_2}$$

$$= \frac{a_1 b_1}{p_1 v_1} = \frac{a_2 b_2}{p_2 v_2} \quad \text{(by Proposition I, Corollary I)}$$

$$= \frac{a_1 b_1}{p_1 r_1} = \frac{a_2 b_2}{p_2 r_2} \quad \text{(by (iv))}$$

$= 1 : 1$ (by similarity of the ellipses).

(b) The case of two ellipses with semiaxes, (a, b_1) and $(a, b_2 \neq b_1)$ is illustrated in the adjoining diagram. The sequence of steps is now the following.

(i) $(\text{C.F.})_{B_1} = (\text{C.F.})_{B_2} = b_1 : b_2 = v_1^2/\rho_1 : v_2^2/\rho_2$.

(ii) $\rho_1 : \rho_2 = (a^2/b_1) : (a^2/b_2)$ (by a known formula for the radius of curvature).

(iii) $v_1^2 : v_2^2 = \rho_1 b_1 : \rho_2 b_2 = 1 : 1$ (by (i)).

(iv) $v_1 = v_2$.

(v) $T_1 : T_2 = ab_1/b_1 v_1 : ab_2/b_2 v_2 = 1 : 1$.

Since the corollary is true for the two cases, it is true generally.

The foregoing is no more than a transcription of what Newton writes in explanation:

> For those times in similar ellipses will be equal (by COR. III and VIII, PROP. IV); but in ellipses that have their greater axis common, they are to each other as the whole areas of the ellipses directly, and the parts of the areas described in the same time inversely; that is, as the lesser axes directly, and the velocities of the bodies in their principal vertices inversely; that is, as those lesser axes directly, and the ordinates to the same point of the common axes inversely; and therefore (because of the equality of the direct and inverse ratios) in the ratio of equality, 1:1.

Finally, in the concluding Scholium, Newton considers the parabolic limit of the ellipse,

$$\frac{x^2}{a^2} + \frac{y^2}{b^2} = 1.$$

One obtains by first transferring the origin to the vertex $(-a, 0)$ and then letting

$$a \to \infty \qquad \text{and} \qquad (1 - e) \to 0,$$

keeping the distance from the vertex to the nearby focus,

$$a(1 - e) = \text{constant} = \alpha \text{ (say)}.$$

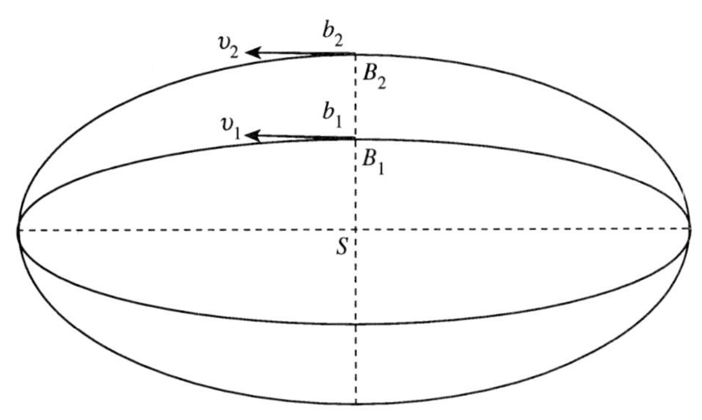

The resulting parabola,

$$y^2 = 4\alpha x,$$

defines the orbit described under the action of a constant attractive force in the direction x. Newton identifies this inference as '*Galileo*'s' theorem!

Newton also considers the hyperbolic orbits described when the centripetal attraction is replaced by centrifugal repulsion.

6

The motion of bodies along conic sections

25. Introduction

The propositions of Section III and most particularly Proposition XI (in which the inverse-square law of attraction is deduced for bodies revolving in ellipses under the action of a 'a centripetal force tending to the focus of the ellipse') are commonly regarded as the apex of the *Principia*. This view, in my judgement, greatly diminishes the *Principia*. Newton himself did not accord Proposition XI any special significance when he came to formulating his universal law of gravitation in Book III: his main references are to Propositions I–IV and especially Proposition IV and its seven ccrollaries (see Chapter 19). Besides, the deepest theoretical insights are revealed in the 'same otherwise' demonstrations of Propositions XI and XII and the emphasis given to Proposition VII, Corollary III in the concluding Scholium—all of which appear for the first time in the second 1713 edition of the *Principia*. (These matters are considered in detail in the Supplement (§§ 31–33) for this chapter.)

26.

Proposition XI. Problem VI

If a body revolves in an ellipse; it is required to find the law of the centripetal force tending to the focus of the ellipse.

Description of figure

S and H are the foci and C is the centre of the ellipse;
CA and CB are the semimajor and semiminor axes of lengths a and b;
P and Q are neighbouring points ($Q \to P$, eventually);
RPZ is the tangent at P;

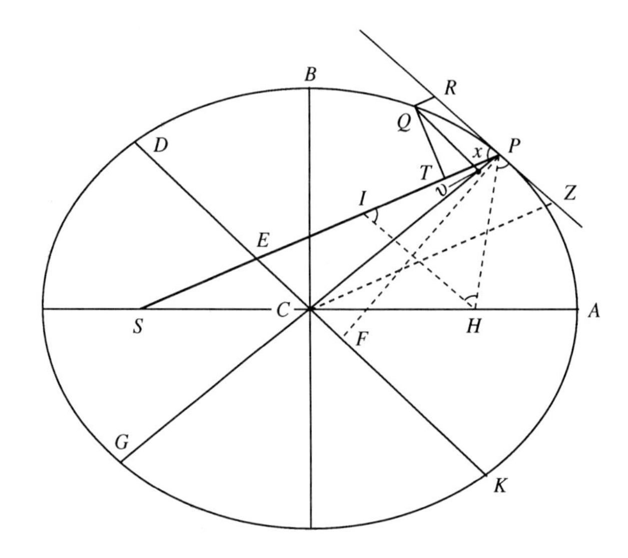

DCK is the diameter, conjugate to PCG and parallel to RPZ;
PF is perpendicular to DCK, and CZ is parallel to SP;
QR and QT are parallel and perpendicular, respectively, to SP;
Qv and IH are parallel to RPZ (and DCK);
v is on PC and x is on PS.

Some properties of the ellipse needed in the solution

(i) $SP + PH = 2a$.
(ii) The latus rectum $L = 2b^2/a = 2BC^2/CA =$ twice the semilatus rectum, l.
(iii) $\angle IPR = \angle HPZ$ implying $\angle PIH = \angle PHI$ and $PI = PH$.
(iv) $Pv\,.\,vG/(Qv)^2 = CP^2/CD^2$ (Proposition X, equation (i), Chapter 5).
(v) $PF^2\,.\,CD^2 = CB^2\,.\,CA^2$ (Proposition X, equation (v), Chapter 5).

A simple consequence of the relations (i)–(v)

Since $SC = CH$ and EC is parallel to IH,

(vi) $SE = EI$.
By (i), (iii), and (vi),

$$2a = PS + PH = PE + SE + PI = PE + EI + PI = 2PE.$$

Therefore
(vii) $PE(=CZ) = a = AC$ ('Newton's theorem').

With these preliminaries out of the way, Newton's solution to the problem is simple, direct, and straightforward. The steps are:

$$\frac{QR}{Pv} = \frac{Px}{Pv} \quad (\because QRPx \text{ is a parallelogram})$$

$$= \frac{PE}{PC} \quad (\text{by the similarity of the } \triangle\text{s } Pxv \text{ and } PEC),$$

or by (vii)

(a) $$\frac{QR}{Pv} = \frac{AC}{PC}.$$

By (iv)

(b) $$\frac{Pv}{(Qv)^2} = \frac{CP^2}{CD^2 . vG}.$$

Multiplication of (a) and (b), gives

(c) $$\frac{QR}{Qv^2} = \frac{AC . PC}{vG . CD^2}.$$

But

$$\frac{(Qv)^2}{QT^2} \asymp \frac{(Qx)^2}{(QT)^2} = \frac{PE^2}{PF^2} \quad (\text{by similarity of } \triangle\text{s } QxT \text{ and } PEF)$$

$$= \frac{AC^2}{PF^2} \quad (\text{by vii}),$$

or, by (v)

(d) $$\frac{(Qv)^2}{QT^2} \asymp \frac{CD^2}{CB^2}.$$

Now by multiplying (c) and (d),

(e) $$\frac{QR}{QT^2} \asymp \frac{AC . PC}{vG . CB^2};$$

or by (ii)

(f) $$L \frac{QR}{QT^2} \asymp 2 \frac{PC}{vG} \asymp 1.$$

Hence

(g) $$L . QR \asymp QT^2;$$

and finally:

(h) $$L . SP^2 \asymp \frac{SP^2 . QT^2}{QR} \asymp \propto (\text{C.F.})^{-1} \quad (\text{by Proposition VI, Corollary I})$$

or

(i) $$\text{C.F.} \propto SP^{-2},$$

that is, an inverse-square law of attraction (Q.E.I)

A fact of some relevance that follows from (Newton's) equation (g) is worth noting. From the equation (Proposition VI, equation (2))

$$SP.QT \asymp 2\frac{\mathrm{d}A}{\mathrm{d}t}\delta t = h\delta t \quad (h = \text{the constant angular momentum})$$

and equation (h), above, it follows that

$$\frac{1}{SP^2} \asymp L\frac{QR}{SP^2QT^2} \asymp \frac{L.QR}{h^2\delta t^2} \asymp \tfrac{1}{2}L\frac{(\text{C.F.})\,\delta t^2}{h^2\delta t^2},$$

the last step following from Galileo's theorem. We thus obtain:

(j) $$\frac{1}{SP^2} \propto \frac{l}{h^2}P(r)$$

where $P(r) = \text{C.F.}$ as defined in Chapter 5, Proposition IX, equation (v).

The same otherwise

By an application of Proposition VII, Corollary III (for which it was apparently intended) Newton shows that *the inverse-square law of attraction tending to S follows from the linear law of attraction* ($\propto r$) *tending to C* (proved in Proposition X); *and conversely.*

Thus with the identifications

$$R \to S, \qquad S \to C, \qquad \text{and} \qquad G \to Z,$$

equation (iv) of Proposition VII, Corollary III gives

$$(\text{C.F.})_S = (\text{C.F.})_C\frac{CZ^3}{PC.SP^2} = (\text{C.F.})_C\frac{AC^3}{PC.SP^2},$$

since $CZ = PE = AC$. From this relation it follows that

$$\text{if} \quad (\text{C.F.})_C \propto PC \quad \text{then} \quad (\text{C.F.})_S \propto SP^{-2},$$

and, conversely,

$$\text{if} \quad (\text{C.F.})_S \propto SP^{-2} \quad \text{then} \quad (\text{C.F.})_C \propto PC. \qquad \text{Q.E.I.}$$

As I have already remarked, neither this 'same otherwise' solution of the celebrated problem nor Proposition VII, Corollary III appear in the first 1687 edition of the *Principia*; but they are included in the second 1713 edition. Further comments are made in the Supplement to this chapter.

27. Proposition XII

Newton prefaces this proposition with the following remark that is a reflection of his joyous attitude at the time of his writing.

> With the same brevity with which we reduced the fifth Problem to the parabola, and hyperbola, we might do the like here; but because of the dignity of the Problem and its use in what follows, I shall confirm the other cases by particular demonstrations.

> *Proposition XII. Problem VII*
>
> *Suppose a body to move in a hyperbola; it is required to find the law of the centripetal force tending to the focus of that figure.*

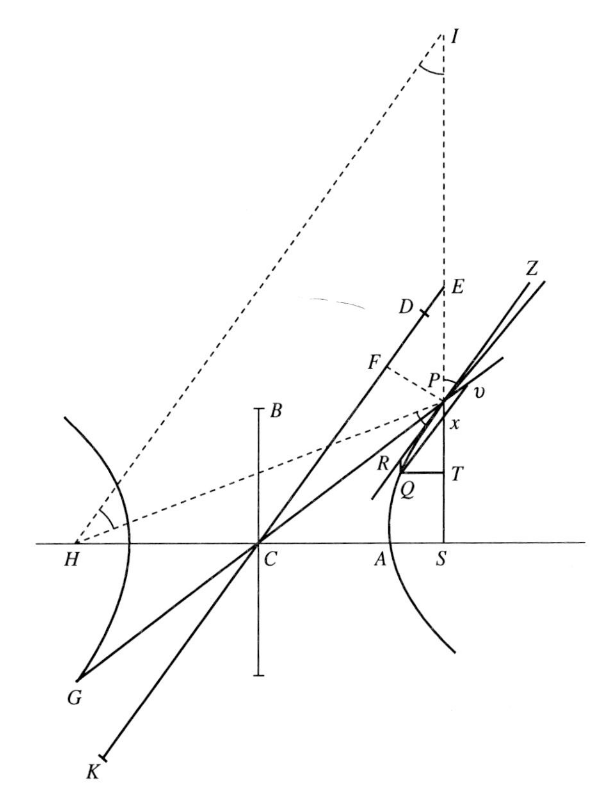

With Newton's careful lettering of the above figure in which the geometrical correspondences that exist between the ellipse (illustrated in Proposition XI, §26) and the hyperbola (illustrated above) are scrupulously observed, all that has been written in the context of the solution of Problem VI literally applies without any alteration, to the solution of the present Problem VII with the *sole* exception of equation (i), $PH + SP = 2a$, which must be altered to read $PH - SP = 2a$. In other words, *all* the equations (ii)–(vii) and (a)–(i)

of §26 are valid as they stand when read in the context of the present figure. The solution to Problem VII (as the solution to Problem VI) is

$$\text{C.F.} \propto SP^{-2}. \qquad \text{(Q.E.I.)}$$

The 'dignity' of the present problem consists then of the *identity* of the two problems if they are viewed correctly!

The same otherwise

The alternative solution is the present analogue of the solution presented in §26 in the context of Proposition XI. The laconic style in which Newton expresses the solution gives us some measure of the depth of his insights.

> Find out the force tending from the centre C of the hyperbola. This will be proportional to the distance CP. But from thence (by Cor. III, Prop. VII) the force tending to the focus S will be as $\dfrac{PE^3}{SP^2}$, that is, because PE is given reciprocally as SP^2. Q.E.I.
>
> And the same way may it be demonstrated, that the body having its centripetal changed into a centrifugal force, will move in the conjugate hyperbola.

The foregoing terse statements of Newton are explicitly verified in §31 (B and C).

28. Proposition XIII: the motion of a body along a parabola

First some preliminary lemmas relating to the geometry of the parabola.

Lemma XIII

> *The latus rectum of a parabola belonging to any vertex is four times the distance of that vertex from the focus of the figure.*

Lemma XIV

> *The perpendicular, let fall from the focus of a parabola on its tangent, is a mean proportional between the distances of the focus from the point of contact, and from the principal vertex of the figure.*

Description of figure

S is the focus and A is the vertex of the parabola,

$$y^2 = 4ax; \qquad \text{(i)}$$

with the origin of the coordinate system at A; and

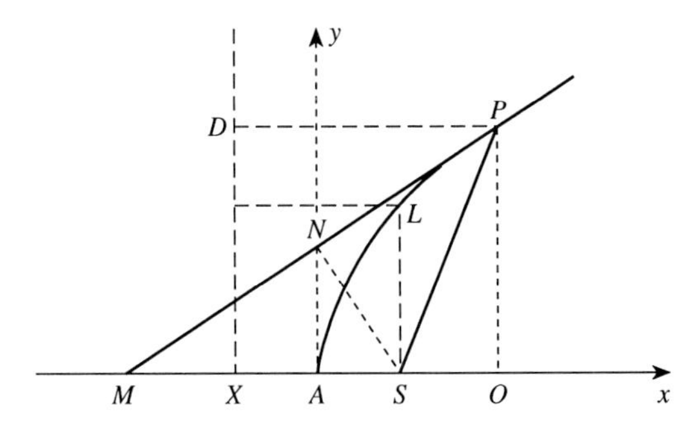

(ii) $$AS = a.$$

DX is the directrix; P is a point on the parabola, and PD is perpendicular to DX. Then (by definition)

(iii) $$SP = PD = OX.$$

PM is the tangent at P; and SN is perpendicular to PM.

$$LS = SX = 2a = \text{semilatus rectum: } L(\text{in figure}) = (a, \tfrac{1}{2}L).$$

(iv) $$L = 4a \qquad \text{(which is Lemma XIII)}$$

The parametric representation of the parabola is:

$$x = a\mu^2 \qquad \text{and} \qquad y = 2a\mu.$$

If

(v) $$P = (AO, OP) = (a\mu^2, 2a\mu), \text{ then,}$$

(vi) $$SO = AO - AS = a(\mu^2 - 1).$$

The equation of the tangent at $P = (a\mu^2, 2a\mu)$ is:

(vii) $$y = \frac{x}{\mu} + a\mu; \qquad \text{when } y = 0,\ x = -a\mu^2.$$

Therefore,

(viii) $$AM = |x|_{y=0} = a\mu^2 = AO \qquad \text{(by (v)),}$$

$$SP^2 = SO^2 + OP^2 = [a(\mu^2 - 1)]^2 + 4a^2\mu^2 \qquad \text{(by (v) and (vi))}$$
$$= [a(\mu^2 + 1)]^2.$$

Hence,

(ix) $$SP = a(\mu^2 + 1) = AM + AS = SM \qquad \text{(by (ii) and (viii)).}$$

By virtue of this last relation,

$$MN = NP \qquad \text{(since } SN \text{ is perpendicular to } MP\text{);}$$

and since $AO = MA$ (by (viii)).

(x) AN is parallel to PO and $\therefore$ perpendicular to OM

(Newton's Cor. III of Lemma (XIV)).

Also since the right-angled $\triangle$s SAN and SNP are similar,

$$\text{(xi)} \qquad \frac{SN}{SA} = \frac{SP}{SN} \qquad \text{or} \qquad SN^2 = SP.SA = aSP.$$

that is, SN is the 'mean proportional' of the distances SP and SA as required by Lemma XIV. It further follows from equation (xi),

$$\text{(xii)} \qquad SN^2/SP = SA = a$$

and

$$\text{(xiii)} \qquad \frac{SP^2}{SN^2} = \frac{SN^2}{SA^2} = \frac{SP}{SA} \qquad \text{(by xii);}$$

and these are, respectively, Newton's Corollaries II and III of Lemma XIV.

It is of interest to recall that Newton was apparently persuaded by Halley to include Lemmas XII, XIII, and XIV for the benefit of the readers of the time (and one might be inclined to add, of the present time, as well!)

Proposition XIII. Problem VIII

If a body moves in the perimeter of a parabola; it is required to find the law of the centripetal force tending to the focus of that figure.

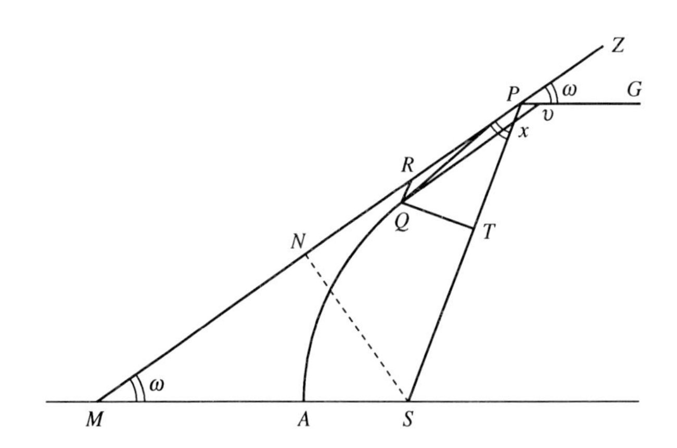

Description of the figure

The lettering of the present figure agrees with that of the preceding figure of the lemmas where they are the same and for the rest of the geometrical correspondence with the figures of Propositions XI and XII is maintained.

S is the focus and A is the vertex of the parabola;
P and Q are neighbouring points on the parabola;
RPZ is the tangent at P;
SN is perpendicular to MRP;

RQ and QT are, respectively, parallel and perpendicular to SP;
PG is parallel to MS.

(xiv) $$\angle GPZ(=\angle Pvx)=\angle SPM(=\angle Pxv)=\omega \quad \text{(say)},$$

(by a property of the parabola). The angle ω defined in (xiv) is related to the parameter μ in equation (vii) for the tangent at P, by

(xv) $$\cot\omega=\mu. \qquad \text{(by (vii))}.$$

The equation of the parabola with respect to the oblique axes PG (the x-axis) and PM (the y-axis) is given by (since $x = Pv$ and $y = Qv$):

$$y^2=4(a\operatorname{cosec}^2\omega)x=4a(1+\mu^2)x \qquad \text{(by (xv))}$$

or by equation (ix)

(xvi) $$y^2=4SPx.$$

With these matters of definition out of the way, Newton's solution to the problem follows easily and directly.

In the chosen oblique system of coordinates

$$P\equiv(0,0) \qquad \text{and} \qquad Q=(Pv,Qv).$$

Accordingly by equation (xvi)

(xvii) $$(Qv)^2=4SP\,.\,Pv=4SP\,.\,QR,$$

because $Pv = Px$ (the $\triangle Pxv$ being isosceles by (xiv)) and $Px = QR$ ($RQPx$ being a parallelogram). Therefore, since $Qv \asymp Qx$,

(xviii) $$(Qx)^2 \asymp 4SP\,.\,QR.$$

On the other hand, by the similarity of the right-angled triangles QxT and SPN,

$$\frac{(Qx)^2}{QT^2}=\frac{SP^2}{SN^2}=\frac{PS}{SA} \qquad \text{(by (xiii))}$$

$$=\frac{4PS\,.\,QR}{4SA\,.\,QR}=\frac{(Qx)^2}{4SA\,.\,QR} \qquad \text{(by (xviii))}.$$

Hence

(xix) $$QT^2=4AS\,.\,QR=L\,.\,QR,$$

where $L = 4AS$ is the latus rectum of the parabola (by (iv)). We observe that equation (xix) is formally the same as equation (g) of Proposition XI (and also Proposition XII); and the solution to the problem follows as before:

$$L\frac{QR}{SP^2QT^2} \asymp \frac{1}{SP^2} \propto \text{C.F.} \qquad \text{by Proposition VI, Corollary 1.} \qquad \text{Q.E.I.}$$

With the inverse-square law of attraction established for the motion along ellipses, hyperbolae, and parabolae, Newton recapitulates the principal results of his major Propositions XI, XII, and XIII in the form of two corollaries.

COR. I. From the three last Propositions it follows, that if any body P goes from the place P with any velocity in the direction of any right line PR, and at the same time is urged by the action of a centripetal force that is inversely proportional to the square of the distance of the places from the centre, the body will move in one of the conic sections, having its focus in the centre of force; and conversely. *For the focus, the point of contact, and the position of the tangent, being given, a conic section may be described*, which at that point shall have a given curvature. But the curvature is given from the centripetal force and velocity of the body being given; and two orbits touching one the other, cannot be described by the same centripetal force and the same velocity.

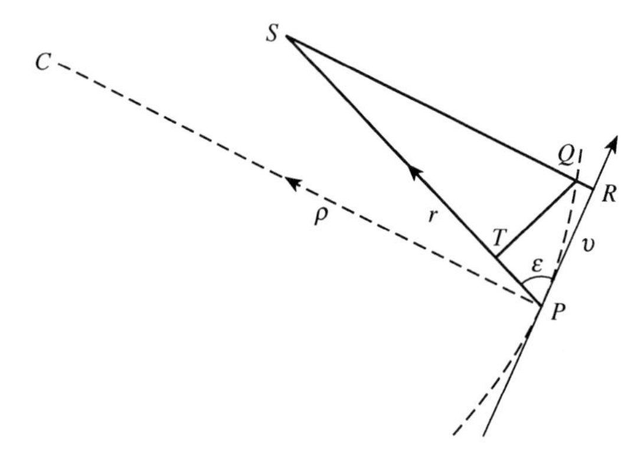

Newton's arguments in this corollary can be paraphrased as follows.

Let S be the centre of attraction with a force k/r^2, where k is some given constant; P the position of a particle at a distance r from S and PR, inclined at an angle ϵ to SP, its instantaneous direction of motion with a velocity v. At an instant of time δt later, by inertia alone, the particle will find itself at R (say) at a distance $v\delta t$ from P. But because of the centripetal attraction, k/r^2, towards S, the particle will find itself at Q along SR where (by Galileo's theorem),

$$QR = \frac{\kappa}{2r^2}(\delta t)^2.$$

Now draw the circle of contact passing through P and Q; and let ρ be the radius of curvature. Then the normal acceleration towards the centre of curvature is v^2/ρ; and this must equal $(\kappa/r^2)\sin\epsilon$. Therefore,

$$\rho = \frac{v^2r^2}{\kappa}\operatorname{cosec}\epsilon,$$

determining ρ in terms of given quantities. And Newton asserts: 'the focus, the point of contact, and the position of the tangent, being given, a conic section may be described, which at that point shall have a given curvature.' But how it 'may be described' is postponed to Proposition XVII to follow.

With respect to the uniqueness of the solution found, Newton states: 'two orbits, touching one the another, cannot be described by the same centripetal force and the same velocity.'

Corollary II

In the diagram for Corollary I, QT (as usual!) is drawn normal to SP. Then, as established in Propositions XI, XII, and XIII,

$$L \asymp \frac{QT^2}{QR},$$

where L is the latus rectum of the ellipse, the hyperbola, or the parabola, as the case may be. This is a relation that is central to Newton's further developments.

29. Kepler's third law: Propositions XIV and XV

After having established the inverse-square law of attraction for motions of bodies along conic sections, Newton turns to proving Kepler's third law (though Newton never associates Kepler's name (as one commonly does today) to the first two of the three laws of planetary motion).

Proposition XIV. Theorem VI

If several bodies revolve about one common centre, and the centripetal force is inversely as the square of the distance of places from the centre: I say, that the principal latera recta of their orbits are as the squares of the areas, which the bodies by radii drawn to the centre describe in the same time.

This proposition is a recapitulation of the principal relations established in the earlier propositions. We have (in the notation, now standard),

$$QR = \tfrac{1}{2}(\text{C.F.})(\delta t)^2 = \kappa\frac{(\delta t)^2}{2SP^2} \qquad \text{(Galileo's theorem)},$$

$$SP.QT = 2A_{,t}\,\delta t \qquad \text{(Proposition VI, Corollary I, equation (2))},$$

where κ is a constant of proportionality and

$$L.QR \asymp QT^2 \qquad \text{(Proposition XIII, Corollary II)}.$$

Therefore

$$L \asymp \frac{QT^2}{QR} = \frac{2SP^2QT^2}{\kappa(\delta t)^2} = \frac{8A_{,t}^2}{\kappa}. \qquad \text{(Q.E.D.)}$$

The corollary for this proposition is a partial statement of Proposition XV.

Proposition XV. Theorem VII

The same things being supposed, I say, that the periodic times in ellipses are as the 3/2th power (in ratione sesquiplicata) *of their greater axes.*

By definition, the latus rectum,

$$L = 2b^2/a \qquad \text{or} \qquad b = \sqrt{(aL/2)}.$$

Therefore,

$$\text{the area of the ellipse} = \pi ab = \pi a^{3/2}\sqrt{(L/2)}.$$

The periodic time, T, of revolution is the area divided by the rate at which the area is described, namely $A_{,t}$ that is,

$$T = \frac{\pi a^{3/2}\sqrt{(L/2)}}{A_{,t}}.$$

But by Proposition XIV,

$$A_{,t} = \tfrac{1}{2}\sqrt{(L/2)}.$$

Therefore,

$$T = 2\pi a^{3/2}. \qquad \text{(Q.E.D.)}$$

By comparison of this last result with Proposition IV, Corollary VI, it follows:

COR. I. Therefore the periodic times in ellipses are the same as in circles whose diameters are equal to the greater axes of the ellipses.

30. Amplifications: Proposition XVI

Proposition XVI with its nine corollaries summarizes and amplifies the principal results of the major propositions that have been established.

Proposition XVI. Theorem VIII

The same things being supposed, and right lines being drawn to the bodies that shall touch the orbits, and perpendiculars being let fall on those tangents from the common focus: I say, that the velocities of the bodies vary inversely as the perpendiculars and directly as the square roots of the principal latera recta.

By the similarity of the right-angled $\triangle$s PQT and PSY in the limit $Q \to P$,

$$\frac{PQ}{QT} \asymp \frac{PS}{SY},$$

or

$$PQ \asymp SP.QT/SY.$$

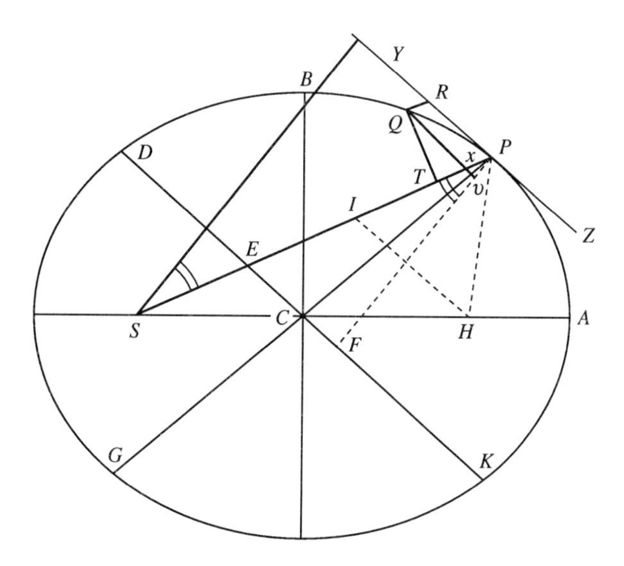

But by Proposition XIV,

$$SP \, . \, QT = \kappa \delta t \sqrt{(L/2)}.$$

where κ is a constant of proportionality. Therefore,

$$PQ \asymp \kappa \frac{\delta t \sqrt{(L/2)}}{SY}.$$

Now,

$$PQ \asymp RP = v\delta t.$$

Combining these last two results we obtain,

(o) $$v = \kappa \sqrt{(L/2)}/SY. \qquad \text{(Q.E.D.)}$$

A series of corollaries making use of this relation follows:

Corollary I

An alternative form of the relation established is

(i) $$\tfrac{1}{2}L = (vSY/\kappa)^2.$$

Corollary II

At A and B, 'SY' (the perpendicular distance from S to the tangent at the point considered) is, respectively, SA and BC. Therefore

(ii) $$v_A = \kappa \sqrt{(L/2)}/SA \qquad \text{and} \qquad v_B = \kappa \sqrt{(L/2)}/BC.$$

Corollary III

The velocity $v_{\odot}(R)$ in a circular orbit of radius R and latus rectum $2R$ is, by (o),

(iii$_a$): $$v_{\odot} = \kappa R^{-1/2}.$$

Therefore, by (ii),

(iii$_b$): $$v_A/v_{\odot}(SA) = \sqrt{(\tfrac{1}{2}L/SA)} \qquad \text{and} \qquad v_B/v_{\odot}(BC) = \sqrt{(\tfrac{1}{2}L/BC)}.$$

Corollary IV

The 'mean distance' from either focus is defined by

$$BS = \sqrt{(BC^2 + SC^2)} = \sqrt{(b^2 + a^2e^2)} = a.$$

Therefore,

$$v_{\odot}(BS) = \kappa a^{-1/2};$$

while according to (o)

$$v(\text{ellipse}; B) = \frac{\kappa}{b}\sqrt{\frac{b^2}{a}} = \kappa a^{-1/2}.$$

Hence,

(iv) $$v_B(\text{ellipse}; B) = v_{\odot}(BS) = \kappa a^{-1/2}.$$

Corollary V

Along any *given* ellipse

(v) $$v_p \propto (SY)^{-1};$$

and this relation obtains also when we compare different ellipses with the same latera recta.

It is possible that when Newton wrote down this corollary, he might have had in mind the earlier common interpretation (by Hooke and by Wren?) of the area-law as implying $v_p \propto (SP)^{-1}$.

Corollary VI

Along a parabola, $L = 4a$ and 'SY' $= SN$ (see the figure for Lemmas XIII and XIV)

$$v = \kappa\sqrt{(2a)}/SN.$$

But by equation (xi) of Lemmas XIII and XIV,

$$SN = \sqrt{(aSP)}.$$

Hence,

(vi) $$v_p(\text{parabola}) = \kappa\sqrt{(2/SP)},$$

that is, the velocity is inversely proportional to the square root of the distance from the focus.

Corollary VII

By ($\mathrm{iii_a}$) and (vi)

($\mathrm{vii_a}$) $$v(\text{parabola}; SP)\colon v_{\odot}(SP) = \sqrt{2}:1,$$

and

($\mathrm{vii_b}$) $$v(\text{parabola}; SP)\colon v_{\odot}(\tfrac{1}{2}SP) = 1:1.$$

(Notice the independence of these ratios on P.)

Corollary VIII

By relations already established,

(viii) $$v(\text{ellipse}; SP)\colon v_{\odot}(\tfrac{1}{2}L) = \tfrac{1}{2}L/SY.$$

Corollary IX

Quite generally:

($\mathrm{ix_a}$) $$v(\text{ellipse}; SP)\colon v_{\odot}(SP) = \sqrt{(\tfrac{1}{2}L\,.\,SP)}\colon SY,$$

and

($\mathrm{ix_b}$) $$v_{\odot}(R_1)\colon v_{\odot}(R_2) = \sqrt{(R_2/R_1)}.$$

The listing of all the foregoing special cases is characteristic of Newton's scientific attitude: to explore *all* matters with thoroughness.

Proposition XVII. Problem IX

Supposing the centripetal force to be inversely proportional to the squares of the distances of places from the centre, and that the absolute value of that force is known; it is required to determine the line which a body will describe that is let go from a given place with a given velocity in the direction of a given right line.

The problem to be solved is this: given that a body P, under the action from S of an inverse-square law of centripetal attraction, k/SP^2 (where k is a known constant of proportionality)* is projected in the direction ZPR with a velocity v, to determine the conic section that it will describe.

* Newton replaces the information of knowing κ by the knowledge of the ratio, $\Lambda/(V\,.\,Sr)^2$ (see the equation below).

For convenience of reference we draw an ellipse assuming (for the present) that the orbit will be found to be an ellipse. The focus about which the conic section will be described is S. The line XPX' on which the other focus H will be found can be readily drawn: we have only to make the angle XPZ equal to the angle SPR. Draw SK and SR perpendicular to XPX' and ZPR, respectively.

The latus rectum L of the conic section to be determined is given in terms of v and SP by

$$L = 2(vSR/\kappa)^2 \qquad \text{(by Proposition XVI, Corollary I).}$$

But this formula was derived by Newton on the assumption that the law of attraction is $(SP)^{-2}$, ignoring an allowed constant of proportionality k. Newton allows for this factor by considering an arbitrary conic section described under the same law of attraction (including the factor k) and about the same focus S. Let pq be an element of arc of this

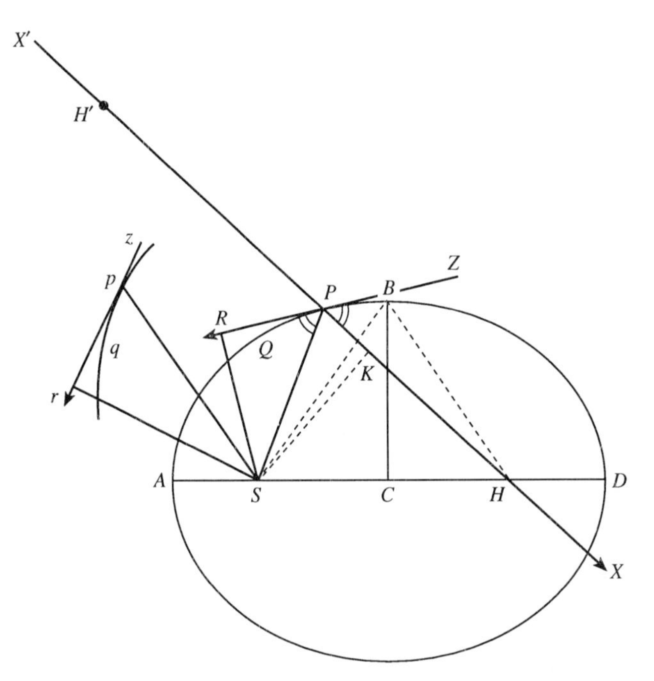

chosen orbit; and let the velocity at p in the direction zpr be V. If Sr is perpendicular to zpr, then by Proposition XVI, Corollary I,

$$\Lambda = 2(V.Sr/\kappa)^2$$

where Λ is the latus rectum of the chosen conic section. By combining the two foregoing relations, we obtain

$$L = \Lambda \frac{(v.SR)^2}{(V.Sr)^2},$$

a relation which determines L.

With the focus S, the line $X'PX$ on which the second focus lies, and the latus rectum L known, it remains to determine the location of H on $X'PX$ and the semiaxes a and b. Newton proceeds on the assumption (to be discarded if need be) that the orbit to be determined will be found to be an ellipse.

$$
\begin{aligned}
SP^2 + PH^2 - 2PH.PK = SH^2 = 4CH^2 &= 4(BH^2 - BC^2)\\
&= 4a^2 - 2a(2b^2/a) \qquad \text{(since } BH = BS = a\text{)},\\
&= (SP + PH)^2 - (SP + PH)L \qquad \text{(since } SP + PH = 2a\text{)},\\
&= SP^2 + PH^2 + 2SP.PH - (SP + PH)L.
\end{aligned}
$$

Hence

(a) $$L(SP + PH) = 2PH(SP + PK);$$

for, alternatively

(b) $$L(1 + SP/PH) = 2(SP + PK).$$

Since SP, PK, and L are given quantities, it is apparent from equation (b) that we have to distinguish three cases:

1. $L < 2(SP + PK)$. In this case, equation (b) can be solved for PH, determining the location of the second focus, H, on PX. The orbit to be determined is an ellipse; and the semimajor axis of the ellipse follows from the relation,

$$SP + PH = 2a,$$

while the semiminor axis follows from the definition of L:

$$L = 2b^2/a \qquad \text{or} \qquad b = \sqrt{(La/2)}.$$

2. $L = 2(SP + PK)$. Then PH is infinite and the orbit will be a parabola, with its axis SH parallel to PK. Since L for a parabola is $4a$ where a is the distance of the vertex from S, the relation (b) gives

$$a = \tfrac{1}{2}(SP + PK).$$

3. $L > 2(SP + PK)$. Clearly in this case the orbit is a hyperbola; the conjugate focus H will be found in the opposite direction along PX' (see figure) and equation (a) will take the form

$$L(PH' - SP) = 2PH'(SP + PK)$$

and

$$L(1 - SP/PH') = 2(SP + PK).$$

The solution can be completed as in case 1 remembering that in this case

$$PH' - SP = 2a$$

Newton concludes:

For if the body, in these cases, revolves in a conic section so found, it is demonstrated in Prop. XI, XII, and XIII, that the centripetal force will be inversely as the square of the distance of the body from the centre of force S; and therefore we have rightly determined the line PQ, which a body let go from a given place P with a given velocity, and in the direction of the right line PR given by position, would describe with such a force. Q.E.F.

The following observation with respect to this proposition may be made.

In Proposition XIII, Corollary I, Newton asserts, as we have noted: 'the focus, the point of contact, and the position of the tangent being given, a conic section may be described which at that point shall have a given curvature'. And as we remarked in the context, how 'the conic section may be described' is considered only in this proposition. But in this proposition, it is the latus rectum L, not the radius of curvature ρ, that is given. The two are, however, very simply related. Equation (viii) of Proposition IX (§24, Chapter 5) for an inverse-square law gives

$$\rho \propto \operatorname{cosec}^3 \epsilon \qquad (\epsilon \text{ is the inclination of } SP \text{ to } RP).$$

The constant of proportionality follows from equation (xv) of the same section. With the known polar equation of the elliptical orbit, namely

$$u = \frac{1}{r} = \frac{2}{L}(1 + e \cos \varphi) \qquad (e = \text{the eccentricity}),$$

equation (xv) of §24 gives

$$\rho = \tfrac{1}{2} L \operatorname{cosec}^3 \epsilon.$$

That Newton must have known this relation requires no argument!

Corollary I

Newton establishes the identity,

$$L = \frac{4DS \, . \, DH}{DS + DH},$$

in the notation of the figure. It follows from the known relations

$$DS + DH = 2a, \qquad DH = a(1 - e) \qquad \text{and} \qquad DS = a(1 + e).$$

By virtue of these relations,

$$\frac{4DS \, . \, DH}{DS + DH} = \frac{4a^2(1 - e^2)}{2a} = 2\frac{b^2}{a} = L$$

as required. An alternative form of the identity is

$$DS : DH = (4DS - L) : L,$$

while a comparison with equation (a) shows that

$$2DS \,.\, DH = PH(PS + PK).$$

Corollary II

The particular simplicity of the case, when the velocity at the principal vertex D is given, is noted. By Proposition XVI, Corollary III (equation (iii)),

$$\frac{\frac{1}{2}L}{DS} = \left[\frac{v(\text{ellipse}; SD)}{v_{\odot}(SD)}\right]^2,$$

a relation which immediately determines DS; and DH (locating the second focus) follows from the relation

$$\frac{DS}{DH} = \frac{4DS - L}{L}.$$

And finally the Corollaries III and IV show how far into the future Newton sees already at this stage of writing the *Principia*.

> COR. III. Hence also if a body moves in any conic section, and is forced out of its orbit by any impulse, you may discover the orbit in which it will afterwards pursue its course. For by compounding the proper motion of the body with that motion, which the impulse alone would generate, you will have the motion with which the body will go off from a given place of impulse in the direction of a right line given in position.
>
> COR. IV. And if that body is continually disturbed by the action of some foreign force, we may nearly know its course, by collecting the changes which that force introduces in some points, and estimating the continual changes it will undergo in the intermediate places, from the analogy that appears in the progress of the series.

It may be useful to summarize the essential content of this proposition.

Newton, in effect, solves the problem of the motion of bodies governed by the equation,

$$\frac{\mathrm{d}^2\vec{r}}{\mathrm{d}t^2} = -\frac{k\vec{r}}{r^3}; \qquad (r = |\vec{r}|)$$

and finds solutions that satisfy given initial conditions and shows how, depending on initial conditions, the orbit described can be an ellipse, a parabola, or a hyperbola. And as Newton explains in Corollary I of Proposition XIII, 'Given the focus, the point of contact and the position of the tangent, a conic section may be described' having a given latus rectum or, equivalently, having at the point of contact, a given curvature; and as he states further, in Corollary I of Proposition XIII, 'two orbits, touching one the other, cannot be described by the same centripetal force and the same velocity'. Besides, Newton establishes the uniqueness of the solution of the initial-value problem quite generally in Proposition XLI.

Nevertheless the adequacy of Newton's treatment was questioned by Bernoulli (and by others since). The following comments by the distinguished mathematician V. I. Arnold (in '*Huygens and Barrow*; *Newton and Hooke*', Birkhäuser Verlag, Basel, 1990, pp. 30–31), should, in my judgement, dispose of these criticisms once and for all.

> Modern mathematicians actually distinguish existence theorems and uniqueness theorems for differential equations and even give examples of equations for which the existence theorem is satisfied but the uniqueness theorem is not. So various troubles can arise, and if Newton's equation were troublesome, it would actually be impossible to make any deductions. A mistaken point of view arises because of the unwarranted extension of the class of functions under consideration. The fact is that in modern mathematics the concepts of function, vector field, differential equation have acquired a different meaning in comparison with classical mathematics. Speaking of a function, we can have in mind a rather nasty object—something differentiable once or even not at all—and we must think about the function class containing it, and so on. But at the time of Newton the word function meant only very good things. Sometimes they were polynomials, sometimes rational functions, but in any case they were all analytic in their domain of definition and could be expanded in Taylor series. In this case the uniqueness theorem is no problem, and at that time nobody gave it a thought.
>
> But in reality Newton proved everything, to a higher standard.

Scholium

As if to draw attention to Corollaries II and III of Proposition VII and to 'the same otherwise'—demonstrations of Propositions XI and XII, newly added to the second 1713 edition of the *Principia*—Newton in the concluding Scholium restates Corollary III of Proposition VII, for the special case when the orbit is an ellipse described about its centre C by a centripetal attraction towards an arbitrary point R in the interior. Identifying S with C in equation (iv) of Proposition VII, Corollary III (Chapter 5, p. 82) we

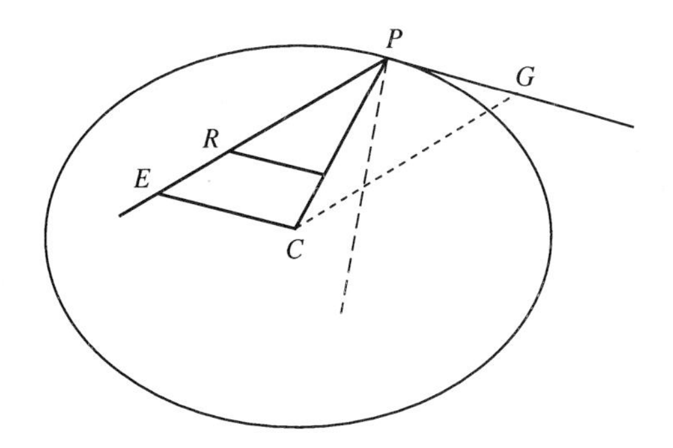

have

$$(\text{C.F.})_R = (\text{C.F.})_C \cdot \frac{CG^3}{CP \, . \, RP^2} \qquad (GC = EP).$$

But, by Proposition X, Corollary III, we know that an ellipse described about the centre C requires a law of centripetal attraction proportional to the distance CP. Consequently

$$(\text{C.F.})_R \propto \frac{CG^3}{RP^2}.$$

This brings us to the end of a remarkable chapter of the *Principia.*

A personal reflection

To repeat once again, what I have remarked on more than one occasion: the Corollaries I, II, and III of Proposition VII and 'the same otherwise'—demonstrations of Propositions XI and XII and the concluding Scholium—were all new additions to the second 1713 edition of the *Principia*, that is, long after the first 1687 edition, during Newton's London years. As will be apparent from the Supplement to this chapter the Corollaries II and III of Proposition VII are of profound theoretical significance and display Newton's deep insight. And it is intriguing to speculate on how Newton might have come to make this discovery.

It is known that during his London years Newton was wont to spend time turning the pages of his personal interleaved copy of the *Principia*—mostly, it appears, for detecting minor errors and misprints that might have been overlooked. Perhaps on one such occasion he noticed (as he must have on numerous previous occasions) that a body describes the same ellipse under two different centres. But on this occasion the thought occurred to him 'Clearly one *must* be deducible from the other'. No sooner had the thought occurred, Newton apparently had no difficulty in devising the proof (as in the Scholium). Perhaps, during those lonely years in London, with the *Principia* in front of him, Newton continued to voyage

'through strange seas of thought alone'.

Supplement: on dual laws of centripetal attraction

31. A recapitulation

As we have remarked on several occasions, Newton in his Corollary III of Proposition VII formulates the problem of a body (with a given angular momentum) describing the same orbit under two different centres. The questions suggested by the formulation of this problem and the illustrations Newton provided are of considerable depth. One can, for example, ask the question: *given a power law of centripetal attraction, is there a dual law for which a body with the same constant of areas will describe the same orbit*? Only recently has this question been raised and answered by V. I. Arnold and T. Needham. We shall present these solutions; but first we shall recapitulate Newton's results on this problem.

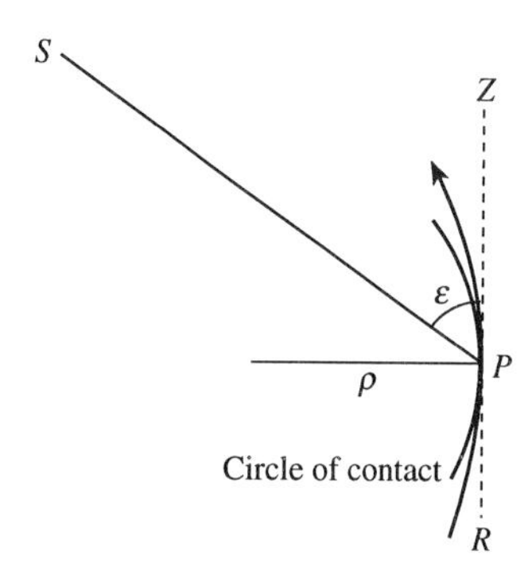

We shall base our recapitulation on the formula

$$\text{C.F.} \propto \frac{1}{2\rho \sin^3 \epsilon} \frac{1}{SP^2}, \tag{1}$$

derived in Chapter 5 (Proposition IX, equation (viii)), where ρ is the radius of curvature at a point P and ϵ is the inclination of the direction of motion at P to the line joining it to the centre of attraction at S.

That bodies describe a conic section under an inverse-square law of attraction emanating from the focus, follows directly from equation (1), since for *any* conic section the radius of curvature ρ is related to the latus rectum L by (cf. Proposition XVII, p. 110)

$$\rho \sin^3 \epsilon = \tfrac{1}{2}L; \tag{2}$$

and equation (1) gives

$$\text{C.F.} \propto \frac{1}{L} (SP)^{-2}. \tag{3}$$

(Q.E.D!)

A digression

Both relations (1) and (2) are included in the *Principia*, albeit implicitly. Newton must have known them: relation (1) follows very directly from several of the relations included in Sections II and III (already of the first 1687 edition); and relation (2) cannot have escaped the Master Geometer of the conic sections. Why then, one may ask, did Newton not include in the *Principia* so simple and direct a derivation of Propositions XI, XII, and XIII that results from combining the two relations (1) and (2)? The answer to this question must be that these simple relations must have escaped him during the two years when he was under the extreme pressure of writing the *Principia*; and that they became transparent to him during his 'lonely voyages', in the leisure of his London years. Be it noted in this connection that the italicized part of Corollary I of Proposition XIII, quoted on page 102, and the concluding sentence of Proposition XVII, quoted on page 107, do not appear in the first 1687 edition of the Principia; they are included for the first time in the second (1713) and the third (1726) editions, respectively. Was Newton, in these later years, 'stooping' a little, contrary to what perhaps was his earlier motto:

> Learn to understand and you shall hear it. But in other terms—no. If you would not rise to us, we cannot stoop to you.
>
> John Ruskin; in *Sesame and Lilies*, 1865

But to the purpose!

If the same orbit is described with respect to two centres S and C, then it follows from equation (1), since ρ is the same for the two orbits,

$$\frac{(\text{C.F.})_S}{(\text{C.F.})_C} \propto \left(\frac{\sin \epsilon_C}{\sin \epsilon_S}\right)^3 \frac{CP^2}{SP^2}. \tag{4}$$

A. The orbit described is an ellipse

Let the orbit be described under the centripetal attraction derived from an arbitrary point S in the interior. Then

$$\epsilon_S = \angle RPS = \angle PEC,$$

and

$$\epsilon_C = \angle RPC = \pi - \angle ECP. \tag{5}$$

Hence

$$\frac{\sin \epsilon_C}{\sin \epsilon_S} = \frac{\sin \angle ECP}{\sin \angle PEC} = \frac{EP}{PC} \tag{6}$$

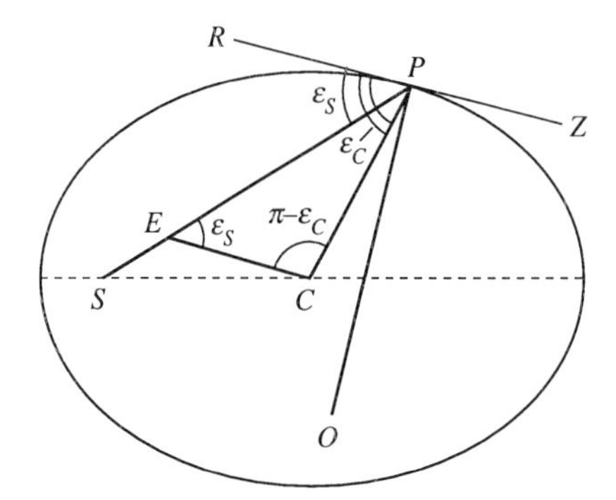

and it follows from equation (4) that

$$\frac{(\text{C.F.})_S}{(\text{C.F.})_C} \propto \left(\frac{EP}{CP}\right)^3 \frac{CP^2}{SP^2}. \tag{7}$$

We know by Proposition X, the centripetal force towards C is

$$(\text{C.F.})_C \propto CP. \tag{8}$$

Therefore,

$$(\text{C.F.})_S \propto \frac{EP^3}{SP^2}, \tag{9}$$

which is the result stated by Newton in his concluding Scholium.

If S is identified with the focus it follows that $EP = a$ (the semimajor axis) is a constant; and we recover Newton's 'the same otherwise' demonstration of Proposition XI.

B. The orbit described is a hyperbola

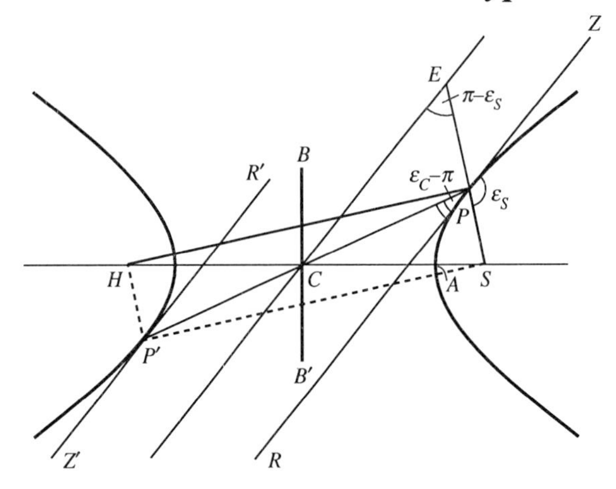

On the right half of the figure the lettering is the same as on the figure for Proposition XII; on the left half the lettering is that appropriate for a body P' orbiting the conjugate hyperbola.

We first compare the forces acting on P centred at S or C. It is clear that

$$\epsilon_C = \angle CPZ = \pi + \angle CPR = \pi + \angle PCE \tag{10}$$

while

$$\epsilon_S = \angle SPZ = \pi - \angle EPZ = \pi - \angle PEC. \tag{11}$$

Therefore

$$\frac{\sin \epsilon_C}{\sin \epsilon_S} = -\frac{\sin \angle PCE}{\sin \angle PEC} = -\frac{PE}{CP}. \tag{12}$$

From equation (4) it now follows:

$$\frac{(\text{C.F.})_S}{(\text{C.F.})_C} \propto -\frac{EP^3}{CP \,.\, SP^2} = -\frac{a^3}{CP \,.\, SP^2}. \tag{13}$$

But we know that the attraction towards S is proportional to SP^{-2}. Hence

$$(\text{C.F.})_C \propto -CP; \tag{14}$$

in other words: a particle will orbit the *same* hyperbola under a *repulsive* force proportional to the distance from the centre C. This conclusion is in accord with Newton's statement, following the 'same otherwise' demonstration of Proposition XII, and quoted on p. 98.

C. A body orbiting the conjugate hyperbola with the centre of attraction at S

It is clear that

$$\angle SP'R' = \angle SPR, \tag{15}$$

while the radii of curvature at P' and P are of opposite signs. Hence

$$\frac{(\text{C.F. acting on } P)_S}{(\text{C.F. acting on } P')_S} = -\frac{(SP')^2}{(SP)^2}; \tag{16}$$

in other words: *under an inverse-square law of repulsion centred at S, a body will describe the conjugate hyperbola*, confirming Newton's statement quoted on p. 98.

By similar considerations (or, more directly by symmetry) it follows that under the action of a repulsive force proportional to the distance from the centre C, a particle can describe either branch of the hyperbola. In this sense *a linear repulsive law of force is self-dual.*

D. The self-duality of the inverse-fifth power law of attraction

We finally turn to Corollary I of Proposition VII.

Consider a circular orbit described under centripetal attraction emanating from two

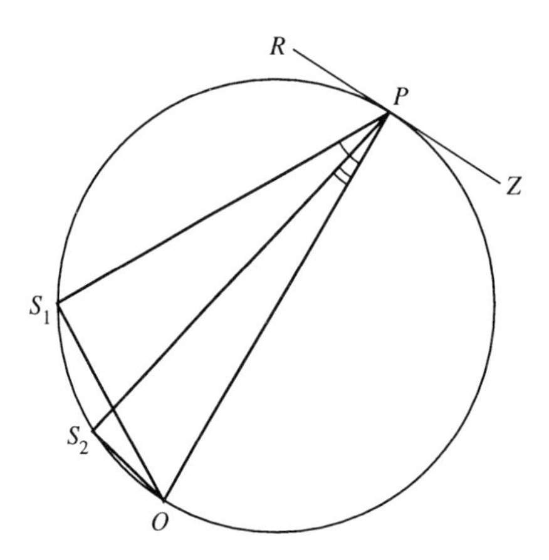

different points S_1 and S_2 on the circumference of the circle. Then, by relation (4),

$$\frac{(\mathrm{C.F.})_{S_1}}{(\mathrm{C.F.})_{S_2}} \propto \left(\frac{\sin \angle S_2PR}{\sin \angle S_1PR}\right)^3 \left(\frac{S_2P}{S_1P}\right)^2 = \left(\frac{\cos \angle S_2PO}{\cos \angle S_1PO}\right)^3 \left(\frac{S_2P}{S_1P}\right)^2$$

$$= \left(\frac{S_2P/PO}{S_1P/PO}\right)^3 \left(\frac{S_2P}{S_1P}\right)^2 \tag{17}$$

or

$$\frac{(\mathrm{C.F.})_{S_1}}{(\mathrm{C.F.})_{S_2}} \propto \frac{(S_2P)^5}{(S_1P)^5}; \tag{18}$$

in other words, the same inverse-fifth power law. One may say, on this account, that the inverse-fifth power law of attraction is *self-dual* for motion in a circle.

To summarize: during the course of proving the various propositions and corollaries of Section III, Newton establishes the following laws of centripetal force as dual:

attractive, r^{-2} and attractive, r	(elliptical orbits);
attractive, r^{-2} and repulsive, r	(hyperbolic orbits);
repulsive, r and repulsive, r	(the conjugate branches of a hyperbola);
repulsive, r^{-2} and attractive r^{-2}	(hyperbolic orbits with the centre of force at the focus of the conjugate hyperbola);
attractive, r^{-5} and attractive, r^{-5}	(a circle with the centre of attraction on any point of the circumference).

(The last two are self-dual.)

32. The mapping of orbits described in the complex plane

Consider the motion of a body described in the complex z-plane under a linear law of attraction. The equation governing such motion is

$$\frac{d^2 z}{dt^2} = -z \tag{1}$$

and a general enough (!) solution of this equation is:

$$z = x + iy = p e^{+it} + q e^{-it}, \tag{2}$$

where p and q ($< p$) are constants, which, without loss of generality, may be considered as real. The motion represented by the solution (2) is a superposition of the uniform, circular motions described in opposite senses and as such describes an elliptical orbit. Thus, rewriting the solution in the form

$$z = (p + q) \cos t + i(p - q) \sin t. \tag{3}$$

and separating the real and the imaginary parts, we have

$$x = (p + q) \quad \cos t \qquad \text{and} \qquad y = (p - q) \sin t \tag{4}$$

which is the parametric representation of an ellipse with semiaxes $a = (p + q)$ and $b = (p - q)$. The eccentricity of the ellipse is given by

$$(p + q)e = \sqrt{[(p + q)^2 - (p - q)^2]} = 2\sqrt{(pq)}. \tag{5}$$

The foci of the ellipse are, therefore, at

$$[\pm 2\sqrt{(pq)}, 0]. \tag{6}$$

Also we may note that the radial distance from the centre is given by

$$r = \sqrt{(p^2 + q^2 + 2pq \cos 2t)} \tag{7}$$

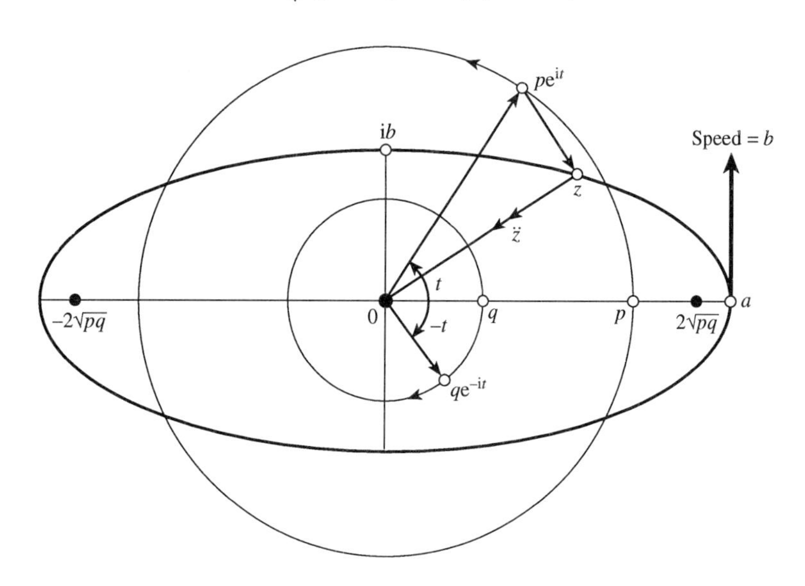

Consider now the mapping

$$\tilde{z} \rightsquigarrow z^2 = p^2 \mathrm{e}^{+2it} + q^2 \mathrm{e}^{-2it} + 2pq$$
$$= (p^2 + q^2)\cos 2t + \mathrm{i}(p^2 - q^2)\sin 2t + 2pq. \tag{8}$$

Separating the real and the imaginary parts, we now have

$$\tilde{x} - 2pq = (p^2 + q^2)\cos 2t \qquad \text{and} \qquad \tilde{y} = (p^2 - q^2)\sin 2t, \tag{9}$$

which represents an ellipse of semiaxes, $\tilde{a} = p^2 + q^2$ and $\tilde{b} = (p^2 - q^2)$ centred at

$$\tilde{x} = 2pq, \tag{10}$$

that is, at the left-hand focus; for its eccentricity $\tilde{e}$, we have

$$(p^2 + q^2)\tilde{e} = \sqrt{[(p^2 + q^2)^2 - (p^2 - q^2)^2]} = 2pq. \tag{11}$$

The polar equation of this mapped ellipse is (cf. equation (7)),

$$\tilde{r} = \frac{(p^2 - q^2)^2}{p^2 + q^2 + 2pq\cos\varphi}. \tag{12}$$

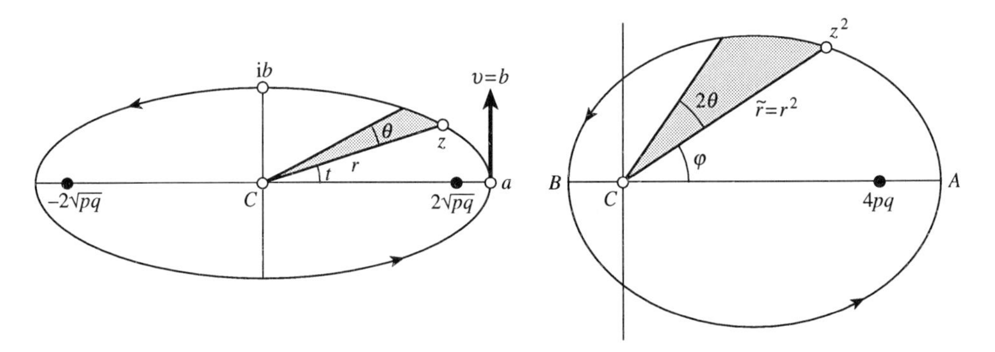

We have shown that by the mapping $z \rightsquigarrow z^2$, the ellipse described under a linear law of attraction is mapped on to another ellipse having the left-hand focus at the centre of the original ellipse. In §33, we shall verify that the mapping does provide a solution for the ellipse described under an inverse-square law of centripetal attraction. Meantime, we shall consider the hyperbola described under a linear law of *repulsion* governed by the equation

$$\frac{\mathrm{d}^2 z}{\mathrm{d}t^2} = +z. \tag{13}$$

Writing the solution of this equation in the form

$$z = \lambda \mathrm{e}^{+t} + \bar{\lambda}\mathrm{e}^{-t} = (\lambda + \bar{\lambda})\cosh t + (\lambda - \bar{\lambda})\sinh t \tag{14}$$

and separating the real and the imaginary parts, we obtain,

$$x = (\lambda + \bar{\lambda})\cosh t \qquad \text{and} \qquad y = -\mathrm{i}(\lambda - \bar{\lambda})\sinh t. \tag{15}$$

The solution represents a hyperbola with semiaxes,

$$a = \lambda + \bar{\lambda} \qquad \text{and} \qquad \mathrm{i}b = \lambda - \bar{\lambda}; \tag{16}$$

and eccentricity $e(>1)$ given by

$$(\lambda + \bar{\lambda})e = \sqrt{[(\lambda + \bar{\lambda})^2 - (\lambda - \bar{\lambda})^2]} = 2|\lambda|. \tag{17}$$

The foci of the two branches of the hyperbola are at

$$[\pm 2|\lambda|, 0]. \tag{18}$$

Now consider, as before, the mapping

$$\begin{aligned} \tilde{z} \rightsquigarrow z^2 &= \lambda^2 e^{+2t} + \bar{\lambda}^2 e^{-2t} + 2|\lambda|^2 \\ &= (\lambda^2 + \bar{\lambda}^2)\cosh 2t + (\lambda^2 - \bar{\lambda}^2)\sinh 2t + 2|\lambda|^2, \end{aligned} \tag{19}$$

which represents the one branch of the hyperbola

$$(\tilde{x} - 2|\lambda|^2) = (\lambda^2 + \bar{\lambda}^2)\cosh 2t \qquad \text{and} \qquad y = -\mathrm{i}(\lambda^2 - \bar{\lambda}^2)\sinh 2t, \tag{20}$$

with semiaxes,

$$(\lambda^2 + \bar{\lambda}^2) \qquad \text{and} \qquad |\lambda^2 - \bar{\lambda}^2| \tag{21}$$

centred at $(\tilde{x} = 2|\lambda|^2, \tilde{y} = 0)$ with eccentricity

$$(\lambda^2 + \bar{\lambda}^2)\tilde{e} = \sqrt{[(\lambda^2 + \bar{\lambda}^2)^2 + |\lambda^2 - \bar{\lambda}^2|^2]} = 2|\lambda|^2. \tag{22}$$

We observe that both branches of the hyperbola

$$\frac{x^2}{(\lambda + \bar{\lambda})^2} - \frac{y^2}{|\lambda - \bar{\lambda}|^2} = 1 \tag{23}$$

are mapped on to the one branch of the hyperbola represented by equation (19). (The case $b^2 > a^2$, when the energy E is positive, is illustrated for reasons explained in §33, B and C.) This fact requires elucidation. It derives from the fact that the same branch of the hyperbola describes the motion of a particle under an inverse-square law repulsion from the conjugate focus H. The equivalent statement, that a particle under an inverse-square law of repulsion from S describes the conjugate hyperbola with its focus at H was established in §31, C. It was further established in §31, B, that under a repulsive force proportional to the distance from the centre, hyperbolic orbits are described.

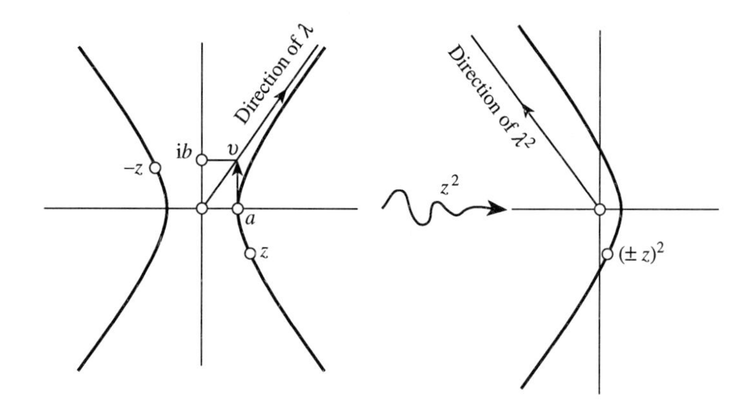

And finally we may note that the linear trajectory,

$$z = a + ibt, \tag{24}$$

in the complex plane (described in the absence of any external force), by the same mapping, $z \rightsquigarrow z^2$, becomes the parabola,

$$\tilde{y}^2 = 4a^2(a^2 - \tilde{x}), \tag{25a}$$

where

$$\tilde{x} = (a^2 - b^2t^2) \qquad \text{and} \qquad \tilde{y} = 2abt. \tag{25b}$$

33. The dual laws of centripetal forces

In §32, we showed that an elliptical (or a hyperbolic) orbit, in the complex z-plane, described about its centre, C, under the action of an attractive (or a repulsive) force proportional to the distance from C, is, by the mapping $z \rightsquigarrow z^2$, transformed into a conic section with its focus at C. Is this fact a mere geometrical curiosity, or has it a deeper physical base? It has: for, we shall show that the trajectory one obtains by the mapping $z \rightsquigarrow z^2$, is, in fact, in accord with the motion of a body under the action of an inverse-square law of attraction (or repulsion, in case the centre of force is at the conjugate focus of the hyperbola that is described).

The basic theorem in the subject is due to K. Bohlin (Bulletin Astronomique, Paris, Vol. 28, p. 8, 1911); see also V. I. Arnold in *Huygens and Barrow, Newton and Hooke* (Birkhäuser Verlag, 1990, pp. 95–100, whose exposition we shall, in effect, follow).

Consider, then, the equation of motion,

$$\frac{d^2\omega}{dt^2} = \mp\omega, \tag{1}$$

in the complex ω-plane, where the upper and the lower signs (here and in the sequel) distinguish the attractive and the repulsive cases, respectively. Equation (1) admits of two integrals: the angular momentum integral,

$$\frac{dA}{dt} = |\omega|^2 \frac{d\varphi}{dt} = \text{constant}; \tag{2}$$

and the energy integral

$$E = \frac{1}{2}\left(\left|\frac{d\omega}{dt}\right|^2 \pm |\omega|^2\right) = \text{constant}. \tag{3}$$

We now determine the equation of motion governing

$$Z = \omega^2 \tag{4}$$

with respect to a time τ determined by the constraint,

$$\frac{\mathrm{d}A}{\mathrm{d}\tau} = |Z|^2 \frac{\mathrm{d}\varphi}{\mathrm{d}\tau} = \frac{\mathrm{d}A}{\mathrm{d}t} = |\omega|^2 \frac{\mathrm{d}\varphi}{\mathrm{d}t}, \tag{5}$$

ensuring the equality of the constants of area for the ω- and the Z-orbits (cf., the remarks following equation (iii) of Proposition VII, Chapter 5, p. 81). An alternative form of equation (5) is

$$\frac{\mathrm{d}}{\mathrm{d}\tau} = \frac{|\omega|^2}{|Z|^2} \frac{\mathrm{d}}{\mathrm{d}t} = \frac{1}{\omega\bar{\omega}} \frac{\mathrm{d}}{\mathrm{d}t}. \tag{6}$$

With these definitions, we find

$$\begin{aligned} \frac{\mathrm{d}^2 Z}{\mathrm{d}\tau^2} &= \frac{1}{\omega\bar{\omega}} \frac{\mathrm{d}}{\mathrm{d}t}\left(\frac{1}{\omega\bar{\omega}} \frac{\mathrm{d}\omega^2}{\mathrm{d}t}\right) = \frac{2}{\omega\bar{\omega}} \frac{\mathrm{d}}{\mathrm{d}t}\left(\frac{1}{\bar{\omega}} \frac{\mathrm{d}\omega}{\mathrm{d}t}\right) \\ &= \frac{2}{\omega\bar{\omega}}\left(-\frac{1}{\bar{\omega}^2}\left|\frac{\mathrm{d}\omega}{\mathrm{d}t}\right|^2 \mp \frac{\omega}{\bar{\omega}}\right) \\ &= -\frac{2}{\omega\bar{\omega}^3}(|\dot{\omega}|^2 \pm |\omega|^2), \end{aligned} \tag{7}$$

Or, making use of the energy integral (3), we have

$$\frac{\mathrm{d}^2 Z}{\mathrm{d}\tau^2} = -4E \frac{Z}{|Z|^3}. \tag{8}$$

The two principal cases to be distinguished are:

$$E > 0 \qquad \text{and} \qquad E < 0. \tag{9}$$

If the orbit of ω is an ellipse (as is the case if the centripetal force is attractive)

$$E > 0. \tag{10}$$

On the other hand, if the orbit of ω is a hyperbola (as is the case when the centripetal force is repulsive) both cases, $E > 0$ and $E < 0$, can arise depending on whether

$$a^2 < b^2 \qquad \text{or} \qquad a^2 > b^2, \tag{11}$$

where a and b are the semiaxes of the hyperbola,

$$\frac{x^2}{a^2} - \frac{y^2}{b^2} = 1, \tag{12}$$

representing the orbit.

I. $E > 0$. In this case, equation (8) is that appropriate for an inverse-square law of attraction.

II. $E < 0$. In this case, equation (8) is that appropriate for an inverse-square law of repulsion.

The foregoing results which follow from our present considerations are in complete accord with Newton's discoveries summarized in §31.

Finally, we consider the case

$$E = 0 \qquad \text{and} \qquad b^2 = a^2, \tag{13}$$

when ω describes straight lines and no external force is operative; and as we have seen in §32, the mapping $z \rightsquigarrow z^2$ produces the 'androgynous' parabola (to borrow a felicitous description by T. Needham).

We shall now proceed to the more general case when the centripetal force is proportional to the ath power of the distance:

$$\text{C.F.} \propto \mp|Z|^a. \qquad (a \text{ an integer}). \tag{14}$$

In this case the equation of motion in the complex z-plane is

$$\frac{\mathrm{d}^2\omega}{\mathrm{d}t^2} = \mp\omega|\omega|^{a-1}, \tag{15}$$

with the two integrals,

$$\frac{\mathrm{d}A}{\mathrm{d}t} = |\omega|^2 \frac{\mathrm{d}\varphi}{\mathrm{d}t} = \text{constant}, \tag{16}$$

and

$$E = \frac{1}{2}\left[\left|\frac{\mathrm{d}\omega}{\mathrm{d}t}\right|^2 \pm \frac{2}{a+1}|\omega|^{a+1}\right] = \text{constant}. \tag{17}$$

We shall now show (following V. I. Arnold) that

$$Z = \omega^\alpha, \tag{18}$$

where

$$\alpha = \tfrac{1}{2}(a+3) \qquad \text{or} \qquad a + 1 = 2(\alpha - 1), \tag{19}$$

satisfies an equation of motion (with respect to a time τ determined by the requirement

$$\left. A_{,\tau} = |Z|^2 \frac{\mathrm{d}\varphi}{\mathrm{d}\tau} = A_{,t} = |\omega|^2 \frac{\mathrm{d}\varphi}{\mathrm{d}t}\right) \tag{20}$$

appropriate to a centripetal force proportional to

$$Z^A, \tag{21}$$

where

$$A - 1 = \frac{2}{\alpha}(1 - 2\alpha) \qquad \text{or} \qquad A + 3 = \frac{2}{\alpha} = \frac{4}{a+3}. \tag{22}$$

Making use of the relation,

$$\frac{\mathrm{d}\tau}{\mathrm{d}t} = \left|\frac{Z}{\omega}\right|^2 = |\omega^{\alpha-1}|^2 = |\omega|^{a+1}, \tag{23}$$

we find successively,

$$\begin{aligned}
\frac{d^2Z}{d\tau^2} &= \frac{1}{|\omega|^{a+1}}\frac{d}{dt}\left[\frac{1}{|\omega|^{a+1}}\frac{dZ}{dt}\right] \\
&= \frac{\alpha}{|\omega|^{a+1}}\frac{d}{dt}\left[\frac{\omega^{\alpha-1}}{|\omega|^{a+1}}\frac{d\omega}{dt}\right] \\
&= \frac{\alpha}{|\omega|^{a+1}}\frac{d}{dt}\left[\frac{1}{\bar{\omega}^{(a+1)/2}}\frac{d\omega}{dt}\right] \qquad \text{(by equations (19))} \\
&= \frac{\alpha}{|\omega|^{a+1}}\left[\frac{1}{\bar{\omega}^{(a+1)/2}}\frac{d^2\omega}{dt^2} - \tfrac{1}{2}(a+1)\frac{1}{\bar{\omega}^{(a+3)/2}}\left|\frac{d\omega}{dt}\right|^2\right] \\
&= -\frac{\alpha(\alpha-1)\omega^{\alpha}}{|\omega|^{4\alpha-2}}\left[\frac{\pm 2}{a+1}|\omega|^{a+1} + \left|\frac{d\omega}{dt}\right|^2\right]; \qquad (24)
\end{aligned}$$

or by virtue of the energy integral (17) and the relations (22) among the coefficients a and α, we obtain,

$$\frac{d^2Z}{d\tau^2} = -2E\alpha(\alpha-1)Z|Z|^{A-1}, \qquad (25)$$

establishing that Z satisfies the required equation.

In summary, we have shown that two bodies with equal angular momenta, revolving under the action of centripetal forces proportional to the ath and the Ath powers of the distance, describe the same orbit with two different centres of action provided (cf. equation (22))

$$(a+3)(A+3) = 4; \qquad (26)$$

and further that the orbits, in the complex plane, belonging to ω can be mapped on to the orbits belonging to Z by the map (cf. equation (18)),

$$z \rightsquigarrow z^{\alpha}, \qquad \text{where } \alpha = \tfrac{1}{2}(a+3) = \frac{2}{A+3}. \qquad (27)$$

The following table lists the dual pairs (a, A) for integral a's together with the required mapping in the complex plane.

a	A	map
$+1$	-2	$z \rightsquigarrow z^2$
-1	-1	$z \rightsquigarrow z$
-4	-7	$z \rightsquigarrow z^{-1/2}$
-5	-5	$z \rightsquigarrow z^{-1}$

The case $(-1, -1)$ is generally excluded on physical grounds; of the remaining cases, only the pair $(-4, -7)$ is not included in the *Principia* (cf. the summary at the end of §31).

❖7❖

Kepler's equation and its solution

Analyzing Kepler's law in two dimensions, Newton discovered an astonishingly modern topological proof of the transcendence of Abelian integrals. Newton's theorem was not really understood by mathematicians at that time, since it was based on the topology of Riemann surfaces. Thus, it was incomprehensible both for Newton's contemporaries and for 20th century mathematicians who were bred on set theory and the theory of functions of a real variable, and who were afraid of multivalued functions.

V. I. Arnold and V. A. Vasil'ev (Notices of the American Mathematical Society, Vol. 38, 1148, 1989)

34. Introduction

A presentation of Section III of the *Principia* was completed in Chapter 6. Sections IV and V which follow are of the nature of an intermezzo: in them we are led into the realm of the conic sections that forms the background against which the rest of the play is to be enacted. There is much of interest in these sections. But we shall pass them by since they are not essential to the subsequent developments; and besides there is an excellent account of them by J. J. Milne that is already available (*Isaac Newton, 1642–1727, a memorial volume*, W. J. Greenstreet, London 1927, pp. 96–114).

We turn then to Section VI which is addressed to the question:

How the motions are to be found in given orbits.

The problem that is considered is the following. We know that a body describes a conic section under the action of an inverse-square law of centripetal attraction towards the focus, in a plane normal to the constant angular momentum. The question is: how is it described *in time*? We already have a built-in clock in the law of areas,

$$r^2(\varphi)\,\frac{\mathrm{d}\varphi}{\mathrm{d}t} = h = \text{constant} \tag{1}$$

where $r(\varphi)$ is the radial distance of the revolving body from the centre of attraction (the focus) in a direction making an angle φ with a fixed direction in the orbital plane. We have only to paint the numerals on the face of the clock by integrating the law of areas:

$$ht = \int^{\varphi} r^2(\varphi)\,\mathrm{d}\varphi.$$

For elliptic orbits we are led to *Kepler's equation*. And Section VI is in the main devoted to the solution of this equation. To appreciate Newton's depth of understanding of this problem and the manner of his solution, we shall begin by providing a conventional treatment (by the standards of today) of the same problem.

35. Kepler's equation

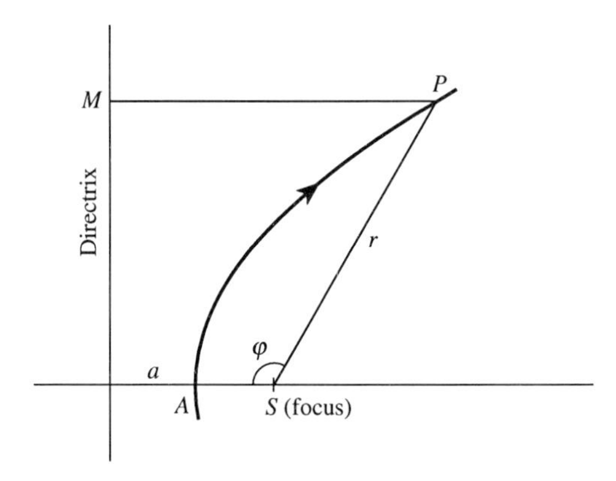

We shall first consider the simpler case of motion along a parabolic orbit. The polar equation of the parabola is

$$\frac{2a}{r} = 1 + \cos\varphi = 2\cos^2\varphi/2, \tag{1}$$

where a, the distance between the vertex A and the focus S, is one-half the semilatus rectum $l = 2a$. The equation to be integrated is

$$ht = a^2 \int_0^{\varphi} \sec^4 \frac{\varphi}{2}\, \mathrm{d}\varphi = 2a^2 \int_0^{\tan\varphi/2} (1 + \tan^2\varphi/2)\, \mathrm{d}(\tan\varphi/2), \tag{2}$$

or

$$ht = 2a^2(\tan\varphi/2 + \tfrac{1}{3}\tan^3\varphi/2). \tag{3}$$

Since

$$r = a(1 + \tan^2\varphi/2), \tag{4}$$

the explicit form of equation (3) is

$$ht = 2a^2\left(\frac{r-a}{a}\right)^{1/2}\left(1 + \frac{1}{3}\frac{r-a}{a}\right). \tag{5}$$

Relations essentially equivalent to (3) and (4) are stated by Newton in Lemmas X and XI of Book III of the *Principia* in the context of his consideration of cometary motions.

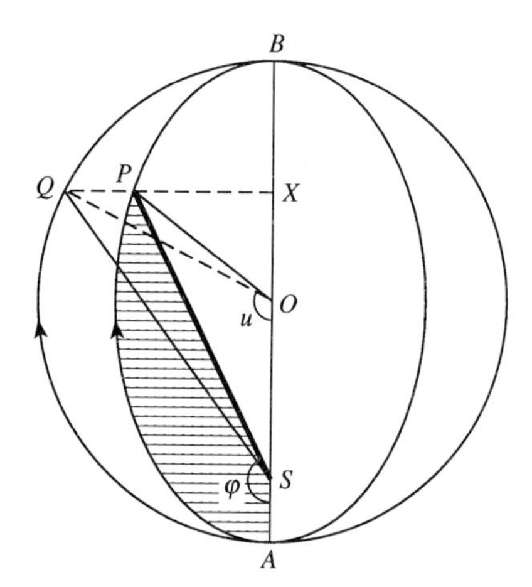

We consider next the motion along an elliptic orbit. Let APB be an ellipse of eccentricity e, and semiaxes a and b. And let P be a point on the ellipse whose image on the auxiliary circle AQB is Q. By the definitions customary in this subject,

$$\begin{aligned} \angle ASP &= \varphi = \text{the true anomaly, and} \\ \angle AOQ &= u = \text{the eccentric anomaly.} \end{aligned} \tag{6}$$

Then,

$$QX = a \sin u, \qquad PX = b \sin u, \qquad \text{and} \qquad OX = -a \cos u. \tag{7}$$

By a known property of the ellipse,

$$SP = r = a + eOX = a(1 - e \cos u), \tag{8}$$

while the polar equation of the ellipse gives

$$l/r = 1 + e \cos \varphi, \tag{9}$$

where

$$l = \text{the semilatus rectum} = a(1 - e^2). \tag{10}$$

Combining equations (8) and (9), we have

$$\begin{aligned} 1 - e^2 &= (1 + e \cos \varphi)(1 - e \cos u) \\ &= (1 - e + 2e \cos^2 \varphi/2)(1 + e - 2e \cos^2 u/2). \end{aligned} \tag{11}$$

On simplifying equation (11), we find

$$(1 + e) \cos^2 \varphi/2 - (1 - e) \cos^2 u/2 = 2e \cos^2 \varphi/2 \cos^2 u/2, \tag{12}$$

or

$$(1 + e) \sec^2 u/2 - (1 - e) \sec^2 \varphi/2 = 2e. \tag{13}$$

An alternative form of equation (13), as may be readily verified, is

$$\tan \varphi/2 = \left(\frac{1+e}{1-e}\right)^{1/2} \tan u/2, \tag{14}$$

an equation which relates the true and the eccentric anomalies.

Turning next to the equation which relates the eccentric anomaly with the time, we first observe,

$$\frac{h}{ab} = \frac{2\pi}{T} = n \quad \text{(by definition)} \qquad \text{or} \qquad h = nab. \tag{15}$$

Since the constant of areas,

$$A_{,t} = \tfrac{1}{2}h, \tag{16}$$

it follows that

$$t = \frac{2}{h} \times \text{area of the elliptic sector } APS, \tag{17}$$

or, since $PX = (b/a)QX$ independently of the position of P on the ellipse,

$$t = \frac{2}{nab}\left(\frac{b}{a} \text{ area of the sector } AQS\right)$$

$$= \frac{2}{na^2}(\text{area of the circular sector } AQO - \text{area of } \triangle SQO)$$

$$= \frac{2}{na^2}(\tfrac{1}{2}a^2u - \tfrac{1}{2}ae\,.\,a \sin u). \tag{18}$$

We thus, finally, obtain

$$nt = u - e \sin u, \tag{19}$$

which is Kepler's equation*—an equation which must be solved to determine the eccentricity e from observations on the image of P (i.e., Q) on the director circle.

36.

Proposition XXX. Problem XXII

To find at any assigned time the place of a body moving in a given parabola.

Let A be the vertex, and S the focus of the parabola

$$y^2 = 4ax = 4AS\,.\,x; \tag{1}$$

* While this equation is commonly ascribed to Kepler, its correct derivation, with its underpinnings in the law of areas, is due to Newton; the derivation we have given, in fact, follows his (see §38, below).

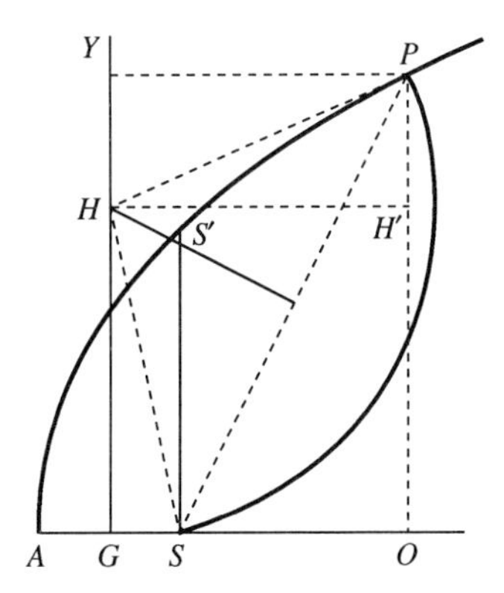

and let the instantaneous position of a body describing the parabola be P. From the mid-point G of AS, draw the perpendicular, GY, to the x-axis AO. Let the normal to SP, from its mid-point, intersect GY at H. With these constructions (which passes understanding) Newton proves:

> *As the point P moves along the parabola, under the action of an inverse-square law of attraction centred at S, the point H moves along GY with a uniform velocity equal to* 3/8th *of the velocity of P at the vertex.*

Proof:

$$\underline{AG^2 + GH^2} = GS^2 + GH^2 = HS^2 = \underline{HP^2}. \tag{2}$$

But

$$\begin{aligned}\underline{HP^2} &= H'H^2 + H'P^2 = (AO - AG)^2 + (PO - GH)^2 \\ &= AO^2 - 2AO.AG + PO^2 - 2PO.GH + \underline{AG^2 + GH^2};\end{aligned} \tag{3}$$

and we conclude that

$$2GH.PO = AO^2 + PO^2 - 2AO.AG. \tag{4}$$

Since AO and PO are the x- and the y-coordinates of P, by equation (1) of the parabola

$$AO = \frac{PO^2}{4AS} \qquad \text{and} \qquad 2AG.AO = ax = \tfrac{1}{4}PO^2. \tag{5}$$

Inserting these results in equation (4), we obtain,

$$2GH = AO\,\frac{PO}{4AS} + \tfrac{3}{4}PO. \tag{6}$$

Alternatively,

$$\begin{aligned}\tfrac{4}{3}GH.AS &= \tfrac{1}{6}PO(AO + 3AS) = \tfrac{1}{6}PO[4AO - 3(AO - AS)] \\ &= \tfrac{1}{6}PO(4AO - 3OS)\end{aligned} \tag{7}$$

or, finally,

$$\tfrac{4}{3}GHa = \tfrac{2}{3}(xy)_P - \text{area of } \triangle SOP. \tag{8}$$

On the other hand,

$$\int_0^x y\,\mathrm{d}x = 2a^{1/2}\int_0^x x^{1/2}\,\mathrm{d}x = \tfrac{4}{3}a^{1/2}x^{3/2} = \tfrac{2}{3}x(4ax)^{1/2}$$

$$= \tfrac{2}{3}(xy)_P = \text{area of segment } ASOP. \tag{9}$$

Equation (8) can, therefore, be rewritten in the form

$$\tfrac{4}{3}aGH = \text{area of the parabolic segment } ASP \tag{10}$$

or, by the law of areas,

$$\tfrac{4}{3}GHa = \tfrac{1}{2}ht, \tag{11}$$

where t is measured from perihelion passage at A. We have thus shown that

$$GH = \frac{3h}{8a}t \quad \text{or} \quad \frac{\mathrm{d}}{\mathrm{d}t}(GH) = \frac{3h}{8a} = \text{constant}. \tag{12}$$

Since the velocity of P at the vertex A is

$$v_A = h/a, \tag{13}$$

it follows that

$$\frac{\mathrm{d}}{\mathrm{d}t}(GH) = \tfrac{3}{8}v_A \qquad \text{(Corollary II).} \quad \text{Q.E.D.}$$

Still another form of equation (9) is obtained by noting that (cf. equation (9))

$$\text{area of segment } ASS' = \tfrac{2}{3}a.2a = \tfrac{4}{3}a^2; \tag{14}$$

and that, therefore,

$$\frac{GH}{a} = \frac{\text{area of segment } ASP}{\text{area of segment } ASS'} \qquad \text{(Corollary I).}$$

And Newton concludes:

> COR. III. Hence, also, on the other hand, the time may be found in which the body has described any assigned arc AP. Join SP, and on its middle point erect a perpendicular meeting the right line GH in H.

A closing remark: since Newton most certainly knew the polar equation of the parabola, it is hard to imagine that he had not deduced for himself the solution to the problem given by equations (3) and (4) of §35: they are, in fact, in essence, stated in Lemmas X and XI of Book III. Newton clearly preferred the present demonstration since it provides for the motion along a parabola a geometrical construction similar to the one he devises for the motion along an ellipse in Proposition XXXI to follow.

37.

Lemma XXVIII

There is no oval figure whose area, cut off by right lines at pleasure, can be universally found by means of equations of any number of finite terms and dimensions.

This lemma is a striking manifestation of Newton's mathematical insight which enabled him, as in this instance, to surpass the level of scientific understanding of his time by two hundred years; and it behooves us to reproduce the lemma in its entirety. But the brevity of Newton's exposition, in which he takes for granted many facts that were obvious to him, makes it difficult to follow even for a modern reader. But once the ideas are explained (as we shall attempt, following V. I. Arnold), Newton's reasoning becomes transparent.

First, some definitions:

1. A plane curve is said to be *algebraic* (or, as Newton calls '*geometrically rational*') if it satisfies an equation of the form $P(x, y) = 0$ where P is some non-zero polynomial.

   Algebraic curves can be *smooth*,* like the conic sections,

   $$x^2 + y^2 = 1, \qquad y^2 = 4ax, \qquad \text{and} \qquad \frac{x^2}{a^2} \pm \frac{y^2}{b^2} = 1;$$

   or *singular* as the lemniscates,

   $$y^m = x^{(n-1)m}(a^2 - x^2),$$

   which are singular (non-differentiable) at $x = y = 0$. The special case

   $$y^2 = x^2(a^2 - x^2)$$

   when $m = n = 2$, is the lemniscate of Huygens.

   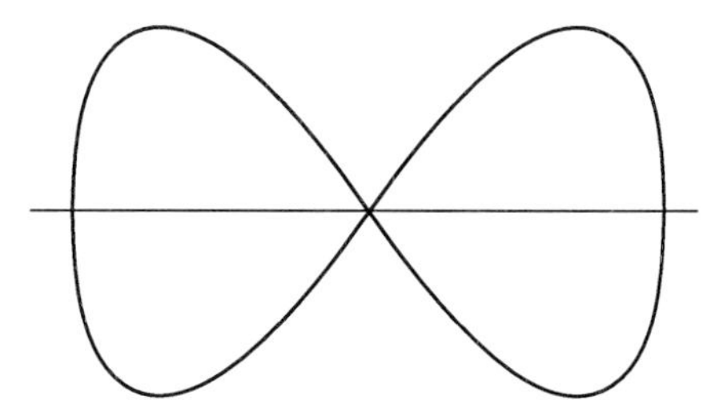

   An example of a *non-algebraic* (or '*geometrically irrational*') curve is the cycloid:

   $$x = a(t - \sin t) \qquad \text{and} \qquad y = a(1 - \cos t).$$

2. An *oval* (or, a closed convex plane curve) is *algebraically integrable*, if the area of the segment S, cut off by the right line $ax + by = c$, is of the form $P(S; a, b, c) = 0$, where P is a non-zero polynomial.

* A curve is smooth if its tangent exists everywhere, is unique, and varies continuously along the curve.

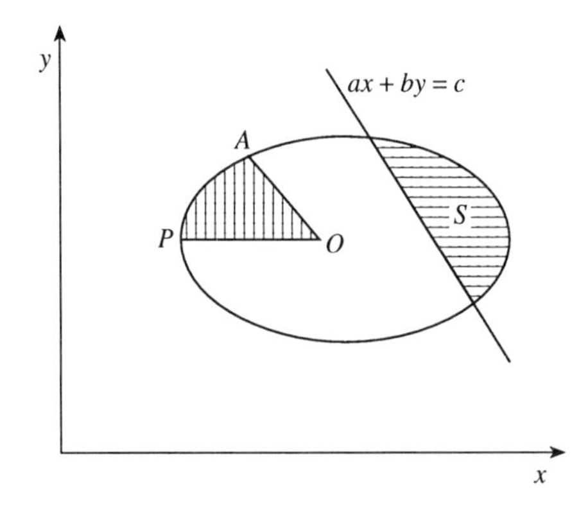

From these definitions, it follows that the area of the triangular sector OPA is an algebraic function of the lines OA and OP bounding the sector since it is the intersection of the segments cut off by the lines OA and OB (extended to cut the oval).

The main theorem that the lemma establishes is:

> *Every algebraically integrable oval has singular points: all smooth ovals are algebraically non-integrable*

Proof:

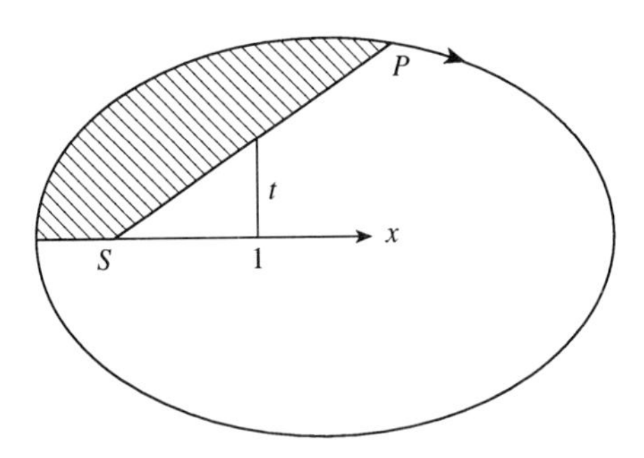

Let the radius vector SP, joining a fixed point S inside the oval, to a point P on the perimeter, rotate (in the positive sense, say). If the oval is algebraically integrable, then the area swept out by SP must be an algebraic function of the tangent, t, of the angle of inclination of SP and the x-axis. If SP is allowed to rotate about S for indefinitely many revolutions, the area swept out by SP increases by the area of the oval once every revolution. Consequently, the area swept out, regarded as a multivalued function of t has infinitely many different values for the same position of P. But an algebraic function cannot be multivalued since the number of roots of a non-zero polynomial cannot exceed its degree. Therefore, the area swept out is not an algebraic function and the oval is not algebraically integrable. Q.E.D.

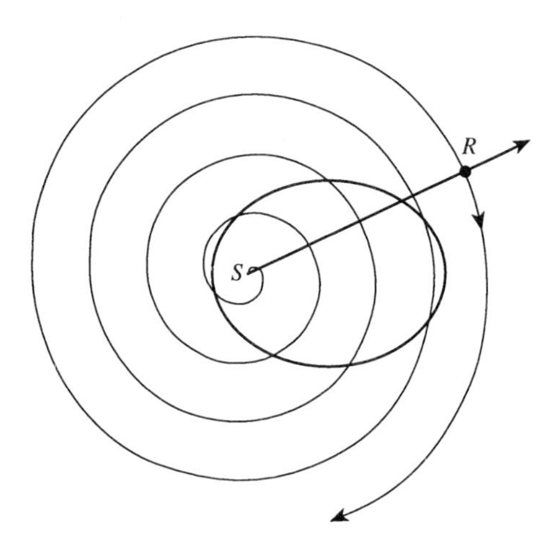

Newton's proof of his theorem is slightly different though in essence equivalent. He considers the radius vector as continually rotating about S with a constant angular velocity Ω while R continually increases at the rate

$$\frac{\mathrm{d}R}{\mathrm{d}t} = \frac{\mathrm{d}A}{\mathrm{d}t} \qquad \text{or} \qquad R = A(t),$$

where $A(t)$ denotes the area swept out by S in a time t. Thus R increases by the area of the oval after each revolution. Note that when P orbits about S under centripetal attraction,

$$A_{,t} = \text{constant} = \tfrac{1}{2}h,$$

and

$$R = \tfrac{1}{2}ht \qquad \text{while } \varphi = \Omega t \quad \text{or} \quad R = \text{constant} \times \varphi.$$

In any event, a body describing an oval will occupy the same position P for a given direction of the ray ($0 \leqslant \varphi \leqslant 2\pi$) for an infinity of times at intervals of the orbital period; and all these infinity of times cannot be obtained as roots of an algebraic equation of a finite degree. The argument is now completed as before.

And in a concluding paragraph Newton adds that by the same arguments *the length of an arc of the oval is not algebraic.*

With the foregoing explanation, if one now reads the parts underlined in the text reproduced below, one finds to one's surprise how transparent Newton's reasoning is!

Suppose that within the oval any point is given, about which as a pole a right line is perpetually revolving with a uniform motion, while in that right line a moveable point going out from the pole moves always forward with a velocity proportional to the square of that right line within the oval. By this motion that point will describe a spiral with infinite circumgyrations. Now if a portion of the area of the oval cut out by that right line could be found by a finite equation, the distance of the point from the pole, which is proportional to this area, might be found by the same equation, and therefore all the points of the spiral might be found by a finite equation also; and therefore the intersection of a right line given in position with the spiral might also be found by a finite equation. But every right line infinitely produced cuts a spiral in an infinite number of points; and the equation by which any intersection of two lines is found at the same time exhibits all their intersections by as many roots, and therefore rises to as many dimensions as there are intersections. Because two circles mutually cut one another in two points, one of these intersections is not to be found but by an equation of two dimensions, by which the other intersection may also be found. Because there may be four intersections of two conic sections, any one of them is not to be found universally, but by an equation of four dimensions, by which they are all found together. For if these intersections are severally sought, because the law and condition of all is the same, the calculus will be the same in every case, and therefore the conclusion always the same, which must therefore comprehend all those intersections at once within itself, and exhibit them all indifferently. Hence it is that the intersections of the conic sections with the curves of the third order, because they may amount to six, come out together by equations of six dimensions; and the intersections of two

Two algebraic curves of degrees m and n intersect in at most mn points (Bézout's theorem)

curves of the third order, because they may amount to nine, come out together by equations of nine dimensions. If this did not necessarily happen, we might reduce all solid to plane problems, and those higher than solid to solid problems. But here I speak of curves irreducible in power. For if the equation by which a curve is defined may be reduced to a lower power, the curve will not be one single curve, but composed of two, or more, whose intersections may be severally found by different calculusses. After the same manner the two intersections of right lines with the conic sections come out always by equations of two dimensions; the three intersections of right lines with the irreducible curves of the third order by equations of three dimensions; the four intersections of right lines with the irreducible curves of the fourth order, by equations of four dimensions, and so on *in infinitum*. Wherefore the innumerable intersections of a right line with a spiral, since this is but one simple curve, and not reducible to more curves, require equations infinite in number of dimensions and roots, by which they may all be exhibited together. For the law and calculus of all is the same. For if a perpendicular is let fall from the pole upon that intersecting right line, and that perpendicular together with the intersecting line revolves about the pole, the intersections of the spiral will mutually pass the one into the other; and that which was first or nearest, after one revolution, will be the second; after two, the third; and so on: nor will the equation in the mean time be changed but as the magnitudes of those quantities are changed, by which the position of the intersecting line is determined.

The notion of reducibility of curves leads to the concepts of local algebraicity and local algebraic integrability.

Theorems that are implied are:

1. No analytic oval is algebraically integgrable even locally.
2. Any locally algebraically integrable oval is algebraic.
3. The total area bounded by a self-intersecting closed locally algebraically integrable curve (taking into account signs) is zero.

Wherefore since those quantities after every revolution return to their first magnitudes, the equation will return to its first form; and consequently one and the same equation will exhibit all the intersections, and will therefore have an infinite number of roots, by which they may all be exhibited. And therefore the intersection of a right line with a spiral cannot be universally found by any finite equation; and of consequence there is no oval figure whose area, cut off by right lines at pleasure, can be universally exhibited by any such equation.

By the same argument, if the interval of the pole and point by which the spiral is described is taken proportional to that part of the perimeter of the oval which is cut off, it may be proved that the length of the perimeter cannot be universally exhibited by any finite equation. But here I speak of ovals that are not touched by conjugate figures running out *in infinitum.*

Corollary. Hence the area of an ellipsis, described by a radius drawn from the focus to the moving body, is not to be found from the time given by a finite equation; and therefore cannot be determined by the description of curves geometrically rational. Those curves I call geometrically rational, all the points whereof may be determined by lengths that are definable by equations; that is, by the complicated ratios of lengths. Other curves (such as spirals, quadratrixes, and cycloids) I call geometrically irrational. For the lengths which are or are not as number to number (according to the tenth Book of Elements) are arithmetically rational or irrational. And therefore I cut off an area of an ellipsis proportional to the time in which it is described by a curve

geometrically irrational, in the following
manner.

The remaining (non-underlined) parts of the text generalize the basic concepts. We have indicated on the margins what these concepts are and the theorems (as stated by Arnold) that are implied. For a fuller account of these matters, the reader is referred to V. I. Arnold in *Huygens and Barrow, Newton and Hooke* (Birkhäuser Verlag, Basel 1990) Chapter 5, pp. 83–94, and Appendix 2, pp. 101–105.

We close (as we began) with a quotation from Arnold:

> Comparing today the texts of Newton with the comments of his successors, it is striking how Newton's original presentation is more modern, more understandable and richer in ideas than the translation due to commentators of his geometrical ideas into the formal language of the calculus of Leibniz.

38.

Proposition XXXI. Problem XXIII

To find the place of a body moving in a given ellipse at any assigned time.

Newton's instructions to find 'the place of the body' are given in the imperial style that on occasion he adopts:

> Suppose A to be the principal vertex, S the focus, and O the centre of the ellipse APB; and let P be the place of the body to be found. Produce OA to G so that $OG:OA = OA:OS$. Erect the perpendicular GH; and about the centre O, with the radius OG, describe the circle GEF; and on the ruler GH, as a base, suppose the wheel GEF to move forwards, revolving about its axis, and in the meantime by its point A describing the cycloid ALI. This done, take GK to the perimeter $GEFG$ of the wheel, in the ratio of the time in which the body proceeding from A described the arc AP, to the time of a whole revolution in the ellipse. Erect the perpendicular KL meeting the cycloid in L; then LP drawn parallel to KG will meet the ellipse in P, the required place of the body.

The proof of the construction he dictates consists of two parts: first the derivation of what has come to be called Kepler's equation; and second its geometric parametrization from which the construction follows.

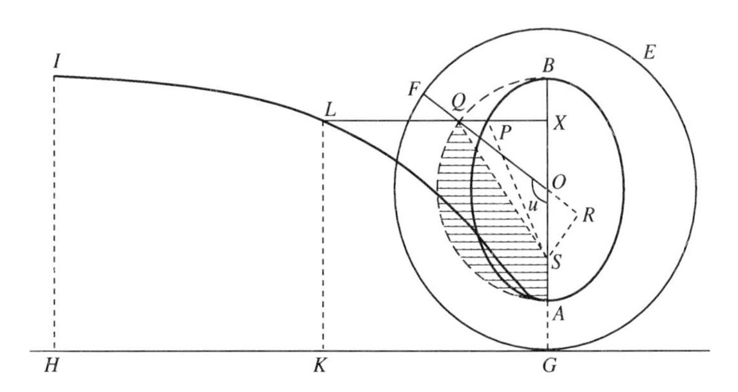

The accompanying diagram is the same as the one in §35 (p. 129) except for the shaded parts and the additional circle *GEF*, concentric to the auxiliary circle *AQB*, of radius *OG*, which according to instruction is a/e: for,

$$OG : a = a : OS = a : ae \qquad \text{or} \qquad OG = a/e.$$

Newton's derivation starts as in §35:

$$\begin{aligned}(\tfrac{1}{2}ht)\frac{a}{b} &= \frac{a}{b}\text{ area of elliptic sector } APS\\ &= \text{area of circular sector } AQS\\ &= \tfrac{1}{2}(OA \text{ arc } AQ) - \text{area of } \triangle SOQ\\ &= \tfrac{1}{2}(OA \text{ arc } AQ - OA.SR)\\ &= \tfrac{1}{2}a(\text{arc } AQ - SR).\end{aligned}$$

Multiplying this equation by $OG/a(=a/OS)$ and remembering that $h = nab$ (by equation (15) of §35), we obtain,

$$\begin{aligned}\tfrac{1}{2}nt(OGa) &= \tfrac{1}{2}a\left(\frac{OG}{a}\text{ arc } AQ - a\frac{SR}{OS}\right)\\ &= \tfrac{1}{2}a(\text{arc } GF - a \sin u);\end{aligned}$$

or, finally,

$$nt(OG) = OG.u - a \sin u[=OG(u - e \sin u)].$$

We now recall that the curve traced by a fixed point on the radius of a circle, which rolls along the x-axis, is described parametrically by the equations,

$$x = OGu - a \sin u \qquad \text{and} \qquad y = OG - a \cos u,$$

where *OG* is the radius of the rolling circle and *a* is the distance of the fixed point from the centre. Accordingly, if we identify *Q* as the fixed point on radius *OF*, then *Q* will trace the cycloid described. To relate the cycloidal motion of *Q* with the orbital motion of *P*,

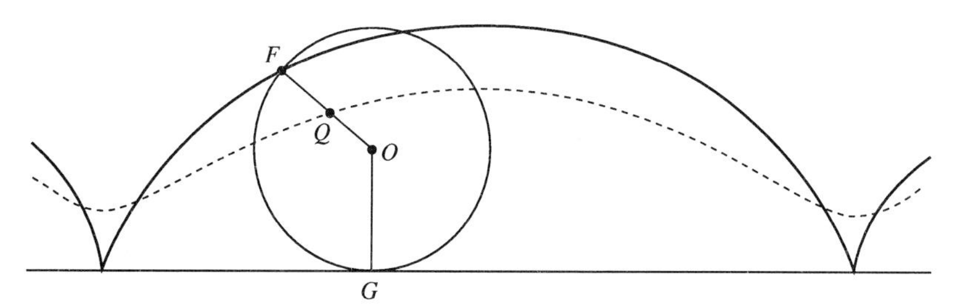

the circle must roll along the x-axis with a uniform constant velocity equal to nOG; and Q is the image on the director circle of the position P that is sought. Q.E.D.

Scholium

This Scholium is particularly difficult to read since Newton describes in connected prose an analytic method of solving Kepler's equation by an approximative iterative procedure; and it remained un-understood for 200 years until it was rescued by J. C. Adams (Monthly Notices of the Royal Astronomical Society, Vol. 53, p. 43, 1882) who presented a very readable account. We shall follow Adams's presentation.

The problem is to solve the equation,

$$\tau(u) = u - e \sin u,$$

where $\tau(u)$ is the known mean anomaly, e the eccentricity, and u the eccentric anomaly. Suppose that by any construction (or even by conjecture) we know an approximate value u_0. Let

$$\tau_0 = u_0 - e \sin u_0 \qquad \text{and} \qquad \tau - \tau_0 \text{ is small.}$$

We obtain an improved approximation by writing

$$u = u_0 + \delta u_0,$$

and determining δu_0 by the iterative procedure:

$$\begin{aligned}\tau &= u_0 + \delta u_0 - e(\sin u_0 + \delta u_0 \cos u_0) + O(\delta u_0)^2 \\ &= \tau_0 + \delta u_0(1 - e \cos u_0) + O(\delta u_0)^2\end{aligned}$$

which gives

$$\delta u_0 = \frac{\tau - \tau_0}{1 - e \cos u_0} = \frac{\delta\tau_0}{1 - e \cos u_0}.$$

We can improve this approximation by including terms of the second order in the expansion of sin u about u_0; thus:

$$\begin{aligned}\tau &= u_0 + \delta u_0 - e[\sin u_0 + \delta u_0 \cos u_0 - \tfrac{1}{2}(\delta u_0)^2 \sin u_0] \\ &= u_0 + \delta u_0 - e[\sin u_0 + \delta u_0 \cos(u_0 + \tfrac{1}{2}\delta u_0)] \qquad \text{(to the same order);}\end{aligned}$$

and we obtain:

$$\delta\tau_0 = \delta u_0[1 - e\cos(u_0 + \tfrac{1}{2}\delta u_0)],$$

or

$$\delta u_0 = \frac{\delta\tau_0}{1 - e\cos(u_0 + \tfrac{1}{2}\delta u_0)}$$

and the iteration can be continued.

An alternative procedure if $e \ll 1$ is to set

$$u_0 = \tau \qquad \text{and} \qquad u = u_0 + \delta u_0,$$

and obtain, as before, in the second approximation,

$$\begin{aligned}\tau &= u_0 + \delta u_0 - e[\sin u_0 + \delta u_0 \cos u_0 - \tfrac{1}{2}(\delta u_0)^2 \sin u_0] \\ &= u_0 + \delta u_0 - e[\sin u_0 + \delta u_0 \cos(u_0 + \tfrac{1}{2}\delta u_0)],\end{aligned}$$

or

$$\delta u_0 = \frac{e\sin u_0}{1 - e\cos(u_0 + \tfrac{1}{2}\delta u_0)} \simeq \frac{e\sin u_0}{1 - e\cos(u_0 + \tfrac{1}{2}e\sin u_0)}.$$

As Adams concludes:

> We need not be surpised that Newton should have employed this method of solving the transcendental equation
>
> $$x - c\sin x = z,$$
>
> since the method is identical in principle with his well-known method of approximation to the roots of algebraic equations.

The Scholium also includes a brief discussion of the hyperbolic case when the appropriate form of Kepler's equation is

$$t - t_0 \propto e\sinh u - u.$$

❖8❖

The rectilinear ascent and descent of bodies

39. Introduction

In the concluding sentence of Section VI, Newton introduces Section VII on rectilinear motion thus:

> And so far concerning the motion of bodies in curved lines. But it may also come to pass that a moving body shall ascend or descend in a right line; and I shall now go on to explain what belongs to such kind of motions.

This is an inversion of the common procedure: rectilinear motion in a central field of force is generally (if not always) considered as a special (and a simpler) case that requires a *separate* treatment since the angular momentum of such motions is zero and there is no constant of areas as a (non-trivial) constraint. Instead, Newton considers rectilinear motion as a limiting case of motion along conic sections and shows how it becomes determinate by an 'invariant' form of the law of areas which applies to orbits independently of their eccentricity. In fact, he reduces the solution for rectilinear motion to a simplified version of Kepler's equation. Besides, he relates rectilinear motion to circular motion reminiscent of his geometrical constructions to describe parabolic and elliptic motion in Propositions XXX and XXXI (see §§36 and 38). And, finally, in the last Proposition XXXIX of this section, he formulates and solves the 'initial-value problem' for rectilinear motion as a model for the solution to the general problem in Proposition XLI and XLII in Section VIII.

We shall begin with a present-day treatment of rectilinear motion that will allow a direct comparison with Newton's treatment.

40. An *ab initio* treatment of rectilinear motion

With the centre of attraction at the origin, the equation of motion governing rectilinear motion along the (vertical) z-direction is

$$\frac{d^2z}{dt^2} = -\frac{\mu}{z^2}, \tag{1}$$

where μ is a constant. (In the *Principia*, the constant μ is normally suppressed; but Newton restores it when occasion demands.) Equation (1) admits of the integral

$$\frac{1}{2}\left(\frac{dz}{dt}\right)^2 = \mu\left(\frac{1}{z} + Q\right) = \tfrac{1}{2}v^2 \qquad \text{(say)}, \tag{2}$$

where Q is a constant of integration. The three cases,

$$Q < 0, \qquad Q > 0, \qquad \text{and} \qquad Q = 0, \tag{3}$$

must be distinguished: they correspond, respectively, to motions along elliptic, hyperbolic, and parabolic orbits in the non-rectilinear case.

(a) *The elliptic case, $Q < 0$*

In this case, we shall write

$$\frac{1}{2}\left(\frac{dz}{dt}\right)^2 = \mu\left(\frac{1}{z} - \frac{1}{2a}\right). \tag{4}$$

At $z = 2a$, the velocity vanishes and the motion is one of descent.

In the accompanying diagram, the origin is at B, and A is at $z = 2a$ where the velocity vanishes. With the mid-point O of AB as centre draw a semicircle of radius $OA = OB = a$. The points on the z-axis are imaged on points on the semicircle at the same height. Thus, the coordinates of z are:

$$z = BC = a(1 + \cos\theta) \qquad \text{and} \qquad x = CD = a\sin\theta. \tag{5}$$

With these definitions,

$$\left(\frac{dz}{dt}\right)^2 = \frac{\mu}{a}\tan^2\frac{\theta}{2}, \tag{6}$$

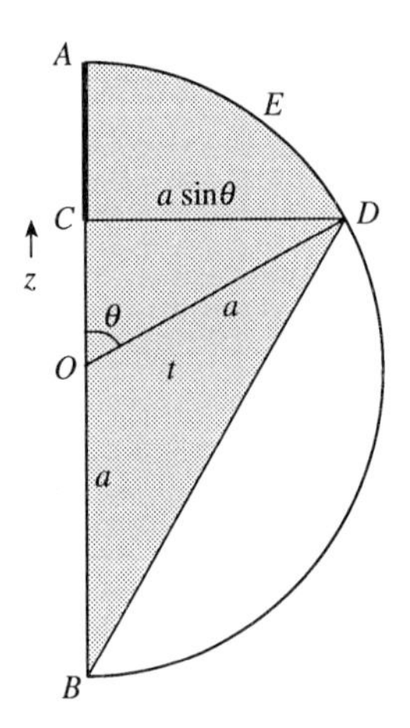

or

$$\frac{\mathrm{d}z}{\mathrm{d}t} = -\left(\frac{\mu}{a}\right)^{1/2} \tan\frac{\theta}{2}, \tag{7}$$

where we have chosen the negative sign appropriate for descending motions. Together with the equation

$$\mathrm{d}z = -a\sin\theta\,\mathrm{d}\theta = -2a\sin\frac{\theta}{2}\cos\frac{\theta}{2}\,\mathrm{d}\theta, \tag{8}$$

equation (7) gives

$$\mathrm{d}t = 2a\left(\frac{a}{\mu}\right)^{1/2}\cos^2\frac{\theta}{2}\,\mathrm{d}\theta = a\left(\frac{a}{\mu}\right)^{1/2}(1+\cos\theta)\,\mathrm{d}\theta. \tag{9}$$

Integrating this equation, we obtain

$$t = \left(\frac{a}{\mu}\right)^{1/2} a(\theta + \sin\theta), \tag{10}$$

where $t = 0$ when the body starts its descent at $z = 2a$. It will be observed that equation (10) is a version of Kepler's equation (cf. Chapter 7, §35, equation (19)): the emergence of Kepler's equation in this context will become transparent from Newton's treatment of this same problem in Proposition XXXII (§41).

As in the case of Kepler's equation, we can rewrite equation (10) in the form

$$t = \left(\frac{a}{\mu}\right)^{1/2}(\text{arc } AD + CD) \tag{11}$$

or, alternatively,

$$\begin{aligned} t &= \frac{2}{(\mu a)^{1/2}}(\text{area of sector } OAD + \text{area of } \triangle OBD) \\ &= \frac{2}{(\mu a)^{1/2}}(\text{area of segment } (BDEAB)). \end{aligned} \tag{12}$$

This is Newton's result (see §41).

(*b*) *The hyperbolic case, Q > 0*

In this case, we shall write

$$\frac{1}{2}\left(\frac{\mathrm{d}z}{\mathrm{d}t}\right)^2 = \mu\left(\frac{1}{z} + \frac{1}{2a}\right). \tag{13}$$

The body now has a finite velocity $(=(\mu/a)^{1/2})$ at infinity; and we shall consider the case of descent starting at infinity in the remote past.

In the accompanying diagram BED is the right half of the rectangular hyperbola,

$$(OC)^2 - (CD)^2 = a^2 \tag{14}$$

where a is its semiaxis with the origin of z at the vertex B. The points on the z-axis are imaged on points, at the same height, on the rectangular hyperbola.

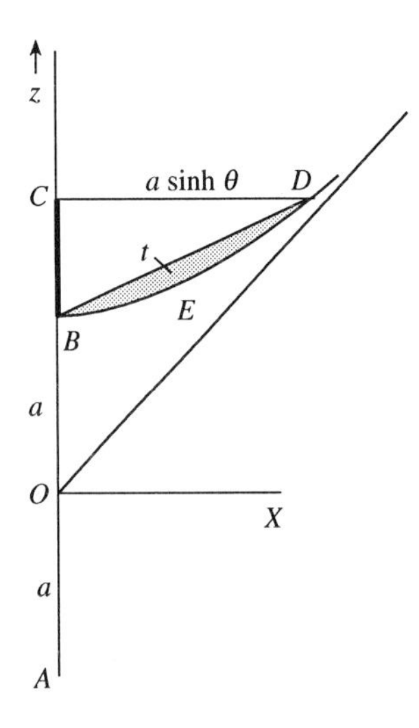

Parametrizing the rectangular hyperbola in the manner,

$$OC = a\cosh\theta \qquad \text{and} \qquad CD = a\sinh\theta \tag{15}$$

we have

$$z = BC = a(\cosh\theta - 1) = 2a\sinh^2\frac{\theta}{2}, \qquad \mathrm{d}z = 2a\sinh\frac{\theta}{2}\cosh\frac{\theta}{2}\,\mathrm{d}\theta, \tag{16}$$

and

$$\frac{\mathrm{d}z}{\mathrm{d}t} = -\left(\frac{\mu}{a}\right)^{1/2}\coth\frac{\theta}{2}. \tag{17}$$

where we have chosen the negative sign appropriate for descent from infinity. From equations (16) and (17), we find,

$$\mathrm{d}t = -\left(\frac{a}{\mu}\right)^{1/2}\tanh\frac{\theta}{2}\,\mathrm{d}z = -a\left(\frac{a}{\mu}\right)^{1/2}(\cosh\theta - 1)\,\mathrm{d}\theta; \tag{18}$$

or, after integration,

$$t = -\left(\frac{a}{\mu}\right)^{1/2} a(\sinh\theta - \theta), \tag{19}$$

where $t = 0$ is the time of arrival at B. Equation (19) is now a version of Kepler's equation for hyperbolic orbits (cf. the last equation of Chapter 7, p. 142).

We readily verify that

$$\text{the area of the segment } BEDB = \tfrac{1}{2}a^2(\sinh\theta - \theta); \tag{20}$$

and therefore,

$$|t| = -t = \frac{2}{(\mu a)^{1/2}} \text{ area of segment } BEDB, \tag{21}$$

which is formally the same as equation (12) for the elliptic case.

(c) *The parabolic case, Q = 0*

In this case, the relevant equations are:

$$\frac{\mathrm{d}^2 z}{\mathrm{d}t^2} = -\frac{\mu}{z^2} \tag{22}$$

and

$$\frac{1}{2}\left(\frac{\mathrm{d}z}{\mathrm{d}t}\right)^2 = \frac{\mu}{z}, \tag{23}$$

appropriate for a body having zero velocity at infinity.

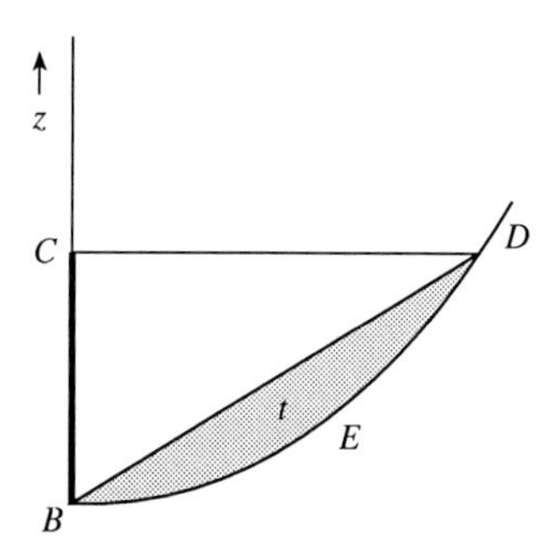

In the accompanying diagram, *BED* is the parabola,

$$x^2 = 4az, \tag{24}$$

with its vertex at *B*. For a body starting its ascent from $z = 0$ at $t = 0$,

$$\mathrm{d}t = \frac{1}{(2\mu)^{1/2}} z^{1/2}\,\mathrm{d}z, \tag{25}$$

and

$$t = \frac{1}{3}\left(\frac{2}{\mu}\right)^{1/2} z^{3/2}. \tag{26}$$

Since (as we may readily verify; cf. equation (9), p. 132)

$$\text{area of segment } BEDB = \tfrac{1}{3}a^{1/2}z^{3/2}, \tag{27}$$

we, once again, have

$$t = \left(\frac{2}{\mu a}\right)^{1/2} (\text{area of segment } BEDB) \tag{28}$$

The velocity of the body at C

The velocity of the body at C at time t, follows directly from equations (7), (17), and (23). Thus, considering equation (7), for the elliptic case, we have

$$|v(C)| = \left(\frac{\mu}{a}\right)^{1/2} \tan\frac{\theta}{2}, \tag{29}$$

while the velocity $v_{\odot}(BC)$ in a circular orbit of radius BC, is

$$v_{\odot}(BC) = \frac{\mu^{1/2}}{(BC)^{1/2}} = \left(\frac{\mu}{a}\right)^{1/2} \frac{1}{\sqrt{(1+\cos\theta)}} = \left(\frac{\mu}{2a}\right)^{1/2} \sec\frac{\theta}{2}. \tag{30}$$

Therefore

$$\frac{|v(C)|}{v_{\odot}(BC)} = 2^{1/2} \sin\frac{\theta}{2} = \sqrt{(1-\cos\theta)} = \left(\frac{AC}{a}\right)^{1/2} = \left(\frac{AC}{AO}\right)^{1/2}. \tag{31}$$

It is manifest that the same relation also obtains in the hyperbolic case. And for the parabolic case, we similarly find from equation (23) that

$$\frac{|v(C)|}{v_{\odot}(\frac{1}{2}BC)} = 1. \tag{32}$$

41.

Proposition XXXII. Problem XXIV

Supposing that the centripetal force is inversely proportional to the square of the distance of the places from the centre; it is required to define the spaces which a body, falling directly, describes in given times.

Newton first considers the case of a body, initially at rest at A, descending directly towards the centre of attraction at B. The problem is to determine the times at which the body will find itself at various heights. Towards this end, Newton considers a sequence of elliptical orbits of different eccentricities but with the same major axis AB. Let $ARPB$ be one of these ellipses with the centre of attraction at its focus S (not at B) and P a point on it at some fixed height. We know that a body at A (the aphelion point) will traverse the arc AP in a time proportional to the area of the sector $SPRA$ which in turn is proportional to the area of the sector $SDEA$ (cf. equations (7) of §35) where D

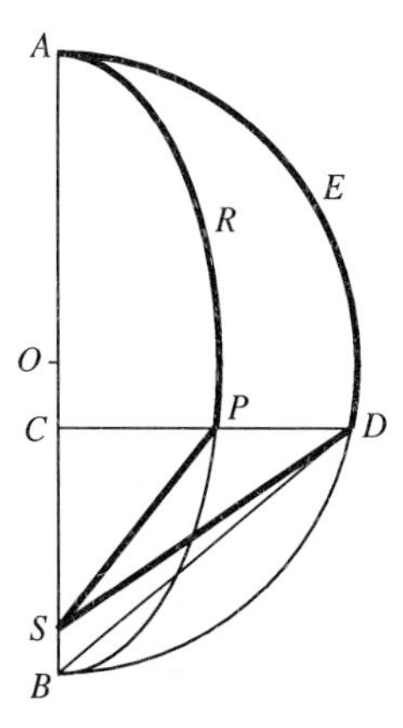

is the image of P on the auxiliary semicircle $BDEA$. This latter proportionality holds independently of the eccentricity of the ellipse. Therefore, passing to the limit when the ellipse collapses on to the right line AB and P coincides with C and S with B, we conclude that the *time of descent from A to C is proportional to the area of the segment $BDEAB$.*

Q.E.I.

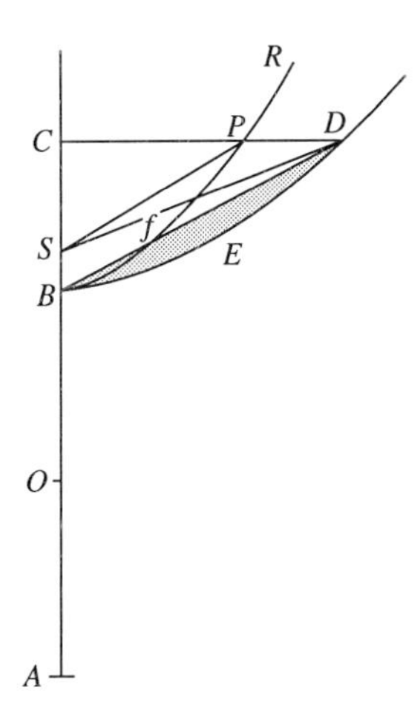

Turning next to the case of a body ascending from B towards infinity, Newton considers a sequence of hyperbolae of different eccentricities but the same major axis AB. Let $BfPR$ be one of these hyperbolae with the centre of attraction at S (not at B) and P a point on it at some fixed height. We know that a body starting at B (the perihelion point) will traverse the arc PfB in a time proportional to the area of the segment $BfPS$ which in turn is proportional to the area of the segment $BEDS$ where D is the image of P on the auxiliary rectangular hyperbola BED.* This latter proportionality holds independently of

* The role of the *auxiliary circle* in the geometry of the ellipse is well described in standard books on conic sections. But I can find no reference to the similar role of the '*auxiliary*' *rectangular hyperbolae* for the geometry of the hyperbola; for example, it is manifest that $CP:CD$ = ratio of the semiaxes, $b:a$.

the eccentricity of the hyperbola. Therefore, passing to the limit when the hyperbola collapses on to the z-axis and P coincides with C and S with B, we conclude that the *time of ascent from B to C is proportional to the area of the segment BEDB.*

Q.E.I.

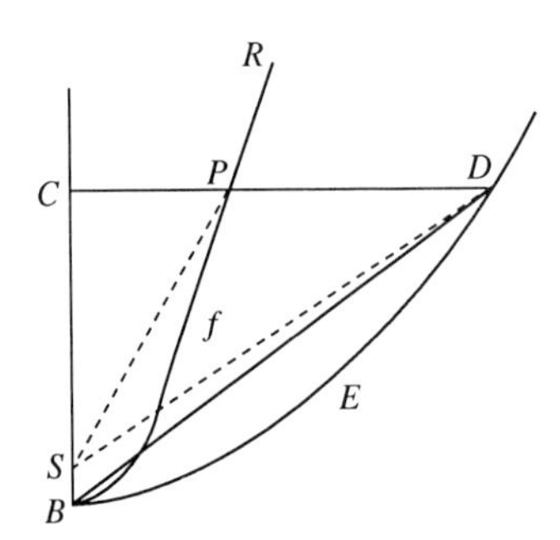

Finally, turning to the parabolic case, Newton considers a sequence of parabolae with different latera recta with their vertices at the same point B; and, in addition, an arbitrary but fixed parabola BED with its vertex also at B. If P is a point, at some fixed height on one of these parabolae, $BfPR$, then from the proportionality of the distances CP and CD (where D is a point at the same height on the fixed parabola BED), the proportionality of the areas of the segments $SBfP$ and $SBED$ follows. And as before, passing to the limit when the parabola $BfPR$ collapses on to the axis, we conclude that *the time of travel from C to B is proportional to the area of the parabolic segment BEDB.*

Q.E.I.

In contrast to the pedestrian derivation of Newton's results in §40, the elegance and the simplicity of Newton's demonstrations are startling.

42.

Proposition XXXIII. Theorem IX

The things above found being supposed, I say, that the velocity of a falling body in any place C is to the velocity of a body, describing a circle about the centre B at the distance BC, as the square root of the ratio of AC, the distance of the body from the remoter vertex A of the circle or rectangular hyperbola, to $\frac{1}{2}AB$*, the principal semidiameter of the figure.*

To prove his assertion, Newton continues with the limiting procedure described in Proposition XXXII. And he is able to treat the elliptic and the hyperbolic cases simultaneously by his scrupulous lettering of the diagrams that retains the geometrical correspondences between the ellipse and the hyperbola in conformity with the 'dignity of the problem'.

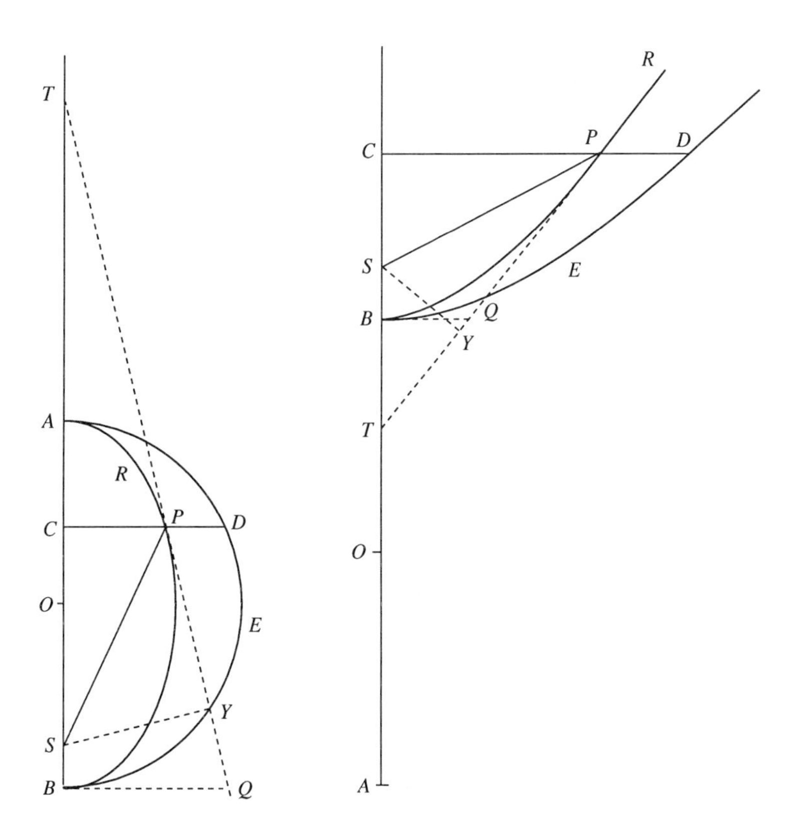

The accompanying diagrams are virtually the same as the ones that accompany the demonstrations of Proposition XXXII. The only additions are the tangent TP at P, the perpendicular SY to the tangent, and BQ parallel to CPD intersecting the tangent at Q.

The proof consists of three steps:

(a) By Corollary IX of Proposition XVI, the ratio of the velocity, $v(P)$, in the elliptic (or hyperbolic) orbit to the velocity $v_{\odot}(SP)$ in a circular orbit of radius SP is given by

$$\left[\frac{v(P)}{v_{\odot}(SP)}\right]^2 = \tfrac{1}{2}L\,\frac{SP}{SY^2}, \tag{1}$$

where L is the latus rectum of the conic section.

(b) The formula,

$$L = 2\,\frac{AO\,.\,CP^2}{AC\,.\,CB} \tag{2}$$

follows from the known relation in the geometry of conic sections:

$$\frac{AC\,.\,CB}{CP^2} = \frac{a^2}{b^2} = \frac{2a}{2b^2/a} = 2\,\frac{AO}{L}. \tag{3}$$

(c) A relation that follows, again, from the known formula is:

$$CO\,.\,OT = OB^2 = OA^2 = a^2. \tag{4}$$

From this formula, we obtain:

$$\frac{CO}{BO} = \frac{BO}{OT} = \frac{BO + CO}{BO + OT} = \frac{BC}{BT}. \tag{5}$$

We also have:

$$\begin{aligned}\frac{AC}{AO} &= 1 - \frac{OC}{AO} = 1 - \frac{OC}{OB} = 1 - \frac{BC}{BT}, \qquad \text{by (5)}\\ &= \frac{BT - BC}{BT} = \frac{TC}{BT} \qquad \text{(Corollary I)},\end{aligned} \tag{6}$$

or, since $\triangle$s CPT and BQT are similar,

$$\frac{AC}{AO} = \frac{TC}{BT} = \frac{CP}{BQ} \qquad \text{or} \qquad CP = \frac{BQ\,.\,AC}{AO}. \tag{7}$$

By combining the relations (1), (2), and (7), we have

$$\begin{aligned}\left[\frac{v(P)}{v_{\odot}(SP)}\right]^2 &= \tfrac{1}{2}L\frac{SP}{SY^2} = \frac{SP}{SY^2}\frac{AO\,.\,CP^2}{AC\,.\,CB}\\ &= \frac{SP}{SY^2}\frac{AO}{AC\,.\,CB}\cdot\frac{BQ^2\,.\,AC^2}{AO^2} = \frac{BQ^2}{SY^2}\frac{SP}{BC}\frac{AC}{AO};\end{aligned} \tag{8}$$

and passing to the limit when the ellipse or the hyperbola, BPR, collapses on to the z-axis, P coincides with C and S with B, we obtain

$$\frac{v(C)}{v_{\odot}(BC)} = \left(\frac{AC}{AO}\right)^{1/2}. \tag{9}$$

Q.E.D.

It follows from equation (9) that

$$v(O) = v_{\odot}(BO). \tag{10}$$

At O the particle ascending the right line BA has still to ascend to A before it comes to rest, or as Newton states:

> Cor. II. A body revolving in any circle at a given distance from the centre, by its motion converted upwards, will ascend to double its distance from the centre.

A personal reflection

Newton's derivation of equation (9) is admittedly less direct than the one given in §40 based on the integral

$$\frac{1}{2}\left(\frac{\mathrm{d}z}{\mathrm{d}t}\right)^2 = \frac{\mu}{z} + \text{constant}.$$

Not that Newton was not aware of this integral: indeed, it occurs explicitly in Proposition XXXIX for more general laws of centripetal attraction and under even more general conditions in Propositions XL–XLII in Section VIII to follow (see Chapter 9). Newton clearly preferred his present demonstration as a sequel to Proposition XXXII. One may surmise that the reason for this preference is to demonstrate that *all* aspects of rectilinear motion are in reality limiting cases of curvilinear motion; *and* to preserve the architectural unity of the subject—a motivation one can discern throughout the *Principia.*

Proposition XXXIV. Theorem X

> *If the figure BED is a parabola, I say, that the velocity of a falling body in any place C is equal to the velocity by which a body may uniformly describe a circle about the centre B at half the interval BC.*

The elliptic and the hyperbolic cases have been dealt with in Proposition XXXIII. The parabolic case is taken up in this proposition. The derivation of the result (equation (32) of §40) by the limiting procedure restores to it its full effulgence. The precision of Newton's natural style, not obscured by the need to explain details of mathematical reasoning, becomes transparent, as in this instance:

> For (by Cor. VII, Prop, XVI) the velocity of a body describing a parabola *RPB* about the centre *S*, in any place *P*, is equal to the velocity of a body uniformly describing a circle about the same centre *S* at half the interval *SP*. Let the breadth *CP* of the parabola be diminished *in infinitum*, so that the parabolic arc *PfB* may come to coincide with the right line *CB*, the centre *S* with the vertex *B*, and the interval *SP* with the interval *BC*, and the Proposition will be manifest.

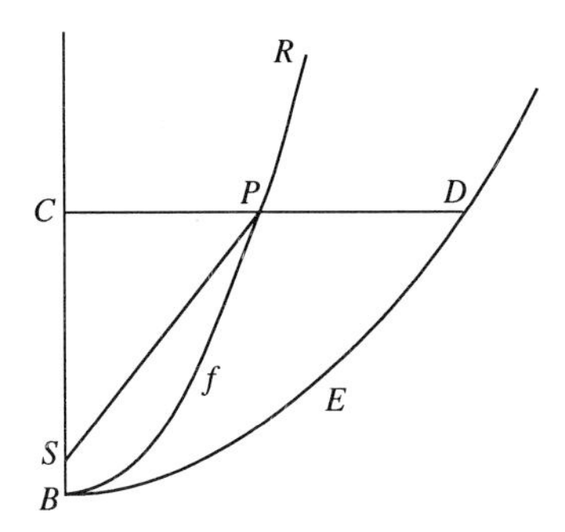

43. The reduction of the problem of rectilinear motion to one in circular motion

Newton's reduction of the solution of the problem of rectilinear motion, in the elliptic case, to the form,

$$t \propto a(\theta + \sin\theta),$$

in Proposition XXXII (p. 149)—Newton must have been fully aware that the area of *ABD* is $a(\theta + \sin\theta)$!—suggests, at once, cycloidal motion of a fixed point *on* the radius of a uniformly rolling circle with the parametrization,

$$x = a(\tau + \sin\tau) \qquad \text{and} \qquad y = a(1 - \cos\tau).$$

This is the real meaning of Proposition XXXV. The geometric construction suggested is no different from the geometric construction described in Proposition XXXI (see Chapter 7) for the solution of Kepler's equation in terms of the cycloidal motion of a fixed *interior* point of a uniformly rolling circle. Newton's apparent reluctance to draw attention to his earlier construction can perhaps be related to the fact that his present simultaneous reduction to cycloidal motion for the hyperbolic and the parabolic cases requires a different approach. Besides, in these cases, it is not meaningful (both physically and mathematically) to consider the cycloid for more than the two half-periods, $(0, \pi)$ and $(\pi, 0)$. One can understand Newton's impatience to stop and explain: there is nothing to be gained and time will be lost—and there is no time to lose!

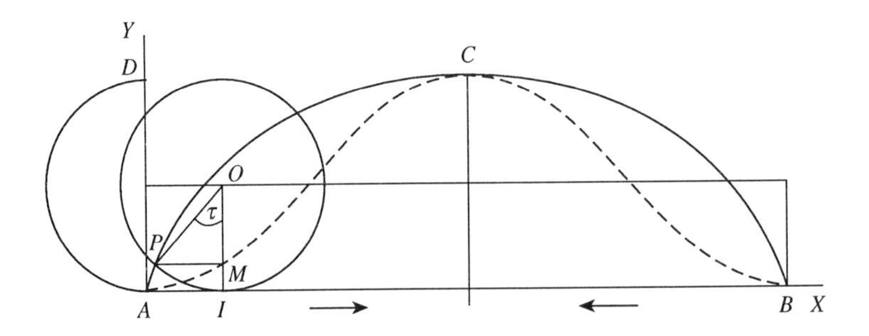

Proposition XXXV. Theorem XI

The same things supposed, I say, that the area of the figure DES, described by the indefinite radius SD, is equal to the area which a body with a radius equal to half the latus rectum of the figure DES describes in the same time, by uniformly revolving about the centre S.

The main features of the accompanying diagrams are the same as in those in §40. Rectilinear motion along the z-axis is considered:

C is the position of the ascending or the descending body at some instant of time, t;

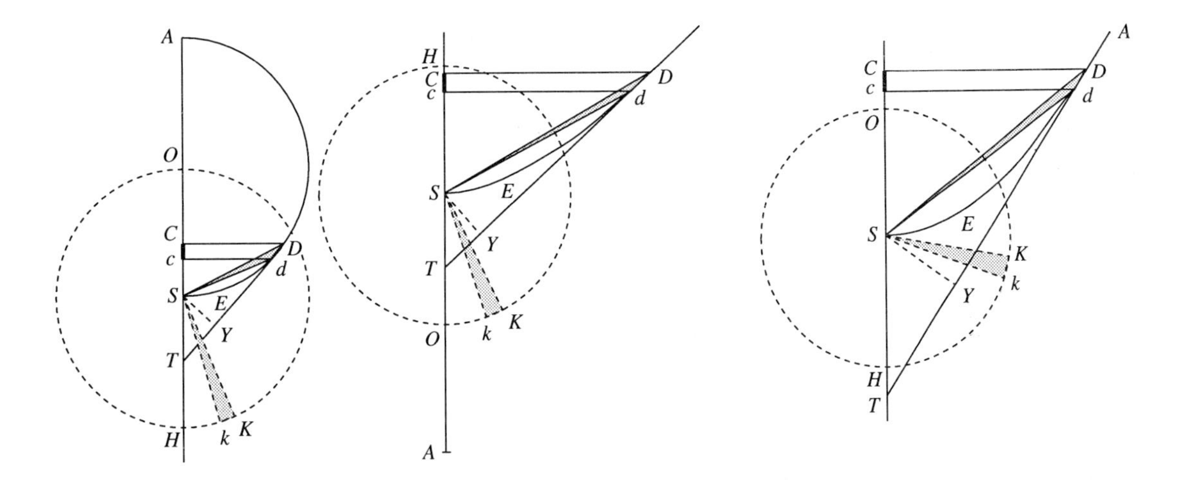

D is the image of C, at the same height, on $SEDA$ (a semicircle, a rectangular hyperbola with semiaxis $AO = OS$, or a fixed parabola in the three respective cases);

c and d the positions of C and its image D at an infinitesimal time δt later;

OKH is a circle of radius equal to the semilatus rectum of $SEDA$;

K is the position at time t of a body orbiting the circle OKH with the uniform circular velocity $v_{\odot}(SO = SK) \propto (SK)^{-1/2}$; and k is its position at a time δt later;

TD is the tangent at D to $SEDA$; and SY is perpendicular to the tangent.

Case 1. Newton considers the elliptic and the hyperbolic cases simultaneously. In both these cases the semilatus rectum of $SEDA = \frac{1}{2}AS = SO = SK$.

The proof consists of three elementary steps:

(a) We note that

$$\frac{TC}{TD} = \frac{Cc}{Dd} \qquad \text{(by construction)},$$

and

$$\frac{TD}{TS} = \frac{CD}{SY} \qquad \text{(by the similarity of } \triangle\text{s } TDC \text{ and } TSY\text{)}.$$

By multiplying the last two equations, we obtain:

$$\frac{TC}{TS} = \frac{Cc \, . \, CD}{SY \, . \, Dd}.$$

But by equation (7) of Proposition XXXIII (p. 152)

$$\frac{TC}{TS} = \frac{AC}{AO} = \frac{AC}{SK}.$$

Hence:

$$\frac{AC}{SK} = \frac{Cc\,.\,CD}{SY\,.\,Dd}. \tag{1}$$

(b) By Proposition XXXIII (equation (9))

$$\left[\frac{v(C)}{v_{\odot}(CS)}\right]^2 = \frac{AC}{AO} = \frac{AC}{SK};$$

and since

$$\left[\frac{v_{\odot}(CS)}{v_{\odot}(SK)}\right]^2 = \frac{SK}{CS},$$

we have

$$\left[\frac{v(C)}{v_{\odot}(SK)}\right]^2 = \frac{AC}{CS}. \tag{2}$$

(c) By using the known relation,

$$AC\,.\,CS = CD^2,$$

in the geometry of conic sections, we obtain

$$\frac{v(C)}{v_{\odot}(SK)} = \frac{AC}{CD}. \tag{3}$$

But

$$Cc = v(C)\delta t \qquad \text{and} \qquad Kk = v_{\odot}(SK)\delta t \tag{4}$$

Hence, by (3)

$$\frac{AC}{CD} = \frac{Cc}{Kk} \qquad \text{or} \qquad AC\,.\,Kk = CD\,.\,Cc.$$

Substituting this last result in (1), we obtain

$$\frac{AC}{SK} = \frac{AC\,.\,Kk}{SY\,.\,Dd},$$

or, finally,

$$SK\,.\,Kk = SY\,.\,Dd. \tag{5}$$

Equation (5) implies that

$$\text{area of sector } SKk = \text{area of segment } SDd; \tag{6}$$

and both areas are described in the same interval of time δt. And as Newton concludes,

> Therefore, if the magnitude of the equal areas SKk and SDk generated in an interval of time δt is diminished, and their number increased *in infinitum*, obtain the ratio of equality, and consequently (by Cor., Lem. IV) the whole areas together generated are always equal. Q.E.D.

Case 2. Turning to the parabolic case, we first observe that equation (1) of case 1, namely,

$$\frac{TC}{TS} = \frac{Cc\,.\,CD}{SY\,.\,Dd},$$

is equally applicable. But, by a well-known geometrical property of the parabola

$$CS = ST = \tfrac{1}{2}TC.$$

Therefore,

$$SY\,.\,Dd = \tfrac{1}{2}CD\,.\,Cc. \tag{7}$$

Next, by Proposition XXXIV (or by equation (32) of §40)

$$v(C) = v_{\odot}(\tfrac{1}{2}SC). \tag{8}$$

And therefore,

$$\frac{v_{\odot}(\frac{1}{2}SC)}{v_{\odot}(SK)} = \left(\frac{SK}{\frac{1}{2}SC}\right)^{1/2} = \left(\frac{4SK^2}{2SK\,.\,SC}\right)^{1/2} = \frac{2SK}{CD}, \tag{9}$$

since, by the equation of the parabola,

$$\mathrm{CD}^2 = 4aSC = 2SK\,.\,SC \qquad (4a = \text{latus rectum}).$$

By equations (8) and (9)

$$\frac{2SK}{CD} = \frac{v(C)}{v_{\odot}(SK)} = \frac{Cc}{Kk} \qquad \text{(by equation (4))},$$

or

$$SK\,.\,Kk = SY\,.\,Dd \qquad \text{(by equation (7))}.$$

We have thus established that the same equality (6) applies in this case as well; and the argument can be completed as before. Q.E.D.

Proposition XXXVI. Problem XXV

To determine the times of the descent of a body falling from a given place A.

In Proposition XXXII it was shown that the time of descent, from rest at A to C, is

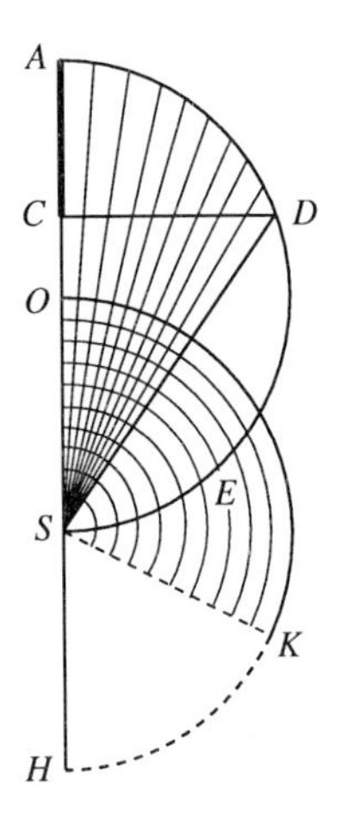

proportional to the area of the segment SAD; and as was shown in Proposition XXXIII, this is the same time in which a body uniformly revolving about S in a circular orbit of radius $SO(=\frac{1}{2}AS)$ describes the arc OK bounding a sector OSK of the same area as SAD.

Newton's demonstration is:

> Upon the diameter *AS*, the distance of the body from the centre at the beginning, describe the semicircle *ADS*, as likewise the semicircle *OKH* equal thereto, about the centre *S*. From any place *C* of the body erect the ordinate *CD*. Join *SD*, and make the sector *OSK* equal to the area *ASD*. It is evident (by Prop. XXXV) that the body in falling will describe the space *AC* in the same time in which another body, uniformly revolving about the centre *S*, may describe the arc *OK*.
>
> Q.E.F.

Proposition XXXVII. Problem XXVI

To define the times of the ascent or descent of a body projected upwards or downwards from a given place.

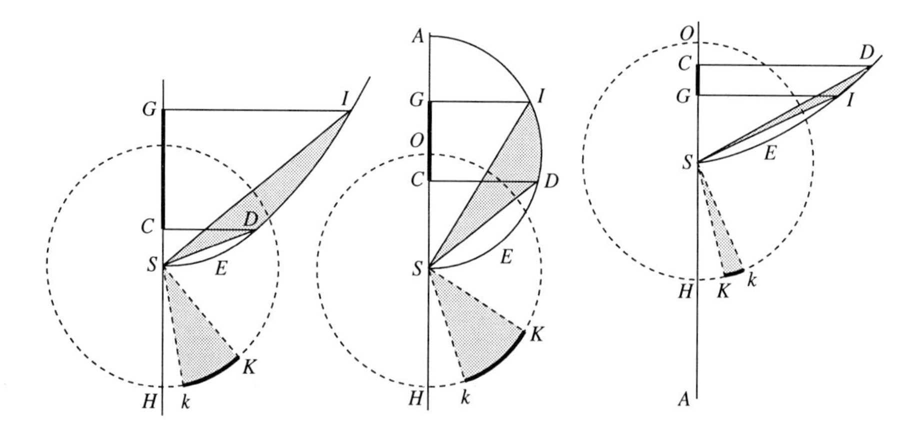

In Proposition XXXV, it was shown that the problem of rectilinear motion allows solutions that are limiting forms of the solutions for curvilinear motions along elliptic, hyperbolic, or parabolic orbits. It remains to prescribe a criterion by which one may decide which of the three classes of solution is appropriate for given initial conditions of height and velocity; and then to define the times of ascent or descent to another given height.

The first question is: given that a body is 'projected upwards or downwards' from a given height G with an assigned velocity $v(G)$, to which of the three classes of solutions does its subsequent motion belong? The answer to this question will depend on whether the body so projected will be or has been in a state of rest at some determinate height A; in which case the solution will belong to the elliptic class. If the solution does belong to the elliptic class, then by Proposition XXXIII, (equation (9))

$$\left[\frac{v(G)}{v_{\odot}(SG)}\right]^2 = \frac{AG}{\frac{1}{2}AS} = \frac{AS - SG}{\frac{1}{2}AS} = 2 - 2\frac{SG}{AS}, \tag{10}$$

where $v_{\odot}(SG)$ is the velocity of uniform motion in a circle of radius SG (where S is the known centre of attraction). Equation (10) rewritten in the form,

$$2\frac{SG}{AS} = 2 - \left[\frac{v(G)}{v_{\odot}(SG)}\right]^2, \tag{11}$$

makes it manifest that a finite positive solution for AS exists if, and only if,

$$\left[\frac{v(G)}{v_{\odot}(SG)}\right]^2 < 2. \tag{12}$$

This then is the necessary and sufficient condition that the motion, subsequent to the initial state—velocity $v(G)$ at height SG—will belong to the elliptic class.

By equation (11), A is 'infinitely remote' if

$$\left[\frac{v(G)}{v_{\odot}(SG)}\right]^2 = 2, \tag{13}$$

which then is the condition for the solution to belong to the parabolic class and the body comes to rest at infinity. It is now clear that the subsequent motion will belong to the hyperbolic class if,

$$\left[\frac{v(G)}{v_{\odot}(SG)}\right]^2 > 2; \tag{14}$$

for, in this case, in place of equation (11), we now have

$$2\frac{SG}{AS} = \left[\frac{v(G)}{v_{\odot}(SG)}\right]^2 - 2, \tag{15}$$

where A is the remote vertex of the rectangular hyperbola SED; and equation (15) will determine the location of A whenever condition (14) is satisfied. (Newton's discussion of the criterion expressed in terms of the discriminant, $[v(G)/v_{\odot}(SG)]^2$, is so brief that the important condition (14) for 'escape' is not explicitly mentioned—though, of course, it is implicit and obvious.)

With the class of solutions to which the subsequent motion of a body, initially with a velocity $v(G)$ at height G, will belong ascertained and the location of A in the elliptic and the hyperbolic cases determined, we are required to 'define' the time of ascent or descent to another specified height C.

By Proposition XXXVI the required time is equal to the time in which a body revolving in a circular orbit with a radius $SH(=\frac{1}{2}AS$ in the elliptic and the hyperbolic cases and the semilatus rectum of the parabola SEI in the parabolic case) will describe the arc Kk which bounds an area SKk which is equal to the shaded area SDI (which is the difference of the areas of the segments SDA and SIA in the elliptic case and of the areas $SEDS$ and $SEIS$ in the hyperbolic and the parabolic cases). Q.E.F.

It is important to draw attention to the significance of this proposition in the larger context of physics: it is the very first instance of the formulation of an *initial-value problem* in physics. Increasing generalizations of the underlying problem are considered in Propositions XXXIX, XLI, and XLII. (For further comments, see §45 below.)

44. A pause

All the propositions in this section so far have been concerned with the inverse-square law of attraction. As a prelude to the consideration of general laws of centripetal attraction in Proposition XXXIX, Newton considers in this proposition the law of attraction proportional to the distance which he deduced in Proposition X and showed that under this law of attraction a body will describe an elliptical orbit about its centre.

Proposition XXXVIII. Theorem XII

Supposing that the centripetal force is proportional to the altitude or distance of places from the centre, I say, that the times and velocities of falling bodies, and the spaces which they describe, are respectively proportional to the arcs, and the sines and versed sines of the arcs.

This proposition is proved by the same limiting process described in the context of Proposition XXXII. Towards the solution of the problem of a body, initially at rest at A, descending directly towards the centre of attraction at S, Newton considers a sequence of ellipses of different eccentricities and with the same semimajor axis SA $(=a)$. Let APB be the first quadrant of one of these ellipses and P a point on it at some fixed height. We know (by Proposition X) that a body at A will describe the arc AP in a time proportional to the area of the sector SPA which in turn is proportional to the area of the sector SDA. This latter proportionality holds independently of the eccentricity of the ellipse. Therefore, passing to the limit when the ellipse collapses on to the right line AS and P coincides with C, we conclude that the time of descent from A to C is proportional to the area of the

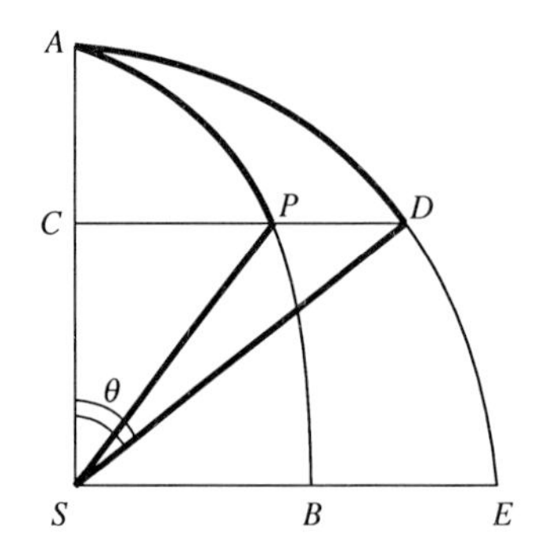

sector SDA, that is, $a\theta$. And since, further, the velocity v along the elliptical orbit is proportional to CD, again, independently of the eccentricity of the ellipse, it follows that the velocity acquired by the body during its descent from A to C is also proportional to CD that is, $a \sin \theta$.

The following corollaries follow in similar fashion:

> COR. I. Hence the times are equal in which one body falling from the place A arrives at the centre S, and another body revolving describes the quadrantal arc ADE.
>
> COR. II. Therefore all the times are equal in which bodies falling from whatsoever places arrive at the centre. For all the periodic times of revolving bodies are equal (by Cor. III, Prop. IV).

45. The initial-value problem

One of the principal tenets of physics is that any valid physical theory must allow an *initial-value formulation.* In general terms what this requirement implies is that given a well-defined initial state of a physical system, one should be able to predict uniquely its future development. To make precise what meaning is to be attached to a 'well-defined state of a physical system' or to its 'future development' are not simple matters; but they are beyond our scope. There is however no ambiguity in the present context: for the initial state of a body under the action of a known centripetal force is no more than its location relative to the centre of force and its velocity at a given instant of time; and we are required to show that its subsequent motion is uniquely predictable. Proposition XXXVII was devoted precisely to this end for rectilinear motion under an inverse-square law of attraction. In Proposition XXXIX, this same problem of rectilinear motion under a general law of centripetal attraction is considered. In Propositions XLI and XLII (in Chapter 9) the general planar problem of curvilinear motion is considered. Newton was the first to recognize that these problems require formulation and solution.

Proposition XXXIX. Problem XXVII

> *Supposing a centripetal force of any kind, and granting the quadratures of curvilinear figures; it is required to find the velocity of a body, ascending or descending in a right line, in the several places through which it passes, as also the time in which it will arrive at any place; and conversely.*

Newton's manner of proving this proposition is astonishing in contrast with earlier propositions: they were proved by a rare combination of physical and geometric insights grafted on to original ideas fashioned for the occasion. In this proposition, Newton simply states, at the outset, the required solutions as indefinite integrals (written out as areas) 'granting quadratures'; and then verifying them by differentiation! And the 'moral of that is' (as the Duchess would say): Newton was second to none when it came to using the integral and differential calculus; but he becomes solicitous (and sometimes secretive) when he does use them. We shall encounter this characteristic on several future occasions.

The problem that requires solution is: given the equation

$$\frac{d^2 z}{dt^2} = -f(z), \tag{1}$$

where $f(z)$ is a positive function of the argument, to find the velocity, dz/dt, and the time, t, when the body is at a given location z. The common manner of solving this problem is to note first that the velocity follows from the integral

$$v^2 = \left[\frac{dz}{dt}\right]^2 = 2\int_z f(z)\,dz, \tag{2}$$

or,

$$\frac{dz}{dt} = \pm\left[2\int_z f(s)\,ds\right]^{1/2}, \tag{3}$$

where the constant of integration is subsumed in expressing the solution (2) as an *indefinite* integral. And the solution for t now follows by a further integration:

$$t = \text{constant} \pm \int^z \frac{dz}{[2\int_z f(s)\,ds]^{1/2}}. \tag{4}$$

Newton expresses this same solution in the context of the accompanying diagram. The diagram more fully lettered and annotated is self-explanatory.

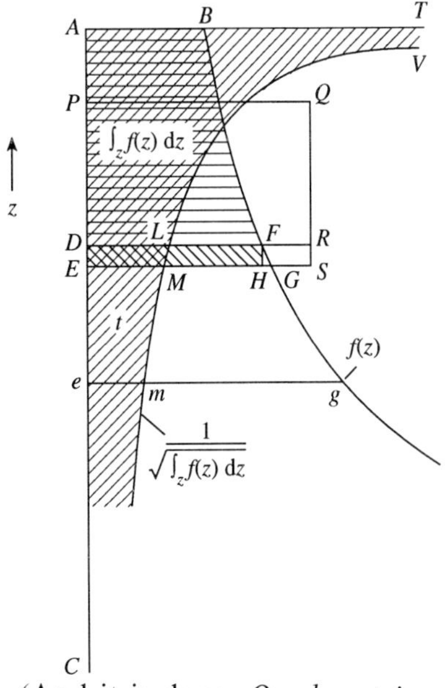

AEC is the z-axis;
$EG = f(z)$ at E;
BFG, the locus of G, represents $f(z)$:

$$[v(E)]^2 = 2 \text{ area of } ABGE = 2\int_E^A f(z)\,dz: \tag{5}$$

Q.E.I.

$$EM \propto [\text{area of } ABGE]^{-1/2} = \left[\int_E^A f(z)\,dz\right]^{-1/2};$$

VLM is the locus of M;
$t \propto$ area of $ABTVME$

$$= \int_E^A \left[\int_P^A f(z)\,dz\right]^{-1/2} dP. \tag{6}$$

Q.E.I.

(And it is done: *Quod erat inveniendum*).

The proof now consists of verifying the solutions by differentiation.

Proof of equation (1): Letting DE be a 'particle' of displacement, we have

$$\text{area of } DFGE = f(E)DE = f(E)\frac{DE}{\Delta t}\Delta t$$
$$= f(E)v(E)\Delta t.$$

But by the second Law ('acceleration = force')

$$\Delta[v(E)] = f(E)\Delta t.$$

Hence

$$\text{area of } DFGE = v(E)\Delta[v(E)] = \tfrac{1}{2}\Delta[v(E)]^2.$$

Q.E.D.

Corollary I

If a body is subject to a force $g(z)$ different from $f(z)$, then the velocity acquired during descent, from rest at P, to D, is given by

$$[v_g(D; P)]^2 = 2\int_D^P g(z)\,\mathrm{d}z.$$

And if this should equal $[v_f(D; A)]^2$, the condition is

$$\int_D^P g(z)\,\mathrm{d}z = \int_D^A f(z)\,\mathrm{d}z.$$

In the particular case, g = constant = PQ (say), then

$$PQ\,.\,PD = \int_D^A f(z)\,\mathrm{d}z,$$

or

$$\text{area of rectangle } PQRD = \textit{area of } ABFD.$$

Corollary II

If the body should be projected from a height D with a velocity V, the velocity at some other point P is given by

$$[v(P)]^2 = V^2 + 2\int_P^D f(z)\,\mathrm{d}z.$$

Corollary III

'The same things being supposed', the time required for descent to a different point e is given by

$$t \propto \int_e^D \left[V^2 + \int_P^D f(z)\,\mathrm{d}z \right]^{-1/2} \mathrm{d}P.$$

Proposition XXXIX together with Corollaries II and III complete the solution of the initial-value problem for rectilinear motion under centripetal attraction.

❖9❖

The conservation of energy and the initial-value problem

The surprising discovery of Newton's is just this, the clear separation of laws of nature on the one hand and initial conditions on the other.

Eugene P. Wigner

46. Introduction

Section VIII, including three Propositions XL, XLI, and XLII, represents a watershed in the development of the *Principia*: it completes the treatment of 'the motion of bodies in immovable orbits'. But it is very much more than that: in deriving the energy integral (and implicitly the notion of potential energy) and in formulating and solving the initial-value problem in the framework of his dynamics, Newton established for the first time two of the basic tenets of physics. Newton must have been fully aware of the significance of these propositions. But his presentation of them manifests a strange impatience—of this later! Meantime, it will suffice to remark that superficially, Propositions XLI and XLII are presented as straightforward generalizations of Proposition XXXIX and its corollary. And there are, of course, examples of his insight and his craftsmanship.

47. The energy integral

The standard derivation of the energy integral will be found in Chapter 4 (§18(c), equation (18)). Newton's route is different:

Proposition XL. Theorem XIII

If a body, acted upon by any centripetal force, is moved in any manner, and another body ascends or descends in a right line, and their velocities be equal in any one case of equal altitudes, their velocities will be also equal at all equal altitudes.

Newton compares the kinetic energies of two bodies, arriving at the same distance from the centre of attraction, C, from a common point V with the same velocity, one of them, D, directly toward C, along a radial trajectory, ADC, and the other, I, along a curvilinear orbit VIK. By Chapter 4 (§18, equation (25)) and Chapter 8 (§45, equation (1)),

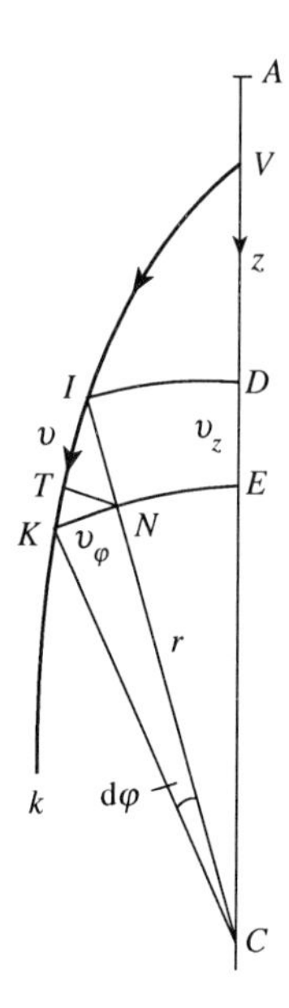

$$\frac{d^2r}{dt^2} = -f(r) + \frac{h^2}{r^3} \tag{1}$$

and

$$\frac{d^2z}{dt^2} = -f(z). \tag{2}$$

From these equations, it follows that,

$$\frac{1}{2}\left(\dot{r}^2 + \frac{h^2}{r^2}\right) = -\int^r f(r)\, dr \tag{3}$$

and

$$\tfrac{1}{2}\dot{z}^2 = -\int^z f(z)\, dz. \tag{4}$$

If v_r and $v_\phi(=r\dot{\phi})$ denote the radial and the transverse velocities, along the orbit VI, at I,

$$v_r^2 + v_\phi^2 = \dot{r}^2 + h^2/r^2 \tag{5}$$

and

$$v_z^2 = \dot{z}^2. \tag{6}$$

Then, by equations (3) and (4),

kinetic energy at I, along the curvilinear orbit,

$$= \int_I^V f(r)\, dr = \int_D^V f(z)\, dz$$

$=$ kinetic energy at D, along the rectilinear orbit, (7)

since by assumption $CI = CD$.

Newton's manner of proof is, however, different. He argues as follows: at I and D, the two bodies, being at the same distance from the centre of attraction at C, are subjected to the same centripetal attraction. They must, therefore, suffer equal increments in their velocity in equal times. When they arrive at E and K (from D and I, respectively) both still equidistant from C but by the infinitesimal distances, $DE = IN$, closer, the increments in their velocities will be given by the accelerations they are subjected to in the directions VDE and ITK and the times during which they act. The particle at D experiences, along the radial trajectory VDE, the increment in velocity

$$\begin{aligned} \Delta v_{ED} = v_E - v_D &= \text{acceleration at } D \times \text{time during which it acts} \\ &= \text{acceleration at } D \times \frac{DE}{v_D}. \end{aligned} \tag{8}$$

On the other hand, since the component of the acceleration along TN, normal to IK, is ineffective in changing the velocity in the direction of motion,

$$\Delta v_{IK} = v_K - v_I = \left[(\text{acceleration at } I) \times \frac{IT}{IN}\right] \times \frac{IK}{v_I}, \tag{9}$$

or

$$\Delta v_{IK} = (\text{acceleration at } I) \times \frac{IN}{v_1} = \Delta v_{ED}, \tag{10}$$

since

$$\left.\begin{array}{c} IT.IK = IN^2, \\ DE = IN, \quad \text{and} \quad v_D = v_I, \end{array}\right\} \tag{11}$$

by assumption.

As Newton concludes:

> and therefore the accelerations generated in the passage of the bodies from D and I to E and K are equal. Therefore the velocities of the bodies in E and K are also equal: and by the same reasoning they will always be found equal in any subsequent equal distances. Q.E.D.
>
> By the same reasoning, bodies of equal velocities and equal distances from the centre will be equally retarded in their ascent to equal distances. Q.E.D.

Newton's demonstration has the clear advantage since it establishes that the *change in velocity* in either body is effected solely by the *change in the distance* from the centre of attraction, thus separating the contributions of the potential energy and the kinetic energy to the total energy: the notion of potential energy is implicit in the statement. Corollary I, as Newton states it, follows directly from this separation of the two forms of energy.

COR. I. Therefore if a body either oscillates by hanging to a string, or by any polished and perfectly smooth impediment is forced to move in a curved line; and another body ascends or descends in a right line, and their velocities be equal at any one equal altitude, their velocities will be also equal at all other equal altitudes. For by the string of the pendulous body, or by the impediment of a vessel perfectly smooth, the same thing will be effected as by the transverse force NT. The body is neither accelerated nor retarded by it, but only is obliged to leave its rectilinear course.

Corollary II is an explicit statement of the proposition for the case,

$$f(r) = -r^{n-1}.$$

In this case,

$$\tfrac{1}{2}(v_A^2 - v_P^2) = -\int_P^A r^{n-1}\,\mathrm{d}r = \frac{1}{n}(P^n - A^n)$$

where P is a fixed point, distant P from the centre and A is a variable point, distant A, the path of the orbit from P to A being irrelevant.

48.

Proposition XLI. Problem XXVIII

Supposing a centripetal force of any kind, and granting the quadratures of curvilinear figures; it is required to find as well the curves in which bodies will move, as the times of their motions in the curves found.

The generalization of Proposition XXXIX, which this proposition represents, is emphasized by the identity of their phrasing. And the demonstration, following parallel lines, is also not different from what one might give today. We should naturally start with the radial equation,

$$\frac{\mathrm{d}^2 r}{\mathrm{d}t^2} = -f(r) + \frac{h^2}{r^3}, \tag{1}$$

and the angular equation,

$$\frac{\mathrm{d}\phi}{\mathrm{d}t} = \frac{h}{r^2}, \tag{2}$$

representing the law of areas. As we have seen in §47, equation (1) allows the energy integral:

$$\dot{r}^2 = -2\int^r f(r)\,\mathrm{d}r - \frac{h^2}{r^2}, \tag{3}$$

where we have left the lower bound of the integral unspecified for later choice in Proposition XLII. Equation (3) rewritten in the form,

$$\frac{\mathrm{d}r}{\mathrm{d}t} = \pm\left[-2\int^{r} f(r)\,\mathrm{d}r - \frac{h^2}{r^2}\right]^{1/2}, \tag{4}$$

together with equation (2), provides the additional equation,

$$\frac{\mathrm{d}\phi}{\mathrm{d}r} = \frac{\pm h}{r^2[-2\int^{r} f(r)\,\mathrm{d}r - h^2/r^2]^{1/2}}. \tag{5}$$

In their integrated forms,

$$t = \pm\int^{r} \frac{\mathrm{d}r}{[-2\int^{r} f(r)\,\mathrm{d}r - h^2/r^2]^{1/2}} \tag{6}$$

and

$$\phi = \pm\int^{r} \frac{h\,\mathrm{d}r}{r^2[-2\int^{r} f(r)\,\mathrm{d}r - h^2/r^2]^{1/2}}, \tag{7}$$

we have the required solution to the problem ‘granting quadratures’.

The following annotated version of Newton’s demonstrations shows that it is the same as the one given above.

Let any centripetal force tend to the centre C, and let it be required to find the curve $VIKk$. Let there be given the circle VR, described from the centre C with any radius CV; and from the same centre describe any other circles ID, KE, cutting the curve in I and K, and the right line CV in

Newton’s notation

$h = Q$; $\quad r = A$;

$$\frac{h}{r} = \frac{Q}{A} = Z.$$

$$\therefore\quad BLFG\colon -2f(r);$$

$$IN = \mathrm{d}r;\qquad \angle XCY = \mathrm{d}\phi,$$

$$KN = r\,\mathrm{d}\phi.$$

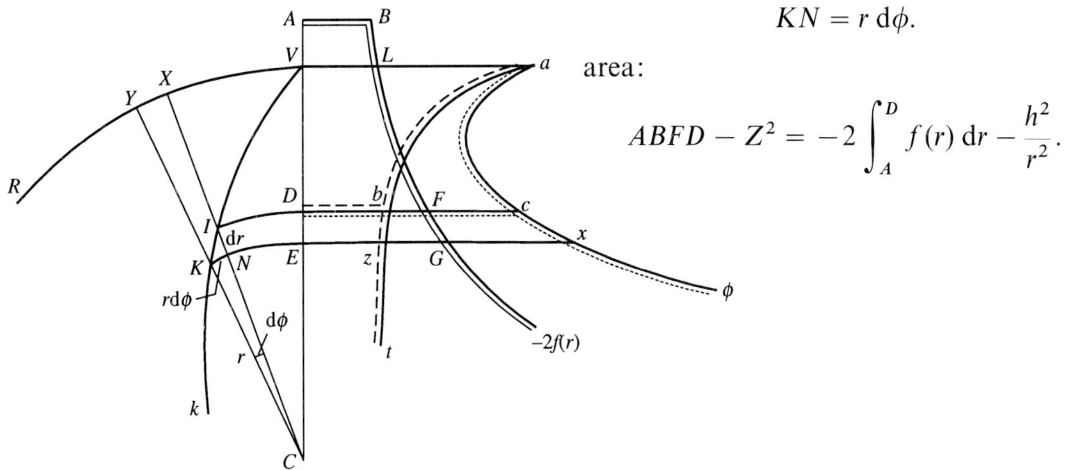

area:

$$ABFD - Z^2 = -2\int_{A}^{D} f(r)\,\mathrm{d}r - \frac{h^2}{r^2}.$$

D and E. Then draw the right line $CNIX$ cutting the circles KE, VR in N and X, and the right line CKY meeting the circle VR in Y. Let the points I and K be indefinitely near; and let the body go on from V through I and K to k; and let the point A be the place from which another body is to fall, so as in the place D to acquire a velocity equal to the velocity of the first body in I. And things remaining as in Prop. XXXIX, the short line IK, described in the least given time, will be as the velocity, and therefore as the right line whose square is equal to the area $ABFD$, and the triangle ICK proportional to the time will be given, and therefore KN will be inversely as the altitude IC; that is (if there be given any quantity Q, and the altitude IC be called A), as Q/A. This quantity Q/A call Z, and suppose the magnitude of Q to be such that in some one case

$$\sqrt{ABFD}:Z = IK:KN, \qquad \rightarrow \frac{\sqrt{ABFD}}{h/r} = \frac{\sqrt{(v_r^2 + v_\phi^2)}}{r\dot{\phi}} = \frac{v_S}{v_\phi} = \frac{IK}{IN}$$

and then in all cases

$$\sqrt{ABFD}:Z = IK:KN,$$

and

$$ABFD:Z^2 = IK^2:KN^2,$$

and by subtraction,

$$ABFD - ZZ:ZZ = IN^2:KN^2, \qquad \rightarrow \frac{ABFD - h^2/r^2}{h^2/r^2} = \frac{v_r^2}{v_\phi^2} = \frac{IN^2}{KN^2}$$

and therefore

$$\sqrt{(ABFD - ZZ)}:Z \quad \text{or} \quad \frac{Q}{A} = IN:KN,$$

and

$$A.KN = \frac{Q.IN}{\sqrt{(ABFD - ZZ)}}, \qquad \rightarrow r^2\,\mathrm{d}\phi = \pm\frac{h\,\mathrm{d}r}{[-2\int^r f(r)\,\mathrm{d}r - h^2/r^2]^{1/2}} = h\,\mathrm{d}t$$

Since

$$YX.XC:A.KN = CX^2:AA, \qquad \rightarrow YX.XC/(r^2\,\mathrm{d}\phi) = CX^2/r^2$$

it follows that

$$YX.XC = \frac{Q.IN.CX^2}{AA\sqrt{(ABFD - ZZ)}}. \qquad \rightarrow YX.XC = CX^2\,\mathrm{d}\phi = \pm\frac{CX^2h\,\mathrm{d}r}{r^2[-2\int^r f(r)\,\mathrm{d}r - h^2/r^2]^{1/2}}$$

Therefore in the perpendicular DF let there be taken continually Db, Dc equal to

$$\frac{Q}{2\sqrt{(ABFD - ZZ)}}, \quad \text{(i)}$$

(i): $$\frac{h}{2[-2\int^r f(r)\,\mathrm{d}r - h^2/r^2]^{1/2}} = \tfrac{1}{2}h\frac{\mathrm{d}t}{\mathrm{d}r}.$$

$$\frac{Q.CX^2}{2AA\sqrt{(ABFD - ZZ)}} \quad \text{(ii)}$$

(ii): $$\frac{hCX^2}{2r^2[-2\int^r f(r)\,\mathrm{d}r - h^2/r^2]^{1/2}} = \tfrac{1}{2}CX^2\frac{\mathrm{d}\phi}{\mathrm{d}r}.$$

respectively, and let the curved lines ab, ac, the loci of the points b and c, be described; and from the point V let the perpendicular Va be erected to the line AC, cutting off the curvilinear areas $VDba$, $VDca$, and let the ordinates Ez, Ex, be erected also. Then because the rectangle $Db.IN$ or $DbzE$ is equal to half the rectangle $A.KN$, or to the triangle ICK; and the rectangle $Dc.IN$ or $DcxE$ is equal to half the rectangle $YX.XC$, or to the triangle XCY; that is, because the nascent particles $DbzE$, ICK of the areas $VDba$, VIC are always equal; and the nascent particles $DcxE$, XCY of the areas $VDca$, VCX are always equal: therefore the generated area $VDba$ will be equal to the generated area VIC, and therefore proportional to the time; and the generated area $VDca$ is equal to the generated sector VCX. If, therefore, any time be given during which the body has been moving from V, there will be also given the area proportional to it $VDba$; and thence will be given the altitude of the body CD or CI; and the area $VDca$, and the sector VCX equal thereto, together with its angle VCI. But the angle VCI, and the altitude CI being given, there is also given the place I, in which the body will be found at the end of that time. Q.E.I.

(i): $\rightarrow$ area $VDba = \tfrac{1}{2}ht$;

(ii): $\rightarrow$ area $VDca = \tfrac{1}{2}h\phi$.

Corollaries I and II

The line of the apsides is determined by the vanishing of $\dot{r} = 0$, that is, by

$$2\int^{r} f(r)\,\mathrm{d}r + h^2/r^2 = 0,$$

or, in Newton's notation,

$$ABFD - Z^2 = 0.$$

Equivalent forms of the same condition are:

$$IN = 0 \qquad \text{and} \qquad \cos\angle KIN = 0.$$

49. The mystery of the missing corollary

While general laws of centripetal force are considered when the occasion demanded (as in Proposition XXXIX and in the propositions of this section) the prime focus in the *Principia* has always been the inverse-square law of attraction. One is therefore puzzled why in this important instance, Newton passed over the inverse-square law of attraction and chose the inverse-cube law for illustrating the proposition. It can hardly be doubted that he did consider the case of the inverse-square law of attraction—it cannot have been more than 'child's play for Newton' (to quote J.E. Littlewood in another context). Why then did he abstain from including it in the *Principia*? In my view the reason is not far to seek. But, first, it is useful to write out the few lines in which Newton would have solved this problem though he might have cloaked his demonstration in the manner of Proposition XLI (described in §48).

Newton would naturally have started with the energy integral ($ABFD - Z^2$ in his notation)

$$\dot{r}^2 = \frac{2}{r} - \frac{h^2}{r^2} + C, \tag{1}$$

appropriate for the inverse-square law, $f(r) \propto r^{-2}$, where C is a constant of integration. By Corollaries I and II of this proposition XLI, the apsides will occur when

$$Cr^2 + 2r - h^2 = 0 \qquad \text{and} \qquad \dot{r} = 0. \tag{2}$$

Considering the case when the quadratic equation (2) allows two real roots,

$$r_1 = a(1 - e) \qquad \text{and} \qquad r_2 = a(1 + e) \tag{3}$$

where a and e are unspecified constants, we find from equation (2) that

$$h^2 = \frac{2r_1r_2}{r_1 + r_2} = a(1 - e^2) \qquad \text{and} \qquad C = -\frac{1}{a}. \tag{4}$$

Inserting these values in equation (1), we find

$$\dot{r}^2 = 2\frac{(r_2 - r)(r - r_1)}{r^2(r_1 + r_2)} = \frac{1}{ar^2}[-r^2 + 2ar - a^2(1 - e^2)]. \tag{5}$$

With this expression for $\dot{r}^2$, the solutions (6) and (7) given in the preceding section take the forms

$$\pm t = a^{1/2}\int\frac{r\,\mathrm{d}r}{[-r^2 + 2ar - a^2(1 - e^2)]^{1/2}}, \tag{6}$$

and

$$\pm\phi = ha^{1/2}\int\frac{\mathrm{d}r}{r[-r^2 + 2ar - a^2(1 - e^2)]^{1/2}}. \tag{7}$$

Evaluating these integrals ('child's play' for Newton!), we find that

$$t = a^{1/2}\left\{a\left(\tfrac{1}{2}\pi - \sin^{-1}\frac{a - r}{ae}\right) + [a(1 + e) - r]^{1/2}[r - a(1 - e)]^{1/2}\right\}. \tag{8}$$

and

$$\phi = \tfrac{1}{2}\pi + \sin^{-1}\frac{r - a(1 - e^2)}{re}, \tag{9}$$

with the initial conditions,

$$t = 0 \quad \text{and} \quad \phi = 0 \quad \text{at} \quad r = a(1 - e). \tag{10}$$

The alternative form of equation (9),

$$\frac{a(1 - e^2)}{r} = 1 + e\cos\phi, \tag{11}$$

makes it manifest that we have indeed recovered the orbit, that is, the converse of Proposition XI.

With the solution to this problem so readily obtained from what had gone before, why did Newton abstain from including it as 'Corollary III' to Proposition XLI? The clue resides in the integrals (6) and (7). Not that these integrals were in any way obstacles: Corollary III of Proposition XCI includes in fact the explicit evaluation of an exactly similar integral:

$$F = 2b - \sqrt{(1 - e^2)}\int_{R-b}^{R+b}\frac{y\,\mathrm{d}y}{\sqrt{[-e^2y^2 + 2Ry - (R^2 - b^2)]}} \tag{12}$$

which expresses the gravitational attraction at an external point on the minor axes of an oblate spheroid. But the solution is given as a 'conundrum' which baffled Newton's contemporaries (including Roger Cotes). To understand Newton's apparent reluctance in dealing with the integrals (6) and (7) in similar fashion, one must bear in mind two facts: First, in 1685 Newton was unable to refer to any printed equivalent to his 1671 'table of areas of curves' later published in his *Tractatus de Quadrature Curvarum*

(London, 1700); and second, Corollary III of Proposition XCI was added out of sequence in the *Principia* in August 1686 at which time he was involved in his theory of the oblateness of the terrestrial planets in Book III, Proposition XIX. (All these matters are discussed in greater detail in Chapter 20.)

One may still ask, why did Newton overcome his reluctance in August 1686 and not some months earlier when he was writing Proposition XLI. The answer, in my view, is clearly that in the former instance he *absolutely* needed the explicit evaluation of the integral (12) to make quantitative his theory of the oblateness of terrestrial planets. In the latter instance, it was all so obvious (to him!) that he did not bother. And later, when criticisms of Bernoulli and others were drawn to his attention, he mostly ignored them with perhaps the attitude:

> Small have continual plodders ever won,
> Save base authority from others' books.

50. Motion under an inverse-cube law of centripetal attraction

In this corollary, Newton illustrates Proposition XLI by considering the inverse-cube law of attraction. Newton gives scant details. Besides, there are apparent misprints (or words like 'body' not having their standard meanings) and alterations of figures from the first and second to the third edition of the *Principia* that add to the obscurity. But enough details are given that it is not difficult to reconstruct Newton's analysis. The following treatment of the problem is what Newton would have given had he the time or the patience; for, as he confesses at the end:

> All these things follow from the foregoing Proposition, by the quadrature of a certain curve, the invention of which, as being easy enough, for brevity's sake I omit.

We start with the energy integral:

$$\dot{r}^2 + \frac{h^2}{r^2} = -2\int \frac{\mathrm{d}r}{r^3} = \frac{1}{r^2} + \text{constant.} \tag{1}$$

In its integrated form we have

$$\dot{r}^2 = (1 - h^2)\frac{1}{r^2} + C \tag{2}$$

where C is a constant of integration. We also have

$$\frac{\mathrm{d}\phi}{\mathrm{d}t} = \frac{h}{r^2}. \tag{3}$$

Following Newton, we shall consider solutions of equations (2) and (3) with the initial conditions,

$$\dot{r} = 0 \qquad \text{and} \qquad t = 0 \tag{4}$$

for some finite

$$r = \beta > 0. \tag{5}$$

(The case when $\dot{r} = 0$ at infinity must be considered separately.)

From equation (2) it is clear that two cases must be distinguished:

$$h^2 < 1 \qquad \text{and} \qquad h^2 > 1. \tag{6}$$

Case 1, $h^2 < 1$. For this case, we shall write equation (2) in the form,

$$\dot{r}^2 = (1 - h^2)\left(\frac{1}{r^2} - \frac{1}{\beta^2}\right) \qquad (r \leqslant \beta) \tag{7}$$

consistently with the requirement (5); and it is also convenient to require (without loss of generality):

$$\dot{r} < 0 \qquad \text{and} \qquad \dot{\phi} > 0 \qquad \text{for } r < \beta. \tag{8}$$

From equations (7) and (3) we readily find:

$$\frac{\mathrm{d}t}{\mathrm{d}r} = -\frac{\beta}{\sqrt{(1-h^2)}}\frac{r}{\sqrt{(\beta^2 - r^2)}}, \tag{9}$$

and

$$\frac{\mathrm{d}\phi}{\mathrm{d}r} = -\frac{h\beta}{\sqrt{(1-h^2)}}\frac{1}{r\sqrt{(\beta^2 - r^2)}}. \tag{10}$$

The required solutions of these equations are:

$$t = +\frac{\beta}{\sqrt{(1-h^2)}}\sqrt{(\beta^2 - r^2)}, \tag{11}$$

and

$$\phi = \frac{h}{\sqrt{(1-h^2)}}\log\frac{\beta + \sqrt{(\beta^2 - r^2)}}{r} = \frac{h}{\sqrt{(1-h^2)}}\cosh^{-1}\frac{\beta}{r}. \tag{12}$$

Rewriting equation (11) in the form

$$\frac{r^2}{\beta^2} + \frac{t^2}{\alpha^2} = 1, \qquad \text{where } \alpha = \frac{\beta^2}{\sqrt{(1-h^2)}}, \tag{13}$$

we conclude that in the (t, r)-plane, the orbit is an ellipse. In the accompanying diagrams, the solution curves in the (t, r)-, (ϕ, r)-, and the (r, ϕ)-planes are illustrated. It is clear, quite generally, that when $h^2 < 1$, the body descends to the centre in a spiral curve.

In addition, Newton derives the following remarkable identity.

Letting

$$\tan\theta = t/r, \tag{14}$$

we have,

$$\sec^2\theta\,\frac{\mathrm{d}\theta}{\mathrm{d}t} = \frac{1}{r} - \frac{t}{r^2}\frac{\mathrm{d}r}{\mathrm{d}t}, \tag{15}$$

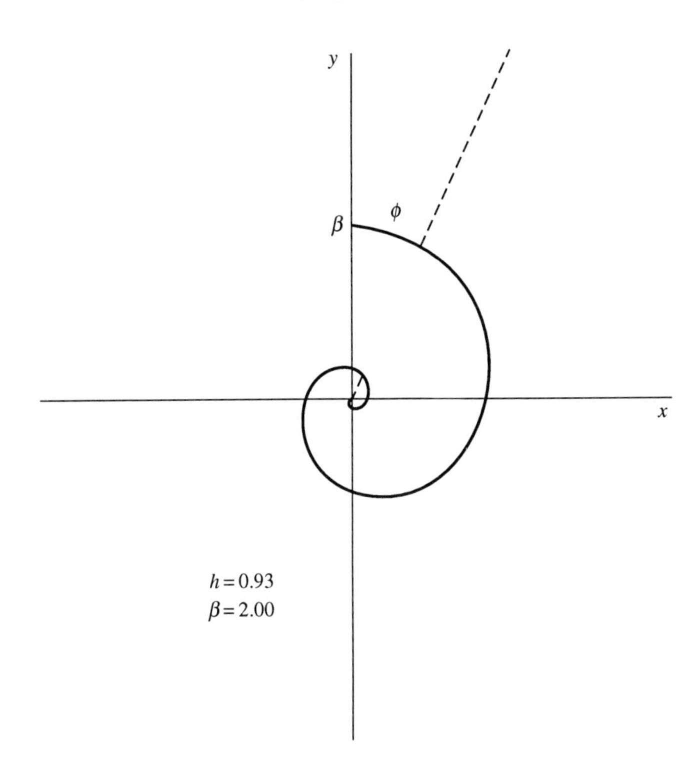

while by equation (13),

$$\frac{\mathrm{d}r}{\mathrm{d}t} = -\frac{\beta^2}{\alpha^2}\frac{t}{r}. \tag{16}$$

Eliminating $\mathrm{d}r/\mathrm{d}t$ from equation (15), we find

$$\frac{\beta}{r}(r^2 + t^2)\,\mathrm{d}\theta = \frac{\beta^3}{h}\,\mathrm{d}\phi. \tag{17}$$

If

$$\mathrm{d}A = \tfrac{1}{2}(r^2 + t^2)\,\mathrm{d}\theta, \tag{18}$$

denotes the element of area in the (t, r)-plane, then

$$\frac{\beta}{r}\,\mathrm{d}A = \frac{1}{2}\frac{\beta^3}{h}\,\mathrm{d}\phi; \tag{19}$$

or in its integrated form

$$\int_\beta^r \frac{\beta}{r}\,\mathrm{d}A = \frac{1}{2}\frac{\beta^3}{h}\,\phi \qquad (\phi = 0 \text{ for } r = \beta). \tag{20}^*$$

* Newton states this result differently:

$$\int_\beta^r CT.\mathrm{d}A = \tfrac{1}{2}\beta^4\phi/h,$$

where $CT = \beta^2/r$ is the distance from the centre C to the intersection T, of the tangent at r (on the ellipse) with the principal axis. (See illustrations on p. 177).

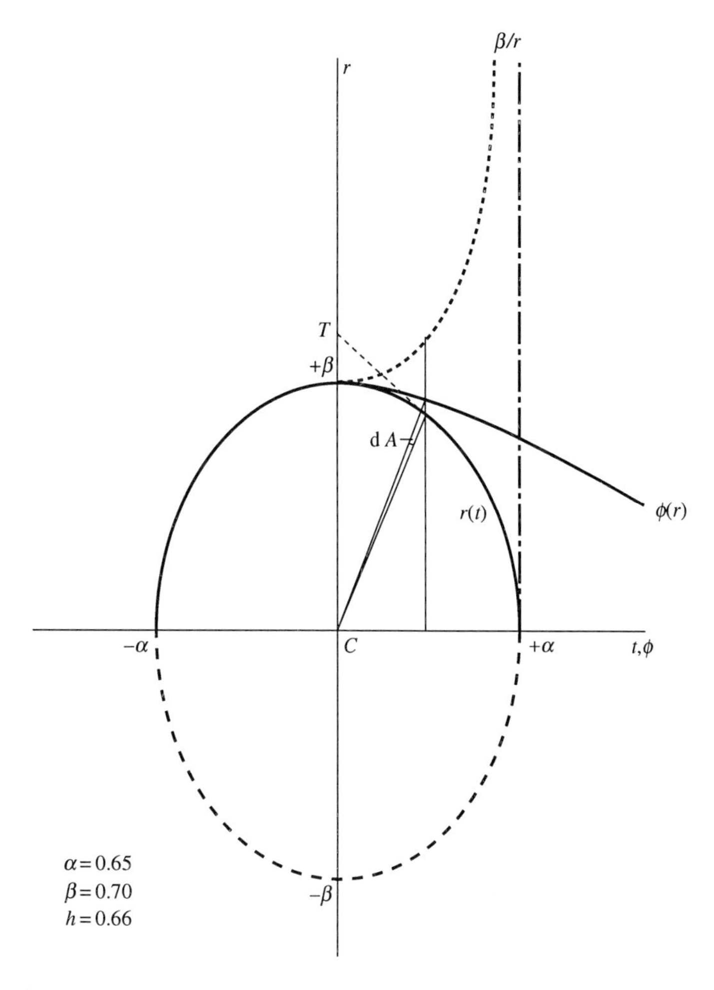

Alternatively, we also have

$$A = \frac{1}{2}\frac{\beta^2}{h}\int_0^{\phi} r\,\mathrm{d}\phi = \frac{1}{2}\frac{\beta^2}{h} \times \text{arc length along the orbit of the particle.} \tag{21}$$

Equations (20) and (21) are equivalent to two of Newton's statements in this corollary. We have called them Newton's identities.

In the accompanying illustration (with $\alpha = 0{\cdot}65$, $\beta = 0{\cdot}70$, and $h = 0{\cdot}66$) the dotted curve represents β/r for various points on the ellipse.

Case 2, $h^2 > 1$. In this case, equations (9) and (10) are replaced by

$$\frac{\mathrm{d}t}{\mathrm{d}r} = +\frac{\beta}{\sqrt{(h^2-1)}}\frac{r}{\sqrt{(r^2-\beta^2)}}, \tag{22}$$

and

$$\frac{\mathrm{d}\phi}{\mathrm{d}r} = +\frac{h\beta}{\sqrt{(h^2-1)}}\frac{1}{r\sqrt{(r^2-\beta^2)}}, \tag{23}$$

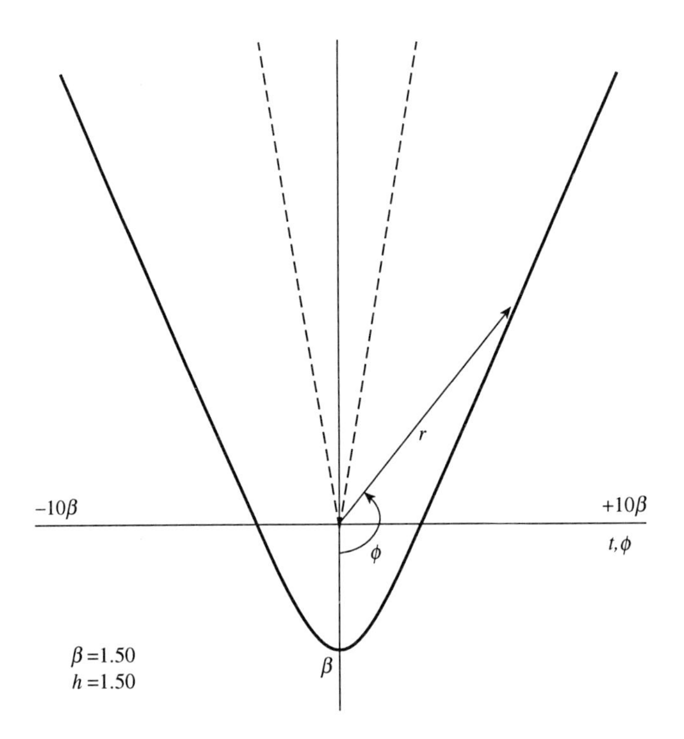

where we have required

$$\dot{r} \geqslant 0 \quad \text{and} \quad \dot{\phi} \geqslant 0. \tag{24}$$

The required solutions of equations (22) and (23) along which $\phi = 0$ when $r = \beta$, are:

$$\frac{r^2}{\beta^2} - \frac{t^2}{\alpha^2} = 1 \quad \text{and} \quad \phi = \frac{h}{\sqrt{(h^2-1)}} \cos^{-1}\frac{\beta}{r} \; [\alpha = \beta^2/\sqrt{(h^2-1)}]. \tag{25}$$

In the (t, r)-plane the orbit is a hyperbola; and along the asymptote, $r/t = \beta/\alpha$,

$$\phi \to \phi_0 = \frac{1}{2}\frac{\pi h}{\sqrt{(h^2-1)}}; \tag{26}$$

the curve therefore ascends indefinitely while ϕ tends to a finite limiting value.

We readily verify that the Newtonian identities (20) and (21) are valid as they stand.

In the accompanying diagrams* the solution curves in the (t, r)-, (ϕ, r)-, in the (r, ϕ)-plane are illustrated.

We have yet to consider the case when the body comes to rest at infinity.

Case 3, $\dot{r} = 0$ at $r = \infty$. In this case the energy integral gives

$$\dot{r}^2 = \frac{1-h^2}{r^2} \quad (h^2 > 1 \text{ is not allowed}). \tag{27}$$

* These diagrams for Case 2 as well as those for Case 1 were kindly prepared for me (for this book) by Drs. Valeria Ferrari and Andrea Malagoli.

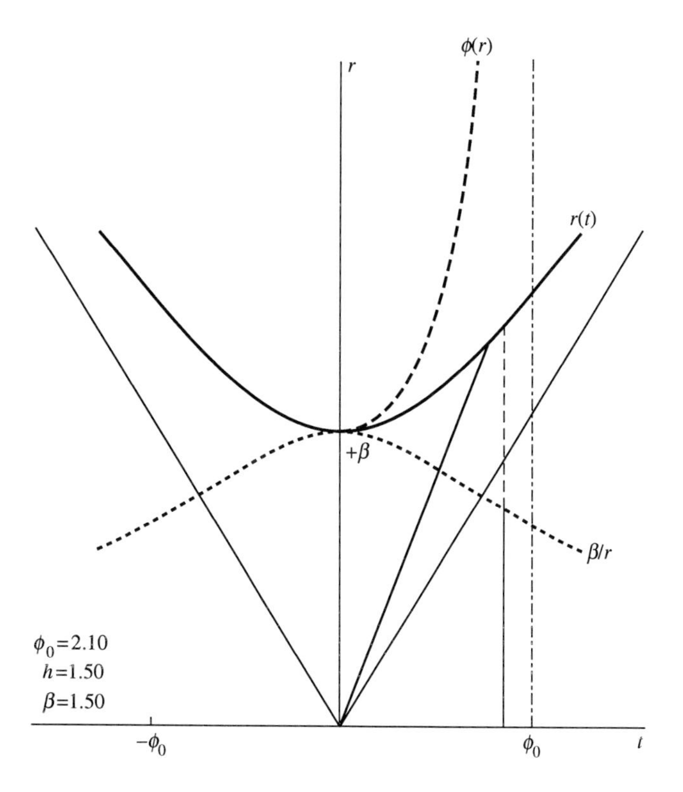

We also have the equation

$$\dot{\phi} = h/r^2. \tag{28}$$

The solutions of these equations are

$$r = \exp\left[\frac{\sqrt{(1-h^2)}}{h}(\phi - \phi_0)\right] = [2(t - t_0)\sqrt{(1-h^2)}]^{1/2}. \tag{29}$$

In the (r, ϕ)-plane the solution represents an equiangular spiral, a case that Newton has already considered in Proposition IX.

Some observations

In the opening paragraph of this section, I stated that misprints (that have survived all three editions of the *Principia* to the present day) add to the difficulty of deciphering this corollary. Consider, for example, the following quotation from the text:

> therefore if the conic section *VRS* be a hyperbola, the body will descend to the centre; but if it be an ellipse, it will ascend continually, and go farther and farther off *in infinitum*. And, on the contrary,

On the face of it, one would conclude that the words 'hyperbola' and 'ellipse' (underlined) have been interchanged by a simple oversight. Certainly, an orbit which is a hyperbola in the (r, t)-plane ascends to infinity while ϕ tends to a finite limit, while an orbit which is an ellipse on the (r, t)-plane descends to the centre in a spiral orbit. How does it happen that this oversight escaped the 'critical eyes' of Roger Cotes? What is more puzzling is that the accompanying illustrations, which are the same in the first and the second editions of the *Principia*, were 'tampered' with (by Pemberton(?)) who apparently sensed that something was amiss): not only in interchanging the order of the illustrations but substantively in details as well (which appear to derive from a misunderstanding of the character of the solution).

I do not, however, have any doubts that Newton himself had a complete grasp of the solution. How else can one explain the remarkable identities (equations (20) and (21)) that he discovered and which bear the stamp of the 'lion's paw'!

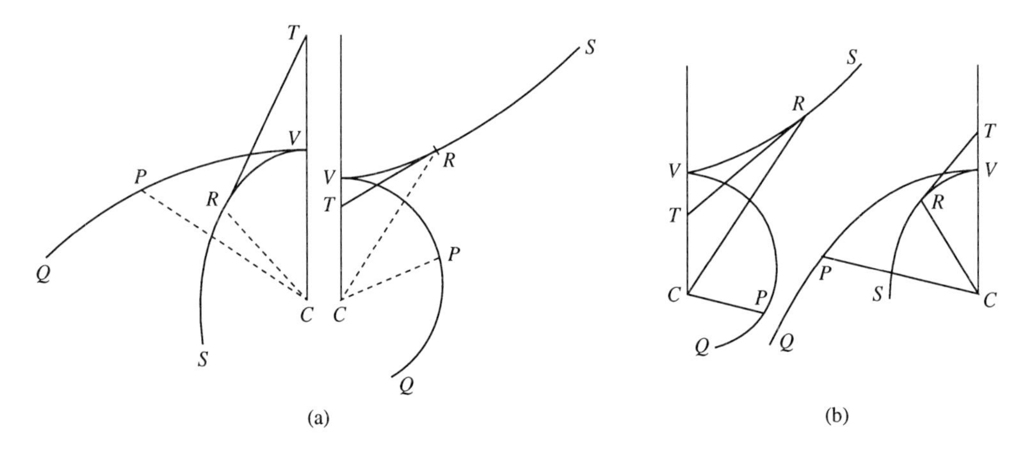

(a) The illustrations as they appeared in the first and the second editions of the *Principia*; (b) as they appeared in the third edition.

51.

Proposition XLII. Problem XXIX

The law of centripetal force being given, it is required to find the motion of a body setting out from a given place, with a given velocity, in the direction of a given right line.

This is Newton's formulation of the initial-value problem. Stated differently, the problem is: at a given instant of time, a body is projected from a point P, at a distance r_0 from the centre S of a known centripetal force of attraction, with a prescribed velocity V in some specified direction; and we are required to find the motion that will ensue. Newton

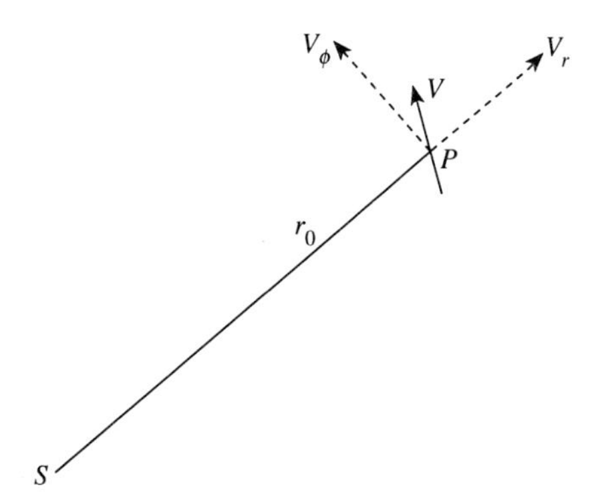

constructs the solution to this problem analogously to his solution of the simpler problem of ascending and descending bodies in Corollaries II and III of Proposition XXXIX. The following is a paraphrase of Newton's solution.

The solution to the problem requires only the specification of the lower bound of the integral in equation (3) of §48:

$$\dot{r}^2 = -2\int^r f(r)\,\mathrm{d}r - h^2/r^2, \tag{1}$$

consistently with the stated initial conditions. The conditions are clearly satisfied by writing

$$\frac{1}{2}\left(\dot{r}^2 + \frac{h^2}{r^2}\right) = -\int_{r_0}^{r} f(r)\,\mathrm{d}r + \tfrac{1}{2}(V_r^2 + V_\phi^2), \tag{2}$$

where V_r and V_ϕ are the radial and the transverse components of V. On the other hand, since the constant of areas h during the subsequent motion of the body must retain the value it had at the beginning,

$$h = r_0 V_\phi. \tag{3}$$

With this value for h, equation (2) gives

$$\dot{r}^2 = -2\int_{r_0}^{r} f(r)\,\mathrm{d}r - \frac{r_0^2}{r^2}V_\phi^2 + V^2. \tag{4}$$

The solution for the initial-value problem follows from inserting this expression (4) in place of (1) in equations (6) and (7) of §48. Thus,

$$t = \pm\int^r \frac{\mathrm{d}r}{[-2\int_{r_0}^{r} f(r)\,\mathrm{d}r + V^2 - r_0^2 V_\phi^2/r^2]^{1/2}} \tag{5}$$

and

$$\phi = \pm\int^r \frac{\mathrm{d}r}{r^2[-2\int_{r_0}^{r} f(r)\,\mathrm{d}r + V^2 - r_0^2 V_\phi^2/r^2]^{1/2}}. \tag{6}$$

Q.E.I.

10

On revolving orbits

52. Introduction

Newton introduces the subject of this Section IX at the conclusion of Section VIII thus:

> I have hitherto considered the motions of bodies in immovable orbits. It remains now to add something concerning their motions in orbits which revolve round the centres of force.

The problem he considers is the following. We know that under a centripetal law of attractive force a body is constrained to move in a fixed plane; and that it describes in this plane equal areas in equal times. Let $r = f(\phi)$ be such an orbit described under a centripetal force $P(r)$. Consider the locus $r = f(\alpha\phi)$ where α is some prescribed constant. *The revolving orbit* so constructed will again describe equal areas in equal times (as Newton shows in Proposition XLIII); and, therefore, by Proposition II, is described under the action of a centripetal force; and the problem is: how is this centripetal force related to the centripetal force acting on the original 'immovable' orbit? The relation that Newton derives in Proposition XLIV is called 'Newton's theorem of revolving orbits'. (Strangely, I have not been able to discover a treatment of this theorem in any current book on dynamics or celestial mechanics with the exception of E. T. Whittaker's *Analytical Dynamics*, Cambridge University Press, p. 83; and it is not generally known either.)

Newton's interest in this theorem was that it enabled him to investigate the apsidal motion of nearly circular orbits (which is the major concern of Proposition XLV with its three examples and two corollaries). And we are introduced at the same time to what was to be his life-long concern with the lunar theory.

53. The theorem of revolving orbits

As an introduction to Propositions XLIII–XLV we shall present a treatment of Newton's theorem by the methods of today. Analytically, the problem that was formulated in §52 is the following.

Given that $r(\phi)$ is the polar equation of an orbit described under the action of a centripetal force $P(r)$ with a constant of areas h, what is the centripetal force $P'(r)$ under which the orbit $r(\alpha\phi)$ with a constant of areas αh would be described where α is some assigned constant?

We recall that the equation governing $u(\phi) = 1/r(\phi)$ is (Chapter 5, §24, equation (ii))

$$\frac{\mathrm{d}^2 u}{\mathrm{d}\phi^2} + u = \frac{P(r)}{h^2 u^2}, \tag{1}$$

or

$$P(r) = h^2 u^3 + h^2 u^2 \frac{\mathrm{d}^2 u}{\mathrm{d}\phi^2}. \tag{2}$$

We now require the centripetal force $P'(r)$ under which the orbit

$$r(\alpha\phi) = r(\phi'), \quad \text{say}, \tag{3}$$

will be described. The constant of areas associated with this 'revolving orbit' is

$$h' = \alpha h. \tag{4}$$

Applying equation (3), for the orbit $u(\alpha\phi) = u(\phi')$, we have,

$$P'(r) = h'^2 u^2 \left(u + \frac{\mathrm{d}^2 u}{\mathrm{d}\phi'^2} \right) = \alpha^2 h^2 u^2 \left(u + \frac{1}{\alpha^2} \frac{\mathrm{d}^2 u}{\mathrm{d}\phi^2} \right)$$

$$= \alpha^2 h^2 u^3 + h^2 u^2 \frac{\mathrm{d}^2 u}{\mathrm{d}\phi^2}; \tag{5}$$

or again using equation (2), now with respect to the orbit $u(\phi)$, we find:

$$P'(r) = \alpha^2 h^2 u^3 - h^2 u^3 + P(r). \tag{6}$$

Hence

$$P'(r) - P(r) = \frac{h^2(\alpha^2 - 1)}{r^3}. \tag{7}$$

Q.E.I.

We now turn to Newton's demonstrations. As usual they provide far greater insights.

54.

Proposition XLIII. Problem XXX

It is required to make a body move in a curve that revolves about the centre of force in the same manner as another body in the same curve at rest.

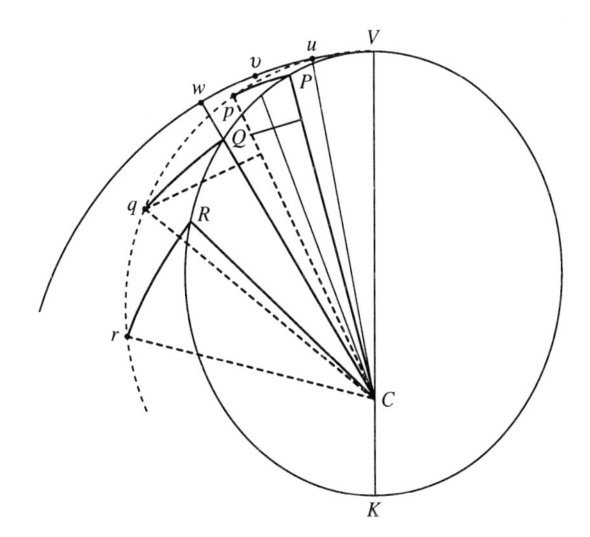

We shall first paraphrase Newton.

Let $VPQR \ldots K$ be the fixed orbit described by a body under an attractive force centred at C and P, Q, R, etc., the positions occupied by the body at certain arbitrarily chosen instants of time. The 'revolving orbit' is obtained by imaging the points, P, Q, R, etc. on to the fixed orbit at the same chosen instants of time by the transformation,

$$Cp = CP \quad \text{and} \quad \angle VCp = \alpha \angle VCP;$$

$$Cq = CQ \quad \text{and} \quad \angle VCq = \alpha \angle VCQ, \quad \text{etc.,} \tag{1}$$

where α is a constant. By this transformation,

$$\begin{aligned} \text{area of sector } VCp &= \tfrac{1}{2}(Cp)^2 \times \angle VCP = \tfrac{1}{2}\alpha(CP)^2 \times \angle VCP \\ &= \alpha \times \text{area of sector } VCP; \end{aligned} \tag{2*}$$

and since both the sectors are described in the same time,

$$\frac{\text{area of sector } VCp}{\text{time in which the area is swept}} = \alpha \frac{\text{area of sector } VCP}{\text{time in which the area is swept}} = \tfrac{1}{2}\alpha h, \tag{3}$$

where h is the constant of areas along the fixed orbit. Since relation (3) obtains for *all* the chosen sectors, it follows that along the revolving orbit, as imaged, equal areas are swept

* The validity of equation (2) becomes clear by observing that for any two infinitesimally close pair of points (P, Q) and (p, q),

$$\text{area of sector } QCP = \tfrac{1}{2}(CP)^2 \, \mathrm{d}\phi$$

and

$$\text{area of sector } qCp = \tfrac{1}{2}\alpha(Cp)^2 \, \mathrm{d}\phi;$$

and because $CP = Cp$ (by construction)

$$\text{area of sector } QCP : \text{area of sector } qCp = \alpha.$$

Since this constant ratio obtains for every infinitesimally close pair of points, it obtains generally. This is effectively Newton's argument.

in equal times; and the constant of areas is αh. And therefore, by Proposition II, the revolving orbit is described under the action of a centripetal force.

We may note the following relations which are implied by the transformation described:

$$\angle VCp = \alpha \angle VCP; \qquad \angle VCq = \alpha \angle VCQ; \qquad \text{(i)},$$

$$\angle PCp = \angle VCu; \qquad \angle QCq = \angle VCv; \qquad \text{(ii)},$$

$$\left.\begin{aligned} \angle PCp &= \angle VCp - \angle VCP = (\alpha - 1) \angle VCP; \\ \angle QCq &= \angle VCq - \angle VCQ = (\alpha - 1) \angle VCQ; \end{aligned}\right\} \qquad \text{(iii)},$$

and

$$\angle PCQ = \angle VCQ - \angle VCP = \frac{1}{\alpha}[\angle VCq - \angle VCp] = \frac{1}{\alpha} \angle pCq,$$

or,

$$\angle pCq = \alpha \angle PCQ. \qquad \text{(iv).} \quad (4)$$

We now give Newton's version in his meticulously phrased language:

> In the fixed orbit *VPK*, let the body *P* revolve, proceeding from *V* towards *K*. From the centre *C* let there be continually drawn *Cp*, equal to *CP*, making the angle *VCp* proportional to the angle *VCP*; and the area which the line *Cp* describes will be to the area *VCP*, which the line *CP* describes at the same time, as the velocity of the describing line *Cp* to the velocity of the describing line *CP*; that is, as the angle *VCp* to the angle *VCP*, therefore in a given ratio, and therefore proportional to the time. Since, then, the area described by the line *Cp* in a fixed plane is proportional to the time, it is manifest that a body, being acted upon by a suitable centripetal force, may revolve with the point *p* in the curved line which the same point *p*, by the method just now explained, may be made to describe in a fixed plane. Make the angle *VCu* equal to the angle *PCp*, and the line *Cu* equal to *CV*, and the figure *uCp* equal to the figure *VCP*, and the body being always in the point *p*, will move in the perimeter of the revolving figure *uCp*, and will describe its (revolving) arc *up* in the same time that the

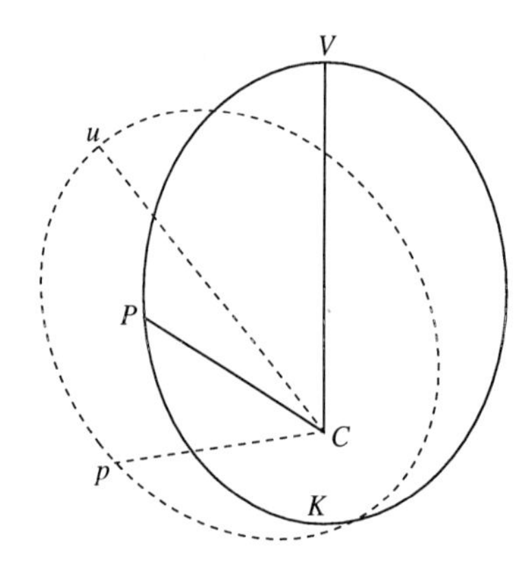

other body P describes the similar and equal arc VP in the fixed figure VPK. Find, then, by Cor. v, Prop. vi, the centripetal force by which the body may be made to revolve in the curved line which the point p describes in a fixed plane, and the Problem will be solved. Q.E.F.

55.

Proposition XLIV. Theorem XIV

The difference of the forces, by which two bodies may be made to move equally, one in a fixed, the other in the same orbit revolving, varies inversely as the cube of their common altitudes.

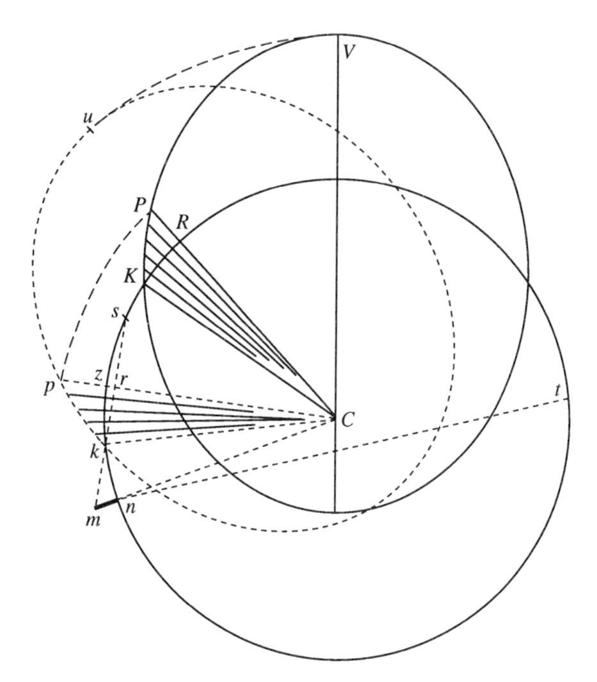

The accompanying diagram, similar to the ones included in Proposition XLIII, is in the same context: VPK is the fixed orbit; the vertex V and the position P in it being imaged on u and p in the congruent revolving orbit upk. By equations (1) and (4, ii) of the preceding section,

$$Cu = CV, \qquad Cp = CP, \qquad \text{and} \qquad \angle VCu = \angle PCp. \tag{1}$$

Let K be a neighbouring point, infinitely close to P; and if n is its image

$$Cn = CK; \tag{2}$$

and by equation (4, iv) of §54,

$$\angle pCn = \alpha \angle PCK. \tag{3}$$

Let k be the intersection of the circle of radius $CK(=Cn)$ with the orbit passing through u and p congruent to VPK. By this construction, the sectors CPK and Cpk are congruent:

$$\triangleleft CPK \equiv \triangleleft Cpk. \tag{4}$$

Draw kr perpendicular to Cp and extend it to intersect the circle $CKkn$ at s; then

$$kr = rs. \tag{5}$$

On the fixed orbit, P arrives at K (after an interval of time Δt, say). In the direction normal to the orbit at P, it will have traversed a distance equal to the perpendicular distance of K from PC, that is, equal to kr (to the first order) by the congruence of the figures CPK and Cpk. The imaged point, p, on the 'revolving orbit' will arrive at m, the continuation of kr, where by equation (3)

$$rm = \alpha rk. \tag{6}$$

If mn be now continued to intersect the circle Kkn at t, then by elementary geometry

$$ms\,.\,mk = mn\,.\,mt, \tag{7}$$

or

$$mn = \frac{ms\,.\,mk}{mt}. \tag{8}$$

Clearly, mn is the distance which p travels during Δt by the *difference* in the centripetal forces acting on p and P. It remains to relate this distance to other known quantities.

By the area theorem,

$$\triangle pCk = \triangle PCK = \tfrac{1}{2}rk\,.\,PC = \tfrac{1}{2}h\Delta t, \tag{9}$$

where h is the constant of areas along the fixed orbit; and by equation (6),

$$\triangle pCm = \tfrac{1}{2}rm\,.\,PC = \tfrac{1}{2}\alpha rk\,.\,PC = \tfrac{1}{2}\alpha h\Delta t. \tag{10}$$

Now by equations (5), (6), (9), and (10)

$$ms = rm + rs = rm + rk = (\alpha + 1)rk = \frac{h(\alpha + 1)}{PC}\Delta t, \tag{11}$$

$$mk = rm - rs = rm - rk = (\alpha - 1)rk = \frac{h(\alpha - 1)}{PC}\Delta t. \tag{12}$$

Hence,

$$ms\,.\,mk = \frac{h^2(\alpha^2 - 1)}{PC^2}(\Delta t)^2, \tag{13}$$

or, since

$$mt \simeq 2PC, \tag{14}$$

it follows from equation (8)

$$mn = \frac{1}{2}\frac{h^2(\alpha^2 - 1)}{PC^3}(\Delta t)^2. \tag{15}$$

We conclude from the meaning of mn already given,

the difference in the centripetal forces acting on p and P, at the same distance from C, is $h^2(\alpha^2 - 1)/PC^3$ (16)

Q.E.D.

COR. I. The difference of the forces, with which the body P revolves in a fixed orbit, and the body p in a movable orbit, will be to the centripetal force, with which another body by a radius drawn to the centre can uniformly describe that sector in the same time as the area VPC is described, as $GG - FF$ to FF.

Abbreviating centripetal force to C.F., consider

$$\frac{(\text{C.F. at } p \text{ in revolving orbit} - \text{C.F. at } P \text{ in fixed orbit}) \times (\Delta t)^2}{(\text{C.F. in circular orbit of radius } PC = pc)(\Delta t)^2}$$

$$= \frac{2mn}{(v^2/PC) \times (\Delta t)^2} = \frac{2mn}{(kr)^2/PC} \quad \text{(By the meaning of } mn\text{)}$$

$$= \frac{mk\,.\,ms}{PC} : \frac{(kr)^2}{PC} \quad \text{(By equation (8) and } mt \simeq 2PC\text{)}$$

$$= mk\,.\,ms : (kr)^2 = \alpha^2 - 1 \quad \text{(By equations (11) and (12))}. \tag{17}$$

Letting

$$\frac{G}{F} = \frac{h'}{h} = \alpha, \tag{18}$$

where h' and h denote the constant of areas in the revolving and in the fixed orbits, respectively, we conclude:

$$\frac{\text{C.F. at } p \text{ in revolving orbit} - \text{C.F. at } P \text{ in fixed orbit}}{\text{C.F. in circular orbit of radius } pc(= PC)} = \frac{G^2 - F^2}{F^2}. \tag{19}$$

Corollary II

What is sought in this corollary is the explicit expression for the centripetal force acting on a body describing the revolving orbit when the fixed orbit is an ellipse.

By equation (16) of this proposition,

C.F. at a distance $pC(=PC = A$, say) in the revolving orbit

$=$C.F. at a distance $PC(=pC = A)$ along the fixed elliptic orbit $+ (G^2 - F^2)/A^3$, (20)

where (as in Corollary I) $G = h\alpha = h'$ and $F = h$. But by Proposition XI, equation (h)

$$\text{C.F. at } P \text{ along the fixed elliptic orbit} = h^2/lA^2 = F^2/RA^2 \tag{21}$$

since in our present notation $l = R =$ semilatus rectum and $h = F$. Hence,

$$\begin{array}{c}\text{C.F. at a distance } A \text{ from the centre}\\ \text{of attraction along the revolving orbit}\end{array} = \frac{1}{R}\left[\frac{F^2}{A^2} + \frac{R(G^2 - F^2)}{A^3}\right]. \tag{22}$$

This is Newton's result. (Newton seems to have absorbed the factor $1/R$ in the constant of proportionality.)

Corollary II includes a special case of equation (22) derived directly from equation (19) of Corollary I.

Quite generally,

$$\text{Force acting at a point } P \text{ in the direction of the inward normal} = v^2/\rho, \tag{23}$$

where v is the velocity in the tangential direction and ρ is the radius of curvature of the orbit at P. Applying this result at V where the inward normal is directed towards the centre,

$$\begin{aligned}\frac{\text{C.F. at } V \text{ along a circular orbit of radius } CV}{\text{C.F. at } V \text{ along the fixed elliptic orbit}} &= \frac{\text{radius of curvature of an ellipse at the vertex } V}{\text{radius of curvature of circular orbit of radius } CV}\\ &= R/CV,\end{aligned} \tag{24}$$

for equal velocities, v, along both orbits. Equation (19) now gives

$$\begin{aligned}\text{C.F. at } V \text{ along the revolving orbit}&\\ &= \text{C.F. at } V \text{ along fixed elliptic orbit} \times \left[1 + \frac{R}{CV}\,\frac{G^2 - F^2}{F^2}\right]\\ &= \frac{F^2}{R(CV)^2}\left[1 + \frac{R}{CV}\,\frac{G^2 - F^2}{F^2}\right] \qquad \text{(by equation (21))}\\ &= \frac{1}{R}\left[\frac{F^2}{(CV)^2} + \frac{R}{(CV)^3}(G^2 - F^2)\right].\end{aligned} \tag{25}$$

Newton derives equation (22) by generalizing this last result by noting that 'this difference at any other altitude A is to itself at the altitude CV as A^{-3} to $(CV)^{-3}$'.

Corollary III

In this corollary, the explicit form of equation (16) is sought when the fixed orbit is an ellipse described about its centre in accordance with Proposition X.

By equation (viii) of Proposition X (p. 88)

$$\frac{QR}{QT^2 \,.\, PC^2} = \frac{PC}{2a^2b^2} \tag{26}$$

But,

$$\left.\begin{aligned} QR &= \tfrac{1}{2}\,\text{C.F.}\,(\Delta t)^2 \\ QT \,.\, PC &= 2A_{,t}\Delta t = h\Delta t. \end{aligned}\right\} \tag{27}$$

Hence,

$$\text{C.F. acting at } P \text{ in fixed orbit} = \frac{h^2}{a^2b^2}\, PC. \tag{28}$$

Since

$$R = \text{semilatus rectum} = b^2/a, \tag{29}$$

and $h = F$ in our present notation,

$$\text{C.F. acting at } P \text{ in fixed orbit} = h^2\left(\frac{1}{R}\,\frac{b^2}{a}\right)\frac{PC}{a^2b^2} = \frac{F^2PC}{Ra^3}. \tag{30}$$

Equation (20) now gives

$$\begin{aligned} \text{C.F. acting at the same distance in a revolving orbit} &= \frac{F^2PC}{Ra^3} + \frac{G^2 - F^2}{PC^3} \\ &= \frac{1}{R}\left[\frac{F^2PC}{a^3} + R\,\frac{G^2 - F^2}{PC^3}\right]; \end{aligned} \tag{31}$$

or, in Newton's notation,

$$PC = A \qquad \text{and} \qquad a = T, \tag{32}$$

we have

$$\text{C.F. at altitude } A \text{ in the revolving orbit} = \frac{1}{R}\left[\frac{F^2A}{T^3} + R\,\frac{G^2 - F^2}{A^3}\right]. \tag{33}$$

Note that by definition the fixed and the revolving orbits both describe a complete revolution in the same time.

Corollary IV

Consider 'universally' equation (16). By equation (viii) of Proposition IX,

$$\text{C.F. at } P \text{ in fixed orbit} = \frac{1}{\rho \sin^3 \epsilon}\,\frac{F^2}{A^2}. \tag{34}$$

We thus obtain,

C.F. at the same distance from C in the revolving orbit

$$= \frac{1}{\rho \sin^3 \epsilon}\left[\frac{F^2}{A^2} + (\rho \sin^3 \epsilon)\,\frac{G^2 - F^2}{A^3}\right], \tag{35}$$

where it may be recalled that ρ denotes the radius of curvature at P and ϵ is the inclination of PC to the direction of motion at P. Since, along an elliptic orbit $\rho \sin^3 \epsilon = R$ (by equation (2) of §31), equation (35) reduces to equation (22) in this case.

There is some confusion in terminology in the statement of this corollary in the *Principia*; it seems to have escaped the scrutiny of Roger Cotes!

> COR. V. Therefore the motion of a body in a fixed orbit being given, its angular motion round the centre of the forces may be increased or diminished in a given ratio; and thence new fixed orbits may be found in which bodies may revolve with new centripetal forces.

This corollary establishes for the first time the notion of *centrifugal potential.*

Corollary VI

Let VX be perpendicular to CV; and consider a point P moving along VX with a uniform velocity, v (say). Since P moves by inertia only, it is not subject to any other force acting on it. Hence the image point p in the revolving orbit (at a distance $Cp = CP$) will describe this orbit under a centripetal force inversely proportional to $(pC)^3$:

$$\text{C.F. acting on } p \text{ at a distance } pC = h^2(\alpha^2 - 1)/(pC)^3, \tag{36}$$

where the constant of areas along the linear trajectory is

$$h = vVC. \tag{37}$$

Hence,

$$\text{C.F. acting on } p \text{ at a distance } r \text{ is} = (vVC)^2(\alpha^2 - 1)/r^3. \tag{38}$$

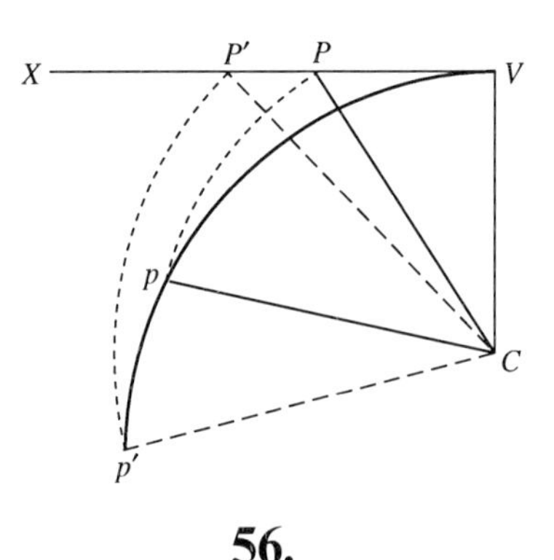

56.

Proposition XLV. Problem XXXI

To find the motion of the apsides in orbits approaching very near to circles.

The problem that Newton considers in this proposition—which was probably the underlying motivation for this entire Section IX at this early stage—is to relate nearly circular orbits described under arbitrary centripetal forces to nearly circular *elliptical* orbits described under inverse-square force; and to determine the precession of the line of apsides under the circumstances. As it will appear, fairly delicate considerations are involved in the solution of this problem; and it is not an easy exercise to follow Newton's account written in connected prose. For this reason, we shall paraphrase Newton. But we shall quote him when his explanations in their lucidity add to our insight.

By Proposition XLIV (cf. equations (7), §52 and (16), §55)

$$P'(r) = P(r) + \frac{h'^2 - h^2}{r^3}. \tag{1}$$

For an elliptical orbit described under an inverse-square law of attraction (cf. equation (21), §55)

$$P(r) = \frac{h^2}{Rr^2}, \tag{2}$$

where R, the semilatus rectum of an ellipse of eccentricity e and semiaxes a and b, is

$$R = b^2/a = a(1 - e^2). \tag{3}$$

Equation (1) now gives

$$P'(r) = \frac{1}{R}\left[\frac{h^2}{r^2} + \frac{R(h'^2 - h^2)}{r^3}\right] = \frac{1}{R}\left[\frac{R(h'^2 - h^2) + rh^2}{r^3}\right]. \tag{4}$$

For a nearly circular orbit, we may write,

$$\left.\begin{aligned} r = r_{\max} - X = T - X \text{ (say)},\\ \text{where}\qquad\qquad\\ r_{\max} = a(1 + e) = T \quad (= CV \text{ of the diagrams in §54 and §55}) \end{aligned}\right\} \tag{5}$$

We therefore have

$$RP'(r) = \frac{1}{r^3}[R(h'^2 - h^2) + (T - X)h^2]. \tag{6}$$

Let the centripetal force, under the action of which the revolving orbit is described, be

$$RP'(r) = C(r)r^{-3} = C(T - X)r^{-3}, \tag{7}$$

where $C(r)$ is at our disposal. Since we are interested in nearly circular orbits, we may expand $C(T - X)$ in a Taylor series and retain only the zero and the first-order terms. We thus obtain

$$RP'(r) = [C(T) - XC'(T)]r^{-3}, \qquad [C'(T) = \mathrm{d}C/\mathrm{d}T]. \tag{8}$$

Equation (6) now gives,

$$R(h'^2 - h^2) + (T - X)h^2 = C(T) - XC'(T), \tag{9}$$

or alternatively,

$$\frac{1}{C(T)}[R(h'^2 - h^2) + Th^2] - \frac{Xh^2}{C(T)} = 1 - \frac{XC'(T)}{C(T)}. \tag{10}$$

The first terms on both sides of this equation are of zero order while the second terms are of the first order. We may therefore equate them separately. We thus obtain,

$$\frac{1}{C(T)}[R(h'^2 - h^2) + Th^2] = 1 \qquad \text{and} \qquad h^2 = C'(T). \tag{11}$$

Since R and T to zero-order are both equal to the semimajor axis a, we find,

$$\frac{Th'^2}{C(T)} = 1 \qquad \text{and} \qquad h^2 = C'(T). \tag{12}$$

We thus obtain

$$\frac{h'^2}{h^2} = \frac{C(T)}{TC'(T)}, \tag{13}$$

where on the right-hand there is no loss of generality in evaluating it for $T = 1$; thus

$$\frac{h'^2}{h^2} = \alpha^2 = \left[\frac{C(T)}{TC'(T)}\right]_{T=1}. \tag{14}$$

This is Newton's basic relation; as he explains:

> And this comes to pass by reason of the likeness of this orbit which a body acted upon by [a uniform centripetal force $C(r)$ describes] and of that orbit which a body performing its circuits in a revolving ellipse will describe in a fixed plane. By this collation of the terms, these orbits are made similar; not universally, indeed, but then only when they approach very near to a circular figure.

It may be recalled that α is the constant ratio of the angles VCp and VCP in the illustrations in §§54 and 55 and p is the image on the revolving orbit of the point P on a fixed orbit at the same distance.

Newton considers several special cases of equation (14).

Example 1

A revolving orbit described under a uniform centripetal force corresponds to (cf. equation (7))

$$C(T) = T^3. \tag{15}$$

By equation (14)

$$\alpha^2 = \tfrac{1}{3} \qquad \text{or} \qquad \angle VCp = \frac{1}{\sqrt{3}} \angle VCP. \tag{16}$$

To continue with Newton's explanation:

> Therefore since the body, in a fixed ellipse, in descending from the upper to the lower apse, describes an angle, if I may so speak, of 180°, the other body in a

movable ellipse, and therefore in the fixed plane we are treating of, will in its descent from the upper to the lower apse, describe an angle VCp of $180°/\sqrt{3}$. ...... A body, therefore, revolving with a uniform centripetal force in an orbit nearly circular, will always describe an angle of $180°/\sqrt{3}$, or 103° 55′ 23″ at the centre; moving from the upper apse to the lower apse when it has once described that angle, and thence returning to the upper apse when it has described that angle again; and so on *in infinitum.*

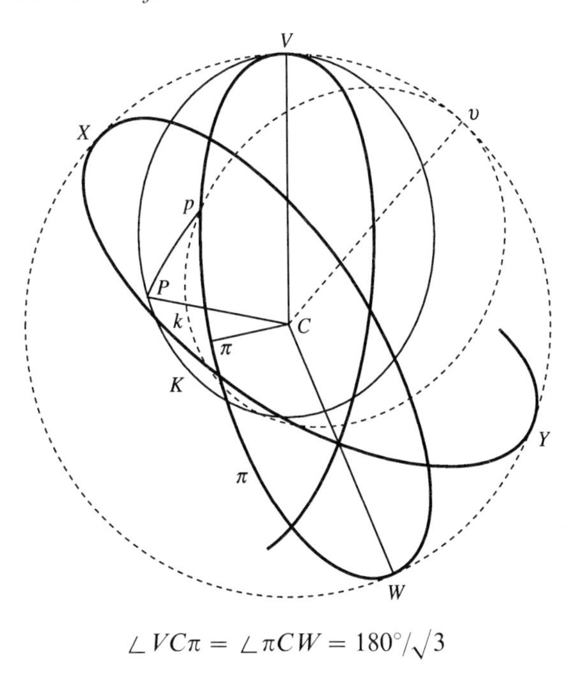

$\angle VC\pi = \angle \pi CW = 180°/\sqrt{3}$

Example 2

More generally, if we consider a law of force proportional to r^{n-3}, which corresponds to

$$C(T) = T^n, \tag{17}$$

we obtain from equation (14),

$$\alpha^2 = \frac{1}{n} \qquad \text{and} \qquad \angle VCp = \frac{1}{\sqrt{n}} \angle VCP. \tag{18}$$

And Newton explains this result in an almost identical phrasing.

Therefore since the angle VCP, described in the descent of the body from the upper apse to the lower apse in an ellipse, is of 180°, the angle VCp, described in the descent of the body from the upper apse to the lower apse in an orbit nearly circular which a body describes with a centripetal force proportional to the power A^{n-3}, will be equal to an angle of $180°/\sqrt{n}$, and this angle being repeated, the body will return from the lower to the upper apse; and so on *in infinitum*.

In particular:

n	Law of force	$180°/\sqrt{n}$
0	r^{-3}	∞
$\frac{1}{4}$	$r^{-11/4}$	360°
1	r^{-2}	180°
2	r^{-1}	127° 16′ 45″
3	constant	103° 55′ 23″
4	r	90°

Newton picks out for special comment the cases $n = 2$ and $n = \frac{1}{4}$. The origin of the 'singularity' for the inverse-cube law is considered in Corollary I below.

Example 3

The case,

$$C(T) = bT^m \pm cT^n, \tag{19}$$

considered in this example plays a crucial role in Newton's later considerations in formulating his universal law of gravitation (see Proposition III, Book III).

For the chosen form of $C(T)$, equation (14) gives

$$\alpha^2 = \left[\frac{C(T)}{TC'(T)}\right]_{T=1} = \left[\frac{bT^m \pm cT^n}{mbT^m \pm ncT^n}\right]_{T=1} = \frac{b \pm c}{mb \pm nc}. \tag{20}$$

> And therefore, since the angle VCP between the upper and the lower apse, in a fixed ellipse, is 180°, the angle VCp between the same apsides in an orbit which a body describes with a centripetal force, that is, as $(bA^m \pm cA^n)/A^3$, will be equal to an angle of $180°/\sqrt{(b \pm c)/(mb \pm nc)}$.

And Newton concludes:

> After the same manner the Problem is solved in more difficult cases.
>
> COR. I. Hence if the centripetal force be as any power of the altitude, that power may be found from the motion of the apsides; and conversely.

If a body describing the revolving orbit returns to the same apse m times when the fixed orbit completes n complete revolutions, then

$$\angle VCp = \frac{m}{n} \angle VCP \quad \text{or} \quad \alpha = \frac{m}{n}. \tag{21}$$

By equations (17) and (18),

$$C(A) = A^{n^2/m^2}; \tag{22}$$

and the centripetal force follows the power law,

$$\text{C.F.} \propto A^{(n^2/m^2)-3}. \tag{23}$$

Conversely, if

$$\text{C.F.} \propto A^q, \tag{24}$$

then

$$\frac{m}{n} = \frac{1}{\sqrt{(q+3)}} \tag{25}$$

We conclude that

$$q > -3. \tag{26}$$

> Hence it is plain that the force in its recess from the centre cannot decrease in a greater than a cubed ratio of the altitude.

Newton explains this limit on the inverse power of the law of centripetal attraction with utmost clarity:

> A body revolving with such a force, and parting from the apse, if it once begins to descend, can never arrive at the lower apse or least altitude, but will descend to the centre, describing the curved line treated of in Cor. III, Prop. XLI. But if it should, at its parting from the lower apse, begin to ascend ever so little, it will ascend *in infinitum*, and never come to the upper apse; but will describe the curved line spoken of in the same Cor., and Cor. VI, Prop. XLIV. So that where the force in its recess from the centre decreases in a greater than a cubed ratio of the altitude, the body at its parting from the apse, will either descend to the centre, or ascend *in infinitum*, according as it descends or ascends at the beginning of its motion. But if the force in its recess from the centre either decreases in a less than a cubed ratio of the altitude, or increases in any ratio of the altitude whatsoever, the body will never descend to the centre, but will at some time arrive at the lower apse; and, on the contrary, if the body alternatively ascending and descending from one apse to another never comes to the centre, then either the force increases in the recess from the centre, or it decreases in a less than a cubed ratio of the altitude; and the sooner the body returns from one apse to another, the farther is the ratio of the forces from the cubed ratio.

It will be noted that Newton here refers to both Proposition XLI, Corollary III (in which he gives a full account of motion under an inverse-cube law of force) and Proposition XLIV, Corollary VI. And it will also be recalled that the inverse-cube law is the first one that Newton considers in Proposition IX even before the inverse-square law in the more celebrated Proposition XI. It looks very much as if Newton was aware from the outset of the crucial role that the inverse-cube law was to play in future considerations! Examples of such prescience abound throughout the *Principia.* A further example follows immediately.

The last example that is considered in this corollary relates to his constant concern with the lunar theory (as we shall presently see).

Consider a revolving body that describes 363° in returning to the same apse while the body in the fixed orbit completes one revolution so that

$$\alpha = m:n = 363^\circ : 360^\circ = 121:120. \tag{27}$$

The index of the corresponding power law is

$$q = (n/m)^2 - 3 = -29\,523:14\,641 \simeq -2\tfrac{4}{243}. \tag{28}$$

Newton quotes this result in Proposition III, Book III (Chapter 19, p. 356):

> ... from the very slow motion of the Moon's apogee; which in every single revolution amounting but to 3° 3′ forwards, ... it appears, that, if the distance of the Moon from the Earth's centre is to the semidiameter of the Earth as D to 1, the force, from which such a motion will result, is inversely as $D^{2+(4/243)}$. i.e., inversely as the power of D, whose exponent is $2\frac{4}{243}$; that is to say, in the proportion of the distance somewhat greater than the inverse square, but which comes $59\frac{3}{4}$ times nearer to the proportion according to the square than to the cube.
>
> COR. II. Hence also if a body, urged by a centripetal force which is inversely as the square of the altitude, revolves in an ellipse whose focus is in the centre of the forces; and a new and foreign force should be added to or subtracted from this centripetal force, the motion of the apsides arising from that foreign force may (by the third Example) be known; and conversely.

In other words, the law of centripetal force considered is:

$$\text{C.F.} = \frac{1}{r^2} - cr, \tag{29}$$

where c is some assigned constant. By the table of values listed in Example 2, the law (29) corresponds to

$$C(T) = T - cT.^4 \tag{30}$$

This is a special case of equation (19) for

$$b = 1, \qquad m = 1, \qquad \text{and} \qquad n = 4. \tag{31}$$

Therefore, by equation (20),

$$\alpha^2 = (1 - c)/(1 - 4c); \tag{32}$$

and we conclude that the angle of revolution between the apsides is

$$180^\circ\sqrt{\frac{1-c}{1-4c}}. \tag{33}$$

The application of this result Newton considers is:

> Suppose that foreign force to be 357·45 times less than the other force with which the body revolves in the ellipse; that is, c to be $\frac{100}{35\,745}$, A or T being equal to 1; and then $180^\circ/\surd(1-c)/(1-4c)$ will be $180^\circ+\sqrt{\frac{35\,645}{35\,345}}$ or $180^\circ\!\cdot\!7623$, that is, $180^\circ\,45'\,44''$. Therefore the body, parting from the upper apse, will arrive at the lower apse with an angular motion of $180^\circ\,45'\,44''$, and this angular motion being repeated, will return to the upper apse; and therefore the upper apse in each revolution will go forward $1^\circ\,31'\,28''$. The apse of the Moon is about twice as swift.

It will be noticed that the rotation of the line of apsides of 3° assumed in the example considered in Corollary I (equation (26)) is 'twice as swift' as the value, $1^\circ\,31'\,28''$, derived in this example!

As always, Newton looks to the future in concluding this Section IX:

> So much for the motion of bodies in orbits whose planes pass through the centre of force. It now remains to determine those motions in eccentric planes.

❖11❖

A pause

57. Introduction

By the end of Section IX, we again reach a watershed in the development of the *Principia*; and again, as at the end of Section III, we have a pause; but this time a shorter one addressed to '... "oscillating pendulous motions of bodies"'. There is much of great interest in this section particularly in those parts dealing with cycloidal pendulums. But we shall pass them by as they are not in the mainstream of the development. But Propositions XLVI and XLVII are exceptions: Proposition XLVI provides an important generalization of Proposition XLII on the initial-value problem; and Proposition XLVII is a beautiful theorem which is at the base of another beautiful Proposition LXIV established in Section XI (see Chapter 12).

58.

Proposition XLVI. Problem XXXII

> *Any kind of centripetal force being supposed, and the centre of force, and any plane whatsoever in which the body revolves, being given, and the quadratures of curvilinear figures being allowed; it is required to determine the motion of a body going off from a given place with a given velocity, in the direction of a given right line in that plane.*

In contrast to Propositions XLII and XLIII, the centre of attraction S is out of the plane (X, Y) in which the motion of P is to take place.

Draw SC perpendicular to the plane (X, Y). The centripetal force at a point Q,

$$-f(r)\,\vec{r}/r \quad \text{(say)},$$

acts in the direction $\overrightarrow{QS}$ and depends only on the distance QS $(=r)$. Resolve this force

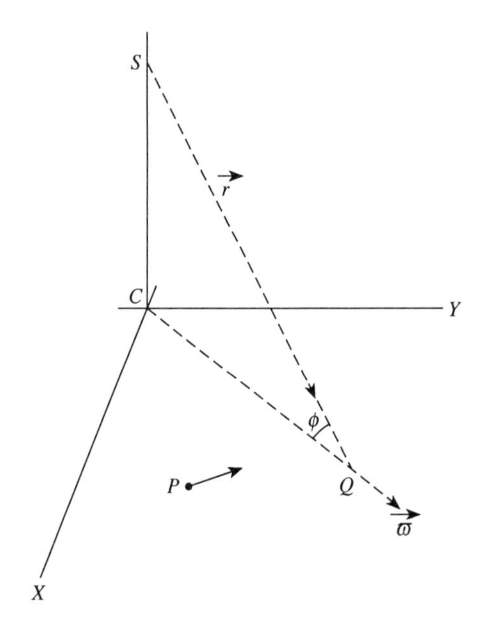

along $\overrightarrow{CQ}$ in the plane (X, Y) and in the direction $\overrightarrow{CS}$ normal to the plane (X, Y). The latter force, of magnitude

$$-f(r)\sin\phi,$$

will clearly not affect any motion confined to the plane (X, Y); only the force in the direction $\overrightarrow{CQ}$ will affect the motion. This force in the direction $\overrightarrow{CQ}$ is

$$-f(r)\cos\phi\,\frac{\vec{\varpi}}{\varpi} = -f(\sqrt{(SC^2+\varpi^2)})\,\frac{\varpi}{\sqrt{(SC^2+\varpi^2)}}\,\frac{\vec{\varpi}}{\varpi},$$

where ϖ denotes the radial distance, CQ, in the plane (X, Y) and $\vec{\varpi}/\varpi$ is a unit vector in the radial direction in the same plane. And since SC is a constant, the force

$$-f(\sqrt{(SC^2+\varpi^2)})\,\frac{\varpi}{\sqrt{(SC^2+\varpi^2)}},$$

is a function only of the radial distance in the plane (XY). The conclusion as Newton states is

> But the action of the other force CQ, coinciding with the position of the plane itself, attracts the body directly towards the given point C in that plane; and therefore causes the body to move in the plane in the same manner as if the force SQ were taken away, and the body were to revolve in free space about the centre C by means of the force CQ alone.

Therefore, Proposition XLII applies to the determination of the motion of P in the plane (X, Y) given its initial position and velocity in that plane. Q.E.I.

Proposition XLVII. Theorem XV

Supposing the centripetal force to be proportional to the distance of the body from the centre; all bodies revolving in any planes whatsoever will describe ellipses, and complete their revolutions in equal times; and those which move in right lines, running backwards and forwards alternately, will complete their several periods of going and returning in the same times.

If the force acting towards S is

$$-k\vec{r} = -kr\frac{\vec{r}}{r} \quad [f(r) = kr],$$

where k is a constant, the component of this force directed towards C in the plane (X, Y) normal to $\overrightarrow{CS}$ is

$$-f(\sqrt{(SC^2 + \varpi^2)})\frac{\varpi}{\sqrt{(SC^2 + \varpi^2)}}\frac{\vec{\varpi}}{\varpi} = -k\vec{\varpi}$$

and this is a centripetal force proportional to the distance in the same plane.

Therefore the forces with which bodies found in the plane (X, Y) are attracted towards the point C, are in proportion to the distances equal to the forces with which the same bodies are attracted every way towards the centre S; and therefore the bodies will move in the same times, and in the same figures, in any plane (X, Y) about the point C, as they would do in free spaces about the centre S; and therefore (by Cor. II, Prop. X, and Cor. II, Prop. XXXVIII) they will in equal times either describe ellipses in that plane about the centre C, or move to and fro in right lines passing through the centre C in that plane; completing the same periods of time in all cases. Q.E.D.

12

The two-body problem

59. Introduction

As we have stated in the introduction to the last chapter (§57), a watershed was reached in the development of the *Principia* at the end of Section IX. In resuming the main development, after a brief pause in Section X, Newton writes:

> I have hitherto been treating of the attractions of bodies towards an immovable centre; though very probably there is no such thing existent in nature. For attractions are made towards bodies, and the actions of the bodies attracted and attracting are always reciprocal and equal, by Law III; so that if there are two bodies, neither the attracted nor the attracting body is truly at rest, but both (by COR. IV of the Laws of Motion), being as it were mutually attracted, revolve about a common centre of gravity. And if there be more bodies, which either are attracted by one body, which is attracted by them again, or which all attract each other mutually, these bodies will be so moved among themselves, that their common centre of gravity will either be at rest, or move uniformly forwards in a right line. I shall therefore at present go on to treat of the motion of bodies attracting each other; considering the centripetal forces as attractions; though perhaps in a physical strictness they may more truly be called impulses. But these propositions are to be considered as purely mathematical; and therefore, laying aside all physical considerations, I make use of a familiar way of speaking, to make myself the more easily understood by a mathematical reader.

(The parts that reflect Newton's lofty approach to science are underlined.)

The first seven Propositions, LVII–LXIII of Section XI, are devoted to the formulation and solution of the two-body problem. Proposition LXIV gives the exact solution for the general problem of n bodies mutually attracting each other with a force proportional to their distances from one another. The remaining Propositions, LXV–LXIX, deal with the problem of the perturbation of the relative Keplerian motion of two bodies by the gravitational effect of a third body. Most noteworthy, however, is Proposition LXVI with its 22 corollaries occupying 15 pages and more than one-half of the entire section: in effect a condensed first monograph on planetary perturbation. Of these Propositions LXV–LXIX, F. Tisserand (*Traité de Mécanique Céleste*, Tome III, Chapitre 3, p. 33) has written:

> I am inclined to think that he [Newton] knew all the formulae (a) [the equations generally known as *Lagrange's planetary equations*] but that, instead of publishing them, he preferred deducing a large number of geometric propositions from them which he obtained by considering in each case the effect of just one component [element].

Tisserand devotes an entire chapter in his *Traité de Mécanique Céleste* (Chapitre III, pp. 27–43) under the heading 'Theorie de la lune de Newton' and gives an excellent account of Propositions LXVI–LXIX. And Laplace, who also devotes a chapter to 'Sur la theorie lunaire de Newton' in his *Mécanique Céleste* (Livre XVI, Chapitre II, pp. 409–23) considers:

> The method [Newton's development of the method of the variation of the elements] appears to me as one of the most remarkable things in the *Principia.*

On account of the closely knit character of Newton's exposition of the 22 corollaries of Proposition LXVI, concealing all details of his treatment of the variation of the elements, we shall postpone to Chapters 13 and 14 Propositions LXV–LXIX. The present chapter will be restricted to Propositions LVII–LXIV.

60. The two-body problem: the general theorems

Of the seven Propositions, LVII–LXIII, devoted to the two-body problem, the first three treat the general case with no restriction on the law of attraction between the two bodies.

*Proposition LVII. Theorem XX**

Two bodies attracting each other mutually describe similar figures about their common centre of gravity, and about each other mutually.

* In Cajori's translation of this proposition, 'aequali motu angulari circum' in the original Latin of the first and the second editions (and 'equable angular motion' and 'equal angular motion' in Motte's translation) is incorrectly interpreted in two places as 'uniform angular motion'—Keplerian motion along an ellipse is clearly not one of 'uniform angular motion'.

First, a modern (common?) version of the proof:

If M_1 and M_2 denote the inertial masses of the two bodies, the attractive force, directed from one to the other, must be of the form

$$f(M_1, M_2; r_{12}) \qquad (f > 0),$$

where f is symmetrical in M_1 and M_2 and r_{12} $(=|\vec{r}_1 - \vec{r}_2|)$ is the distance between them. The symmetry of f in M_1 and M_2 is required by the third Law of motion. The equations governing the motions of M_1 and M_2 are:

$$M_1 \frac{\mathrm{d}^2\vec{r}_1}{\mathrm{d}t^2} = -f(M_1, M_2; r_{12}) \frac{\vec{r}_1 - \vec{r}_2}{r_{12}} \tag{1}$$

and

$$M_2 \frac{\mathrm{d}^2\vec{r}_2}{\mathrm{d}t^2} = +f(M_1, M_2; r_{12}) \frac{\vec{r}_1 - \vec{r}_2}{r_{12}}. \tag{2}$$

From these equations the uniform motion in a straight line of the centre of gravity,

$$\vec{C} = \frac{M_1}{M_1 + M_2} \vec{r}_1 + \frac{M_2}{M_1 + M_2} \vec{r}_2 \tag{3}$$

follows directly from the sum of equations (1) and (2):

$$M_1 \frac{\mathrm{d}^2\vec{r}_1}{\mathrm{d}t^2} + M_2 \frac{\mathrm{d}^2\vec{r}_2}{\mathrm{d}t^2} = 0, \tag{4}$$

or

$$(M_1 + M_2)\vec{C} = M_1\vec{r}_1 + M_2\vec{r}_2 = (M_1 + M_2)\vec{V}t + \vec{A}, \tag{5}$$

where $\vec{V}$ and $\vec{A}$ are constant vectors.

The motion of the two bodies 'round one another' (to adopt Newton's manner of speaking) is determined by the equation

$$\frac{\mathrm{d}^2}{\mathrm{d}t^2}(\vec{r}_1 - \vec{r}_2) = -\frac{M_1 + M_2}{M_1 M_2} f(M_1, M_2; r_{12}) \frac{\vec{r}_1 - \vec{r}_2}{r_{12}}. \tag{6}$$

And the motion of the two bodies relative to their centre of gravity is obtained by considering

$$\vec{\xi}_1 = \vec{r}_1 - \frac{1}{M_1 + M_2}(M_1\vec{r}_1 + M_2\vec{r}_2) = +\frac{M_2}{M_1 + M_2}(\vec{r}_1 - \vec{r}_2) \tag{7}$$

and

$$\vec{\xi}_2 = \vec{r}_2 - \frac{1}{M_1 + M_2}(M_1\vec{r}_1 + M_2\vec{r}_2) = -\frac{M_1}{M_1 + M_2}(\vec{r}_1 - \vec{r}_2). \tag{8}$$

Therefore,

$$\frac{M_1 + M_2}{M_2}\vec{\xi}_1 = -\frac{M_1 + M_2}{M_1}\vec{\xi}_2 = (\vec{r}_1 - \vec{r}_2). \tag{9}$$

From this proportionality of the vectors $\vec{\xi}_1, \vec{\xi}_2$, and $(\vec{r}_1 - \vec{r}_2)$ the stated proposition follows.

Newton's proof, couched in words, is essentially the same:

The centre of gravity moves uniformly in a straight line (by Corollary IV of the Laws of Motion). Therefore, the two bodies will always find themselves, in opposite directions, along the straight line at opposite ends and dividing the line $(\vec{r}_1 - \vec{r}_2)$, joining them in the constant ratio $M_2 : M_1$. Hence $\vec{\xi}_1$, $\vec{\xi}_2$, and $(\vec{r}_1 - \vec{r}_2)$ are collinear; and moreover,

$$\frac{|\vec{\xi}_1|}{M_2} = \frac{|\vec{\xi}_2|}{M_1} = \frac{r_{12}}{M_1 + M_2}. \tag{10}$$

And Newton concludes

> But right lines that are in a given ratio to each other, and are carried round their extremities with an equal angular motion, describe upon planes, which either rest together with them, or are moved with any motion not angular, figures entirely similar round those extremities. Therefore the figures described by the revolution of these distances are similar. Q.E.D.

Proposition LVIII. Theorem XXI

If two bodies attract each other with forces of any kind, and revolve about the common centre of gravity: I say, that, by the same forces, there may be described round either body unmoved a figure similar and equal to the figures which the bodies so moving describe round each other.

Consider two bodies P (planet) and S (Sun) of masses M_1 and M_2 revolving about their common centre of gravity C. Then in accordance with equation (6) the equation governing the motion of M_1 round M_2 is given by

$$M_1 \frac{d^2}{dt^2}(\vec{r}_1 - \vec{r}_2) = -\frac{M_1 + M_2}{M_2} f(M_1, M_2; r_{12}) \frac{\vec{r}_1 - \vec{r}_2}{r_{12}}. \tag{11}$$

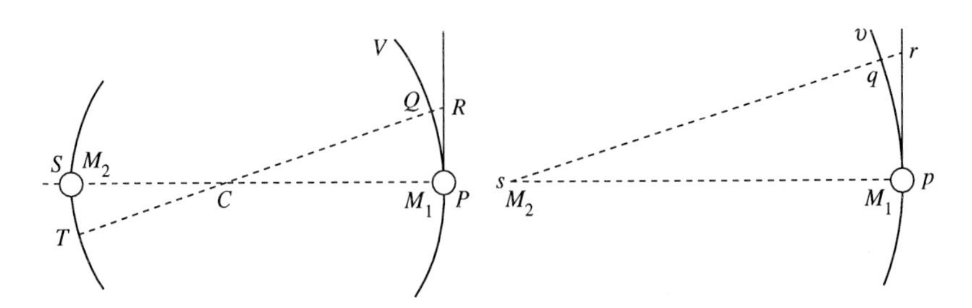

On the other hand, the motion of p (of mass M_1), under the (centripetal) attraction of S (of mass M_2) assumed to be immovable, is governed by

$$M_1 \frac{\mathrm{d}^2\vec{S}}{\mathrm{d}t^2} = -f(M_1, M_2; |\vec{S}|)\frac{\vec{S}}{|\vec{S}|}, \tag{12}$$

where $\vec{S}$ denotes the radius vector joining s and p. A comparison of equations (11) and (12) shows that the orbit of P round S, under their mutual attraction, may be formally considered as described under a law of centripetal attraction towards S which is the same as when S is immovable, but with a force enhanced by the factor $(M_1 + M_2)/M_2 = (S + P)/S$.

It follows that the orbits, similar and congruent for the two cases, exist. Q.E.D.

Newton's proof, again couched in words, is essentially the same; and is based on the observation (deduced by an examination of the diagram) that the acceleration of P towards S, in the revolving orbit, is the same as the acceleration of p towards the immovable S enhanced by the factor $(S + P)/S$.

Three corollaries follow which require a knowledge of the acceleration experienced by P (or S) towards the centre of gravity C in the revolving orbit (which Newton takes for granted, as apparently obvious!).

From equations (6), (7), and (9) it follows:

$$M_1 \frac{\mathrm{d}^2\vec{\xi}_1}{\mathrm{d}t^2} = \frac{M_1 M_2}{M_1 + M_2}\frac{\mathrm{d}^2}{\mathrm{d}t^2}(\vec{r}_1 - \vec{r}_2) = -f\left(M_1, M_2; \frac{M_1 + M_2}{M_2}|\vec{\xi}_1|\right)\frac{\vec{\xi}_1}{|\vec{\xi}_1|}. \tag{13}$$

Now writing

$$\vec{\xi}_1 = \frac{M_2}{M_1 + M_2}\vec{X} \quad \text{and} \quad \tau = t\left(\frac{M_1 + M_2}{M_2}\right)^{1/2} \tag{14}$$

in equation (13), we obtain

$$M_1 \frac{\mathrm{d}^2\vec{X}}{\mathrm{d}\tau^2} = -f(M_1, M_2; |\vec{X}|)\frac{\vec{X}}{|\vec{X}|}, \tag{15}$$

which is identical in form with equation (12). We conclude:

$$\vec{S}(t) \equiv \vec{X}(\tau) = \frac{M_1 + M_2}{M_2}\vec{\xi}_1\left(t\sqrt{\frac{M_2}{M_1 + M_2}}\right). \tag{16}$$

Hence, the orbit described by M_1 under the centripetal attraction of M_2, considered immovable, is similar to the orbit described by M_1, about the common centre of gravity C under their mutual attraction.

> COR I. Hence two bodies attracting each other with forces proportional to their distance, describe (by Prop. X), both round their common centre of gravity, and round each other, concentric ellipses; and, conversely, if such figures are described, the forces are proportional to the distances.

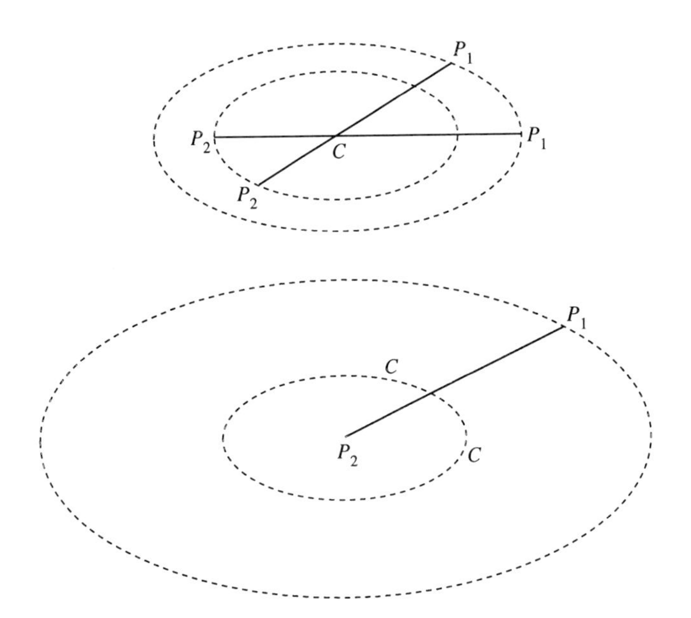

COR II. And two bodies, whose forces are inversely proportional to the square of their distance, describe (by Prop. XI, XII, XIII), both round their common centre of gravity, and round each other, conic sections having their focus in the centre about which the figures are described. And, conversely, if such figures are described, the centripetal forces are inversely proportional to the square of the distance.

$$CP_1 = a_1(1 + e),$$

$$CP_2 = a_2(1 + e),$$

$$CP'_1 = a_1(1 - e),$$

$$CP'_2 = a_2(1 - e);$$

$$P_2P_1 = (a_1 + a_2)(1 + e),$$

$$P'_1P'_2 = (a_1 + a_2)(1 - e),$$

$$CP_1 : CP_2 = M_2 : M_1,$$

$$P'_1P_1 = 2(a_1 + a_2).$$

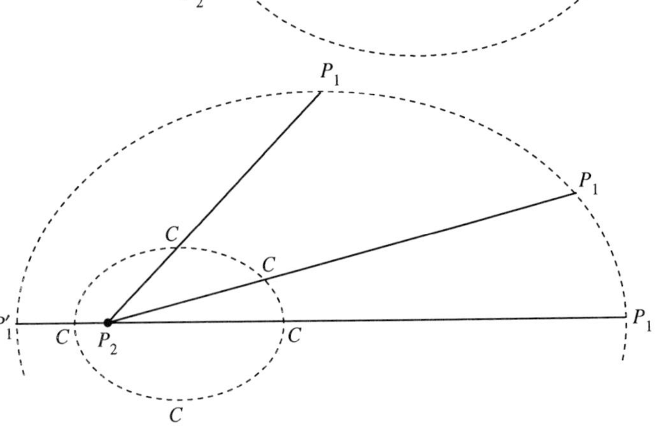

COR. III. Any two bodies revolving round their common centre of gravity describe areas proportional to the times, by radii drawn both to that centre and to each other.

Proposition LIX. Theorem XXII

The periodic time of two bodies S and P revolving round their common centre of gravity C, is to the periodic time of one of the bodies P revolving round the other S remaining fixed, and describing a figure similar and equal to those which the bodies describe about each other, as $\sqrt{S}$ is to $\sqrt{(S+P)}$.

By writing

$$t = \left(\frac{M_2}{M_1 + M_2}\right)^{1/2} \tau \qquad \text{and} \qquad \vec{R} = \vec{r}_1 - \vec{r}_2 \tag{17}$$

equation (6) becomes

$$M_1 \frac{\mathrm{d}^2\vec{R}}{\mathrm{d}\tau^2} = -f(M_1, M_2; |\vec{R}|)\frac{\vec{R}}{|\vec{R}|}, \tag{18}$$

which is identical in form with equation (12),

$$M_1 \frac{\mathrm{d}^2\vec{S}}{\mathrm{d}t^2} = -f(M_1, M_2; |\vec{S}|)\frac{\vec{S}}{|\vec{S}|}, \tag{19}$$

governing the motion of M_1 about an immovable mass M_2. Therefore, by comparing equations (6) and (18), we conclude:

$$\frac{[\text{periodic time of } M_1 \text{ revolving round } M_2 \text{ under their mutual attraction}]}{[\text{periodic time of } M_1 \text{ about an immovable } M_2 \text{ in a congruent orbit}]}$$

$$= \frac{t}{\tau} = \sqrt{[M_2/(M_1 + M_2)]} = \sqrt{S} : \sqrt{(S+P)}. \quad \text{Q.E.D.}$$

Proposition LX. Theorem XXIII

If two bodies S and P, attracting each other with forces inversely proportional to the square of their distance, revolve about their common centre of gravity: I say, that the principal axis of the ellipse which either of the bodies, as P describes by this motion about the other S, will be to the principal axis of the ellipse, which the same body P may describe in the same periodic time about the other body S fixed, as the sum of the two bodies S + P to the first of two mean proportionals between that sum and the other body S.*

* x and y are said to be the mean proportionals between a and b if

$$\frac{a}{x} = \frac{x}{y} = \frac{y}{b} \qquad \text{or} \qquad x^2 = ay \quad \text{and} \quad y^2 = bx.$$

Alternatively,

$$x^3 b = y^3 a \qquad \text{or} \qquad x : y = a^{1/3} : b^{1/3}.$$

For a law of attraction inversely proportional to the square of the distance

$$f(M_1, M_2; r_{12}) = \frac{M_1 M_2}{r_{12}^2}, \tag{20}$$

where, as hitherto, the factor G has been suppressed. For f of the foregoing form, equations (11) and (12) become

$$\frac{d^2\vec{R}}{dt^2} = -(M_1 + M_2)\frac{\vec{R}}{|\vec{R}|^3}, \tag{21}$$

and

$$\frac{d^2\vec{S}}{dt^2} = -M_2\frac{\vec{S}}{|\vec{S}|^3}. \tag{22}$$

By Proposition XV, the periodic times for elliptical orbits, of semiaxes, a_R and a_S, respectively, are given by

$$T_R = 2\pi\frac{a_R^{3/2}}{(M_1 + M_2)^{1/2}} \quad \text{and} \quad T_S = 2\pi\frac{a_S^{3/2}}{M_2^{1/2}}. \tag{23}$$

Hence,

$$\text{if} \quad T_R = T_S, \quad \left(\frac{a_R^3}{M_1 + M_2}\right)^{1/2} = \left(\frac{a_S^3}{M_2}\right)^{1/2} \quad \text{or} \quad a_R : a_S = \sqrt[3]{(M_1 + M_2)} : \sqrt[3]{M_2} \tag{24}$$

while

$$\text{if} \quad a_R = a_S, \quad T_R : T_S = \sqrt{M_2} : \sqrt{(M_1 + M_2)}. \tag{25}$$

Proposition LXI. Theorem XXIV

If two bodies attracting each other with any kind of forces, and not otherwise agitated or obstructed, are moved in any manner whatsoever, those motions will be the same as if they did not at all attract each other, but were both attracted with the same forces by a third body placed in their common centre of gravity; and the law of the attracting forces will be the same in respect of the distance of the bodies from the common centre, as in respect of the distance between the two bodies.

The first part of the proposition is a direct consequence of equations

$$M_1\frac{d^2\vec{\xi}_1}{dt^2} = -f\left(M_1, M_2; \frac{M_1 + M_2}{M_2}|\vec{\xi}_1|\right)\frac{\vec{\xi}_1}{|\vec{\xi}_1|}, \tag{26a}$$

and

$$M_2\frac{d^2\vec{\xi}_2}{dt^2} = -f\left(M_1, M_2; \frac{M_1 + M_2}{M_1}|\vec{\xi}_2|\right)\frac{\vec{\xi}_2}{|\vec{\xi}_2|}. \tag{26b}$$

The second part of the proposition follows from the foregoing equations, equation (12) (or (18)), and the further fact that the law of attraction, $f(M_1, M_2; r_{12})$, between M_1 and M_2, is unrestricted except for the requirement of symmetry in M_1 and M_2. Q.E.D.

We may parenthetically note that if the law of attraction is inversely proportional to the square of the distance, equations (26) become

$$\frac{\mathrm{d}^2\vec{\xi}_1}{\mathrm{d}t^2} = -M_2\left(\frac{M_2}{M_1 + M_2}\right)^2 \frac{\vec{\xi}_1}{|\vec{\xi}_1|^3}. \tag{27}$$

In other words, the motion of M_1, around the centre of gravity C, is formally the same as if at C there was an immovable mass,

$$M_2\left(\frac{M_2}{M_1 + M_2}\right)^2. \tag{28}$$

and M_2 was absent.

Moreover, by writing,

$$\vec{\xi}_1 = \frac{M_2}{M_1 + M_2}\vec{\eta} \quad \text{in equation (26a)} \tag{29}$$

and

$$\vec{\xi}_2 = \frac{M_1}{M_1 + M_2}\vec{\eta} \quad \text{in equation (26b)} \tag{30}$$

both equations take the form,

$$\frac{\mathrm{d}^2\vec{\eta}}{\mathrm{d}t^2} = -\frac{M_1 + M_2}{M_1 M_2} f(M_1, M_2; |\vec{\eta}|) \frac{\vec{\eta}}{|\vec{\eta}|}. \tag{31}$$

It follows that M_1 and M_2 describe similar orbits and

$$|\vec{\xi}_1| : |\vec{\xi}_2| = M_2 : M_1. \tag{32}$$

All the foregoing remarks are implicit in Newton's comments.

61. Initial-value problems

Proposition LXII. Problem XXXVIII

To determine the motions of two bodies which attract each other with forces inversely proportional to the squares of the distance between them, and are let fall from given places.

In Proposition XXXVI (Problem XXV), the solution to the problem of rectilinear descent of a body (initially at rest), under the attraction of an immovable centre with a force inversely proportional to the square of their distance, was given. In this proposition, it is shown how that solution can be adapted to the motion of two bodies, initially at rest, attracting each other with the same inverse-square law of force. Newton presents this adaptation of the solution with clarity and precision.

> The bodies, by the last theorem, will be moved in the same manner as if they were attracted by a third placed in the common centre of their gravity; and by

> the hypothesis that centre will be fixed at the beginning of their motion, and therefore (by Cor. iv of the Laws of Motion) will be always fixed. The motions of the bodies are therefore to be determined (by Prob. xxv) in the same manner as if they were impelled by forces tending to that centre; and then we shall have the motions of the bodies attracting each other. Q.E.I.

The reference to Corollary IV of the Laws of Motion is to remind the reader that 'the common centre of gravity of two or more bodies does not alter its state of motion or rest by the actions of the bodies among themsleves'.

Proposition LXIII. Problem XXXIX

> *To determine the motions of two bodies attracting each other with forces inversely proportional to the squares of their distance, and going off from given places in given directions with given velocities.*

In Proposition XVII (p. 107), it was shown how the motion of a body, under the attraction of an immovable centre with a force inversely proportional to the square of their distance, that ensues an initially assigned position and velocity can be determined. In this proposition, it is shown how the same manner of determination can be adapted to the analogous initial-value problem for two mutually attracting bodies. Again, Newton's explanation is a model of compression and clarity.

> The motions of the bodies at the beginning being given, there is given also the uniform motion of the common centre of gravity, and the motion of the space which moves along with this centre uniformly in a right line, and also the very first, or beginning motions of the bodies in respect of this space. Then (by Cor. v of the Laws, and the last theorem) the subsequent motions will be performed in the same manner in that space, as if that space together with the common centre of gravity were at rest, and as if the bodies did not attract each other, but were attracted by a third body placed in that centre. The motion therefore in this movable space of each body going off from a given place, in a given direction, with a given velocity, and acted upon by a centripetal force tending to that centre, is to be determined by Prob. ix and xxvi, and at the same time will be obtained the motion of the other round the same centre. With this motion compound the uniform progressive motion of the entire system of the space and the bodies revolving in it, and there will be obtained the absolute motion of the bodies in immovable space. Q.E.I.

It may be useful to quote 'Cor. v of the Laws':

> *The motions of bodies included in a given space are the same among themselves, whether that space is at rest, or moves uniformly forwards in a right line without any circular motion.*

Two observations may be relevant:

(1) the statements of Corollaries IV and V of the Laws are exactly right in the context; and
(2) Newton's extraordinary insistence on the importance of initial-value problems.

62. The solution of a many-body problem

It is characteristic of Newton's style that before proceeding directly to the perturbation of the Kepler motions of two bodies about one another by a third body, he should consider the only case of the n-body problem that admits of an exact solution: the case of n bodies attracting each other by a force proportional to their mutual distances.

Proposition LXIV. Problem XL

Supposing forces with which bodies attract each other to increase in a simple ratio of their distances from the centres; it is required to find the motions of several bodies among themselves.

Consider then, N mass points, $M_1, M_2, \ldots, M_N$, the mass M_i being attracted by the mass M_j with a force

$$-k^2 M_i M_j (\vec{r}_i - \vec{r}_j) \qquad (i \neq j), \tag{1}$$

where k is some positive constant. (Newton prefers to write instead

$$-k_i k_j (\vec{r}_i - \vec{r}_j) \qquad (i \neq j), \tag{1'}$$

though he later reverts to the form (1) by identifying k_i with kM_i.)

The relevant equations of motion are:

$$M_i \frac{\mathrm{d}^2 \vec{r}_i}{\mathrm{d}t^2} = -k^2 \sum_{\substack{j=1 \\ j \neq i}}^{N} M_i M_j (\vec{r}_i - \vec{r}_j) \qquad (i = 1, \ldots, N). \tag{2}$$

These equations admit the integral,

$$\sum_{i=1}^{N} M_i \vec{r}_i = (M_1 + M_2 + \cdots + M_N)\vec{V}t + \vec{A}, \tag{3}$$

where $\vec{V}$ and $\vec{A}$ are constant vectors. In a frame of reference in which the centre of gravity is at rest at the origin,

$$\sum_{j=1}^{N} M_j \vec{r}_j = 0 \qquad (\vec{V} = 0 \text{ and } \vec{A} = 0). \tag{4}$$

The equations of motion in this frame are

$$\frac{d^2\vec{r}_i}{dt^2} = -k^2\left(\sum_{j\neq i}^{N} M_j\right)\vec{r}_i + k^2 \sum_{j\neq i}^{N} M_j\vec{r}_j \qquad (i = 1, \ldots, N)$$

$$= -k^2\left(\sum_{j\neq i}^{N} M_j\right)\vec{r}_i - k^2 M_i\vec{r}_i \qquad \text{(by equation (4)).} \tag{5}$$

Hence

$$\frac{d^2\vec{r}_i}{dt^2} = -k^2\left(\sum_{j=1}^{N} M_j\right)\vec{r}_i \tag{6}$$

It follows:

$$\vec{r}_i = \vec{A}_i\, e^{\pm i\sigma t} \quad (\vec{A}_i = \text{constant vector}), \tag{7}$$

where

$$\sigma = k\sqrt{\left[\sum_{j=1}^{N} M_j\right]} = k\sqrt{(\text{mass of the system})}. \tag{8}$$

The foregoing mechanical solution will not be confused as Newton's! Newton proceeds differently: the following is a paraphrase of his method of solution.

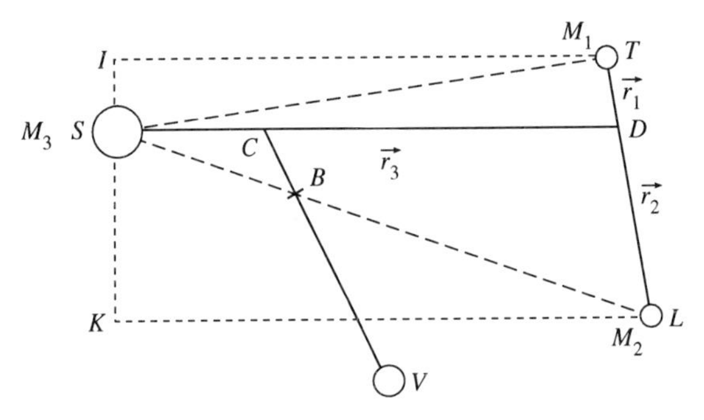

Consider first two bodies T and L of masses M_1 and M_2 attracting each other in accordance with the same basic postulate (1). The equations governing their motions are:

$$\frac{d^2\vec{r}_1}{dt^2} = -k^2 M_2(\vec{r}_1 - \vec{r}_2) \qquad \text{and} \qquad \frac{d^2\vec{r}_2}{dt^2} = -k^2 M_1(\vec{r}_2 - \vec{r}_1). \tag{9}$$

In the frame of reference in which the centre of gravity, D, of the two masses is at rest,

$$D = M_1\vec{r}_1 + M_2\vec{r}_2 = 0. \tag{10}$$

In this frame, the equations are:

$$\frac{d^2\vec{r}_1}{dt^2} = -k^2(M_1 + M_2)\vec{r}_1 \qquad \text{and} \qquad \frac{d^2\vec{r}_2}{dt^2} = -k^2(M_1 + M_2)\vec{r}_2. \tag{11}$$

The masses M_1 and M_2, therefore, describe elliptical orbits with the same period about D as centre.

Now let a third body S of mass M_3 be introduced which attracts T and L and in turn is attracted by them in accordance with the same postulate (1). By including the attraction by S, the motions of T and L (still in the same frame), will be governed by

$$\begin{aligned}\frac{d^2\vec{r}_1}{dt^2} &= -k^2(M_1 + M_2)\vec{r}_1 - k^2M_3(\vec{r}_1 - \vec{r}_3) = -k^2(M_1 + M_2 + M_3)\vec{r}_1 + k^2M_3\vec{r}_3\\ \frac{d^2\vec{r}_2}{dt^2} &= -k^2(M_1 + M_2)\vec{r}_2 - k^2M_3(\vec{r}_2 - \vec{r}_3) = -k^2(M_1 + M_2 + M_3)\vec{r}_2 + k^2M_3\vec{r}_3.\end{aligned} \tag{12}$$

Therefore M_1 and M_2, besides executing elliptic orbits about D more 'swiftly' than in the absence of M_3, are transported parallelly in the direction $\vec{r}_3$ with the same acceleration $k^2M_3\vec{r}_3$. By Corollary VI of the Laws of Motion, T and L

> *urged in the direction of parallel lines by equal accelerative forces, they will all continue to move among themselves, after the same manner as if they had not been urged by those forces.*

By considering next the motion of S, the equation governing it is:

$$\begin{aligned}\frac{d^2\vec{r}_3}{dt^2} &= -k^2M_1(\vec{r}_3 - \vec{r}_1) - k^2M_2(\vec{r}_3 - \vec{r}_2)\\ &= -k^2(M_1 + M_2)\vec{r}_3 + k^2(M_1\vec{r}_1 + M_2\vec{r}_2),\end{aligned} \tag{13}$$

or, by equation (10),

$$\frac{d^2\vec{r}_3}{dt^2} = -k^2(M_1 + M_2)\vec{r}_3 = -k^2(M_1 + M_2 + M_3)\vec{r}_3 + k^2M_3\vec{r}_3. \tag{14}$$

Now by changing the frame of reference from the one in which the centre of gravity of M_1 and M_2 is at rest to the one in which the centre of gravity of M_1, M_2, and M_3 is at rest, when

$$M_1\vec{r}_1 + M_2\vec{r}_2 + M_3\vec{r}_3 = 0 \tag{15}$$

that is, to a frame which is accelerated with respect to the frame (10) by $-k^2M_3\vec{r}_3$, the equations governing the motions of M_1, M_2, and M_3 take the common form,

$$\frac{d^2\vec{r}_i}{dt^2} = -k^2(M_1 + M_2 + M_3)\vec{r}_i \qquad (i = 1, 2, 3). \tag{16}$$

The solution is now completed by induction to obtain the general result (6). As Newton states:

> it will easily be concluded that all the bodies will describe different ellipses, with equal periodic times about their common centre of gravity B, in an immovable plane. Q.E.I.

The superiority of Newton's solution over the one given earlier is that it exhibits the physical basis for the unfolding of the solution. And one is left marvelling at Newton's ability to explain precisely in words involved analytical arguments.

❖13❖

The method of the variation of the elements of a Kepler orbit and Newton's lunar theory: An introduction to Propositions LXV–LXIX

63. Introduction

As we have stated in the introduction to Chapter 12, Tisserand was of the opinion that Newton had derived for himself the equations expressing the variations with time of the elements of the Kepler orbit under the action of an external force $\vec{F}$—elements that would otherwise have been constants; and that Proposition LXVI and its 22 corollaries provide the principal basis for this inference (see Chapter 14). To make Propositions LXV–LXIX more readily understandable, we shall give in this chapter, in the spirit of the *Principia*, an elementary self-contained account of the method of the variation of the elements and its application to the perturbations of the Moon.

The present chapter is in effect a continuation of Chapter 4 whose notations we shall follow.

64. The basic equations, definitions, and the coordinate system adopted

Suppose that at some point P, where the position $\vec{r}$ and the velocity $\vec{v} = \mathrm{d}\vec{r}/\mathrm{d}t$ are known, the perturbing force suddenly ceases. The body will then move in an ellipse defined by the initial values. This is the *osculating ellipse* at P: it is a Kepler orbit having at that point the same position and velocity as the true orbit. The orbital elements of the osculating ellipse will not remain constant but vary with time. The method we shall describe for determining the resulting variation with time is the *method of the variation of the elements.* But first, a few preliminaries setting up the framework of the problem.

Equations (3) of §18 and (2) of §19 in Chapter 4, generalized to include the action of the external force $\vec{F}$ are:

$$\frac{d\vec{r}}{dt} = \vec{v}; \tag{1}$$

and

$$\frac{d\vec{v}}{dt} = -\frac{\mu}{r^3}\vec{r} + \vec{F}. \tag{2}$$

We define the vectors $\vec{h}$ and $\vec{e}$ (no longer constants) by the equations:

$$\vec{h} = \vec{r} \times \vec{v} \tag{3}$$

$$\mu\left(\frac{\vec{r}}{r} + \vec{e}\right) = \vec{v} \times \vec{h} \qquad \text{and} \qquad \vec{e} \cdot \vec{v} = -\frac{1}{r}\vec{r} \cdot \vec{v}. \tag{4}$$

From the foregoing definitions and equations (1) and (2) it follows:

$$\frac{d}{dt}\frac{\vec{r}}{r} = \frac{\vec{h} \times \vec{r}}{r^3} \tag{5}$$

and

$$\frac{d\vec{h}}{dt} = \vec{r} \times \left(-\mu\frac{\vec{r}}{r^3} + \vec{F}\right) = \vec{r} \times \vec{F}. \tag{6}$$

Similarly, from equations (4) and (5), we find:

$$\begin{aligned}\mu\left(\frac{\vec{h} \times \vec{r}}{r^3} + \frac{d\vec{e}}{dt}\right) &= \frac{d\vec{v}}{dt} \times \vec{h} + \vec{v} \times \frac{d\vec{h}}{dt}\\ &= \left(-\mu\frac{\vec{r}}{r^3} + \vec{F}\right) \times \vec{h} + \vec{v} \times (\vec{r} \times \vec{F}),\end{aligned} \tag{7}$$

or

$$\mu\frac{d\vec{e}}{dt} = \vec{F} \times \vec{h} + \vec{v} \times (\vec{r} \times \vec{F}). \tag{8}$$

The coordinate system adopted

We choose the right-handed system of coordinates defined by the unit vectors (see adjoining figure)

$$\vec{1}_h, \quad \vec{1}_r, \quad \text{and} \quad \vec{1}_\alpha \tag{9}$$

along the directions of

$$\vec{h}, \quad \vec{r}, \quad \text{and} \quad \vec{\alpha} \tag{10}$$

of magnitudes,

$$|\vec{h}| = h, \quad |\vec{r}| = r, \quad \text{and} \quad |\vec{\alpha}| = hr. \tag{11}$$

Along these vectors, we have the relations

$$\vec{r} \times \vec{\alpha} = r^2\vec{h}, \quad \vec{\alpha} \times \vec{h} = h^2\vec{r}, \quad \text{and} \quad \vec{\alpha} \cdot \vec{r} = 0 \tag{12}$$

and, we shall resolve $\vec{F}$ along the unit vectors (9) with the components,

$$F_h, \quad F_r, \quad \text{and} \quad F_\alpha. \tag{13}$$

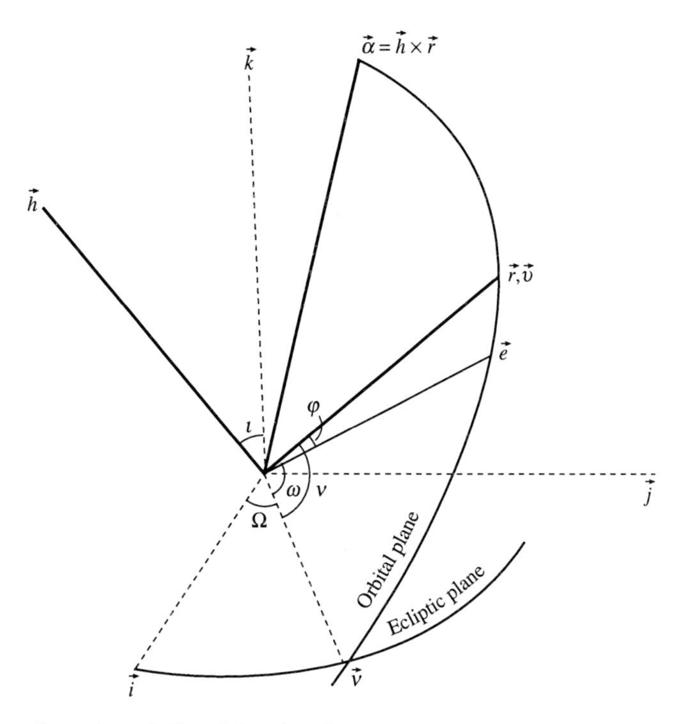

Other vectors and angles defined in the figure are:

$\vec{i}, \vec{j}, \vec{k}$: unit vectors along the axes of a rectangular system of coordinates defined with respect to the ecliptic plane;

$\vec{i}$: direction of the vernal equinox;

$\vec{\nu}$: direction of the ascending node;

$\vec{h}$: angular momentum vector;

$\vec{e}$: Lenz vector;

$\Omega = \angle(\vec{\nu}, \vec{i})$;

$\omega = \angle(\vec{\nu}, \vec{e})$; $\omega + \varphi = \nu$;

$\varphi = \angle(\vec{r}, \vec{e})$;

$\vec{g} = (\vec{h} \times \vec{\nu})$;

$\iota = \angle(\vec{h}, \vec{k})$ = inclination of the orbital plane to the ecliptic.

Let the velocity $\vec{v}$ resolved along the directions of $\vec{r}$ and $\vec{\alpha}$ be

$$\vec{v} = A\vec{r} + B\vec{\alpha}, \tag{14}$$

where A and B are to be determined. By this definition,

$$\vec{h} = \vec{r} \times \vec{v} = B(\vec{r} \times \vec{\alpha}) = Br^2\vec{h}, \tag{15}$$

and therefore,

$$B = r^{-2}. \tag{16}$$

From the disposition of the vectors in the figure, it further follows:

$$\vec{e} \cdot \vec{\alpha} = e\alpha \cos(90^\circ + \varphi) = -e\alpha \sin \varphi = -ehr \sin \varphi \qquad \text{[by equation (11)]}$$

and

$$\vec{e} \cdot \vec{r} = er \cos \varphi; \tag{17}$$

and therefore,

$$\begin{aligned} \vec{e} \cdot \vec{v} = \vec{e} \cdot (A\vec{r} + B\vec{\alpha}) &= Aer \cos \varphi - Behr \sin \varphi \\ &= -\frac{1}{r} \vec{r} \cdot \vec{v} = -Ar \qquad \text{(by equation (4));} \end{aligned} \tag{18}$$

or, with B having the value (16),

$$Ar^2 = -Aer^2 \cos \varphi + eh \sin \varphi. \tag{19}$$

In view of our dealing with an osculating ellipse we can write (cf. Chapter 4, §19, equation (11))

$$r = \frac{h^2}{\mu(1 + e \cos \varphi)}. \tag{20}$$

We thus obtain:

$$A = \frac{\mu e}{rh} \sin \varphi. \tag{21}$$

The solution for $\vec{v}$ in the chosen coordinate system is therefore

$$\vec{v} = \frac{\mu e}{rh} \vec{r} \sin \varphi + \frac{1}{r^2} \vec{\alpha}. \tag{22}$$

Finally, for future reference, we spell out in the accompanying table the relative orientations of the two coordinate systems, $(\vec{h}/h, \vec{r}/r, \vec{\alpha}/\alpha)$ and $(\vec{i}, \vec{j}, \vec{k})$ that we have introduced.

	$\vec{i}$	$\vec{j}$	$\vec{k}$
$\vec{h}/h$	$\sin \iota \sin \Omega$	$-\sin \iota \cos \Omega$	$\cos \iota$
$\vec{r}/r$	$\cos \Omega \cos v - \sin \Omega \sin v \cos \iota$	$\sin \Omega \cos v + \cos \Omega \sin v \cos \iota$	$\sin v \sin \iota$
$\vec{\alpha}/\alpha$	$-\cos \Omega \sin v - \sin \Omega \cos v \cos \iota$	$-\sin \Omega \sin v + \cos \Omega \cos v \cos \iota$	$\cos v \sin \iota$

$(v = \omega + \varphi)$

65. The variation of the elements

The elements of the Kepler orbit whose variations we have to find are:

(a) $\vec{h}$: the angular momentum vector;
(b) ι: the inclination of the orbital plane to the elliptic;
(c) Ω: the angle of the ascending node in the equatorial plane;
(d) $\vec{e}$: the Lenz vector;
(e) ω: the inclination of $\vec{e}$ to the direction of $\vec{v}$;
(f) a: the semimajor axis; and $n = 2\pi/\text{period}$.

A related problem is the effect of $\vec{F}$ on the equation of time, that is, on Kepler's equation.

As we have stated we shall resolve $\vec{F}$ along the directions of $\vec{h}$, $\vec{r}$, and $\vec{\alpha}$ (cf. equation (13) of §64):

$$\vec{F} = F_h \frac{\vec{h}}{h} + F_r \frac{\vec{r}}{r} + F_\alpha \frac{\vec{\alpha}}{\alpha}. \tag{1}$$

(*a*) *Variation of* $\vec{h}$

By equation (6), of §64,

$$\frac{d\vec{h}}{dt} = \vec{r} \times \vec{F} = \frac{\vec{h}}{h} rF_\alpha - \frac{\vec{\alpha}}{h} F_h. \tag{2}$$

Therefore,

$$\vec{h} \cdot \frac{d\vec{h}}{dt} = rhF_\alpha \qquad \text{or} \qquad \frac{dh}{dt} = rF_\alpha. \tag{3}$$

(*b*) *Variation of* ι

Since (see the table on p. 222)

$$\vec{h} \cdot \vec{k} = h \cos \iota, \tag{4}$$

$$\begin{aligned}\frac{d\vec{h}}{dt} \cdot \vec{k} &= \frac{dh}{dt} \cos \iota - h \sin \iota \frac{d\iota}{dt} \\ &= rF_\alpha \cos \iota - h \sin \iota \frac{d\iota}{dt} \qquad \text{(by equation (3))}.\end{aligned} \tag{5}$$

Also, by equation (2),

$$\frac{d\vec{h}}{dt} \cdot \vec{k} = \left(\frac{\vec{h}}{h} rF_\alpha - \frac{\vec{\alpha}}{h} F_h\right) \cdot \vec{k} = rF_\alpha \cos \iota - r \sin \iota \cos(\omega + \varphi) F_h. \tag{6}$$

Combining equations (5) and (6), we obtain

$$\frac{d\iota}{dt} = \frac{r}{h}\cos(\omega + \varphi)F_h. \tag{7}$$

(c) *Variation of* Ω

Multiplying equation (2) scalarly by $\vec{j}$, we have

$$\vec{j}\cdot\frac{d\vec{h}}{dt} = -rF_\alpha \sin\iota\cos\Omega + r[\sin(\omega+\varphi)\sin\Omega - \cos(\omega+\varphi)\cos\Omega\cos\iota]F_h$$

$$= -\frac{dh}{dt}\sin\iota\cos\Omega - h\cos\Omega\cos\iota\frac{d\iota}{dt} + rF_h\sin(\omega+\varphi)\sin\Omega. \tag{8}$$

On the other hand,

$$\vec{j}\cdot\vec{h} = -h\sin\iota\cos\Omega; \tag{9}$$

and since $\vec{j}$ is a constant vector,

$$\vec{j}\cdot\frac{d\vec{h}}{dt} = -\frac{dh}{dt}\sin\iota\cos\Omega - h\cos\iota\cos\Omega\frac{d\iota}{dt} + h\sin\iota\sin\Omega\frac{d\Omega}{dt}. \tag{10}$$

Combining equations (8) and (10), we obtain:

$$h\sin\iota\frac{d\Omega}{dt} = rF_h\sin(\omega+\varphi). \tag{11}$$

(d) *Variation of* $\vec{e}$

By equation (8) of §64,

$$\mu\frac{d\vec{e}}{dt} = \vec{F}\times\vec{h} - (\vec{r}\cdot\vec{v})\vec{F} + (\vec{F}\cdot\vec{v})\vec{r}. \tag{12}$$

Resolving $\vec{F}$ into its components and substituting for $\vec{v}$ from equations (14) and (16) of §64, we obtain, successively,

$$\mu\frac{d\vec{e}}{dt} = F_r\frac{\vec{r}\times\vec{h}}{r} + F_\alpha\frac{\vec{\alpha}\times\vec{h}}{hr} - Ar^2\left(\frac{\vec{h}}{h}F_h + \frac{\vec{r}}{r}F_r + \frac{\vec{\alpha}}{hr}F_\alpha\right) + \left(ArF_r + \frac{h}{r}F_\alpha\right)\vec{r}$$

$$= -\frac{\vec{\alpha}}{r}F_r + \frac{\vec{r}}{r}hF_\alpha - \frac{\vec{h}}{h}Ar^2F_h - \frac{\vec{\alpha}}{h}ArF_\alpha + \frac{\vec{r}}{r}hF_\alpha$$

$$= -\frac{\vec{h}}{h}Ar^2F_h + 2\frac{\vec{r}}{r}hF_\alpha - \left(\frac{1}{r}F_r + A\frac{r}{h}F_\alpha\right)\vec{\alpha}. \tag{13}$$

Multiplying this equation scalarly by $\vec{e}$ and remembering that $\vec{e}$ is orthogonal to $\vec{h}$ and inclined to the direction of $\vec{r}$ by the angle φ (by definition) we find:

$$\mu \frac{\mathrm{d}\vec{e}}{\mathrm{d}t} \cdot \vec{e} = 2heF_\alpha \cos\varphi + \left(\frac{1}{r}F_r + A\frac{r}{h}F_\alpha\right) ehr \sin\varphi$$

$$= heF_r \sin\varphi + r\frac{\mu e}{h}\left(2\frac{h^2}{r\mu}\cos\varphi + e\sin^2\varphi\right)F_\alpha \qquad \text{(equation (21) of §64)}$$

$$= heF_r \sin\varphi + \frac{he}{1 + e\cos\varphi}(e + 2\cos\varphi + e\cos^2\varphi)F_\alpha, \qquad \text{(equation (20) of §64)} \tag{14}$$

or, alternatively,

$$\frac{\mu}{h}\frac{\mathrm{d}e}{\mathrm{d}t} = F_r \sin\varphi + F_\alpha \frac{e + 2\cos\varphi + e\cos^2\varphi}{1 + e\cos\varphi}$$

$$= F_r \sin\varphi + F_\alpha \frac{r\mu}{h^2}(e + 2\cos\varphi + e\cos^2\varphi). \tag{15}$$

(e) *Variation of* $\omega = \angle(\vec{v}, \vec{e})$

Since (cf. figure on p. 221)

$$(\vec{k} \cdot \vec{e}) = e \sin\iota \sin\omega, \tag{16}$$

we have

$$\vec{k} \cdot \frac{\mathrm{d}\vec{e}}{\mathrm{d}t} = \sin\iota \sin\omega \frac{\mathrm{d}e}{\mathrm{d}t} + e\cos\iota \sin\omega \frac{\mathrm{d}\iota}{\mathrm{d}t} + e\sin\iota\cos\omega\frac{\mathrm{d}\omega}{\mathrm{d}t}$$

$$= (\sin\iota\sin\omega)\frac{h}{\mu}\left[F_r \sin\varphi + \frac{r\mu}{h^2}(e + 2\cos\varphi + e\cos^2\varphi)F_\alpha\right]$$

$$+ \frac{re}{h}[\cos\iota \sin\omega \cos(\omega + \varphi)]F_h + e\sin\iota\cos\omega\frac{\mathrm{d}\omega}{\mathrm{d}t}, \tag{17}$$

where we have substituted from equations (7) and (15). Alternatively, substituting for $\mathrm{d}\vec{e}/\mathrm{d}t$ from the last line of equation (13), we have,

$$\vec{k} \cdot \frac{\mathrm{d}\vec{e}}{\mathrm{d}t} = \frac{\vec{k}}{\mu} \cdot \left[-\frac{\vec{h}}{h}Ar^2F_h + 2\frac{\vec{r}}{r}hF_\alpha - \left(\frac{1}{r}F_r + A\frac{r}{h}F_\alpha\right)\vec{\alpha}\right]$$

$$= -\frac{Ar^2}{\mu}(\cos\iota)F_h + \frac{2h}{\mu}[\sin(\omega + \varphi)\sin\iota]F_\alpha$$

$$-\frac{1}{\mu}\left(\frac{1}{r}F_r + \frac{Ar}{h}F_\alpha\right)hr\sin\iota\cos(\omega + \varphi). \tag{18}$$

Combining equations (17) and (18), and substituting for A from equation (21) of §64, we obtain

$$e\frac{d\omega}{dt} = -\left(\frac{h}{\mu}\cos\varphi\right)F_r - \left[\frac{re}{h}\cot\iota\sin(\omega+\varphi)\right]F_h + \left(\frac{h^2+r\mu}{h\mu}\sin\varphi\right)F_\alpha. \tag{19}$$

Before finding the variations of a and n it is useful to assemble some basic relations that obtain along the osculating ellipse. They are:

$$r = \frac{l}{1+e\cos\varphi} = \frac{h^2}{\mu}\frac{1}{1+e\cos\varphi}, \tag{20}$$

where

$$l = \text{semilatus rectum} = h^2/\mu = a(1-e^2). \tag{21}$$

And if u denotes the eccentric anomaly (cf. Chapter 7, §35, equation (8)),

$$r = a(1-e\cos u), \tag{22}$$

or equivalently;

$$e\cos u = 1 - \frac{r}{a} = 1 - \frac{1-e^2}{1+e\cos\varphi} = e\frac{e+\cos\varphi}{1+e\cos\varphi}; \tag{23}$$

and we deduce,

$$\sin u = \frac{r}{h}\left(\frac{\mu}{a}\right)^{1/2}\sin\varphi \qquad \text{and} \qquad a(e-\cos u) = -r\cos\varphi. \tag{24}$$

Also, by equations (22) of §64 and (24),

$$\vec{r}\cdot\vec{v} = \frac{\mu e}{h}r\sin\varphi = (\mu a)^{1/2}e\sin u. \tag{25}$$

And finally, by Kepler's third law of motion (Chapter 4, §19, equation (23)),

$$n = \frac{2\pi}{\text{period}} = \left(\frac{\mu}{a^3}\right)^{1/2}. \tag{26}$$

(*f*) *Variation of a and n*

From equation (21), it follows that

$$2\frac{h_{,t}}{h} = \frac{a_{,t}}{a} - \frac{2ee_{,t}}{1-e^2} = \frac{a_{,t}}{a} - \frac{2\mu aee_{,t}}{h^2}, \tag{27}$$

or, alternatively,

$$\begin{aligned}\tfrac{1}{2}h\frac{a_{,t}}{a} &= h_{,t} + \frac{\mu a}{h}ee_{,t}\\ &= rF_\alpha + ae\left\{F_r\sin\varphi + F_\alpha\frac{r\mu}{h^2}[e(1+\cos^2\varphi)+2\cos\varphi]\right\}\end{aligned}$$

(by equations (3) and (15)) (28)

where the coefficients of F_α combine to give

$$r\left\{1 + \frac{e}{1-e^2}\,[e(1+\cos^2\varphi) + 2\cos\varphi]\right\} = \frac{h^2 a}{r\mu}. \tag{29}$$

Hence,

$$\tfrac{1}{2}h\,\frac{a_{,t}}{a} = (ae\sin\varphi)F_r + \frac{h^2 a}{r\mu}\,F_\alpha; \tag{30}$$

while by equation (26),

$$\frac{n_{,t}}{n} = -\frac{3}{2}\frac{a_{,t}}{a}. \tag{31}$$

(g) *Variation of Kepler's equation*

First, the variation of the defining equation (22) gives

$$r_{,t} = (1 - e\cos u)a_{,t} + (ae\sin u)u_{,t} - (a\cos u)e_{,t}. \tag{32}$$

But, by equation (25)

$$r_{,t} = v = \frac{\sqrt{(\mu a)}}{r}\,e\sin u. \tag{33}$$

Hence,

$$\frac{\sqrt{(\mu a)}}{r}\,e\sin u = \frac{r}{a}\,a_{,t} + (ae\sin u)u_{,t} - (a\cos u)e_{,t}. \tag{34}$$

Next, by Kepler's equation (Chapter 7, §35, equation (19)), the time $(t - T)$, elapsed after passing the vernal equinox at time T, is given by

$$n(t - T) = u - e\sin u. \tag{35}$$

The variation of this equation gives

$$n_{,t}(t - T) + n(1 - T_{,t}) = (1 - e\cos u)u_{,t} - (\sin u)e_{,t}, \tag{36}$$

or, by making use of equations (26) and (31),

$$\frac{\sqrt{\mu}}{a^{3/2}}\left[\frac{3}{2}\frac{a_{,t}}{a}(t - T) + T_{,t}\right] = \frac{\sqrt{\mu}}{a^{3/2}} - \frac{r}{a}\,u_{,t} + (\sin u)e_{,t}. \tag{37}$$

Now, eliminating $u_{,t}$ from equations (34) and (37) we obtain,

$$\left(\frac{\mu}{a}\right)^{1/2}\left[\frac{3}{2}\frac{a_{,t}}{a}(t - T) + T_{,t}\right]e\sin u = \frac{r^2}{a^2}\,a_{,t} + (ae\sin^2 u - r\cos u)e_{,t}; \tag{38}$$

or, after some further simplifications making use of equations (23) and (24),

$$\left(\frac{\mu}{a}\right)^{1/2}\left[\frac{3}{2}\frac{a_{,t}}{a}(t - T) + T_{,t}\right]e\sin u = \frac{r^2}{a^2}\,a_{,t} - (r\cos\varphi)e_{,t}. \tag{39}$$

Making use, once again, of equation (24), we find

$$(\mu e \sin \varphi)T_{,t} = [rh - \tfrac{3}{2}\mu e(t - T)\sin \varphi]\frac{a_{,t}}{a} - (ah \cos \varphi)e_{,t}. \tag{40}$$

Summary

Collecting the various variational equations, we have:

$$\left.\begin{aligned}
&\frac{\mathrm{d}h}{\mathrm{d}t} = rF_\alpha, && \text{(i)}\\
&\frac{\mathrm{d}\iota}{\mathrm{d}t} = \left[\frac{r}{h}\cos(\omega + \varphi)\right]F_h, && \text{(ii)}\\
&\sin \iota \frac{\mathrm{d}\Omega}{\mathrm{d}t} = \left[\frac{r}{h}\sin(\omega + \varphi)\right]F_h, && \text{(iii)}\\
&\frac{\mathrm{d}e}{\mathrm{d}t} = \left(\frac{h}{\mu}\sin \varphi\right)F_r + \frac{r}{h}(e + 2\cos \varphi + e\cos^2 \varphi)^{*}F_\alpha, && \text{(iv)}\\
&e\frac{\mathrm{d}\omega}{\mathrm{d}t} = -\left(\frac{h}{\mu}\cos \varphi\right)F_r - \left[\frac{re}{h}\cot \iota \sin(\omega + \varphi)\right]F_h\\
&\qquad + \left[\left(\frac{h^2 + r\mu}{h\mu}\right)^{\dagger}\sin \varphi\right]F_\alpha, && \text{(v)}\\
&\frac{1}{a}\frac{\mathrm{d}a}{\mathrm{d}t} = \left(\frac{2ae}{h}\sin \varphi\right)F_r + \frac{2ah}{r\mu}F_\alpha = -\frac{2}{3n}\frac{\mathrm{d}n}{\mathrm{d}t}, && \text{(vi)}\\
&(\mu e \sin \varphi)\frac{\mathrm{d}T}{\mathrm{d}t} = [rh - \tfrac{3}{2}\mu e(t - T)\sin \varphi]\frac{1}{a}\frac{\mathrm{d}a}{\mathrm{d}t} - (ah\cos \varphi)\frac{\mathrm{d}e}{\mathrm{d}t}, && \text{(vii)}\\
&\frac{\mathrm{d}\varpi}{\mathrm{d}t} = \frac{\mathrm{d}}{\mathrm{d}t}(\omega + \Omega) = -\left(\frac{h}{\mu e}\cos \varphi\right)F_r + \left[\frac{1}{e}\left(\frac{h^2 + r\mu}{h\mu}\right)^{\dagger}\sin \varphi\right]F_\alpha\\
&\qquad + \left[\frac{r}{h}\sin(\omega + \varphi)\tan\frac{\iota}{2}\right]F_h. && \text{(viii)}
\end{aligned}\right\} \tag{41}$$

66. Application of the method of the variation of the elements to lunar motion

The part of the problem of lunar motion to which Newton gave his principal attention was to find the solution of the three-body problem presented by the system of the Earth,

* An alternative simpler form of the coefficient of F_α is (as can be directly verified),

$$h(\cos \varphi + \cos u)/\mu.$$

† And similarly,

$$\frac{1}{h\mu}(h^2 + r\mu) = \frac{h}{\mu}\left(1 + \frac{r\mu}{h^2}\right) = \frac{h}{\mu}\left(1 + \frac{r}{l}\right).$$

the Moon, and the Sun. Newton's attitude was that the Kepler motion of the Moon about the Earth was disturbed mainly by the tidal action of the Sun. Newton's principal task was then to find the components of the disturbing force; and this problem he solved completely.

(*a*) *The disturbing function*

We start then with the equations of motion of three bodies of masses m_1, m_2, and m_3 under their mutual gravitational attraction. The equations are:

$$\left.\begin{aligned} m_1\frac{\mathrm{d}^2\vec{r}_1}{\mathrm{d}t^2} &= \frac{Gm_1m_2}{r_{12}^3}(\vec{r}_2-\vec{r}_1)+\frac{Gm_1m_3}{r_{13}^3}(\vec{r}_3-\vec{r}_1),\\ m_2\frac{\mathrm{d}^2\vec{r}_2}{\mathrm{d}t^2} &= \frac{Gm_2m_1}{r_{21}^3}(\vec{r}_1-\vec{r}_2)+\frac{Gm_2m_3}{r_{23}^3}(\vec{r}_3-\vec{r}_2),\\ m_3\frac{\mathrm{d}^2\vec{r}_3}{\mathrm{d}t^2} &= \frac{Gm_3m_1}{r_{31}^3}(\vec{r}_1-\vec{r}_3)+\frac{Gm_3m_2}{r_{32}^3}(\vec{r}_2-\vec{r}_3). \end{aligned}\right\} \tag{1}$$

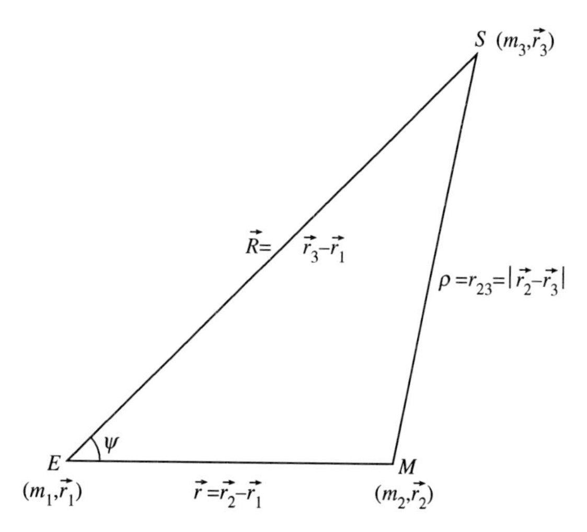

Letting (see the figure)

$$\vec{R}=\vec{r}_3-\vec{r}_1,\quad \vec{r}=\vec{r}_2-\vec{r}_1,\quad |\vec{r}|=r,\quad r_{ij}=|\vec{r}_i-\vec{r}_j|,\qquad \text{and}\qquad \rho=|\vec{r}_3-\vec{r}_2|=r_{23}, \tag{2}$$

we combine equations (1) to give

$$\frac{\mathrm{d}^2\vec{r}_2}{\mathrm{d}t^2} = -Gm_1\frac{\vec{r}}{r^3}+Gm_3\frac{\vec{R}-\vec{r}}{\rho^3} \tag{3}$$

and

$$\frac{\mathrm{d}^2\vec{r}_1}{\mathrm{d}t^2} = +Gm_2\frac{\vec{r}}{r^3}+Gm_3\frac{\vec{R}}{R^3}. \tag{4}$$

Subtracting equation (4) from equation (3), we obtain

$$\frac{d^2\vec{r}}{dt^2} = -\mu\frac{\vec{r}}{r^3} + Gm_3\left(\frac{\vec{R}-\vec{r}}{\rho^3} - \frac{\vec{R}}{R^3}\right) \qquad [\mu = G(m_1+m_2)]. \tag{5}$$

We now identify m_1, m_2, and m_3 with the masses of the Earth, the Moon, and the Sun, respectively. Then $\vec{r}$ denotes the vector joining the Moon and the Earth. The equation governing the motion of the Moon (in which, of course, we are primarily interested), is

$$\frac{d^2\vec{r}}{dt^2} = -\mu\frac{\vec{r}}{r^3} + GM_\odot\left[-\frac{\vec{r}}{\rho^3} + \vec{R}\left(\frac{1}{\rho^3} - \frac{1}{R^3}\right)\right], \tag{6}$$

where we have written $M_\odot$ in place of m_3 (to conform with the conventional symbol for the mass of the Sun) and

$$\mu = G(\text{mass of Earth} + \text{mass of Moon}). \tag{7}$$

The disturbing function, $\vec{F}$, is accordingly given by

$$\vec{F} = GM_\odot\left[-\frac{\vec{r}}{\rho^3} + \vec{R}\left(\frac{1}{\rho^3} - \frac{1}{R^3}\right)\right]. \tag{8}$$

(*b*) *The components of* $\vec{F}(=F_r, F_\alpha, F_h)$

The adjoining figure, which is essentially the same as the one on p. 221, includes some additional information relative to the direction, $\vec{R}$, of the Sun (on the ecliptic plane) relative to the Earth. In particular,

$$\angle v = \angle(\omega+\varphi) = \angle(\vec{v}, \vec{r}), \qquad \angle\psi = \angle(\vec{R}, \vec{r}), \qquad \text{and} \qquad \angle U = \angle(\vec{R}, \vec{v}). \tag{9}$$

From the diagram, it follows that

$$\vec{R} = R[\vec{i}\cos(U+\Omega) + \vec{j}\sin(U+\Omega)] \tag{10}$$

or, substituting for $\vec{i}$ and $\vec{j}$ from the table on p. 222, we have

$$\begin{aligned}\vec{R} &= R[\vec{i}\cos(U+\Omega) + \vec{j}\sin(U+\Omega)]\\ &= R[\cos(U+\Omega)\{(\sin\iota\sin\Omega)\vec{1}_h + [\cos\Omega\cos v - \sin\Omega\sin v\cos\iota]\vec{1}_r\\ &\qquad - [\cos\Omega\sin v + \sin\Omega\cos v\cos\iota]\vec{1}_\alpha\}\\ &\quad + \sin(U+\Omega)\{-(\sin\iota\cos\Omega)\vec{1}_h + [\sin\Omega\cos v + \cos\Omega\sin v\cos\iota]\vec{1}_r\\ &\qquad - [\sin\Omega\sin v - \cos\Omega\cos v\cos\iota]\vec{1}_\alpha\}]\\ &= R[-(\sin U\sin\iota)\vec{1}_h + (\cos U\cos v + \sin U\sin v\cos\iota)\vec{1}_r\\ &\quad + (-\cos U\sin v + \sin U\cos v\cos\iota)\vec{1}_\alpha].\end{aligned} \tag{11}$$

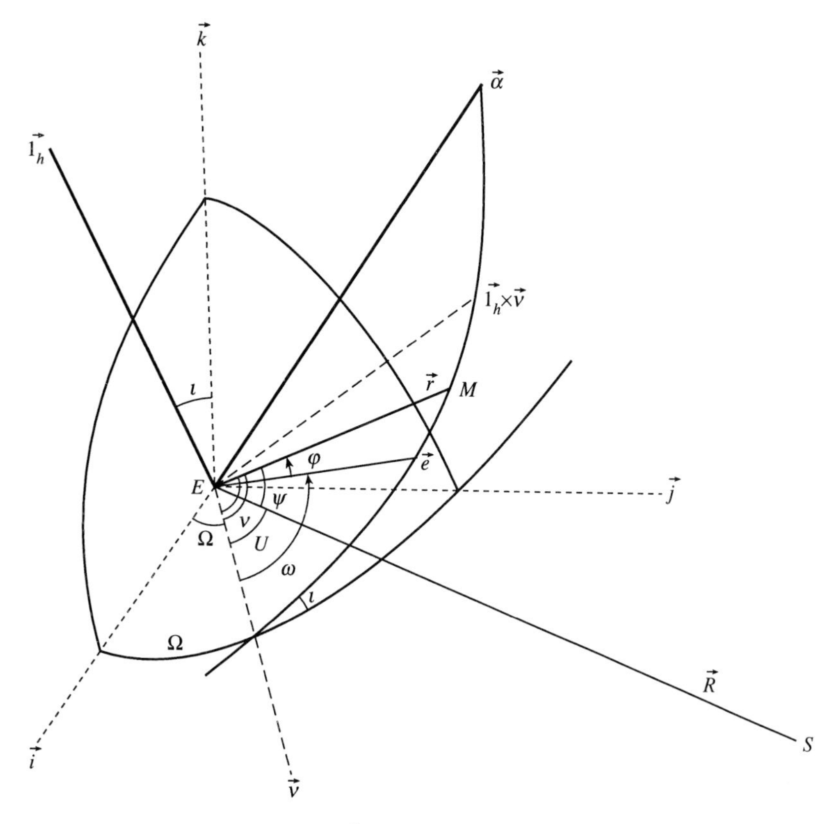

Inserting the foregoing expression for $\vec{R}$ in equation (8) we can read off the components, (F_r, F_α, F_h) of $\vec{F}$. We find:

$$F_r = GM_\odot\left[-\frac{r}{\rho^3} + R(+\cos U \cos v + \sin U \sin v \cos \iota)\left(\frac{1}{\rho^3} - \frac{1}{R^3}\right)\right], \tag{12}$$

$$F_\alpha = GM_\odot\left[\qquad + R(-\cos U \sin v + \sin U \cos v \cos \iota)\left(\frac{1}{\rho^3} - \frac{1}{R^3}\right)\right], \tag{13}$$

$$F_h = GM_\odot\left[\qquad - R \sin U \sin \iota\left(\frac{1}{\rho^3} - \frac{1}{R^3}\right)\right]. \tag{14}$$

Since (see equation (9))

$$\cos \psi = \cos U \cos v + \sin U \sin v \cos \iota = \cos \angle(\vec{R}, \vec{r}), \tag{15}$$

we can rewrite the expression (12) for F_r in the form,

$$F_r = GM_\odot\left[-\frac{r}{\rho^3} + R\left(\frac{1}{\rho^3} - \frac{1}{R^3}\right)\cos \psi\right]. \tag{16}$$

Now we proceed to make approximations appropriate to the problem on hand. First, we observe that since

$$\frac{r}{R} = \frac{\text{radius of the Moon's orbit}}{\text{radius of the Earth's orbit}} \simeq 2{\cdot}6 \times 10^{-3}, \tag{17}$$

we can neglect quantities of $O(r/R)$. From the equation

$$\rho^2 = R^2 + r^2 - 2rR \cos \psi, \tag{18}$$

we deduce,

$$\left.\begin{aligned} \rho &= R - r \cos \psi + O(r^2/R) \\ \text{and} \qquad \frac{1}{\rho^3} - \frac{1}{R^3} &= 3\frac{r}{R^4} \cos \psi + O(r^2/R^5). \end{aligned}\right\} \tag{19}$$

Inserting these approximations in equation (16), we find

$$\begin{aligned} F_r &= GM_\odot\left\{-r\left[\frac{1}{R^3} + 3\frac{r}{R^4}\cos\psi + O\left(\frac{r^2}{R^5}\right)\right] + 3\frac{r}{R^3}\cos^2\psi + O\left(\frac{r^2}{R^4}\right)\right\} \\ &= \frac{GM_\odot}{R^3}\left[r(-1 + 3\cos^2\psi) + O\left(\frac{r^2}{R^4}\right)\right]. \end{aligned} \tag{20}$$

Hence,

$$F_r \simeq \frac{1}{2}\frac{GM_\odot}{R^3} r(1 + 3\cos 2\psi), \tag{21}$$

where terms of $O(r^2/R^4)$ have been omitted. A further approximation that is permissible is to retain the inclination ι only to the first order and set (cf. equations (15) and (19))

$$\left.\begin{aligned} \cos\iota = 1 \qquad \text{when} \qquad & \cos\psi \simeq \cos(v - U) \\ \text{and} \qquad & \frac{1}{\rho^3} - \frac{1}{R^3} \simeq 3\frac{r}{R^4}\cos(v - U). \end{aligned}\right\} \tag{22}$$

Inserting these approximations in equations (12)–(14) we finally obtain, to the order of accuracy needed,

$$F_r = +\frac{1}{2}\frac{GM_\odot}{R^3} r[1 + 3\cos 2(v - U)], \tag{23}$$

$$F_\alpha = -\frac{3}{2}\frac{GM_\odot}{R^3} r \sin 2(v - U), \tag{24}$$

$$F_h = -3\frac{GM_\odot}{R^3} r \sin\iota \sin U \cos(v - U), \tag{25}$$

where by Kepler's third law (cf. equation (26) of §65)

$$GM_\odot/R^3 = N^2, \tag{26}$$

where $N(=2\pi/\text{period of the Earth about the Sun})$ is the mean motion of the Sun (relative to the Earth).

(c) *Application of the variational equations*

By equations (21) and (26) of §65 applied to the Earth–Moon system we have

$$n = (\mu/a^3)^{1/2} \qquad \text{and} \qquad h = [a\mu(1-e^2)]^{1/2} = na^2(1-e^2)^{1/2}. \tag{27}$$

We further define

$$m = N/n. \tag{28}$$

With these definitions and with the expressions for the components of $\vec{F}$ given in equations (23)–(25), we readily find, from equations (41) of §65 the following variational equations as applied to the motion of the Moon.

$$\frac{1}{h}\frac{\mathrm{d}h}{\mathrm{d}t} = \frac{r}{h}[-\tfrac{3}{2}N^2 r \sin 2(v-U)]$$

$$= -\tfrac{3}{2}N^2 \frac{r^2}{na^2\sqrt{1-e^2}} \sin 2(v-U), \tag{i}$$

or,

$$\frac{1}{h}\frac{\mathrm{d}h}{\mathrm{d}t} = -\tfrac{3}{2}m^2 n \frac{r^2}{a^2\sqrt{(1-e^2)}} \sin 2(v-U). \tag{i$'$}$$

$$\frac{\mathrm{d}\iota}{\mathrm{d}t} = \left(\frac{r}{h}\cos v\right)[-3N^2 r \sin\iota \sin U \cos(v-U)]$$

$$= -3N^2 \frac{r^2}{na^2\sqrt{1-e^2}} \cos v \sin\iota \sin U \cos(v-U),$$

or,

$$\frac{\mathrm{d}\iota}{\mathrm{d}t} = -3m^2 n \frac{r^2}{a^2\sqrt{(1-e^2)}} \cos v \sin\iota \sin U \cos(v-U). \tag{ii}$$

$$\frac{\mathrm{d}\Omega}{\mathrm{d}t} = -3m^2 n \frac{r^2}{a^2\sqrt{(1-e^2)}} \sin v \sin U \cos(v-U). \tag{iii}$$

$$\frac{\mathrm{d}e}{\mathrm{d}t} = \frac{h}{\mu}[F_r \sin\varphi + (\cos u + \cos\varphi)F_\alpha]$$

$$= n\frac{a^2\sqrt{(1-e^2)}}{\mu}\{\tfrac{1}{2}N^2 r[1+3\cos 2(v-U)]\sin\varphi - \tfrac{3}{2}N^2 r \sin 2(v-U)(\cos u + \cos\varphi)\}$$

$$= \tfrac{1}{2}N^2 n\frac{a^2 r\sqrt{(1-e^2)}}{\mu}\{[1+3\cos 2(v-U)]\sin\varphi - 3(\cos u+\cos\varphi)\sin 2(v-U)\},$$

or,

$$\frac{\mathrm{d}e}{\mathrm{d}t} = \tfrac{1}{2}m^2 n\frac{r\sqrt{(1-e^2)}}{a}\{[1+3\cos 2(v-U)]\sin\varphi - 3(\cos u+\cos\varphi)\sin 2(v-U)\}. \tag{iv}$$

$$e\frac{\mathrm{d}\omega}{\mathrm{d}t} = \frac{h}{\mu}\left\{-\cos\varphi \,.\, F_r + \left(1+\frac{r}{l}\right)\sin\varphi F_\alpha\right\} \tag{v}$$

or,

$$\frac{\mathrm{d}\omega}{\mathrm{d}t} = n\frac{a^2\sqrt{(1-e^2)}}{\mu e}\left\{-\cos\varphi\,.\,F_r + \left(1+\frac{r}{l}\right)\sin\varphi\,.\,F_\alpha\right\}.$$

The quantity in braces is

$$\{\cdots\} = -\tfrac{1}{2}N^2 r[1 + 3\cos 2(v-U)]\cos\varphi + \left(1+\frac{r}{l}\right)\sin\varphi[-\tfrac{3}{2}N^2 r\sin 2(v-U)]$$

Therefore,

$$\frac{\mathrm{d}\omega}{\mathrm{d}t} = -\tfrac{1}{2}N^2 n\frac{a^2\sqrt{(1-e^2)}}{\mu e}r\left[\cos\varphi[1+3\cos 2(v-U)] + 3\sin\varphi\sin 2(v-U)\left(1+\frac{r}{l}\right)\right]$$

or

$$\frac{\mathrm{d}\omega}{\mathrm{d}t} = -\tfrac{1}{2}m^2 n\frac{\sqrt{(1-e^2)}}{e}\frac{r}{a}\left\{\cos\varphi[1+3\cos 2(v-U)] + 3\sin\varphi\sin 2(v-U)\left(1+\frac{r}{l}\right)\right\}. \quad \text{(vi)}$$

Ignoring the first term in F_r (with the factor e) in the equation for $\mathrm{d}a/\mathrm{d}t$ in (41) of §65, and ignoring the factor $(1-e^2)^{1/2}$ in equation (27), we have

$$\frac{\mathrm{d}a}{\mathrm{d}t} = \frac{2a^2h}{r\mu}F_\alpha = \frac{2a}{rn}F_\alpha;$$

or, substituting for F_α from equation (24), we obtain,

$$\frac{\mathrm{d}a}{\mathrm{d}t} = -3N^2\frac{a}{n}\sin 2(v-U). \quad \text{(vii)}$$

14

The three-body problem: the foundations of Newton's lunar theory

67. Introduction

In Propositions LXV–LXIX, Newton took the first steps in the study of the motions of three bodies under their mutual attractions. His main considerations are devoted to the changes in elements of a Kepler orbit effected by an external perturbation, in particular that of a distant third body. The considerations are clearly intended as a prelude to his lunar theory to be elaborated further, in Book III (Propositions XXV–XXXIII; Chapter 22).

Newton's arguments in the 22 Corollaries of Proposition LXVI are of 'grande finesse' (to quote Tisserand) and often difficult to follow because of the concise prose in which they are couched. However, with the introduction provided in Chapter 13, it is not difficult to give an account which reveals the subtlety and the depth of Newton's considerations.

68.

Proposition LXV. Theorem XXV

Bodies, whose forces decrease as the square of their distances from their centres, may move among themselves in ellipses; and by radii drawn to the foci may describe areas very nearly proportional to the times.

Newton considers two illustrations of this proposition as sufficiently illustrative.

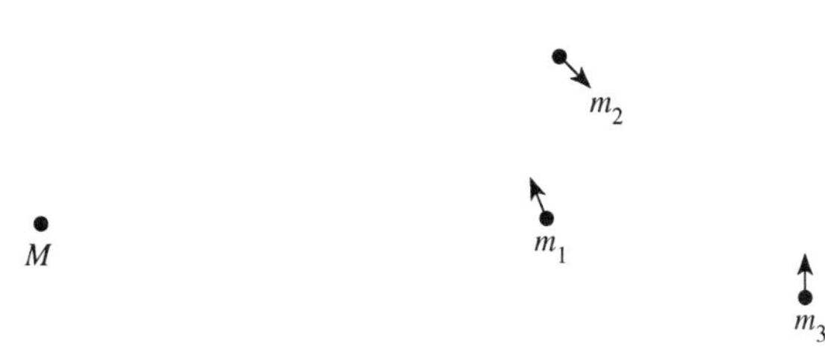

Case I. Consider a 'great body' M, about which several 'lesser bodies' m_1, m_2, m_3, etc., are revolving. The centre of gravity of all these masses should either be at rest or move uniformly forward in a right line. If the masses, m_1, m_2, m_3, etc., are sufficiently tiny (compared to M) then the centre of gravity will not be sensibly different from the location of M, which may, then, be considered to be at rest or moving uniformly forward in a right line; and about which the lesser bodies will revolve. If further, the mutual attractions between m_1, m_2, m_3, etc., are sufficiently feeble compared to the attractive force of M, then the lesser bodies will each revolve about M in ellipses describing equal areas in time.

In considering this example, Newton has the solar system clearly in mind.

Case II. Consider the same system as in case I, except that now the attractive force

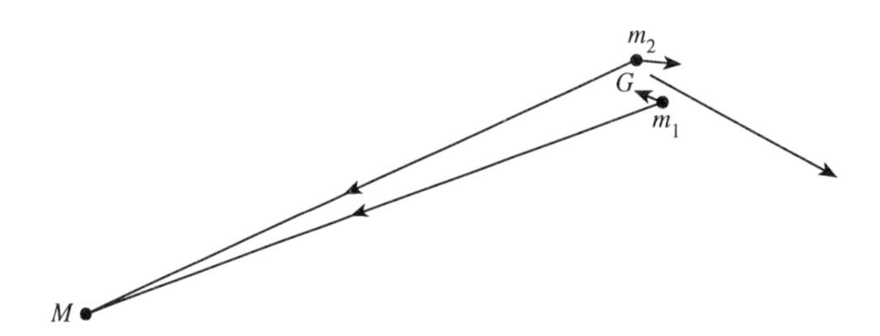

between m_1 and m_2 (say, to restrict ourselves to two 'lesser bodies') is sufficiently strong that they describe elliptic orbits about their common centre of gravity G. If the great mass M is now sufficiently far that the distance between m_1 and m_2 is small compared to the distance MG, then M will attract m_1 and m_2 along (nearly) parallel lines and subject them to equal parallel accelerations. Since equal and parallel accelerations will not affect their relative motions (by Corollary VI of the Laws of Motion, applied earlier in Proposition LXIV) it follows that M will act on m_1 and m_2 as if the two together 'were one body'. Accordingly the centre of gravity of m_1 and m_2 will describe about M a parabola or a hyperbola when 'the attraction is but languid and an ellipse when it is more vigorous' (!).

It is clear that the foregoing arguments are readily generalized to a system of several lesser bodies, m_1, m_2, m_3, etc., provided no three of them are close together.

It should be pointed out that Newton is very careful in emphasizing the limitations of the assumptions that are at the base of the foregoing deductions. In Corollaries I and II he is even more explicit. He concludes with the remarkable corollary:

> COR. III. Hence if the parts of this system move in ellipses or circles without any remarkable perturbation, it is manifest that, if they are at all impelled by accelerative forces tending to any other bodies, the impulse is very weak, or else is impressed very near equally and in parallel directions upon all of them.

69.

Proposition LXVI. Theorem XXVI

> *If three bodies, whose forces decrease as the square of the distances, attract each other; and the accelerative attractions of any two towards the third be between themselves inversely as the squares of the distances; and the two least revolve about the greatest: I say, that the interior of the two revolving bodies will, by radii drawn to the innermost and greatest, describe round that body areas more proportional to the times, and a figure more approaching to that of an ellipse having its focus in the point of intersection of the radii, if that great body be agitated by those attractions, than it would do if that great body were not attracted at all by the lesser, but remained at rest; or than it would do if that great body were very much more or very much less attracted, or very much more or very much less agitated, by the attractions.*

It is characteristic (?) of Newton that he should have begun his magisterial enunciation of the 22 corollaries of this proposition (which Laplace has described as one of the most remarkable chapters in the *Principia*) in this low key:

> This appears plainly enough from the demonstration of the second Corollary of the foregoing Proposition; but it may be made out after this manner by a way of reasoning more distinct and more universally convincing.

Case I. Newton begins with a clear formulation of the context in which the problem of the motion of three bodies (*T*, *P*, and *S*), under their mutual gravitational attractions, is considered in this proposition: it is the perturbation of the elliptic orbit, described by a 'lesser body' *P* about the 'greatest body' *T* by another 'lesser body' *S*. To avoid misunderstanding, it may be well to clarify (as Newton in fact does in Corollary VI) that *S* is a 'lesser body', only in the sense that its influence on the motion of *P* may be considered as a small perturbation of the (unperturbed) orbit that *P* describes about *T*. It is thus not contrary to Newton's assumptions to identify *S* with the Sun, *T* with the Earth (Terram), and *P* with the Moon (Lunam). Indeed, Laplace, Lagrange, and Tisserand describe the contents of this proposition as addressed to lunar theory.

Also, contrary to the impression created by Newton's choice of the same illustration (six times in this proposition and once in Proposition XXV, Book III), his considerations are not limited to the perturbation of circular orbits (described by *P*). This is explicitly not the case in the earlier corollaries of this proposition. (For this reason, we have chosen a slightly modified illustration that is more applicable.)

Newton first considers the case when the orbits described by *P*, *S*, and *T* are coplanar.

In the illustration below, *P*, when undisturbed, will describe an elliptic orbit about *T* as its focus (in accordance with Proposition LVII, Corollaries II and III). This orbit will

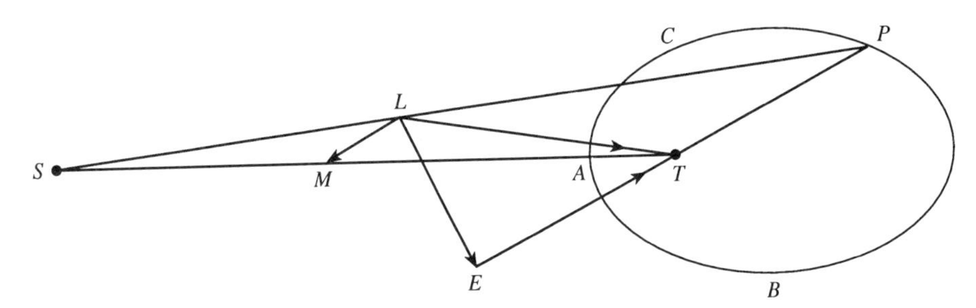

be disturbed by the gravitational attraction of S—it is permissible to add the qualification 'at a far distance'. Let the length TS be a measure of the force $|\overrightarrow{TS}|$ exerted by S on T per unit mass. Then

$$LS = TS\,\frac{ST^2}{SP^2} \tag{1}$$

will be a measure of force $|\overrightarrow{LS}|$ (per unit mass) exerted by S on P. Now resolve $\overrightarrow{LS}$ along LM parallel to PT, and $\overrightarrow{MS}$. Then

$$\overrightarrow{LS} = \overrightarrow{LM} + \overrightarrow{MS}. \tag{2}$$

Of these two forces, $\overrightarrow{LM}$ being parallel to $\overrightarrow{PT}$ can be 'super added' to the gravitational force ($\propto PT^{-2}$) of T already acting in this direction; and the continuing central character of the force acting on P will preserve the law of areas. But since $|\overrightarrow{LM}|$ is not proportional to the inverse square of the distance PT, its action will make the orbit of P depart from an ellipse. But the force $\overrightarrow{MS}$, no longer parallel to $\overrightarrow{PT}$, will affect the law of areas besides making the orbit of P depart even further from an ellipse. However, since equal accelerations imparted to T and P will not affect their relative motion, it follows from

$$\overrightarrow{TS} - \overrightarrow{LS} = \overrightarrow{TL} = \overrightarrow{TM} - \overrightarrow{LM} \tag{3}$$

that only the part $\overrightarrow{TM}$ of $\overrightarrow{TS}$ will affect the relative motion of P and T. Therefore, of the two forces $\overrightarrow{LM}$ and $\overrightarrow{TM}$ that affect the orbit of P about T, $\overrightarrow{LM}$ does not affect the law of areas, while both of them contribute to the departure of the orbit from an ellipse.

Case II. Newton considers next how the non-coplanarity of the orbits of P and T will affect the conclusions arrived at under case I. Newton simply states, as intuitively obvious, what directly follows from the expression for the force F_h, in the direction of the angular momentum and normal to the orbital plane, given in Chapter 13, Section 66 equation (14):

$$F_h \propto \sin \imath \sin U \cos \psi, \tag{4}$$

where $\imath$ denotes the angle of inclination, U the angle between the line of the nodes and the direction of $\overrightarrow{TS}$, and ψ is the angle between the direction $\overrightarrow{TS}$ and $\overrightarrow{TP}$. (See the accompanying figure.) Equation (4) is correct only to the first order in the inclination. Under these same circumstances,

$$\overrightarrow{TM} \cdot \vec{1}_h = F_h \propto -\sin \imath \sin U \cos \psi; \tag{4a}$$

(or, as Newton expresses, F_h is 'as the generating force $\overrightarrow{TM}$' and that it is non-vanishing only when ST is 'without the nodes', that is, when $U \neq 0$). Since,

$$\overrightarrow{TM} \cdot \vec{1}_h = TM \frac{\vec{R}}{R} \cdot \vec{1}_h = -TM\,(\sin \iota \sin U); \tag{4b}$$

it follows that

$$TM \propto \cos \psi. \tag{4c}$$

70. Proposition LXVI (continued): Corollaries I–VI

Following Laplace (*Mécanique Céleste*, Vol. V, Book 6) and Tisserand (*Traité de Mécanique Céleste*, Tome III, Chapitre 3, p. 33) we shall first paraphrase Newton's results in more formal language.

In these corollaries, Newton's considerations are mostly restricted to the case when the unperturbed orbit is circular and coplanar with the orbit of T, and further that S is far away and $PK/ST \ll 1$ (see the illustration below). (Some of the results are, however, valid under less restrictive conditions.)

(*a*) *The perturbing function*

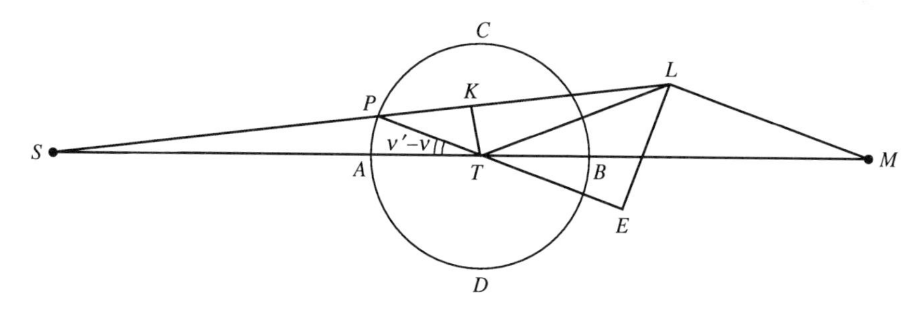

The above figure is the same as that on p. 238 except that the orbit described by P is circular and an additional line TK is drawn perpendicular to SL. Also under the present assumption, $ST \gg PK$,

$$ST \simeq SK. \tag{1}$$

As in §69 the disturbing attractive force on P, in the unit ST measuring the gravitational attraction of T by S, is

$$\overrightarrow{LS} - \overrightarrow{TS} = \overrightarrow{LT}. \tag{2}$$

Resolve $\overrightarrow{LT}$ into its orthogonal components,

$$\overrightarrow{LT} = \overrightarrow{LE} + \overrightarrow{ET}, \tag{3}$$

respectively, perpendicular and parallel to $\overrightarrow{TP}$. By equation (1),

$$ST \simeq SP + PK \qquad \text{and} \qquad SL = SP + PL; \tag{4}$$

and by construction (by §69, equation (1))

$$LS = ST^3/SP^2. \tag{5}$$

Therefore,

$$\begin{aligned} 3SP^2.PK + 3SP.PK^2 + PK^3 &= (SP + PK)^3 - SP^3 \simeq ST^3 - SP^3 \\ &= LS.SP^2 - SP^3 = SP^2(LS - SP) \\ &= SP^2.PL; \end{aligned} \tag{6}$$

or, dividing by SP^2,

$$3PK\left(1 + \frac{PK}{SP} + \frac{1}{3}\frac{PK^2}{SP^2}\right) \simeq PL. \tag{7}$$

Since $PK \ll SP$, we may write,

$$3PK = PL + O(PT/ST). \tag{8}$$

The triangle EPL being right-angled,

$$\begin{aligned} LE &= PL \sin \angle EPL \simeq 3PK \sin \angle TPK \\ \text{and} \qquad TE + TP &= PL \cos \angle TPK \simeq 3PK \cos \angle TPK; \end{aligned} \tag{9}$$

or, alternatively,

$$\begin{aligned} LE &= 3PK\frac{TK}{TP} = 3TP\frac{PK}{TP}\frac{TK}{TP} \simeq 3TP \sin(v' - v)\cos(v' - v) \\ \text{and} \qquad TE &= 3PK\frac{PK}{TP} - TP = TP\left[3\left(\frac{PK}{TP}\right)^2 - 1\right] \simeq TP[3\cos^2(v' - v) - 1]. \end{aligned} \tag{10}$$

Since the forces are measured in the unit ST (standing for gravitational attraction of T by S), the forces, F_α and F_r, in absolute units, perpendicular and parallel to the radius vector TP, are obtained by multiplying LE and TE by

$$\frac{GM_\odot}{ST^2}\frac{1}{ST} = N^2, \tag{11}$$

where $N = 2\pi$/period of the Earth about the Sun. Hence

$$F_\alpha = \tfrac{3}{2}N^2TP \sin 2(v' - v) = -\tfrac{3}{2}N^2TP \sin 2\psi, \tag{12}$$

and

$$F_r = N^2TP[3\cos^2(v' - v) - 1] = \tfrac{1}{2}N^2TP(1 + 3\cos 2\psi), \tag{13}$$

where

$$\angle\psi = \angle(v - v') = \angle(v - U) = (1 - m)\angle v, \tag{14}$$

in the notation of Chapter 13 (equations (9) and (22) of §66). We observe that these expressions for F_r and F_h are in agreement with those derived in Chapter 13, equations (23) and (24) of §66—but by how different routes!

It may be noted parenthetically that at conjunction and opposition (or syzygies) and at quadratures, we have respectively (see the accompanying figure)

$$\psi = 0 \quad \text{and} \quad \pi \qquad \text{and} \qquad \psi = \pi/2 \quad \text{or} \quad 3\pi/2. \tag{15}$$

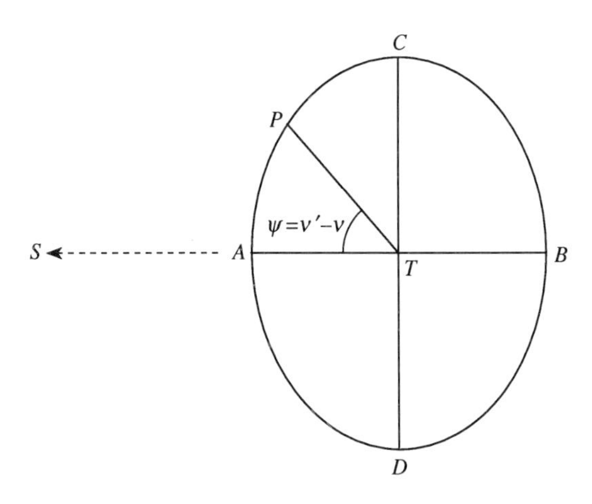

(*b*) *The centripetal attraction*

Turning to the centripetal attraction, we have for the unperturbed circular orbit,

$$\frac{GM_T}{PT^2} = n^2\overline{PT} = k = \text{Constant}, \tag{16}$$

where n (analogous to N) denotes the mean lunar motion. The centripetal attraction along the perturbed orbit can therefore be written as

$$\left.\begin{aligned} k - F_r &= k[1 - \tfrac{1}{2}m^2(1 + 3\cos 2\psi)], \\ &\text{since} \\ F_r &= \tfrac{1}{2}m^2n^2r(1 + 3\cos 2\psi) = \tfrac{1}{2}m^2k\frac{r}{\overline{PT}}(1 + 3\cos 2\psi) \simeq \tfrac{1}{2}km^2(1 + 3\cos 2\psi), \end{aligned}\right\} \tag{17}$$

where (as in Chapter 13)

$$m = \frac{N}{n} = \frac{U}{v}. \tag{18}$$

(Note that F_r is subtracted since, by definition, it acts in the direction of increasing r.)

Measuring r in units of the unperturbed radius of the circular orbit of P, let its values at syzygies and at quadratures (distinguished by subscripts 1 and 0), respectively, be

$$r_1 = 1 - x \qquad \text{and} \qquad r_0 = 1 + x. \tag{19}$$

The centripetal forces, F_1 and F_0, at syzygies and at quadratures are, by equation (17),

$$F_1 = \frac{k}{(1-x)^2}(1 - 2m^2) \qquad \text{and} \qquad F_0 = \frac{k}{(1+x)^2}(1 + m^2). \tag{20}$$

(c) *The perturbed orbit*

Newton assumes that the orbit is a prolate ellipse centred at T; and that its minor and major axes, in the directions of the syzygies ($\psi = 0$ and $\psi = \pi$) and quadratures ($\pi/2$ and $3\pi/2$), are given by (19). Commenting on this assumption, Laplace writes 'it is correct but it requires proof'; but goes on to say 'this hypothesis in calculation, based on insight which is very likely, is permissible to inventors in researches as difficult as these'.

If the orbit is indeed a prolate ellipse, then one readily verifies that in our present approximation,

$$r = 1 - x\cos 2\psi = 1 - x\cos[2(1-m)v]. \tag{21}$$

We shall presently need the radius of the curvature of the orbit (21). The formula given in Chapter 5 (§24, equation (xiv)), rewritten in terms of r (rather than in terms of its reciprocal u) is

$$\rho = \frac{[r^2 + (\mathrm{d}r/\mathrm{d}\psi)^2]^{3/2}}{r^2 + 2(\mathrm{d}r/\mathrm{d}\psi)^2 - r\,\mathrm{d}^2r/\mathrm{d}\psi^2} \simeq \frac{r^2}{r - \mathrm{d}^2r/\mathrm{d}\psi^2}. \tag{22}$$

For r given by equation (21)

$$\rho \simeq \frac{1 - 2x\cos[2(1-m)v]}{1 - [1 + 4(1-m)^2]x\cos[2(1-m)v]}. \tag{23}$$

In particular,

$$\frac{\rho_0}{\rho_1} \simeq \frac{1+2x}{1-2x}\,\frac{1 - x[1 + 4(1-m)^2]}{1 + x[1 + 4(1-m)^2]}, \tag{24}$$

or

$$\frac{\rho_0}{\rho_1} \simeq 1 - 2x[4(1-m)^2 - 1]. \tag{25}$$

(*d*) *The variation of the 'constant of areas'*

To determine the variation of the constant of areas, Newton effectively uses the formula derived in Chapter 13, § 65, equation (i);

$$\frac{\mathrm{d}h}{\mathrm{d}t} = rF_\alpha = -\tfrac{3}{2}N^2r^2 \sin[2(1-m)v], \tag{26}$$

or, by equation (18),

$$\mathrm{d}h = -\tfrac{3}{2}m^2(nr^2)\sin[2(1-m)v](n\,\mathrm{d}t). \tag{27}$$

For an underlying circular orbit,

$$n\,\mathrm{d}t = \mathrm{d}v; \tag{28}$$

and we have

$$\mathrm{d}h = -\tfrac{3}{2}m^2(nr^2)\sin[2(1-m)v]\,\mathrm{d}v. \tag{29}$$

On integrating this equation, we obtain

$$h \simeq \text{constant} + \frac{3}{4}\frac{m^2}{1-m}(n\langle r^2\rangle_{\text{av}})\cos[2(1-m)v]. \tag{30}$$

Since (cf. equation (16))

$$h = r^2\frac{\mathrm{d}v}{\mathrm{d}t} \qquad \text{and} \qquad \langle h\rangle_{\text{av}} = n\langle r^2\rangle_{\text{av}}, \tag{31}$$

when averaged over a synodic month, we can clearly write

$$h = r^2\frac{\mathrm{d}v}{\mathrm{d}t} = \langle h\rangle_{\text{av}}\left\{1 + \frac{3}{4}\frac{m^2}{1-m}\cos[2(1-m)v]\right\}. \tag{32}$$

(*e*) *The determination of x*

We start with the formula,

$$V^2 = \left(\frac{\mathrm{d}r}{\mathrm{d}t}\right)^2 + r^2\left(\frac{\mathrm{d}v}{\mathrm{d}t}\right)^2 \simeq r^2\left(\frac{\mathrm{d}v}{\mathrm{d}t}\right)^2. \tag{33}$$

for the square of the velocity, V, along the orbit. By equation (32)

$$\begin{aligned} V^2 &= \frac{\langle h\rangle_{\text{av}}^2}{r^2}\left\{1 + \frac{3}{4}\frac{m^2}{1-m}\cos[2(1-m)v]\right\}^2 \\ &\simeq \frac{\langle h\rangle_{\text{av}}^2}{r^2}\left[1 + \frac{3}{2}\frac{m^2}{1-m}\cos 2\psi + O(m^4)\right], \end{aligned} \tag{34}$$

it being assumed that m is much less than 1; actually, in the context $m \simeq 0{\cdot}075$; see equation (42) below. In particular at conjunction ($\psi = 0$) and at quadrature ($\psi = \pi/2$)

$$V_1^2 \simeq \frac{\langle h \rangle_{\rm av}^2}{(1-x)^2}\left(1 + \frac{3}{2}\frac{m^2}{1-m}\right) \quad \text{and} \quad V_0^2 \simeq \frac{\langle h \rangle_{\rm av}^2}{(1+x)^2}\left(1 - \frac{3}{2}\frac{m^2}{1-m}\right). \tag{35}$$

On the other hand the centripetal force F in the direction of the inward normal is given by

$$F = V^2/\rho, \tag{36}$$

where ρ denotes the radius of curvature. At syzygies and at quadratures, the inward normals are along the principal axes; and therefore, by equation (36)

$$\frac{F_1}{F_0} = \frac{V_1^2 \rho_0}{V_0^2 \rho_1} \simeq \left(\frac{1+x}{1-x}\right)^2 \left(1 + \frac{3m^2}{1-m}\right)\frac{\rho_0}{\rho_1}, \tag{37}$$

where, by equation (20),

$$\frac{F_1}{F_0} \simeq \left(\frac{1+x}{1-x}\right)^2 (1 - 3m^2). \tag{38}$$

Combining equations (37) and (38) we obtain,

$$\frac{\rho_0}{\rho_1} \simeq 1 - 3m^2\left(1 + \frac{1}{1-m}\right). \tag{39}$$

Now making use of the expression (25) for ρ_0/ρ_1 we have,

$$1 - 3m^2\left(1 + \frac{1}{1-m}\right) = 1 - 2x[4(1-m)^2 - 1]; \tag{40}$$

or, solving for x, we have

$$x = \frac{3}{2}m^2 \frac{1 + \dfrac{1}{1-m}}{4(1-m)^2 - 1} = m^2\left(1 + \frac{19}{6}m\right) + O(m^3). \tag{41}$$

a result known to be correct to $O(m^3)$. For the Earth–Moon system

$$m = 0{\cdot}074803, \qquad x = 0{\cdot}007202, \qquad \text{and} \qquad \frac{1-x}{1+x} = 0{\cdot}98571 \simeq \frac{69}{70}; \tag{42}$$

and this is Newton's result (explicitly given in Proposition XXVIII, Book III; see Chapter 22).

Now to Newton's own manner of enunciating the results derived more expansively in §§ (a)–(e) above:

> COR. I. Hence it may be easily inferred, that if several less bodies *P*, *S*, *R*, etc., revolve about a very great body *T*, the motion of the innermost revolving body *P* will be least disturbed by the attractions of the others.

This corollary follows from the fact that the disturbing forces, F_r and F_α, acting on each of the bodies, is directly proportional to the radius of its orbit about T. (It may be noted, parenthetically, that a derivation of the expressions (12) and (13) for F_r and F_α, explicitly along the same lines, will be found in Proposition XXVI, Book III.)

> COR. II. In a system of three bodies T, P, S, if the accelerative attractions of any two of them towards a third be to each other inversely as the squares of the distances, the body P, by the radius PT, will describe its area about the body T swifter near the conjunction A and the opposition B than it will near the quadratures C and D.

This follows from equation (32) according to which 'the constant of areas' h, at syzygies and at quadratures, are, respectively,

$$h_1 = \langle h \rangle_{\text{av}}\left(1 + \frac{3}{4}\frac{m^2}{1-m}\right) \quad \text{and} \quad h_0 = \langle h \rangle_{\text{av}}\left(1 - \frac{3}{4}\frac{m^2}{1-m}\right). \tag{43}$$

> COR. III. And from the same reasoning it appears that the body P, other things remaining the same, moves more swiftly in the conjunction and opposition than in the quadratures.

This is a statement of the result that follows from equation (35), namely that

$$V_1 > V_0. \tag{44}$$

> COR. IV. The orbit of the body P, other things remaining the same, is more curved at the quadratures than at the conjunction and opposition.

Again, this is a statement of the result,

$$\rho_0^{-1} > \rho_1^{-1}, \tag{45}$$

that follows from equation (39). Its origin in equation (36) is recalled by Newton:

> COR. V. Hence the body P, other things remaining the same, goes farther from the body T at the quadratures than at the conjunction and opposition.

This was initially an assumption (cf. equation (19)) but justified later by deriving equation (41) from the premise. Laplace's statement quoted in § (c) above is precisely in this context.

> COR. VI. The periodical time would be increased and diminished in a ratio compounded of the $\frac{3}{2}$th power of the ratio of the radius, and of the square root of that ratio in which the centripetal force of the central body T was diminished or increased, by the increase or decrease of the action of the distant body S.

This corollary is remarkable for the insight that Newton brings to bear on its solution. In order better to appreciate his reasoning, we shall first consider (as hitherto) the problem directly from the relevant variational equations.

By Chapter 13, § 66(c), equation (vii),

$$\frac{da}{dt} = \frac{2a}{rn} F_\alpha = -3N^2 \frac{a}{n} \sin 2\psi = -3m^2 n a_0 \sin[2(1-m)v], \tag{46}$$

where a is replaced by a 'mean value' a_0; or, after integration.

$$a - a_0 = +\frac{3}{2}\frac{m^2}{1-m} a_0 \cos 2\psi. \tag{47}$$

This expression for $a - a_0$, which one obtains from the first-order variational equation, is *qualitatively* incorrect: it predicts an oblate ellipse for the perturbed orbit contrary to what Newton has already shown to be a prolate ellipse (see equation (42)).

Newton obtains for the period a formula correct to $O(m^2)$ by an ingenious argument. First, he notes what is implied in Corollary II (equation (43)) and more explicitly in Corollary VII, §71, equations (15) and (17), below, namely,

$$\frac{(\text{C.F.})_\text{s}}{(\text{C.F.})_\text{q}} = \frac{r_\text{q}^2}{r_\text{s}^2}(1 - 3m^2 r^3), \tag{48}$$

where the subscripts s and q refer to syzygies and to quadratures. (It will be noted that we have not distinguished between r_s and r_q when r occurs with the small factor m^2). Next, referring to Corollary VI, Proposition IV (p. 74), he notices that

$$\frac{(\text{C.F.})_\text{s}}{(\text{C.F.})_\text{q}} = \frac{r_\text{s}}{r_\text{q}}\left(\frac{P_\text{q}}{P_\text{s}}\right)^2, \tag{49}$$

where P_q and P_s denote the periods at the quadratures and at the syzygies. Combining equations (48) and (49), he obtains

$$\frac{P_\text{q}}{P_\text{s}} = \left(\frac{r_\text{q}}{r_\text{s}}\right)^{3/2} \cdot \sqrt{(1 - 3m^2 r^3)}. \tag{50}$$

More generally, we may write, in accordance with equation (32),

$$P \propto r^{3/2}[1 + \tfrac{1}{4}m^2 r^3(1 + 3\cos 2\psi)]. \qquad \text{Q.E.D.} \tag{51}$$

An equivalent formula for n is

$$n = \left(\frac{\mu}{a^3}\right)^{1/2}[1 - \tfrac{1}{4}m^2 r^3(1 + 3\cos 2\psi)]. \tag{52}$$

An equation for the variation of a follows from inverting equation (52). Thus,

$$a = \left(\frac{\mu}{n^2}\right)^{1/3}[1 - \tfrac{1}{6}m^2 r^3(1 + 3\cos 2\psi)], \tag{53}$$

or, replacing r by 1,

$$a = a_{\text{ellipse}}(1 - \tfrac{1}{6}m^2 - \tfrac{3}{2}m^2 \cos 2\psi), \tag{54}$$

where a_{ellipse} stands for the semimajor axis of the unperturbed orbit. Alternatively, we can write

$$\frac{a - a_{\text{ellipse}}}{a_{\text{ellipse}}} = -(\tfrac{1}{6}m^2 + \tfrac{3}{2}m^2 \cos 2\psi), \tag{55}$$

which differs from Newton's earlier result (41) by including the uniform contraction term $-\frac{1}{6}m^2$ and the replacement of

$$m^2(1 + \tfrac{19}{6}m) \qquad \text{by} \qquad \tfrac{3}{2}m^2. \tag{56}$$

It may be noted here that for $m = 0{\cdot}07480$ the second-order formula replaces $1{\cdot}5m^2$ by $1{\cdot}237m^2$.

In concluding an account of Corollaries I–VI, we may note that their ordering is not the same as that in which they are derived in §§(a)–(e) above. But it is abundantly clear that they all follow directly the variational equation

$$\frac{dh}{dt} = rF_\alpha; \tag{57}$$

and Newton must have known it.

71. Proposition LXVI (continued): Corollaries VII and VIII—the rotation of the line of apsides

In Corollaries VII and VIII, Newton returns to the problem of the rotation of the line of apsides he had considered earlier in Proposition XLV; and makes use in fact of the result obtained in Corollary II of the proposition.

> COR. VII. It also follows, from what was before laid down, that the axis of the ellipse described by the body P, or the line of the apsides, does as to its angular motion go forwards and backwards by turns, but more forwards than backwards, and by the excess of its direct motion is on the whole carried forwards.

But first we shall consider the problem by way of the variational equation for

$$\varpi = \omega + \Omega = \angle(\vec{e}, \vec{v}) + \angle(\vec{v}, \vec{i}), \tag{1}$$

When the inclination ι is zero (as presently assumed)

$$\varpi = \angle(\vec{e}, \vec{i}). \tag{2}$$

Since $\vec{i}$ is a fixed vector, $d\varpi/dt$ describes the rotation of the Lenz vector with respect to a fixed direction and therefore of the rotation of the line of apsides. By the last of the three equations listed in equation (41) of §65 in Chapter 13 for $\iota = 0$,

$$e\frac{d\varpi}{dt} = \frac{h}{\mu}\left[-F_r \cos\varphi + F_\alpha\left(1 + \frac{r}{l}\right)\sin\varphi\right]. \tag{3}$$

Making use of the zero-order relation,

$$\frac{h}{\mu} = \frac{na^2\sqrt{(1-e^2)}}{\mu} = \frac{\sqrt{(1-e^2)}}{na}, \tag{4}$$

and substituting for F_r and F_α from equations (22)–(24) of §66 in Chapter 13, we obtain

$$e\frac{\mathrm{d}\varpi}{\mathrm{d}t} = -\tfrac{1}{2}\,m^2 n\frac{r}{a}\sqrt{(1-e^2)}\left\{(1+3\cos 2\psi)\cos\varphi + 3\left(1+\frac{r}{l}\right)\sin 2\psi \sin\varphi\right\}. \tag{5}$$

We may note here for future reference that at quadratures ($\psi = \pi/2$ and $3\pi/2$) and syzygies ($\psi = 0$ and π), respectively,

$$\left.\begin{aligned} e\left(\frac{\mathrm{d}\varpi}{\mathrm{d}t}\right)_0 &= +m^2 n\frac{r}{a}(1-e^2)^{1/2}\cos\varphi,\\ e\left(\frac{\mathrm{d}\varpi}{\mathrm{d}t}\right)_1 &= -2m^2 n\frac{r}{a}(1-e^2)^{1/2}\cos\varphi. \end{aligned}\right\} \tag{6}$$

Returning to equation (5), we observe that on the average the contributions from the terms with the factors $\cos 2\psi$ and $\sin 2\psi$ will vanish. Therefore, to determine a secular term (if there is one) it will suffice to consider

$$e\langle \mathrm{d}\varpi\rangle = -\tfrac{1}{2}m^2\sqrt{(1-e^2)}\left\langle\left(\frac{r}{a}\cos\varphi\right)n\,\mathrm{d}t\right\rangle. \tag{7}$$

Using the equations defined in the context of Kepler's equation, we have (cf. Chapter 13, §65, equations (24) and (35))

$$\langle e\,\mathrm{d}\varpi\rangle = -\tfrac{1}{2}m^2\sqrt{(1-e^2)}\langle(\cos u - e)(1-e\cos u)\rangle\,\mathrm{d}u, \tag{8}$$

where u denotes the eccentric anomaly. Since

$$\begin{aligned}(\cos u - e)(1 - e\cos u) &= (1+e^2)\cos u - e(1+\cos^2 u)\\ &= -\tfrac{3}{2}e + (1+e^2)\cos u - \tfrac{1}{2}e\cos 2u, \end{aligned} \tag{9}$$

we conclude from equation (8) that

$$\langle \mathrm{d}\varpi\rangle = \tfrac{3}{4}m^2\sqrt{(1-e^2)}\,\mathrm{d}u\,(\sec^{-1}), \tag{10}$$

or

$$\langle \varpi\rangle = \tfrac{3}{4}m^2\sqrt{(1-e^2)}u\,(\sec^{-1}). \tag{11}$$

In other words, the line of apsides will rotate forward (in the direction of rotation) with a mean amplitude

$$\tfrac{3}{4}m^2\sqrt{(1-e^2)}. \tag{12}$$

We shall find that this is in agreement with what Newton had found earlier in Proposition XLV.

Newton derives the result differently. He first observes that the force transverse to the radial direction (i.e., F_α) will have no effect on the average, while the total centripetal force (per unit mass of the Moon) which will have a direct effect, is given by (cf. Chapter 13, §66, equations (6), (16), (22), and (23))

$$\text{C.F.} = -\frac{GM_T}{r^2} + \frac{GM_\odot}{2R^3}r(1 + 3\cos 2\psi). \tag{13}$$

Or, by making use of the standard definitions,

$$\text{C.F.} = -n^2\frac{(\overline{PT})^3}{r^2} + \tfrac{1}{2}N^2 r(1 + 3\cos 2\psi), \tag{14}$$

or, measuring r in units of the average (circular) radius of the Moon ($\overline{PT}$), we have

$$\text{C.F.} = -k\left[\frac{1}{r^2} - \tfrac{1}{2}m^2 r(1 + 3\cos 2\psi)\right], \tag{15}$$

where (cf. §70, equation (16))

$$k = n^2\overline{PT}. \tag{16}$$

From equation (15) it follows that at quadratures and at syzygies,

$$(\text{C.F.})_0 = -k\left(\frac{1}{r^2} + m^2 r\right) \quad \text{and} \quad (\text{C.F.})_1 = -k\left(\frac{1}{r^2} - 2m^2 r\right). \tag{17}$$

Therefore, as Newton states, the centripetal force 'decreases less than' and 'more than the square of the ratio of the distance PT' at quadratures and at syzygies, respectively.

In Proposition XLV, the rotation of the line apsides under a centripetal force, (writing r in place of A)

$$-k\left(\frac{1}{r^2} - cr\right), \tag{18}$$

was considered; and it was shown that the angle of revolution between the apsides is given by

$$180^\circ\sqrt{\frac{1-c}{1-4c}} \simeq 180^\circ(1 + \tfrac{3}{2}c). \tag{19}$$

Therefore, for the laws of force, prevailing at quadratures and at syzygies, the angle of revolution is given

$$180^\circ(1 - \tfrac{3}{2}m^2) \quad \text{and} \quad 180^\circ(1 + 3m^2), \quad \text{respectively,} \tag{20}$$

that is at the quadratures 'the upper apse will go backwards', while at syzygies 'the line of apsides will go forwards'. And therefore since the advance 'in the syzygies is almost twice as great—as the regression in the quadratures—the line of apsides will go forward'.

Q.E.D.(!)

Newton also notices that 'in places between the syzygies and the quadratures' the effect is one of 'going forward'. For example at the octants ($\psi = \pi/4$), the centrifugal force is

$$-k\left(\frac{1}{r^2} - \tfrac{1}{2}m^2 r\right), \tag{21}$$

which will, by equation (19), predict an angle of revolution by the apsides,

$$180^\circ(1 + \tfrac{3}{4}m^2), \tag{22}$$

which is the average of the predictions (20).

Newton had clearly before him the results (20) and (22); but he does not repeat his earlier statement in Proposition XLV that in comparison with observation it is 'half as swift'.

Corollary VIII

This corollary states that 'it is plain that, when the apsides are in syzygies', the line of apsides will 'go forwards swiftly, and in the quadratures... go backwards more slowly'. These are direct consequences of equations (6) at the upper apse where $\cos\varphi = -1$.

72. Proposition LXVI (continued): Corollaries IX–XVII

In Corollaries IX–XI, Newton considers in turn the variation in the eccentricity, inclination, and the direction of the ascending node. In the following Corollaries, XIII–XVII, he considers various related matters: some elaborations and some generalizations.

(a) Corollary IX: the variation of eccentricity

First we shall consider the problem directly from the variational equations derived in Chapter 13, §65, equations (41)(iv), namely,

$$\frac{de}{dt} = \frac{h}{\mu}[F_r \sin\varphi + (\cos\varphi + \cos u)F_\alpha]. \tag{1}$$

Substituting for F_r and F_α from Chapter 13, §66, equations (23) and (24) and making use of the relations,

$$h = na^2\sqrt{(1 - e^2)}; \qquad \frac{h}{\mu} = \frac{\sqrt{(1 - e^2)}}{na}, \tag{2}$$

we obtain (cf. Chapter 13, §66(c), equation (iv));

$$\frac{de}{dt} = \tfrac{1}{2}m^2 n\sqrt{(1 - e^2)}\,\frac{r}{a}\,\{(1 + 3\cos 2\psi)\sin\varphi - 3(\cos u + \cos\varphi)\sin 2\psi\}. \tag{3}$$

To zero order in e, the foregoing equation becomes,

$$\frac{de}{dt} = \tfrac{1}{2}m^2 n[(1 + 3\cos 2\psi)\sin\varphi - 6\sin 2\psi\cos\varphi]; \tag{4}$$

or, after some elementary trigonometry,

$$\frac{de}{dt} = \tfrac{1}{4}m^2 n[2 \sin \varphi - 3 \sin(2\psi + \varphi) - 9 \sin(2\psi - \varphi)]. \tag{5}$$

To bring out the dependence on v explicitly we make use of the definitions (cf. the illustration on p. 231)

$$\left.\begin{aligned} &\varphi = v - \omega; \qquad \psi = (1 - m)v, \\ &\text{whence} \\ &2\psi + \varphi = (3 - 2m)v - \omega; \qquad 2\psi - \varphi = (1 - 2m)v + \omega. \end{aligned}\right\} \tag{6}$$

With these substitutions, equation (5) becomes

$$de = \tfrac{1}{4}m^2\{2 \sin(v - \omega) - 3 \sin[(3 - 2m)v - \omega] - 9 \sin[(1 - 2m)v + \omega]\}n\, dt. \tag{7}$$

Since in our present approximation $n\, dt = dv$, equation (7) can be integrated directly to give

$$\begin{aligned} e - e_0 = \tfrac{1}{4}m^2\Bigg\{ -2 \cos(v - \omega) &+ \frac{3}{3 - 2m} \cos[(3 - 2m)v - \omega] \\ &+ \frac{9}{1 - 2m} \cos[(1 - 2m)v + \omega]\Bigg\}. \end{aligned} \tag{8}$$

Ignoring the m's in the denominators and reverting to the variables φ and ψ, we have:

$$e - e_0 = \tfrac{1}{4}m^2\{-2 \cos \varphi + \cos(2\psi + \varphi) + 9 \cos(2\psi - \varphi)\}. \tag{9}$$

To bring out the dependence on ψ explicitly, we substitute for φ (see the illustration on p. 231):

$$\varphi = \psi - \gamma, \tag{10}$$

where $\gamma = \angle(\vec{R}, \vec{e})$, may be considered to be a constant during a synodic month; and we find

$$e - e_0 = \tfrac{1}{4}m^2[-2 \cos(\psi - \gamma) + \cos(3\psi - \gamma) + 9 \cos(\psi + \gamma)], \tag{11}$$

or, after some elementary trigonometry,

$$(e - e_0) = m^2[(1 + \cos^2 \psi) \cos \psi \cos \gamma - (2 + \sin^2 \psi) \sin \psi \sin \gamma]. \tag{12}$$

In particular,

$$\left.\begin{aligned} &\text{at quadratures } (\psi = \pi/2): \qquad e - e_0 = -3m^2 \sin \gamma, \\ &\text{and} \\ &\text{at syzygies } (\psi = 0): \qquad e - e_0 = +2m^2 \cos \gamma. \end{aligned}\right\} \tag{13}$$

Now to Newton's presentation. Though he must have derived for himself the relations (13) (see below) he chose to argue qualitatively (and secretively). His principal argument relates to the variation of the centripetal force according to the formula (§71, equation (15))

$$\text{centripetal force:} \qquad -k\left[\frac{1}{r^2} - \tfrac{1}{2}m^2 r(1 + 3 \cos 2\psi)\right]; \tag{14}$$

and the 'accession of the new forces'

$$\left.\begin{array}{ll} -km^2r & \text{at the quadratures} \\ +2km^2r & \text{at the syzygies,} \end{array}\right\} \tag{15}$$

and

to the prevailing $-k/r^2$. He repeats essentially the arguments of Corollary VII which led to the 'direct motion of the line of the apsides, as was just now said'. He considers also the dependence of the increments (15) on the radial distances at the upper and the lower apse in case the orbit is an ellipse of finite eccentricity. Since he concentrates primarily on the forces acting at the syzygies and at the quadratures, he does not have to mention the effect of the transverse force F_α because it vanishes at these points. With these brief qualitative remarks he concludes (correctly!)

> COR. IX. ...... Therefore the ratio of the whole increase and decrease [of the centripetal force] in the passage between the apsides is least at the quadratures and greatest at the syzygies; and therefore in the passage of the apsides from the quadratures to the syzygies it [the eccentricity] is continually augmented, and increases the eccentricity of the ellipse; and in the passage from the syzygies to the quadratures it is continually decreasing, and diminishes the eccentricity.

The first part of this statement reiterates the facts (15), while the second part is in accord with equations (13). It will be noted that in the derivation of equations (13) both the forces F_r and F_α play an equal part; and it is hard to believe that Newton did not derive the equations for himself in essentially the manner we have.

(b) Corollary X: the variation of the inclination

> The inclination, therefore, is the greatest of all when the nodes are in the syzygies. In their passage from the syzygies to the quadratures the inclination is diminished at each appulse of the body to the nodes; and becomes least of all when the nodes are in the quadratures, and the body in the syzygies; then it increases by the same degrees by which it decreased before; and, when the nodes come to the next syzygies, returns to its former magnitude.

Since Newton shows complete awareness of the variational equation for the inclination we shall start with the equation (Chapter 13, §65, equation (41)(ii))

$$\frac{d\iota}{dt} = \frac{r}{h} F_h \cos v = \frac{r}{na^2\sqrt{(1-e^2)}} F_h \cos v, \tag{16}$$

or, substituting for F_h, we have (Chapter (13), §66(c), equation (ii))

$$\frac{d\iota}{dt} = -3m^2 n \frac{r^2}{a^2\sqrt{(1-e^2)}} \sin \iota \cos v \sin U \cos \psi. \tag{17}$$

To zero order in the eccentricity, the equation becomes

$$\frac{d\iota}{dt} = -3m^2 n \sin\iota \cos v \sin U \cos\psi; \tag{18}$$

or, after some elementary trigonometry,

$$d\iota = -\tfrac{3}{4}m^2 \sin\iota\{\sin 2v + \sin(2mv) - \sin[2(1-m)v]\}n\,dt. \tag{19}$$

On integrating this equation, we obtain:

$$\iota - \iota_0 = \tfrac{3}{8}m^2 \sin\iota\left(\cos 2v + \frac{1}{m}\cos 2U - \frac{1}{1-m}\cos 2\psi\right). \tag{20}$$

Clearly, the dominant term is $(\cos 2U)/m$. Hence

$$\iota - \iota_0 \simeq \tfrac{3}{8}m \sin\iota \cos 2U. \tag{21}$$

The variation in the inclination is therefore periodic with a period of one-half a nodal year with an amplitude $(3m/8)$ radians $(=1{\cdot}6^\circ)$.

Newton's description of the results derived from the posited variational equation (18) is a *tour de force* leaving no doubt of his awareness of the equation.

The dependence of $d\iota/dt$ on $\sin\iota$ is implied in his remarks that there will be no change in the inclination if initially ι were to vanish identically; and that a change can occur only by the operation of the force (F_h) normal to the orbital plane. And the periodic dependence of the variation on $-\sin U$ is an obvious inference from the statements that the effect vanishes when the nodes are in syzygies (when $U = 0$) and that it is a maximum at quadratures (when $U = \pi/2$); and more fully in the following passage:

> But when the nodes are in the quadratures, it disturbs them very much, and, attracting the body P continually out of the plane of its orbit, it diminishes the inclination of the plane in the passage of the body from the quadratures to the syzygies, and again increases the same in the passage from the syzygies to the quadratures.

Without any further explanation, he slips in ('hence it comes to pass'!) the results of the integration of equation (19):

> Hence it comes to pass that when the body is in the syzygies, the inclination is then least of all, and returns to the first magnitude nearly, when the body arrives at the next node. But if the nodes are situated at the octants after the quadratures, that is, between C and A, D and B, it will appear, from what was just now shown, that in the passage of the body P from either node to the ninetieth degree from thence, the inclination of the plane is continually diminished; then, in the passage through the next 45 degrees to the next quadrature, the inclination is increased; and afterwards, again, in its passage through another 45 degrees to the next node, it is diminished.

The statement says (in accordance with equation (21)) that $(\iota - \iota_0)$ diminishes in the interval, $\pi/4 \leqslant U \leqslant 3\pi/4$, and increases in the following period. The conclusion is repeated in the concluding sentences quoted at the beginning.

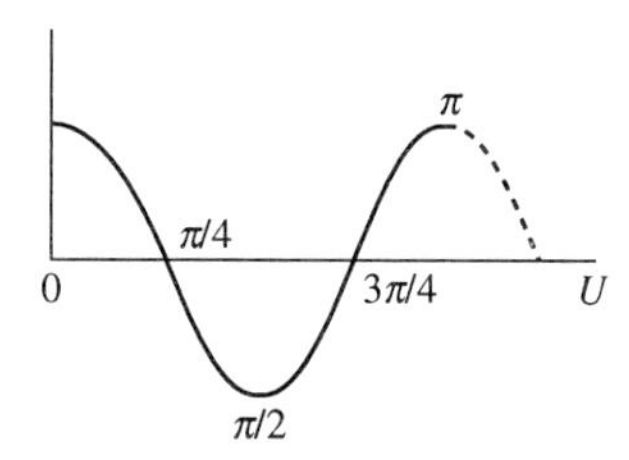

(c) *Corollary XI: the variation of the direction of the ascending node* (Ω)

As in our presentations of Corollaries IX and X we shall start with the variational equations for Ω (Chapter 13, §65 equation (41)(iii)):

$$\sin \iota \frac{d\Omega}{dt} = \frac{r}{h} F_h \sin v. \tag{22}$$

Substituting for F_h and retaining only terms of zero order in e, we find as in §(b) (see equations (16) and (18));

$$\frac{d\Omega}{dt} = -3m^2 n \sin v \sin U \cos \psi; \tag{23}$$

or, after some elementary trigonometry,

$$\frac{d\Omega}{dt} = -\tfrac{3}{4}m^2 n(1 + \cos 2\psi - \cos 2v - \cos 2U). \tag{24}$$

Rewriting this equation in the form,

$$d\Omega = -\tfrac{3}{4}m^2[1 + \cos[2(1-m)v] - \cos 2v - \cos(2mv)]n\, dt, \tag{25}$$

we can integrate it directly to give

$$\Omega = -\tfrac{3}{4}m^2 nt - \tfrac{3}{8}m^2\left[\frac{1}{1-m}\sin 2\psi - \sin 2v - \frac{1}{m}\sin 2U\right]. \tag{26}$$

Of the periodic terms, $(\sin 2U)/m$ is clearly the dominant one and we have

$$\Omega \simeq -\tfrac{3}{4}m^2 nt + \tfrac{3}{8}m \sin 2U. \tag{27}$$

Thus, there is a regression of the nodes of $2\pi(3m^2/4)$ per nodal month besides a periodic variation of amplitude $(3m/8)$ radians $(=1{\cdot}6^\circ)$ (or, $19{\cdot}2^\circ$ per year),* with a period of one-half nodal year.

Newton's explanation of the regression of the nodes is based on the observation that

* This is the value that Newton explicitly derives in Book III, Proposition XXXII.

the variation of the inclination is exactly out of phase with the variation of Ω as follows from equations (18) and (23):

$$\left.\begin{aligned}\frac{\mathrm{d}\iota}{\mathrm{d}t} &\propto \cos v \sin U \cos\psi = -\tfrac{1}{4}\sin 2\psi + \tfrac{1}{2}\cos\psi\sin(v+U),\\ \text{while}\qquad \frac{\mathrm{d}\Omega}{\mathrm{d}t} &\propto \sin v \sin U \cos\psi = +\tfrac{1}{2}\cos^2\psi - \tfrac{1}{2}\cos\psi\cos(v+U).\end{aligned}\right\} \tag{28}$$

The fact that

$$\langle\Delta\iota\rangle_{\mathrm{av}} = 0, \quad \text{while by equation (27),} \quad \langle\Delta\Omega\rangle_{\mathrm{av}} = -\tfrac{3}{4}m^2 \quad \text{per synodic month,} \tag{29}$$

are immediate consequences. Newton does not spell out the equations but concludes that 'by like reasoning'(!)

> the nodes will continue to recede in their passage from this node to the next. The nodes, therefore, when situated in the quadratures, recede continually; and at the syzygies, where no perturbation can be produced in the motion as to latitude, are quiescent; in the intermediate places they partake of both conditions, and recede more slowly; and, therefore, being always either retrograde or stationary, they will be carried backwards, or made to recede in each revolution.

(*d*)

> COR. XII. All the errors described in these corollaries are a little greater at the conjunction of the bodies P, S than at their opposition; because the generating forces NM and ML are greater.

This is the shortest of the 22 corollaries. Its special importance is the insight it gives on what Newton was contemplating already at this stage: after having investigated the consequences, in their entirety, of the variational equations and having also realized their inadequacy for accounting for all effects (e.g. the advance of the line apsides in Corollary VII), Corollary XII reveals, in no doubtful terms, that Newton had already started to assess the second-order effects; and in particular on the centripetal force.

It can be shown that correct to $O(r^2/R^4)$, F_r is given by

$$F_r = \frac{GM_\odot}{R^3}\left[r(3\cos^2\psi - 1) + \frac{3}{2}\frac{r^2}{R}(5\cos^2\psi - 3)\cos\psi\right]. \tag{30}$$

This expression follows from differentiating with respect to r the expression for the potential given in E. W. Brown, *An introductory treatise on the lunar theory* (Cambridge University Press, 1896) equation (1) on p. 79. This expression can also be derived directly by the method used in Chapter 13.

An alternative form of equation (30), comparable to our earlier expression for F_r, is

$$F_r = \tfrac{1}{2}N^2 r\left[(1 + 3\cos 2\psi) + 3\frac{r}{R}(5\cos^2\psi - 3)\cos\psi\right]. \tag{31}$$

It follows that at conjunction ($\psi = 0$) we have an increment in F_r of amount $3N^2r^2/R$ while at opposition ($\psi = \pi$) we have an equal decrement. There is evidence that Newton might have derived this result in his efforts to refine his lunar theory (see Chapter 22); but the essential point of the corollary is to point out that the second-order corrections, relative to the first-order corrections, are odd and that at conjunction and at opposition they will have opposite signs.

A comment (contrary to the traditional view)* that I should like to make at this point is that Newton, after the concentrated mental effort required of him to complete his investigation of the first-order variational equations, must have fully realized the much greater effort that would be needed to evaluate all second-order effects—not that the problem itself, in any measure, was beyond his capacity; and it is not surprising that the very thought of embarking on the project 'made his head ache and kept him awake so often that he would think of it no more'. Besides, the evaluation of all the second-order terms would require 'more muscle than brains' (to quote Henry Norris Russell in another context).

(*e*) *Further elaborations: Corollaries XIII–XVII*

In Corollaries XIII–XVII, Newton fills in certain gaps in his presentation of Corollaries I–XI: in particular the factor $GM_\odot/R^3$ in the expressions for F_r, F_α, and F_h and the factor m^2 in the variational equations $\omega_{,t}/n$, $e_{,t}/n$, $\iota_{,t}/\mathrm{n}$, and $\Omega_{,t}/\mathrm{n}$.

Corollary XIII. In the expression for F_r, F_α, and F_h derived in Chapter 13, §66 (equations (23)–(25)) the factor $GM_\odot/R^3$ was replaced by N^2 in equation (26), as representing the 'Mean motion of the Sun (relative to the Earth)' Strictly, the noted equivalence is valid only if T describes its orbit relative to an 'immovable' Sun. But if N^2 is to represent the true anomalistic mean motion of S around T under their mutual attractions, then by Proposition LIX (the concluding remark)

$$N^2(S \text{ fixed}) = N^2(\text{under the mutual attractions of } S \text{ and } T) \times \frac{M_\odot}{M_T + M_\odot}. \tag{32}$$

Consequently, if N^2 in the variational equations means the 'true' mean square motion of S then the additional factor $[(M_T + M_\odot)/M_\odot]$ must be included. It appears that it is to this enhancement, that the 'consequent increase in the centripetal force' refers (though Newton is not explicit).

* Chapter 22, Page 449)

Corollary XIV. In this corollary Newton states explicitly for the first time (in this proposition) that the factor that occurs in the expressions for the disturbing functions is $GM_{\odot}/R^3$. In Newton's words, the disturbance

> will be very nearly in a ratio compounded of the direct ratio of the absolute force of the body *S*, and the cubed inverse ratio of the distance *ST*... and their effects, will be (by Cor. ii. and vi, Prop. iv) inversely as the square of the periodical time.

Corollary XV. In this corollary Newton infers the proportionality of the disturbing function to the radius r of the orbit of P around T by an argument of similarity:

> the force of the body *S*, which causes the body *P* to deviate from that orbit will act always in the same manner, and in the same proportion.

And he concludes:

> that is, that all the linear errors will be as the diameters of the orbits, the angular errors the same as before; and the times of similar linear errors, or equal angular errors, are as the periodical times of the orbits.

(It may be noted that here and elsewhere Newton means by 'errors' what we call the disturbing function.)

Corollary XVI. In Corollaries VII and XI it was shown that the terms in the mean motion of the line of the apsides and the mean regression of the line of ascending nodes are (apart from sign) the same. By §71 equations (12) and (27)

$$\left.\begin{aligned}\langle\varpi\rangle_{\text{av}} &= +\tfrac{3}{4}m^2nt = +\tfrac{3}{4}(N^2/n)t,\\ \text{and}\quad \langle\Omega\rangle_{\text{av}} &= -\tfrac{3}{4}m^2nt = -\tfrac{3}{4}(N^2/n)t.\end{aligned}\right\} \tag{33}$$

By arguments essentially the same as in Cor. VII and Cor. XI, Newton now concludes:

> And therefore the mean motion of the line of the apsides will be in a given ratio to the mean motion of the nodes; and both those motions will be directly as the periodical time of the body *P*, and inversely as the square of the periodical time of the body *T*. The increase or diminution of the eccentricity and inclination of the orbit *PAB* makes no sensible variation in the motions of the apsides and nodes, unless that increase or diminution be very great indeed.
>
> Cor. XVII. The mean force *SN* or *ST*, by which the body *T* is retained in the orbit it describes about *S*, is to the force with which the body *P* is retained in its orbit about *T* in a ratio compounded of the ratio of the radius *ST* to the radius *PT*, and the squared ratio of the periodical time of the body *P* about *T* to the periodical time of the body *T* about *S*.

The mean force with which 'T is retained in its orbit about S' is

$$GM_{\odot}/R^2, \tag{34}$$

while 'the mean force with which the body P is retained in its orbit' is

$$GM_T/r^2. \tag{35}$$

Therefore, their ratio is

$$\left(\frac{GM_{\odot}}{R^3}\right)R:\left(\frac{GM_T}{r^3}\right)r = \frac{N^2}{n^2}\frac{ST}{PT}, \tag{36}$$

where ST and PT have the meanings they have in §70. (Q.E.D.)

A personal reflection

By Corollary XVII, the analysis of the perturbations of a Kepler orbit by a distant gravitating body has been completed. And the corollaries that follow are devoted to peripheral matters; such as the theory of tides (to be expanded in Proposition XXIV of Book III) and why gravitating bodies tend to be spherical in the absence of external perturbations. It is an appropriate time to pause and reflect.

Two thoughts are uppermost in my mind. First, Newton's astonishing grasp of the entire problem of planetary perturbations and the power of his insight. No surprise that Laplace thought that these corollaries are amongst the most remarkable in the *Principia.* A second thought concerns the paucity of any real explanation and an apparent attempt to conceal details by recourse, too often, to phrases like 'hence it comes to pass', 'by like reasoning', and 'it is manifest that' at crucial points of the argument. This 'secretive style' is nowhere present, to this same extent, in the *Principia.*

It is possible that Newton developed the entire theory in his mind by an unsurpassed concentration* suspected by Lord Keynes in his memorable address at the Newton Tercentenary Celebrations at the Royal Society in 1946.

* Apropos of the quotation from Lord Keynes to follow, I have often thought of Newton's ability for unsurpassed concentration with that of Prince Arjuna of whom the following tale is told:

> Drona assembled them all to test their knowledge of weaponry. He had craftsmen fashion an artificial bird and attach it to a treetop where it was hardly visible, and proceeded to point out the target to the princes.
>
> And he said:
>
> Quickly take your bows, put your arrow to the string, and take your position aiming at this bird. As soon as I give the word, shoot off its head. I shall order you one after the other, and you do it boys!
>
> He first turned to Yudhisthira [the eldest of the princes]: 'Lay on the arrow, invincible prince', he said, 'and as soon as I have ceased talking, let go of it!' Yudhisthira then first took his loud-sounding bow and at his guru's command stood aiming at the bird. Drona said to him after a while, 'Do you see the bird in the treetop, prince?' 'I see it', Yudhisthira replied to his teacher. After a while Drona again said to him, 'Now can you see the tree or me, or your brothers?' 'Yes', he said to each question, 'I see the tree, and yourself, and my brothers, as well as the bird'. Then Drona, dissatisfied, scolded him: 'You won't be able to hit that target'. Then the famous teacher questioned Duryodhana and the other sons of Dhrtarástra one after the other in the same way, to put them to the test; and also the other pupils and the foreign kings. They all said that they could see everything, and were scolded.

> I believe that the clue to his mind is to be found in his unusual powers of continuous concentrated introspection.... His peculiar gift was the power of holding continuously in his mind a purely mental problem until he had seen straight through it. I fancy his pre-eminence is due to his muscles of intuition being the strongest and most enduring with which a man has ever been gifted....
>
> I believe that Newton could hold a problem in his mind for hours and days and weeks until it surrendered to him its secret. Then being a supreme mathematical technician he could dress it up, how you will, for purposes of exposition, but it was his intuition which was pre-eminently extraordinary—'so happy in his conjectures', said de Morgan, 'as to seem to know more than he could possibly have any means of proving'. The proofs, for what they are worth, were, as I have said, dressed up afterwards—they were not the instrument of discovery.

And, if indeed the entire content of Proposition LXVI (inclusive of the Corollaries I–XVII) was developed by Newton in his mind and the presentation was 'dressed up afterwards', then his reluctance to work out the second-order corrections is even more understandable (see the comments at end of Corollary XII).

73. Proposition LXVI (continued): Corollaries XVIII–XXII

In the last five corollaries of this proposition (better described as a 'monograph') Newton turns his attention to the relevance of the results established in the context of the perturbations of a Kepler orbit to the theory of tides—a topic that he considers much later in Book III, Proposition XIV in its 'proper place'. It is significant that already at this stage (even before[†] he had established his 'superb' theorems on the gravitational attraction of spherical and non-spherical bodies in Section XII) Newton was thinking of accounting for the 'motions... of our sea' (as he was to state at the conclusion of the penultimate paragraph of the *Principia*).

> Thereupon Drona spoke smilingly to Arjuna, 'Now you must shoot at the target'. 'Listen'. 'As soon as I give the word you must shoot the arrow'. 'Now first stand there for a little while, son, and keep the bow taut'. The left-handed archer stretched the bow until it stood in a circle and kept aiming at the target as his guru had ordered. After a while Drona said to him in the same way, 'Do you see the bird sitting there? And the tree? And me?' 'I see the bird', Arjuna replied; 'but I don't see the tree or you'. And Drona again asked, 'if you see the bird, describe it to me'. 'I see its head, not its body'.
>
> (From J. A. B. van Buitenen's translation of the *Mahábhárata* Vol. I, pp. 272–3; University of Chicago Press, 1974)

† As I have stated earlier in Chapter 1 (p. 11), Newton proved his 'superb theorem' in the spring of 1685. And he had started on writing the *Principia* only in December 1684. When, then, had he written the first 17 corollaries of Proposition LXVI?

COR. XVIII. By the same laws by which the body P revolves about the body T, let us suppose many fluid bodies to move round T at equal distances from it; and to be so numerous, that they may all become contiguous to each other, so as to form a fluid annulus, or ring, of a round figure, and concentric to the body T; and the several parts of this ring, performing their motions by the same law as the body P, will draw nearer to the body T,[1] and move swifter[2] in the conjunction and opposition of themselves and the body S, than in the quadratures. And the nodes of this ring or its intersections with the plane of the orbit of the body S or T, will rest at the syzygies; but out of the syzygies they will be carried backwards, or in a retrograde direction, with the greatest swiftness in the quadratures, and more slowly in other places.[3] The inclination of this ring also will vary, and its axis will oscillate in each revolution, and when the revolution is completed will return to its former situation, except only that it will be carried round a little by the precession of the nodes.[4]

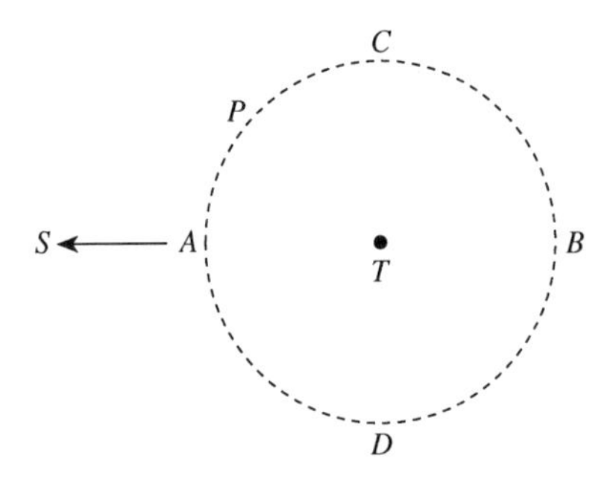

Newton's statement of the corollary is reproduced *in extenso* for two reasons. First, for the clarity with which he formulates the enlargement of the basis of his considerations to a continuous ring of particles, each of which revolves, like P, about T in accordance with the same laws. And second, for Newton's recalling (and thus summarizing) the corollaries in the enlarged context, in a sequence that is perhaps indicative of his priorities. The sequence, distinguished by superior numerals in the text, are correlated below with the corollaries.

1: COR. V: $TA = r(1 - x) < TC = r(1 + x)$, (§70, equation (19));

2: COR. III: $V_1(\psi = 0) > V_0(\psi = \pi/2)$, (§70, equation (35));

3: COR. XI: §72, equations (24) and (27); $\langle\Omega\rangle_{av} = -\frac{3}{4}m^2nt$;

4: COR. X: §72, equation (21); §65, equation (7).

COR. XIX. Suppose now the spherical body T, consisting of some matter not fluid, to be enlarged, and to extend itself on every side as far as that ring, and that a channel were cut all round its circumference containing water; and that this sphere revolves uniformly about its own axis in the same periodical time.

> This water being accelerated and retarded by turns (as in the last Corollary), will be swifter at the syzygies, and slower at the quadratures, than the surface of the globe, and so will ebb and flow in its channel after the manner of the sea.

The foregoing assertions are of course valid. But Newton could have made their truth more obvious if he could have stated at this point that replacing the point mass at T by a uniform 'spherical globe* consisting of some matter' is entirely permissible. But the underlying justifying propositions are proved only in Section XII to follow. Or, did Newton know of the justification already? (This question renews interest in the chronological sequence of the great discoveries of Sections XI and XII.)

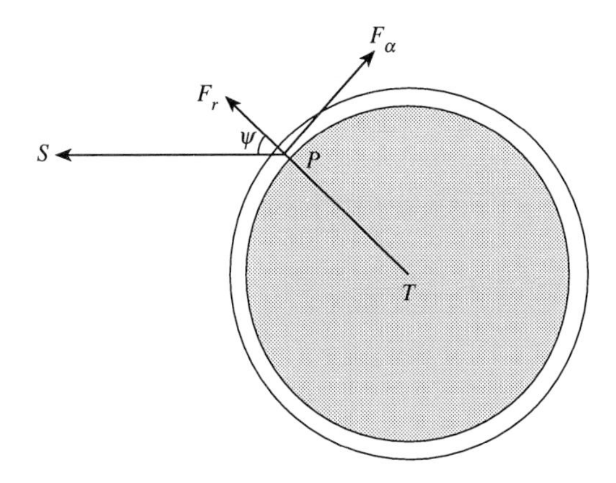

In the second part of the corollary, Newton, recalling Corollaries V and VI of the Laws of Motion (Chapter 2, §8), reminds us that all the conclusions reached up to this point will in no way be affected if all three bodies, S, T, and P partake of the same uniform rectilinear motion.

In the final part of the corollary Newton notes that the principal forces that act on the water in the channel are the perturbations in the centripetal force and the force in the direction towards the Sun. These forces are:

$$\frac{1}{2}\frac{GM_\odot}{R^3}r(1+3\cos 2\psi) \qquad \text{and} \qquad +2\frac{GM_\odot}{R^3}r\cos\psi. \tag{1}$$

The first of these forces attains the values,

$$-\frac{GM_\odot}{R^3}r \ \text{ at quadratures} \qquad \text{and} \qquad +2\frac{GM_\odot}{R^3}r \ \text{ at syzygies,} \tag{2}$$

and 'will attract the water downwards at the quadratures, and depress it as far as the syzygies'; while the second of the forces in the direction PS is

$$\pm 2\frac{GM_\odot}{R^3}r \ \text{ at syzygies; and vanishes at quadratures,} \tag{3}$$

* Or, more generally, any spherically symmetric distribution of matter having the same mass as the spherical globe.

and 'will attract it upward in the syzygies, and withhold its descent and make it rise as far as the quadratures'.

It is implied that the ring of fluid bodies of Corollary XVIII is coplanar with the orbital plane of P about T; and similarly that the channel of Corollary XIX lies nearly in the equator.

Corollary XX. In this corollary, Newton effectively considers the application of Corollary XI on the regression of the nodes and Corollary X on the variation of the inclination to the problem considered in Corollary XIX.

By Corollary XI, equation (27),

$$\Omega \simeq -\tfrac{3}{4}m^2nt + \tfrac{3}{8}m \sin 2U. \tag{4}$$

Newton formulates the content of this equation for the circumstances of Corollary XIX as follows:

> Let the globe have the same axis with the ring, and perform its revolutions in the same times, and at its surface touch the ring within, and adhere to it; then the globe partaking of the motion of the ring, this whole body will oscillate, and the nodes will go backwards.

He next considers the correlated variations of the node and of the inclination as exemplified by equations (18) and (23) of §72

$$\left.\begin{aligned} \frac{d\iota}{dt} &= -3m^2n \sin \iota \cos v \cos \psi \sin U, \\ \text{and} \qquad \frac{d\Omega}{dt} &= -3m^2n \sin v \cos \psi \sin U; \end{aligned}\right\} \tag{5}$$

and the solution (§72, equation (21))

$$\iota - \iota_0 = \tfrac{3}{8}m \sin \iota \cos 2U. \tag{6}$$

According to these equations

$$\iota - \iota_0 \begin{cases} \text{is a maximum at syzygies } (U = 0), \\ \text{is a minimum at quadratures } (U = \pi/2), \\ \text{and vanishes at octants } (U = \pi/4 \text{ and } 3\pi/4). \end{cases} \tag{7}$$

Further, when the nodes are at quadratures ($U = \pi/2$),

$$\text{both } \left|\frac{d\Omega}{dt}\right| \text{ and } \left|\frac{d\iota}{dt}\right| \text{ are at maximum,} \tag{8}$$

while at octants ($U = 0$ and $3\pi/4$),

$$\iota - \iota_0 = 0. \tag{9}$$

Newton's principal object in this corollary is to apply the foregoing conclusions to a globe that is 'a little higher or a little denser in the equatorial than in the polar regions';

and he concludes:

> yet the phenomena of this and the preceding Corollary would scarce be altered; except that the places of the greatest and least height of the water will be different; for the water is now no longer sustained and kept in its orbit by its centrifugal force, but by the channel in which it flows.

Newton then reiterates the conclusions arrived at, at the end of Corollary XIX.

Finally in Corollary XXI, Newton further elaborates on the same theme.

> COR. XXI. For the same reason that redundant matter in the equatorial regions of a globe causes the nodes to go backwards, and therefore by the increase of that matter that retrograde motion is increased, by the diminution is diminished, and by the removal quite ceases; it follows, that, if more than that redundant matter be taken away, that is, if the globe be either more depressed, or of a rarer consistence near the equator than near the poles, there will arise a direct motion of the nodes.

While one is astonished at Newton's consideration of the theory of tides in such detail at this early stage, it is important to point out that he does not seem to have realized that the pressure at the bottom of the channel balances the vertical component F_r of the gravitational force and that the only tide-generating force is the tangential component F_α. This fact vitiates some of the conclusions of Corollary XX. Laplace was to correct them a century later in his definitive equilibrium theory of tides. Nevertheless, the results of Corollary XIX are entirely correct providing the depth of water be less than about thirteen miles. We shall return to these matters when considering Newton's theory of the tides as set out in Book III, Propositions XXIV, XXXVI, and XXXVIII.

The last Corollary XXII of this Proposition LXVI is a remarkable restatement of the ideas outlined in Corollaries XVIII–XXI and of the directions in which one may think beyond. We shall not therefore attempt to paraphrase the original text.

> COR. XXII. And thence from the motion of the nodes is known the constitution of the globe. That is, if the globe retains unalterably the same poles, and the motion (of the nodes) is retrograde, there is a redundance of the matter near the equator; but if that motion is direct, a deficiency. Suppose a uniform and exactly spherical globe to be first at rest in a free space; then by some impulse made obliquely upon its surface to be driven from its place, and to receive a motion partly circular and partly straight forward. Since this globe is perfectly indifferent to all the axes that pass through its centre, nor has a greater propensity to one axis or to one situation of the axis than to any other, it is manifest that by its own force it will never change its axis, or the inclination of its axis. Let now this globe be impelled obliquely by a new impulse in the same part of its surface as before; and since the effect of an impulse is not at all

> changed by its coming sooner or later, it is manifest that these two impulses, successively impressed, will produce the same motion, as if they had been impressed at the same time; that is, the same motion, as if the globe had been impelled by a simple force compounded of them both (by Cor. ii of the Laws), that is, a simple motion about an axis of a given inclination. And the case is the same if the second impulse were made upon any other place of the equator of the first motion; and also if the first impulse were made upon any place in the equator of the motion which would be generated by the second impulse alone; and therefore, also, when both impulses are made in any places whatsoever; for these impulses will generate the same circular motion as if they were impressed together, and at once, in the place of the intersections of the equators of those motions, which would be generated by each of them separately. Therefore, a homogeneous and perfect globe will not retain several motions distinct, but will unite all those that are impressed on it, and reduce them into one; revolving, as far as in it lies, always with a simple and uniform motion about one single given axis, with an inclination always invariable. And the inclination of the axis, or the velocity of the rotation, will not be changed by centripetal force. For if the globe be supposed to be divided into two hemispheres, by any plane whatsoever passing through its own centre, and the centre to which the force is directed, that force will always urge each hemisphere equally; and therefore will not incline the globe to any side with respect to its motion round its own axis. But let there be added anywhere between the pole and the equator a heap of new matter like a mountain, and this, by its continual endeavor to recede from the centre of its motion, will disturb the motion of the globe, and cause its poles to wander about its surface describing circles about themselves and the points opposite to them. Neither can this enormous deviation of the poles be corrected otherwise than by placing that mountain either in one of the poles, in which case, by Cor. xxi, the nodes of the equator will go forwards; or in the equatorial regions, in which case, by Cor. xx, the nodes will go backwards; or, lastly, by adding on the other side of the axis a new quantity of matter, by which the mountain may be balanced in its motion; and then the nodes will either go forwards or backwards, as the mountain and this newly added matter happen to be nearer to the pole or to the equator.

The underlined statements are remarkable for the use that Newton makes of the geometrical *symmetry* of the sphere to reflection and to rotation about its centre to draw *physical conclusions.* They are reminiscent of the arguments that L. Lictenstein (1918; Sitzungsber. Pruess Akad. Wiss. Phys. Math. Kl., p. 1120) was to use some two centuries later to prove that a fluid mass in equilibrium under its own gravitation must necessarily be spherically symmetric.

74. Propositions LXVII–LXIX

In Propositions LXVII, LXVIII, and its corollaries, Newton is concerned with the most appropriate choice for the coordinate frame for considering the perturbations of a three-body problem such as of S, T, and P. His choice is:

> Therefore the perturbation is least when the common centre of the three bodies is at rest; that is, when the innermost and greatest body T is attracted according to the same law as the rest are; and is always greatest when the common centre of the three, by the diminution of the motion of the body T, begins to be moved, and is more and more agitated.

This is precisely the coordinate frame in which the variational equations of Chapter 13, §66 were derived.

In the more general context of the many-body problem (with perhaps the configuration of Jupiter and its satellites in mind) Newton's recommendation is:

> Cor. And hence if several smaller bodies revolve about the great one, it may easily be inferred that the orbits described will approach nearer to ellipses; and the descriptions of areas will be more nearly uniform, if all the bodies attract and agitate each other with accelerative forces that are directly as their absolute forces, and inversely as the squares of the distances, and if the focus of each orbit be placed in the common centre of gravity of all the interior bodies (that is, if the focus of the first and innermost orbit be placed in the centre of gravity of the greatest and innermost body; the focus of the second orbit in the common centre of gravity of the two innermost bodies; the focus of the third orbit in the common centre of gravity of the three innermost; and so on), than if the innermost body were at rest, and was made the common focus of all the oribits.

These coordinates have recently been resurrected under the name 'Jacobi coordinates' (!) (cf. D. Brouwer and G. Clemence, *Celestial mechanics*, Academic Press, New York, p. 588).

Proposition LXIX. Theorem XXIX

In a system of several bodies A, B, C, D, etc., if any one of those bodies, as A, attract all the rest, B, C, D, etc., with accelerative forces that are inversely as the squares of the distances from the attracting body; and another body, as B, attracts also the rest, A, C, D, etc., with forces that are inversely as the squares of the distances from the attracting body; the absolute forces of the attracting bodies A and B will be to each other as those very bodies A and B to which those forces belong.

COR. I. Therefore if each of the bodies of the system *A*, *B*, *C*, *D*, etc., does singly attract all the rest with accelerative forces that are inversely as the squares of the distances from the attracting body, the absolute forces of all those bodies will be to each other as the bodies themselves.

COR. II. By a like reasoning, if each of the bodies of the system *A*, *B*, *C*, *D*, etc., does singly attract all the rest with accelerative forces, which are either inversely or directly in the ratio of any power whatever of the distances from the attracting body; or which are defined by the distances from each of the attracting bodies according to any common law; it is plain that the absolute forces of those bodies are as the bodies themselves.

The problem that Newton considers in this proposition and its corollaries is the following:

We are given a set of bodies, *A*, *B*, *C*, *D*, etc., each of which 'attracts all the rest with accelerative force... which are defined by the distances from each of the attracting bodies according to any common law'. If *P* and *Q* are two of these bodies, the assumption is that between the two,

$$\frac{\mathrm{d}^2\vec{r}_P}{\mathrm{d}t^2} = +\Pi_Q(\vec{r}_Q - \vec{r}_P)F(|\vec{r}_Q - \vec{r}_P|) \tag{1}$$

where F is a 'universal' function of the absolute distance, $|\vec{r}_Q - \vec{r}_P|$, between P and Q and Π_Q is a positive numerical factor dependent only on a scalar property of the body Q. Consider equation (1) for two of the bodies X and $Y\ (\neq X)$. Then,

$$\left.\begin{aligned}\frac{\mathrm{d}^2\vec{r}_X}{\mathrm{d}t^2} &= +\Pi_Y(\vec{r}_Y - \vec{r}_X)F(|\vec{r}_Y - \vec{r}_X|),\\ \frac{\mathrm{d}^2\vec{r}_Y}{\mathrm{d}t^2} &= +\Pi_X(\vec{r}_X - \vec{r}_Y)F(|\vec{r}_X - \vec{r}_Y|).\end{aligned}\right\} \tag{2}$$

From these equations it follows that,

$$\left|\frac{\mathrm{d}^2\vec{r}_X}{\mathrm{d}t^2}\right| : \left|\frac{\mathrm{d}^2\vec{r}_Y}{\mathrm{d}t^2}\right| = \Pi_Y : \Pi_X. \tag{3}$$

But by the third Law of Motion,

$$\left|\frac{\mathrm{d}^2\vec{r}_X}{\mathrm{d}t^2}\right| : \left|\frac{\mathrm{d}^2\vec{r}_Y}{\mathrm{d}t^2}\right| = m_Y : m_X, \tag{4}$$

where m_X and m_Y denote the inertial masses of the two bodies. Since equation (4) applies to all pairs, X and Y, we conclude that

$$\Pi_X = Gm_X \tag{5}$$

where G is a universal constant, that is, it is the same for all of the bodies.

Corollary I is stated explicitly for the case 'When bodies whose forces decrease as the square of the distance', that is, when

$$f(|r_X - r_Y|) \propto |\vec{r}_X - \vec{r}_Y|^{-2}. \tag{6}$$

The constant G in this case is of course Newton's constant of gravitation.

Newton adds this further corollary.

> COR. III. In a system of bodies whose forces decrease as the square of the distances, if the lesser revolve about one very great one in ellipses, having their common focus in the centre of that great body, and of a figure exceedingly accurate; and moreover by radii drawn to that great body describe areas proportional to the times exactly; the absolute forces of those bodies to each other will be either accurately or very nearly in the ratio of the bodies. And so conversely. This appears from Cor. of Prop. XLVIII, compared with the first Corollary of this Proposition.

The addition of this corollary at this point, similarly phrased as Cor. III. of Proposition LXV and with an apparent emphasis, may appear strange. But he must have had in mind a question that will occur when he comes to formulating his universal law of gravitation in Book III (see Proposition VI). This is a further example of Newton's prescience to which we have already referred in Chapter 10 (p. 198).

Again, we have reached a watershed after a long and arduous journey. And as his wont, Newton speaks of what the next destination has to be on his view of Natural Philosophy.

Scholium

> These Propositions naturally lead us to the analogy there is between centripetal forces and the central bodies to which those forces are usually directed; for it is reasonable to suppose that forces which are directed to bodies should depend upon the nature and quantity of those bodies, as we see they do in magnetical experiments. And when such cases occur, we are to compute the attractions of the bodies by assigning to each of their particles its proper force, and then finding the sum of them all. I here use the word *attraction* in general for any endeavor whatever, made by bodies to approach to each other, whether that endeavor arise from the action of the bodies themselves, as tending to each other or agitating each other by spirits emitted; or whether it arises from the action of the ether or of the air, or of any medium whatever, whether corporeal or incorporeal, in any manner impelling bodies placed therein towards each other. In the same general sense I use the word *impulse*, not defining in this treatise the species or physical qualities of forces, but investigating the quantities and mathematical proportions of them; as I observed before in the Definitions.

In mathematics we are to investigate the quantities of forces with their proportions consequent upon any conditions supposed; then, when we enter upon physics, we compare those proportions with the phenomena of Nature, that we may know what conditions of those forces answer to the serveral kinds of attractive bodies. And this preparation being made, we argue more safely concerning the physical species, causes, and proportions of the forces. Let us see, then, with what forces spherical bodies consisting of particles endued with attractive powers in the manner above spoken of must act upon one another; and what kind of motions will follow from them.

15

'The superb theorems'

75. Introduction

In Chapter 1, we have already described the antecedents of Newton's discovery of the theorem that was later to be described as 'superb' by J. W. L. Glaisher. Therefore, on this occasion, we shall, as the saying goes 'cut the cackle and come to the hosses'.

We shall begin with an explanation which Newton provides later:

Scholium

By the surfaces of which I here imagine the solids composed, I do not mean surfaces purely mathematical, but orbs so extremely thin, that their thickness is as nothing; that is, the evanescent orbs of which the sphere will at last consist, when the number of the orbs is increased, and their thickness diminished without end. In like manner, by the points of which lines, surfaces, and solids are said to be composed, are to be understood equal particles, whose magnitude is perfectly inconsiderable.

76. Propositions LXX–LXXII

Proposition LXX. Theorem XXX

If to every point of a spherical surface there tend equal centripetal forces decreasing as the square of the distances from those points, I say, that a corpuscle placed within that surface will not be attracted by those forces any way.

The proof is Newton's: it is the same as one finds in any modern textbook.

'Let $HIKL$ be that spherical surface, and P a corpuscle placed within.' With P as vertex, construct a cone of small solid angle $d\omega$ intersecting the surface in the elements KL and HI. The attractions per unit mass at P of these elements of surface area dS and dS' are

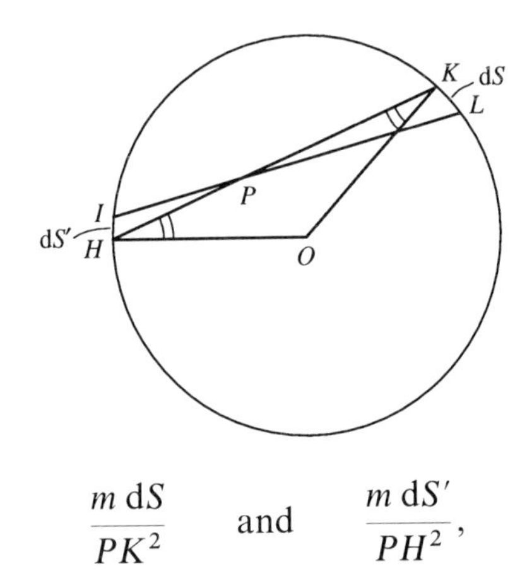

$$\frac{m\,\mathrm{d}S}{PK^2} \quad \text{and} \quad \frac{m\,\mathrm{d}S'}{PH^2},$$

where m is the mass per unit area of the surface; and these attractions are in opposite directions. We have

$$\mathrm{d}S = PK^2 \frac{\mathrm{d}\omega}{\cos OKP} \quad \text{and} \quad \mathrm{d}S' = PH^2 \frac{\mathrm{d}\omega}{\cos OHP}.$$

Since $\angle OKP = \angle OHP$, it follows that the attraction of the two elements are equal and opposite. Since, further, the entire spherical surface can be divided into such similar pairs of elements, the resultant attraction is zero. Q.E.D.

Proposition LXXI. Theorem XXXI

The same things supposed as above, I say, that a corpuscle placed without the spherical surface is attracted towards the centre of the sphere with a force inversely proportional to the square of its distance from that centre.

J. E. Littlewood has conjectured (though I do not share in the conjecture—see below) that Newton had perhaps first constructed a proof based on calculus which 'we can infer with some possibility what the proof was.' And here is his conjectured proof.

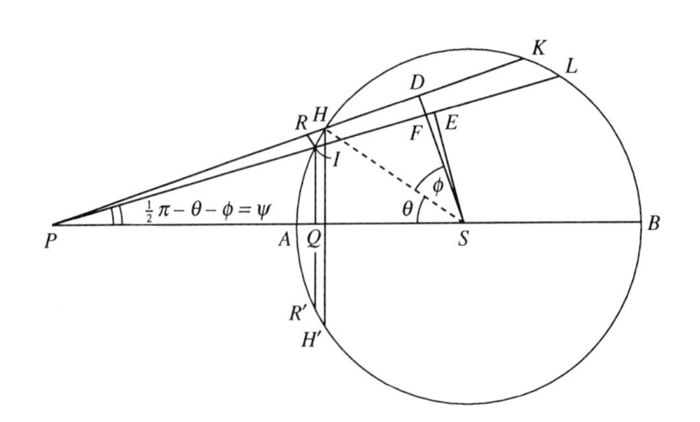

Let $LKHRR'H'B$ be the same spherical surface that was considered in Proposition LXX with the corpuscle P now 'placed without the spherical surface' at a distance $r(=PS)$ from the centre S. We shall divide the spherical surface of radius a into infinitesimal circular caps like $HRR'H$, with their axes along PS.

The direction of attraction is along HP. Only the component along PS will additively contribute to the net attraction: that in the direction normal to PS will cancel by symmetry. We now evaluate the contribution, $\mathrm{d}F$, to the force by the spherical cap $HRR'H$. The successive steps in the evaluation (following Littlewood) are:

$$r^2\,\mathrm{d}F = \frac{PS^2}{PI^2}\cos\psi\,.\,2\pi a\,\mathrm{d}\theta\,.\,IQ \quad (=a\sin\theta)$$

$$= \frac{PS^2}{PI^2}\sin(\theta+\phi)\,.\,2\pi a^2\,\mathrm{d}\theta\,.\sin\theta; \tag{1}$$

$$\frac{r^2}{2\pi a^2}\,\mathrm{d}F = \left(\frac{\sin\angle PIS}{\sin\theta}\right)^2\sin(\theta+\phi)\sin\theta\,\mathrm{d}\theta = \cos^2\phi\,\frac{\sin(\theta+\phi)}{\sin\theta}\,\mathrm{d}\theta. \tag{2}$$

But

$$\frac{a}{r} = \frac{HS}{PS} = \frac{\sin(\frac{1}{2}\pi-\theta-\phi)}{\sin(\frac{1}{2}\pi+\phi)} = \frac{\cos(\theta+\phi)}{\cos\phi}. \tag{3}$$

By differentiating this last expression we obtain,

$$\left.\begin{aligned}&[\cos(\theta+\phi)\sin\phi - \sin(\theta+\phi)\cos\phi]\,\mathrm{d}\phi = \cos\phi\sin(\theta+\phi)\,\mathrm{d}\theta,\\ &\text{and}\\ &a\sin\phi\,\mathrm{d}\phi = r\sin(\theta+\phi)(\mathrm{d}\theta+\mathrm{d}\phi).\end{aligned}\right\} \tag{4}$$

Hence,

$$\frac{\mathrm{d}\theta}{\mathrm{d}\phi} = -\frac{\sin\theta}{\cos\phi\sin(\theta+\phi)}. \tag{5}$$

By combining equations (2) and (5), we have

$$\frac{r^2}{2\pi a^2}\frac{\mathrm{d}F}{\mathrm{d}\phi} = -\cos\phi. \tag{6}$$

Integrating this last equation over ϕ, we obtain

$$\frac{r^2}{2\pi a^2}F = \int_{-\pi/2}^{\pi/2}\cos\phi\,\mathrm{d}\phi = 2, \tag{7}$$

or,

$$F = \frac{4\pi a^2}{r^2}. \tag{8}$$

Now, to Newton's proof. He considers the situation described in the previous illustration together with the illustration of the companion situation in which the corpuscle, now designated, p, is at a different distance ps from the centre s of the spherical surface $lkhrab$.

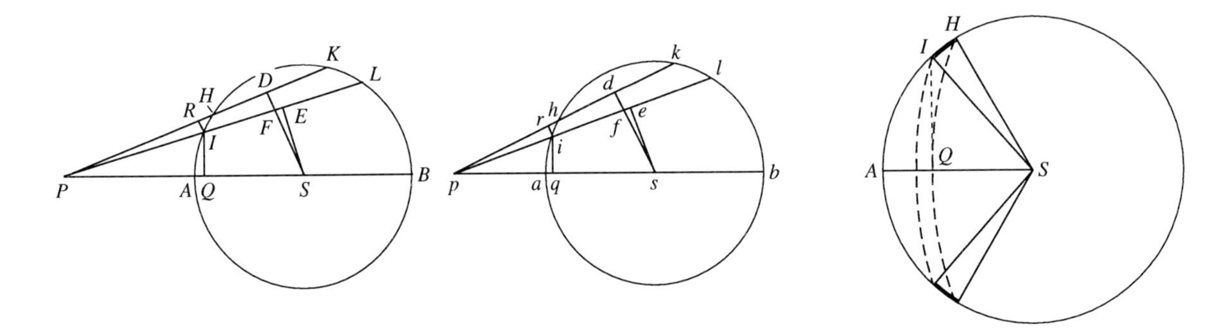

Newton compares the attraction of the two spherical surfaces at the points P and p. The wonder of the proof is the initial geometrical construction: The right lines PHK and phk cut off equal chords HK and hk; and similarly the infinitesimally neighbouring lines PIL and pil cut off equal chords IL and il. As Littlewood has remarked, this construction 'must have left its readers in helpless wonder'. By this construction

$$HK = hk, \quad IL = il; \qquad \therefore \; SD = sd \quad \text{and} \quad se = SE; \tag{9}$$

and, therefore,*

$$DF = SD - SE = sd - se = df. \tag{10}$$

The successive (elementary!) steps are:

$$\frac{\delta F_p}{\delta F_P} = \frac{PI^2}{pi^2} \cdot \frac{(PS)(pf)}{(PF)(ps)} \cdot \frac{hi}{HI} \cdot \frac{iq}{IQ} \tag{11}$$

$$\frac{pf}{pi} \cdot \frac{PI}{PF} = \frac{df}{ri} \cdot \frac{RI}{DF} = \frac{RI}{ri} = \frac{HI}{hi}, \qquad \text{(By the similarity of } \triangle\text{s } RHI \text{ and } rhi) \tag{12}$$

and

$$\frac{PI}{PS} \cdot \frac{ps}{pi} = \frac{IQ}{SE} \cdot \frac{se}{iq} = \frac{IQ}{iq}, \qquad \text{(By the similarity of } \triangle\text{s } PIQ\ (piq) \text{ and } PSE\ (pse)) \tag{13}$$

By multiplication of equations (12) and (13), we obtain,

$$\frac{PI^2}{pi^2} \cdot \frac{(pf)(ps)}{(PF)(PS)} = \frac{HI}{hi} \cdot \frac{IQ}{iq}. \tag{14}$$

Now by combining equations (11) and (14), we find

$$\frac{\delta F_p}{\delta F_P} = \frac{PS^2}{ps^2}. \tag{15}$$

Q.E.D.

Notice the entire absence of any superfluous steps; and how well-knit the proof is!

* These relations have to be understood in the sense of 'last ratios ... when $\angle$s DPE and dpe vanish together.'

A personal reflection

I stated earlier that I did not share Littlewood's conjecture. Littlewood stated his conjecture as follows.

> To the Newton of 1685 the problem was bound to yield in reasonable time: it is possible (though this is quite conjectural) that he tried the approach via a shell of radius r, to be followed by an integration with respect to r; this would of course instantly succeed.

Littlewood was, of course, repeating a legend that has often been told, namely, that Newton first constructed proofs of most (if not all) of his propositions by calculus and then transliterated them into 'his' geometrical language. I do not believe in this legend. First, there are enough propositions that are proved directly by integral calculus, for example, Propositions XXXIX and XLI: and in Proposition LXXI he simply preferred not to give the proofs by calculus. Second, his physical and geometrical insights were so penetrating that the proofs emerged whole in his mind: 'he was happy in his thoughts' (qualifying de Morgan). Besides, where was the time to dissimulate? For my part, I am not surprised that 'to the Newton of 1685' the geometrical construction 'that must have left its readers in helpless wonder' came quite naturally (see the comments at the end of the next proposition).

Proposition LXXII. Theorem XXXII

If to the several points of a sphere there tend equal centripetal forces decreasing as the square of the distances from those points; and there be given both the density of the sphere and the ratio of the diameter of the sphere to the distance of the corpuscle from its centre: I say, that the force with which the corpuscle is attracted is proportional to the semidiameter of the sphere.

Consider two homogeneous spheres of equal density and of radii R_1 and R_2; and two 'corpuscles' P_1 and P_2 of equal masses at distances C_1P_1 and C_2P_2 which are in the ratio of their radii:

$$C_1P_1 : C_2P_2 = R_1 : R_2. \tag{16}$$

And, let p_1 and p_2 be two 'equal' corpuscles (i.e. of equal infinitesimal volumes) in the two spheres at distances from P_1 and P_2 that are in the same ratio of the radii R_1 and R_2, that is,

$$p_1P_1 : p_2P_2 = R_1 : R_2. \tag{17}$$

The attractions, f_1 and f_2, exerted by p_1 and p_2 on P_1 and P_2 are in the ratio,

$$f_1 : f_2 = (p_1P_1)^{-2} : (p_2P_2)^{-2} = R_1^{-2} : R_2^{-2}. \quad \text{(by equation (17))} \tag{18}$$

This ratio obtains for every pair of similarly situated particles in the two spheres. Therefore, the entire attractions, F_1 and F_2 of the whole spheres will be 'directly' as the number of

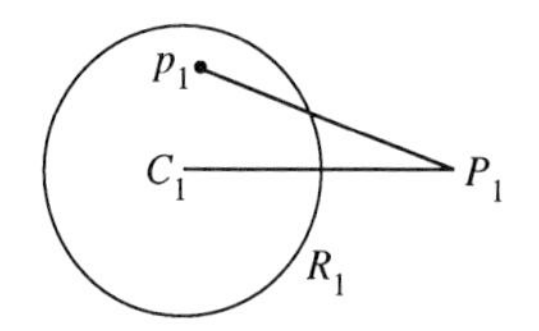

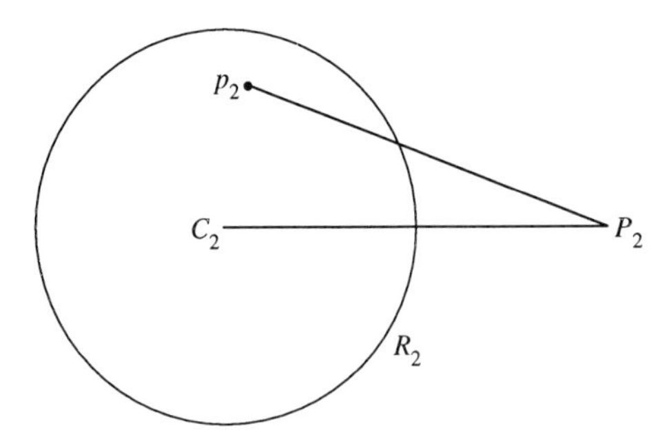

corpuscles, N_1 and N_2 in the two spheres: and since, by assumption, all the corpuscles in the two spheres are of equal infinitesimal volumes,

$$N_1 : N_2 = R_1^3 : R_2^3. \tag{19}$$

Hence, for the total forces,

$$F_1 : F_2 = N_1 f_1 : N_2 f_2 = R_1 : R_2. \tag{20}$$

Q.E.D.

Contrast the foregoing presentation (in the manner of our times) with Newton's concision in his inimitable style:

> Then the attractions of one corpuscle towards the several particles of one sphere will be to the attractions of the other towards as many analogous particles of the other sphere in a ratio compounded of the ratio of the particles directly, and the square of the distances inversely. But the particles are as the spheres, that is, as the cubes of the diameters, and the distances are as the diameters; and the first ratio directly with the last ratio taken twice inversely, becomes the ratio of diameter to diameter. Q.E.D.

The equally 'baffling' nature of Newton's demonstration of this proposition does not seem to have been noted: it is *not* based (as one might have expected) on dividing the sphere into infinitesimal shells and then summing over the result of Proposition LXXI (this comes later in Proposition LXXVI as a 'Lebesgue' integral); it is based, instead, on more general grounds of similarity. In this last respect, it is no different from the proof of Proposition LXXI which is also based on similarity arguments as exemplified by the number of appeals to the similarity of the triangles in the geometrical construction. Indeed it is possible that Newton first 'saw' the proof of this simpler(!) Proposition LXXII; and soon after, the proof of Proposition LXXI.

From the fact that the centripetal forces acting on P_1 and P_2 are directly as the distances from the centre, it follows immediately from Corollary VII, Proposition IV (equations (15) and (16) on p. 74) that the periodic times of circular orbits about the sphere at distances proportional to R_1 and R_2 are equal; and vice versa. This is the content of Corollaries I and II.

The real power of the similarity arguments used in the proof of this proposition emerges in Corollary III.

> Cor. iii. If to the several points of any two solids whatever, of like figure and equal density, there tend equal centripetal forces decreasing as the square of the distances from those points, the forces, with which corpuscles placed in a like situation to those two solids will be attracted by them, will be to each other as the diameters of the solids.

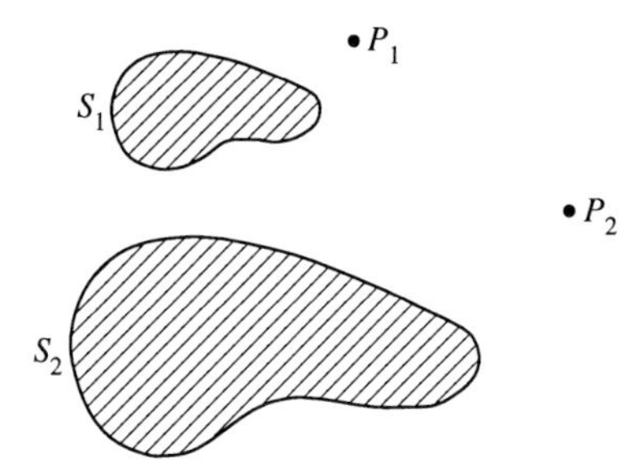

This corollary is used in Proposition XCI, Corollary III (p. 316).

77. Propositions LXXIII–LXXV

Propositions LXX, LXXI, and LXXII proved in §76 are mutually independent and provide the corner stones for the final Proposition LXXVI. Meantime, the three Propositions LXXXIII–LXXV provide intermediate steps.

Proposition LXXIII. Theorem XXXIII

> *If to the several points of a given sphere there tend equal centripetal forces decreasing as the square of the distances from the points, I say, that a corpuscle placed within the sphere is attracted by a force proportional to its distance from the centre.*

Given a homogeneous sphere *ACBD*, each point of which is a centre of centripetal force decreasing as the square of the distance from it, is required to find the net force acting on an interior point *P*. To this end, draw a sphere *PEQF* passing through *P* and concentric

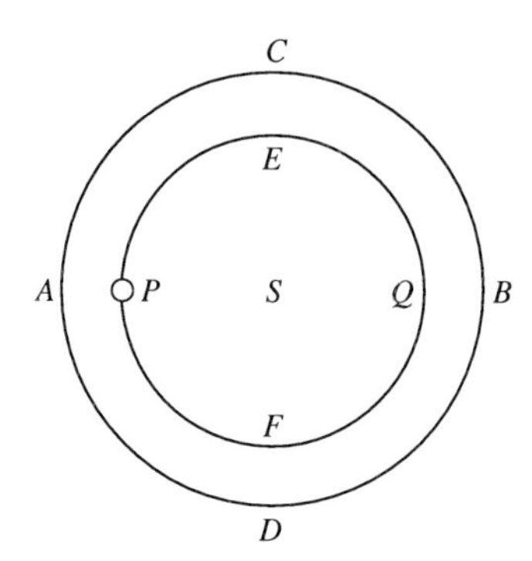

to *ACBD*. Then by Proposition LXX the spherical shell bounded by *ACBD* and *PEQF* will have

> no effect at all on the body *P*, their attractions being destroyed by contrary attractions. There remains therefore, only the attraction of the interior sphere *PEQF*. And by Prop. LXXII this is as the distance *PS*. Q.E.D.

Proposition LXXIV. Theorem XXXIV

> *The same things supposed, I say, that a corpuscle situated without the sphere is attracted with a force inversely proportional to the square of its distance from the centre.*

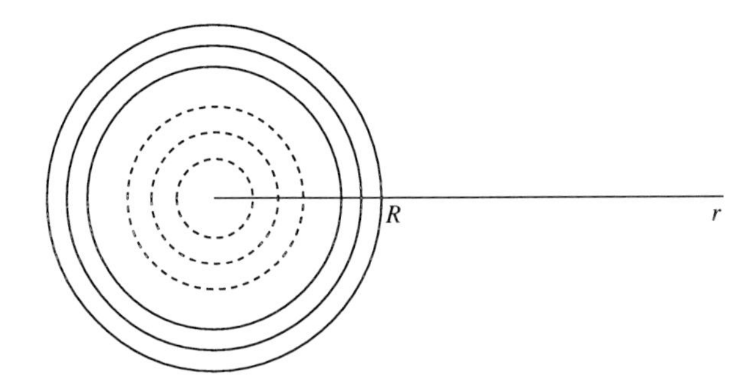

Newton's proof is clear and to the point.

> For suppose the sphere to be divided into innumerable concentric spherical surfaces, and the attractions of the corpuscle arising from the several surfaces will be inversely proportional to the square of the distance of the corpuscle from the centre of the sphere (by Prop. LXXI). And, by composition, the sum of those attractions, that is, the attraction of the corpuscle towards the entire sphere, will be in the same ratio. Q.E.D.

Accordingly, the attraction *F* of any homogeneous spherical body *S*, at an external point, at a distance r_1 from the centre, is of the form

$$F = C(S)r_1^{-2} \tag{1}$$

where $C(S)$ is a scalar property characterizing the *whole* body S. Apply equation (1) to two homogeneous spherical bodies, S_1 and S_2 of radii R_1 and R_2. Their attractions F_1 and F_2 at points, distant r_1 and r_2 from their respective centres are in the ratio,

$$\frac{F(S_1; r_1)}{F(S_2; r_2)} = \frac{C_1(S_1)}{C_2(S_2)}\frac{r_2^2}{r_1^2}. \tag{2}$$

Now, let

$$r_1 : R_1 = r_2 : R_2 = \alpha \quad \text{(say)}. \tag{3}$$

Then

$$\frac{F_1(S_1; \alpha R_1)}{F_2(S_2; \alpha R_2)} = \frac{C_1(S_1)}{C_2(S_2)}\frac{R_2^2}{R_1^2}. \tag{4}$$

But by Proposition LXXII (equation (20))

$$\frac{F_1(S_1; \alpha R_1)}{F_1(S_2; \alpha R_2)} = \frac{R_1}{R_2}. \tag{5}$$

Hence, combining equations (4) and (5) we have

$$\frac{C_1(S_1)}{C_2(S_2)}\frac{R_2^2}{R_1^2} = \frac{R_1}{R_2} \quad \text{or} \quad \frac{C_1(S_1)}{R_1^3} = \frac{C_2(S_2)}{R_2^3}. \tag{6}$$

Since S_1 and S_2 are any two homogeneous spheres we may write quite generally,

$$F = AR^3/r^2, \tag{7}$$

where A is a constant, that is, the same for all homogeneous spheres of the same density. And since matter of uniform diversity can be uniquely characterized by its density ρ, we can also write,

$$F = \text{constant}\ \frac{\rho R^3}{r^2} \tag{8}$$

or

$$F = \mathscr{C}\frac{M}{r^2}, \quad \text{per particle (or, equivalently, per unit mass)}, \tag{9}$$

where M denotes the 'weight' of the sphere (as Newton states at a later point) and $\mathscr{C}$ is a constant that is the same for all homogeneous spheres.

The contents of Corollaries I and II are expressed in the relation (9) though we have been more explicit in stating the conclusions than Newton is in the *Principia*. Newton stops at equation (7), but later assumes (9).

Corollary III draws attention to an important conclusion that follows from Propositions LXXII and LXXIII.

COR. III. If a corpuscle placed without an homogeneous sphere is attracted by a force inversely proportional to the square of its distance from the centre, and the sphere consists of attractive particles, the force of every particle will decrease as the square of the distance from each particle.

This is almost a statement of the universal law of gravitation—'almost' because the inference is in a very particular context. (And Newton of course, knows it!)

Proposition LXXV. Theorem XXXV

If to the several points of a given sphere there tend equal centripetal forces decreasing as the square of the distances from the point, I say, that another similar sphere will be attracted by it with a force inversely proportional to the square of the distance of the centres.

In this proposition and its corollaries, the attraction of homogeneous spheres by homogeneous spheres is considered. And the important distinction is made between what we shall call 'passive' particles which submit to attraction, and 'active' particles which exert attraction. Of course, a particle can be both active and passive subject, however, to the third Law of Motion (see below).

In Proposition LXXII we considered a homogeneous sphere S consisting of active particles each of which is the centre of 'equal centripetal force decreasing as the square of the distance from it'. And in Proposition LXXIV, we showed that such a sphere will attract a passive particle p at an external point (see Fig. (a)) with a force proportional to (by equation (9))

$$F = \mathscr{C}M/r^2 \qquad \text{per unit mass (or per particle)} \tag{10}$$

where r is the distance of p from the centre.

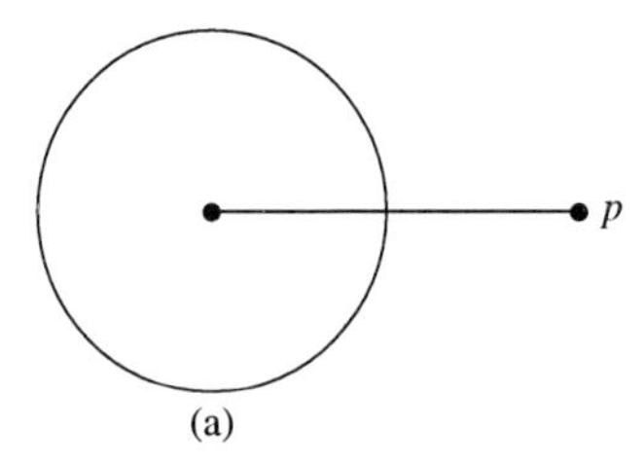

(a)

This attractive force by the whole sphere is, therefore, the same as 'if it issued from a single corpuscle placed in the centre of the sphere'.

Consider now another similar sphere (see Fig. (b)) equal in every way to the one we have considered. Its attraction at the point C will also be proportional to (again by Proposition LXXIV)

$$F = \mathscr{C}\frac{M}{D^2} \qquad \text{per particle of the second sphere} \tag{11}$$

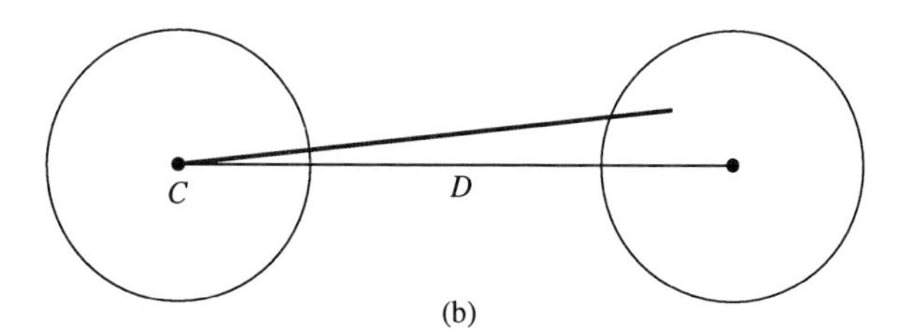

(b)

where D is the distance between the two centres. But this force will act on the particles of the second sphere only if they are all passive. Q.E.D.

In Corollary I attention is called explicitly to the presence of the factor R^3 in equation (7).

In Corollary II the case is considered (see Fig. (c)) in which the particles in the two spheres are both active and passive which in the context means that

> the several points of the one attract the several points of the other with the same force with which they themselves are attracted by the others again; and therefore since in all attractions (by Law III) the attracted and attracting point are both equally acted on, the force will be doubled by their mutual attractions, the proportions remaining.

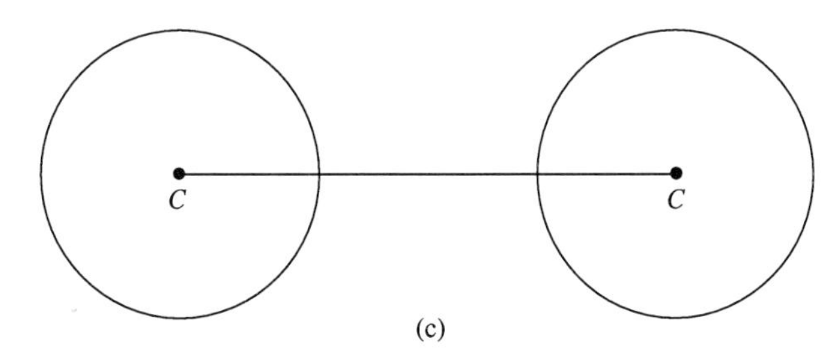

(c)

In other words, the mutual attraction between the two identical spheres will be proportional to

$$M^2/D^2 \tag{12}$$

where D is the distance between the centres of the two spheres.

More generally, the attraction between any two homogeneous spheres of masses M_1 and M_2 will likewise be proportional to

$$M_1M_2/D^2 \tag{13}$$

where D is again the distance between the centres of the two spheres.

In Corollary III Newton (ever conscious of his ultimate objective) draws attention to the relevance of the propositions established so far to the problems considered in Chapters 6 and 7.

COR. III. Those several truths demonstrated above concerning the motion of bodies about the focus of the conic sections will take place when an attracting sphere is placed in the focus, and the bodies move without the sphere.

And Corollary IV follows directly from Propositions X and LXXIII.

COR. IV. Those things which were demonstrated before of the motion of bodies about the centre of the conic sections take place when the motions are performed within the sphere.

78. Proposition LXXVI

'The preparations being made' Newton proceeds to the proof of the main theorem.

Proposition LXXVI. Theorem XXXVI

If spheres be however dissimilar (as to density of matter and attractive force) in the same ratio onwards from the centre to the circumference; but everywhere similar, at every given distance from the centre, on all sides round about; and the attractive force of every point decreases as the square of the distance of the body attracted: I say, that the whole force with which one of these spheres attracts the other will be inversely proportional to the square of the distance of the centres.

Newton's proof of this proposition is so terse that he had to relegate to the Corollaries I–V simple statements of the omitted details. In the presentation which follows, we shall be more expansive and include the details of the first five corollaries at the same time.

Consider first the attraction of two homogeneous spherical shells, $\mathscr{S}_1$ and $\mathscr{S}_2$ (not necessarily of the same density) included, respectively, between concentric spheres of radii R_1 and R_2, and of radii r_1 and r_2. By expressing the spherical shells as the difference of the homogeneous spheres, of radii R_2 and R_1 and masses M_2 and M_1 in one of them, and of radii r_2 and r_1 and masses m_2 and m_1 in the other, we find by successive applications of Proposition LXXV equation (12), that the attraction, $F(\mathscr{S}_1, \mathscr{S}_2)$, between the two spherical shells is

$$F(\mathscr{S}_1, \mathscr{S}_2) = \mathscr{C}(M_2 - M_1)(m_2 - m_1)D^{-2} \tag{i}$$

where D is the distance between the centres C_1 and C_2.

To obtain the attraction between two whole spheres in which the density distributions

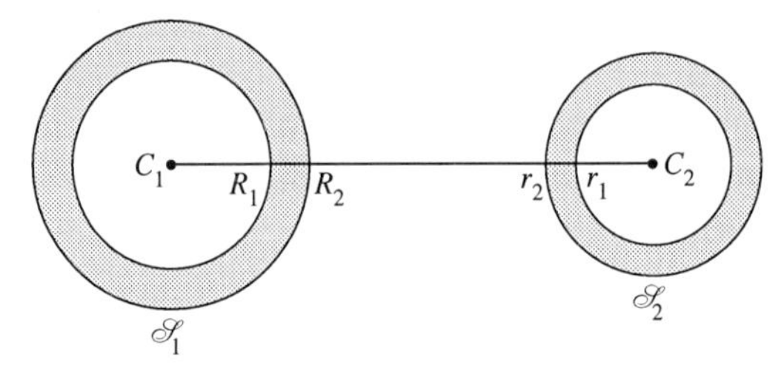

are spherically symmetric (or, as Newton quaintly describes: 'everywhere similar, at every given distance from the centre, on all sides round about') we divide the spheres by n concentric spheres, including in one case, a sphere S_0 at the centre of mass ΔM_0 and spherical shells of masses $\Delta M_1, \Delta M_2, \ldots, \Delta M_n$, and in the other case a central sphere of mass δm_0 and spherical shells of masses $\delta m_1, \delta m_2, \ldots, \delta m_n$. We then obtain the attraction of the two whole spheres, *as divided*, by first finding the attraction of a particular shell ΔM_i of one of the spheres on each of the shells $\delta m_k (k = 0, 1, \ldots, n)$ of the other sphere,

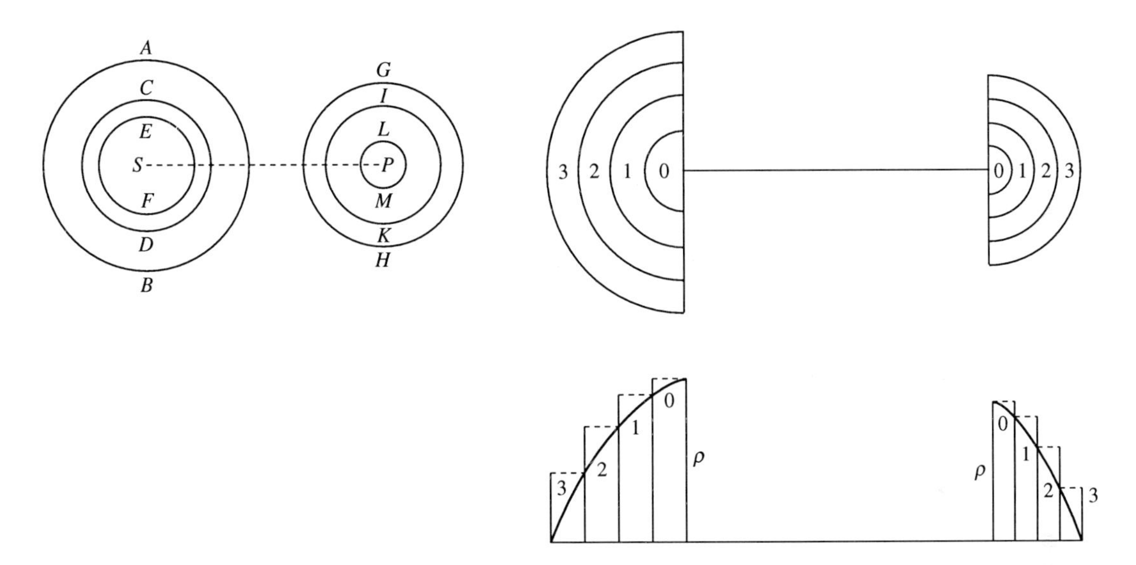

summing over all k, and then summing the resulting expression over all $i (=0, \ldots, n)$, we find (by equation (i)):

$$F\left(\sum_{i=0}^{n} \Delta M_i; \sum_{k=0}^{n} \delta m_k\right) = \frac{\mathscr{C}}{D^2}\left[\sum_{i=0}^{n} \Delta M_i\left(\sum_{k=0}^{n} \delta m_i\right)\right]. \tag{ii}$$

Now letting the number of concentric spheres n increase 'in infinitum' we shall obtain, in the limit:

$$F(M, m) = \mathscr{C}\frac{Mm}{D^2} \tag{iii}$$

where M and m are the masses of the two spheres. Q.E.D.

(Note that the 'sum' is evaluated over a division of the ordinates (M, m) and not on a division of the abscissae (R, r)—in this sense, Newton evaluates a 'Lebesgue sum' and not a 'Riemann sum'!)

Corollaries I–V state, as amplifications, details not included in Newton's proof of the proposition.

Corollaries I and II. If there are a number of 'spheres', the attraction between any two of them is the same; independently of the presence of the others. More explicitly stated, if there are n spheres of masses, $M_1, M_2, \ldots, M_n$, at relative distances R_{ij} between M_i and M_j, the attraction between two of them, M_i and M_j (say), is:

$$F(M_i, M_j) = \mathscr{C}\frac{M_i M_j}{R_{ij}^2} \qquad (i \neq j), \tag{iv}$$

for all distinct pairs, (i, j).

Corollaries III and IV. The distinction between writing

$$F(M_i, M_j) \propto \frac{V_i V_j}{R_{ij}^2} \qquad \text{and} \qquad F(M_i, M_j) = \mathscr{C}\frac{M_i M_j}{R_{ij}^2}, \tag{v}$$

is made, where V_i and V_j are the volumes of the two spheres.

(There is some ambiguity in Newton's statement of these corollaries as it is not always clear whether he means by 'spheres' their volumes or their masses.)

Corollary V. In this corollary, Newton returns to the distinction between 'active' and 'passive' masses made in Proposition LXXV; Corollary II, though by qualifying the spheres as attracting 'each other' or 'one another' he was considering only 'active' masses. But the distinction is important enough to make (see Corollaries VIII and IX below).

> COR. VI. If spheres of this kind revolve about others at rest, each about each, and the distances between the centres of the quiescent and revolving bodies are proportional to the diameters of the quiescent bodies, the periodic times will be equal.
>
> COR. VII. And, again, if the periodic times are equal, the distances will be proportional to the diameters.

In these corollaries Newton returns to Proposition LX, after replacing in it the point masses M_1 and M_2 by finite spheres of the kind considered in this proposition.

In Proposition LX, by comparing the elliptic orbit (of semimajor axis a_s) described by M_1 about a 'quiescent' mass M_2 and the 'revolving' elliptic orbit (of semimajor axis a_R) described by the same mass, M_1, about the common centre of gravity of M_1 and M_2, we showed that for the equality of the periodic times (see equation (24), p. 212):

$$a_S : a_R = \sqrt[3]{M_2} : \sqrt[3]{(M_1 + M_2)} = (\tfrac{4}{3}\pi\rho_2)^{1/3} R_2 : \sqrt[3]{(M_1 + M_2)} \tag{vi}$$

where ρ_2 and R_2 denote the mean density and the radius of M_2; and conversely. Q.E.D.

Corollaries VIII and IX. In these corollaries, Newton repeats Corollary III of Proposition LXXV (in an almost identical phrasing) under the present more general context, distinguishing, further the cases when the orbiting mass is 'passive' or 'active'.

> COR. VIII. All those truths above demonstrated, relating to the motions of bodies about the foci of conic sections, will take place when an attracting sphere, of any form and condition like that above described, is placed in the focus.

COR. IX. And also when the revolving bodies are also attracting spheres of any condition like that above described.

79. Propositions LXXVII and LXXVIII

Having completed his discussion of the attraction of and by spherical bodies when the points of the sphere are centres of attraction with a force proportional to the inverse square of the distances from the attracted bodies, Newton proceeds to the consideration of other laws of attraction. The simplest case, by far, is when the law of centripetal attraction is directly as the distance. Indeed, the case is so simple that Newton disposes of it in two relatively short propositions.

Proposition LXXVII. Theorem XXXVII

If to the several points of spheres there tend centripetal forces proportional to the distances of the points from the attracted bodies, I say, that the compounded force with which two spheres attract each other is as the distance between the centres of the spheres.

CASE 1. Let *AEBF* be a sphere; *S* its centre; *P* a corpuscle attracted; *PASB* the axis of the sphere passing through the centre of the corpuscle; *EF*, *ef* two planes cutting the sphere, and perpendicular to the axis, and equidistant, one on one side, the other on the other, from the centre of the sphere; *G* and *g* the intersections of the planes and the axis; and *H* any point in the plane *EF*.

Strictly, *EF* and *ef* are equally thin circular discs of thickness dz (say) along the axis *PB*. The attraction of the disc *EF* 'upon the corpuscle' *P* is the sum over the attractions of circular rings of radius ϖ $(=HG)$ and width $d\varpi$. The attraction by an element of this

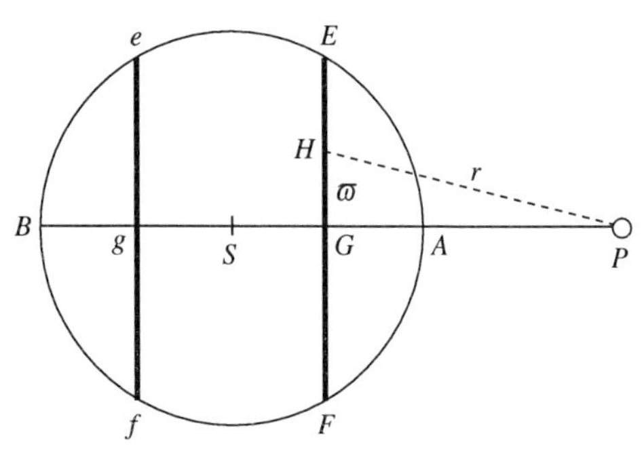

ring is in the direction of *P* and of magnitude proportional to the distance *HP* (the same for all elements of the ring). The component of this force normal to *PS* will cancel by symmetry and only the component of the attraction $\overrightarrow{PH}$ parallel to *PS*, namely $\overrightarrow{PG}$, will survive. Therefore, the attraction per unit mass by the ring considered is

$$\rho k(2\pi\varpi\, d\varpi)(dz)\overrightarrow{PG} \tag{i}$$

where k is the constant of proportionality in the law of the centripetal force assumed and ρ is the density. (Newton argues in similar fashion in Proposition XC.) Integrating the foregoing expression over ϖ from 0 to GE, we find for the attraction by the circular disc EF,

$$\rho[k\pi(GE)^2\,\mathrm{d}z]\overrightarrow{PG}. \tag{ii}$$

Similarly, the attraction by the disc ef is

$$\rho[k\pi(ge)^2\,\mathrm{d}z]\overrightarrow{Pg}. \tag{iii}$$

And since $ge = GE$ (by construction), the attraction of P by the two discs together is

$$\rho k\pi(GE)^2\,\mathrm{d}z(\overrightarrow{PG} + \overrightarrow{Pg}) = 2\rho k\pi(GE)^2\,\mathrm{d}z\overrightarrow{PS}. \tag{iv}$$

Therefore,

> the forces of all the planes in the whole sphere, equidistant on each side from the centre of the sphere, are as the sum of those planes multiplied by the distance *PS*, that is, as the whole sphere and the distance *PS* conjointly:

that is,

$$\text{attraction of } P \text{ by the sphere} = \rho k \text{ (volume pf sphere)}.\overrightarrow{PS}. \tag{v}$$

Q.E.D.

(It should be noted that by 'sphere' Newton means a homogeneous sphere.)

> CASE 2. Let now the corpuscle *P* attract the sphere *AEBF*. And, by the same reasoning, it will appear that the force with which the sphere is attracted is as the distance *PS*. Q.E.D.

Case 3. This is the case of the attraction of two homogeneous spheres, S_1 and S_2 with centres C_1 and C_2, and each particle of S_1 attracting each particle of S_2 with a force as the distance.

Let P be a particle of S_2. Then, as we have shown, the sphere S_1 will attract P with a force given by equation (v); namely,

$$\rho_1 k.\text{(volume of sphere } S_1).\overrightarrow{PC_1} \tag{vi}$$

that is, 'as if the whole attractive force issued from one single corpuscle at the centre of

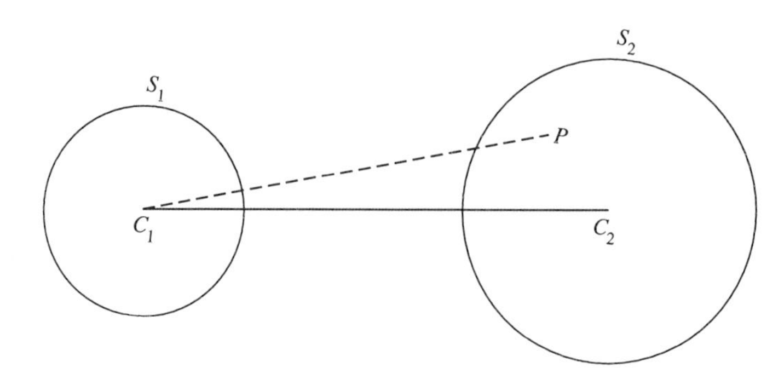

the sphere', C_1. By considering the 'giant' corpuscle at C_1 as attracting the particles of S_2 by the same harmonic law, we can conclude that the force of attraction, by the same proposition, is

$$[\rho_1 k(\text{volume of sphere } S_1)] \times [\rho_2(\text{volume of sphere } S_2)] \, . \, \overrightarrow{C_2C_1}. \qquad \text{(vii)}$$

Case 4. If the attraction of the particles of S_1 and S_2 are mutual, that is, they attract each other with equal and opposite forces, the attraction of S_1 by S_2 will be

$$[\rho_1 k(\text{volume of sphere } S_2)] \times [\rho_2(\text{volume of sphere } S_1)] \, . \, \overrightarrow{C_1C_2}. \qquad \text{(viii)}$$

In our earlier terminology the particles of S_1 and S_2 are both 'active and passive'.

Case 5. We now consider case 1; when the particle P, now designated p, is in the interior of the sphere. The diagram is the same as in case 1 except for the location of p. By the same arguments as in case 1 the attraction of the circular rings EF and ef on p are, respectively,

$$\rho k\pi(GE)^2\overrightarrow{pG}\,\mathrm{d}z \qquad \text{and} \qquad \rho k\pi(ge)^2\overrightarrow{pg}\,\mathrm{d}z. \qquad \text{(ix)}$$

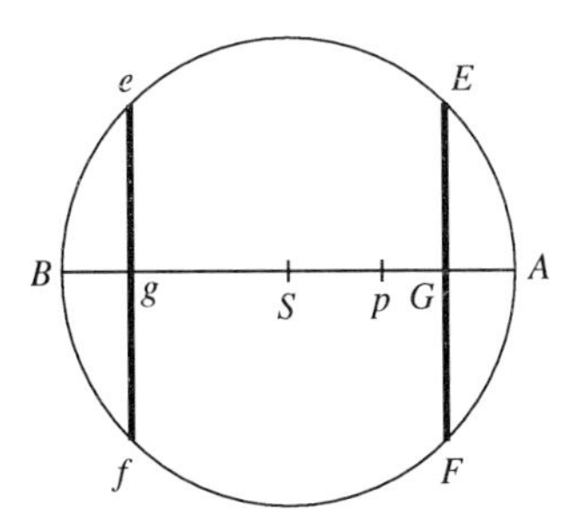

The two together give

$$\begin{aligned}\rho k\pi(GE)^2(\overrightarrow{pG} + \overrightarrow{pg})\,\mathrm{d}z &= \rho k\pi(GE)^2[(\overrightarrow{SG} - \overrightarrow{Sp}) + (\overrightarrow{pS} + \overrightarrow{Sg})]\,\mathrm{d}z \\ &= 2\rho k\pi(GE)^2\overrightarrow{pS}\,\mathrm{d}z, \qquad \text{(x)}\end{aligned}$$

since $\overrightarrow{SG}$ and $\overrightarrow{Sg}$ are equal and opposite. And by the whole sphere, the attraction is

$$\rho k(\text{volume of sphere}) \, . \, \overrightarrow{pS}. \qquad \text{(xi)}$$

Q.E.D.

Case 6. Here we are concerned with the attraction of the sphere S_1 on a sphere S_2 in its interior. By the same arguments as in case 3, the attraction of S_1 by S_2 is given by

$$[k\rho_1(\text{volume of sphere } S_1)] \, . \, [\rho_2(\text{volume of sphere } S_2)] \, . \, \overrightarrow{C_2C_1}. \qquad \text{(xii)}$$

And if the particles of S_1 and S_2 are both active and passive, the attraction of S_2 by S_1 is given by the same expression except that it is in the opposite direction $\overrightarrow{C_1C_2}$.

So far, our considerations have been restricted to spheres of uniform density; the generalization to spherically symmetric distributions of density proceeds as in Propositions LXXIV–LXXVI. Newton's statements are clear and concise; and need no further explanation.

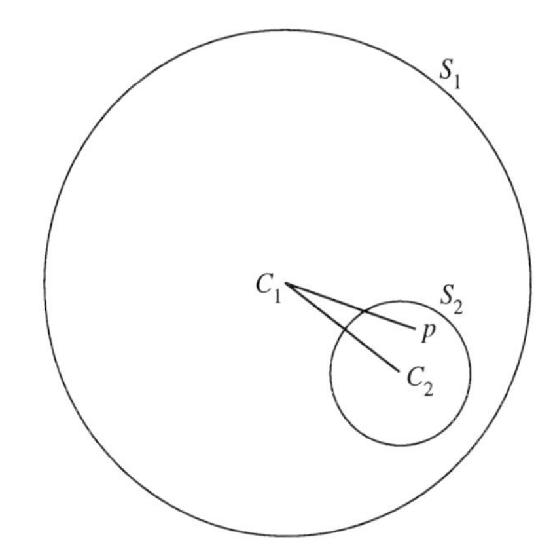

Proposition LXXVIII. Theorem XXXVIII

If spheres in the progress from the centre to the circumference be however dissimilar and unequable, but similar on every side round about at all given distances from the centre; and the attractive force of every point be as the distance of the attracted body: I say, that the entire force with which two spheres of this kind attract each other mutually is proportional to the distance between the centres of the spheres.

This is demonstrated from the foregoing Proposition, in the same manner as Prop. LXXVI was demonstrated from Prop. LXXV.

In the following corollary, Newton points out that, in the solution, obtained in Proposition LXIV (p. 215), for the motion of n point particles which attract each other with forces that increase linearly as their relative distances, mass points can be replaced by finite spheres with arbitrary spherically symmetric density distributions.

COR. Those things that were above demonstrated in Prop. X and LXIV, of the motion of bodies round the centres of conic sections, take place when all the attractions are made by the force of spherical bodies of the condition above described, and the attracted bodies are spheres of the same kind.

Newton concludes with this Scholium:

Scholium

I have now explained the two principal cases of attractions: when the centripetal forces decrease as the square of the ratio of the distances, or increase in a simple ratio of the distances, causing the bodies in both cases to revolve in conic sections, and composing spherical bodies whose centripetal forces observe the same law of increase or decrease in the recess from the centre as the forces of the particles themselves do; which is very remarkable. It would be tedious to run over the other cases, whose conclusions are less elegant and important, so particularly as I have done these. I choose rather to comprehend and determine them all by one general method as follows.

We may draw particular attention to the underlined phrase: nowhere else in the *Principia* has Newton allowed himself a similar expression of surprise.

80. Lemma XXIX and Propositions LXXIX–LXXXI

In Lemma XXIX and in the three propositions that follow, Newton is concerned with the force exerted on an external particle by a homogeneous sphere 'to the several equal particles of which there tend equal centripetal forces' according to some specified law. The treatment is an analytical *tour de force* in which an integral over two variables is reduced to one over only one of them by inverting the order of the integrations and explicitly evaluating the integral over the other. The treatment closely parallels the one by integral calculus but, written in continuous prose, is not easy to unravel. On this account, we shall treat the problem directly by integral calculus noting at points the agreement with Newton's results as he states them.

In the adjoining diagram, S is the centre of a sphere of radius a; P is the external particle at a distance $R(=PS)$ from the centre; EE' is a chord that cuts the axis at right angles at D; and $z(=PD)$ is the distance of P from D. The range of the variable z is clearly $(R-a) \leqslant z \leqslant (R+a)$.

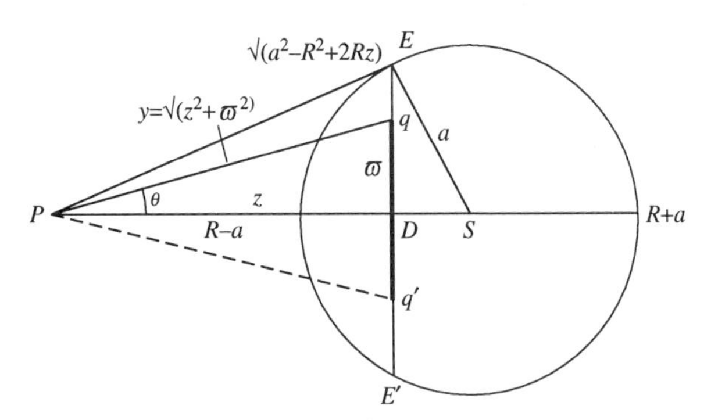

We determine the attraction of P by the sphere by summing over the contributions by circular discs, EDE', of the same thickness dz, at various distances $z(=PD)$ from P; and determine the attraction by the disc EDE', in turn, by summing over the circular rings, qDq' of radii ϖ in the range $0 \leqslant \varpi \leqslant DE$ (in the manner Newton describes later in Proposition XC).

Considering the attraction by the circular disc, EDE', we first observe that only the components of the attraction parallel to SP contribute, while the contributions normal to SP are 'destroyed by contrary attractions'. Accordingly, if $f(r)$ is the centripetal attraction per unit mass at a distance $y(=Pq = (\sqrt{(z^2+\varpi^2)}$ in the context), the attraction of P by the disc EDE' of thickness $\mathrm{d}z$ is given by

$$2\pi\,\mathrm{d}z\int_0^{ED}\mathrm{d}\varpi\,\varpi f(\sqrt{(z^2+\varpi^2)})\cos\theta. \tag{1}$$

Since

$$\cos\theta = z/\sqrt{(z^2+\varpi^2)} \tag{2}$$

we obtain for the attraction by the entire sphere, the double integral,

$$F = 2\pi\int_{R-a}^{R+a}\mathrm{d}z\,z\int_0^{ED}\mathrm{d}\varpi\,\frac{\varpi}{\sqrt{(z^2+\varpi^2)}}f(\sqrt{(z^2+\varpi^2)}); \tag{3}$$

or, alternatively,

$$F = 2\pi\int_{R-a}^{R+a}\mathrm{d}z\,z\int_z^{PE}\mathrm{d}y\,f(y). \tag{4}$$

It is clear, by reference to the illustration, that

$$y^2 = PE^2 = PD^2 + ED^2 = z^2 + a^2 - (R-z)^2 = a^2 - R^2 + 2Rz; \tag{5)*}$$

and we can write:

$$F = 2\pi\int_{R-a}^{R+a}\mathrm{d}z\,z\int_z^{\sqrt{(a^2-R^2+2Rz)}}\mathrm{d}y\,f(y). \tag{6}$$

We now invert the order of the integrations and find, as is manifest from the illustration,

$$F = 2\pi\int_{R-a}^{R+a}\mathrm{d}y\,f(y)\int_{(y^2-a^2+R^2)/2R}^{y}\mathrm{d}z\,z. \tag{7}$$

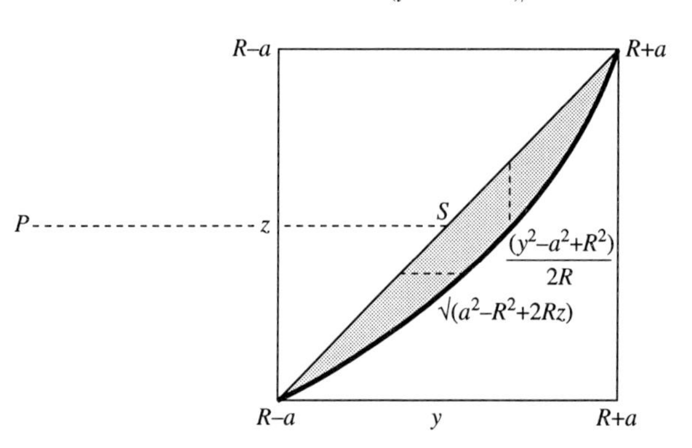

The shaded portion of the illustration is the domain of integration; it is equivalent to the area denoted by ANB in Newton's illustration on p. 292.

In equation (7) the integration over z is readily effected; and we find:

$$\int_{(y-a^2+R^2)/2R}^{y} z\,\mathrm{d}z = \tfrac{1}{2}z^2\Big|_{(y^2-a^2+R^2)/2R}^{y} = \tfrac{1}{2}y^2 - \frac{(y^2-a^2+R^2)^2}{8R^2}. \tag{8}$$

* In this and in the following equations y has the meaning of y_{max}.

Inserting this expression in equation (7) we finally obtain:

$$F = \pi \int_{R-a}^{R+a} f(y)\left[y^2 - \frac{(y^2 - a^2 + R^2)^2}{4R^2}\right] \mathrm{d}y. \tag{9}$$

The expression for the force given by equation (9) is the same as that stated in Proposition LXXX. To show this, we revert to the variable z. By equation (5),

$$y^2 = a^2 - R^2 + 2Rz, \qquad (y^2 - a^2 + R^2)^2 = 4R^2z^2,$$

and (10)

$$y\,\mathrm{d}y = R\,\mathrm{d}z.$$

With these substitutions, equation (9) becomes,

$$F = \pi \int_{R-a}^{R+a} f(\sqrt{(z^2 + \varpi^2)}) \frac{\varpi^2 R}{\sqrt{(z^2 + \varpi^2)}}\,\mathrm{d}z = \pi \int_{R-a}^{R+a} f(y) \frac{\varpi^2 R}{y}\,\mathrm{d}z. \tag{11}$$

Now, by reference to the illustration on p. 287, we observe that

$$y = PE, \qquad \varpi = DE, \qquad \text{and} \qquad R = PS. \tag{12}$$

Accordingly,

$$F = \pi \int_{R-a}^{R+a} f(PE) \frac{DE^2 PS}{PE}\,\mathrm{d}z. \tag{13}$$

Q.E.D.

In his demonstration, Newton inverts the order of the integrations by appealing to the *Book of the Sphere and the Cylinder* by Archimedes. The reference here is to Archimedes's theorem establishing the equality of the area of the surface of a sphere of radius a and that of the enveloping cylinder:

$$\begin{aligned} 4\pi a^2 &= 2\int_0^a \mathrm{d}z \int_0^{2\pi} a\,\mathrm{d}\phi \\ &= 2\int_0^{\pi/2} a\,\mathrm{d}\theta \int_0^{2\pi} a \sin\theta\,\mathrm{d}\phi \\ &= 2\int_0^{2\pi} \mathrm{d}\varphi \int_0^{\pi/2} a^2 \sin\theta\,\mathrm{d}\theta; \end{aligned} \tag{14}$$

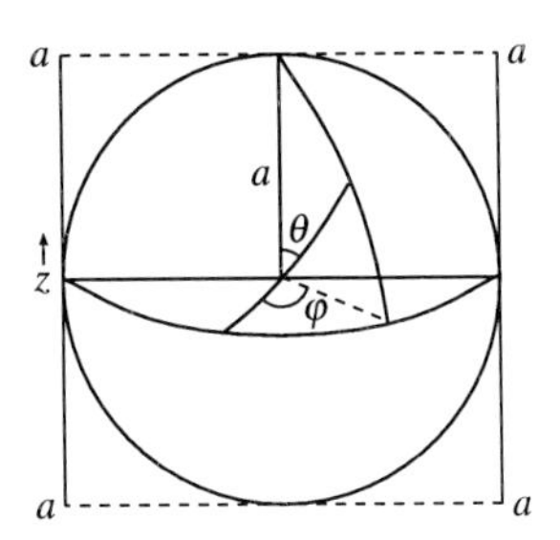

and the 'equality'

$$a\,\mathrm{d}z = a^2 \sin\theta\,\mathrm{d}\theta \tag{15}$$

is the content of Lemma XXIX and Proposition LXXIX.

Corollaries I–IV which conclude Proposition LXXX spell out the integrand of equation (13) for the cases,

$$f(y) \propto \text{constant}, \quad y^{-1}, \quad y^{-3}, \quad \text{and} \quad y^{-4}, \tag{16}$$

skipping the cases,

$$f(y) \propto y \quad \text{and} \quad y^{-2}, \tag{17}$$

to emphasize(?) that these cases have already been treated 'particularly'.

It is curious that Newton, who is concise to a fault when he comes to proving the more difficult of his Propositions (LXVI, for example) should want to write down 'particularly' the integrands of equation (13) for the various cases (16).

In Proposition LXXXI, Newton's main objective is to demonstrate how equation (13) can be 'manipulated' (to be irreverent for once) to obtain the forces explicitly for the last three of the special cases (16). Before turning to these cases, we shall first obtain a simpler expression for F by replacing f in equation (9) by its potential V:

$$f(y) = -\frac{\mathrm{d}V}{\mathrm{d}y}. \tag{18}$$

Equation (9) becomes

$$F = -\pi \int_{R-a}^{R+a} \frac{\mathrm{d}V}{\mathrm{d}y}\left[y^2 - \frac{(y^2 - a^2 + R^2)^2}{4R^2}\right]\mathrm{d}y. \tag{19}$$

Since,

$$y^2 - \frac{1}{4R^2}(y^2 - a^2 + R^2)^2 = 0 \qquad \text{for } y = R + a \quad \text{and} \quad R - a, \tag{20}$$

and integration by parts of the integral yields:

$$F = \pi \int_{R-a}^{R+a} Vy\left[2 - \frac{1}{R^2}(y^2 - a^2 + R^2)\right]\mathrm{d}y \tag{21}$$

or

$$F = \frac{\pi}{R^2}\int_{R-a}^{R+a} Vy[R^2 + a^2 - y^2]\,\mathrm{d}y, \tag{22}$$

an equation which Newton appears to have known (see Examples 2 and 3 below).

The form of the equation (13), on which Newton bases his evaluation, is:

$$F = \pi \int_{R-a}^{R+a}\left[y^2 - \frac{1}{4R^2}(y^2 - a^2 + R^2)^2\right] f(y)\frac{R}{y}\,\mathrm{d}z, \tag{23}$$

with the equivalence,

and,

$$\left.\begin{aligned} f(y) = \frac{1}{y}, \qquad R = PS, \qquad PE = y, \\ DE^2 = y^2 - \frac{1}{4R^2}(y^2 - a^2 + R^2)^2 = \varpi^2. \end{aligned}\right\} \tag{24}$$

Where y now stands for y_{max}. We shall however continue to use equations (9) or (22) with y as the variable of integration.

Example 1. $f(y) = y^{-1}$

Equation (9) now gives,

$$F = \pi \int_{R-a}^{R+a} \left[y - \frac{(y^2 + R^2 - a^2)^2}{4R^2 y} \right] \mathrm{d}y$$

$$= \frac{\pi}{4R^2} \int_{R-a}^{R+a} \left\{ 4R^2 y - \frac{1}{y} [y^4 + 2y^2(R^2 - a^2) + (R^2 - a^2)^2] \right\} \mathrm{d}y; \tag{25}$$

or, on further simplification,

$$F = \frac{\pi}{4R^2} \int_{R-a}^{R+a} \left[-\frac{(R^2 - a^2)^2}{y} + 2y(R^2 + a^2) - y^3 \right] \mathrm{d}y. \tag{26}$$

On evaluating this integral, we find:

$$F = \frac{\pi}{4R^2} \left[-(R^2 - a^2)^2 \log \frac{R + a}{R - a} + 2aR(R^2 + a^2) \right]. \tag{27}$$

Newton reduces the integrand to the form given in equation (26) and distinguishes the first term from the rest by simply saying that it 'describes the area of a hyperbola' (with a graph of the function y^{-1}) without expressing its value in terms of the logarithmic function. (One wonders why.)

Example 2.

In conformity with Newton's definitions,

$$f(y) = \frac{2a^2}{y^3} \qquad \text{and} \qquad V(y) = \frac{a^2}{y^2}. \tag{28}$$

By equation (22),

$$F = \pi \frac{a^2}{R^2} \int_{R-a}^{R+a} \left(\frac{R^2 + a^2}{y} - y \right) \mathrm{d}y, \tag{29}$$

which integrates immediately to give,

$$F = \pi \frac{a^2}{R^2}\left[(R^2 + a^2)\log\frac{R+a}{R-a} - 2aR\right]. \tag{30}$$

The expression (30) for F agrees exactly with the one Newton gives. To show that this is the case, it is proper that we reproduce at this time the illustration which appears four times (with minor variations) in the *Principia*. It is the same as the one on p. 287 with

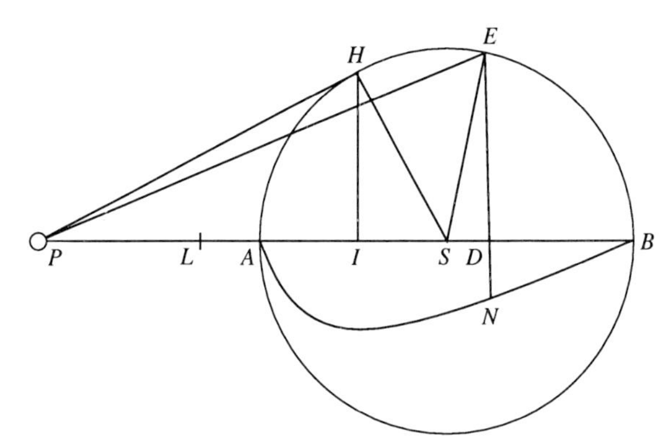

the addition that PH is the tangent to the circle, HI is drawn perpendicularly to PS, and L bisects PI. We readily verify that by this construction,

$$2LS = \frac{R^2 + a^2}{R}, \qquad 2LD = \frac{y^2}{R}, \qquad PS = R, \qquad \text{and} \qquad IS = \frac{a^2}{R}. \tag{31}$$

Returning to the solution (30), Newton gives

$$2LS\,.\,IS = \frac{(R^2 + a^2)a^2}{R^2} \tag{32}$$

for the factor in front of the area 'described by a hyperbola' and

$$-2SA\,.\,IS = -2a^3/R \tag{33}$$

for the remaining term. (Q.E.D.!)

The fact that Newton finds two terms for F, in exact correspondence with the expression for F derived from equation (29), suggests that he had prior knowledge of the simplification of equation (23) which results from replacing $f(y)$ by its potential, and integrating by parts. The suggestion is confirmed (see Example 3 below) by Newton omitting the crucial steps by simply stating 'Whence arises this construction of the problem'.

Example 3

Again in conformity with Newton's definition

$$f(y) = \frac{2a^3}{y^4} \qquad \text{and} \qquad V(y) = \frac{2a^3}{3y^3}. \tag{34}$$

Equation (22) now gives,

$$F = \pi \frac{2a^3}{3R^2} \int_{R-a}^{R+a} \left(\frac{R^2 + a^2}{y^2} - 1 \right) \mathrm{d}y, \tag{35}$$

which on integration gives

$$F = \frac{8\pi a^6}{3R^2(R^2 - a^2)}. \tag{36}$$

Again, this agrees with Newton's result:

$$F \propto (PS^3 PI)^{-1}; \tag{37}$$

for, by equations (31),

$$R^2(R^2 - a^2) = PS^2(PS^2 - PS \, . \, IS) = PS^3(PS - IS) = PS^3 PI. \tag{38}$$

Q.E.D.

The brevity of the demonstration and the omission of the crucial steps by the statement 'these after due reductions come forth . . .' confirms, in my judgment, that Newton had knowledge of the simplification of equation (13) that results from the introduction of the potential.

81. Proposition LXXXII and the discovery of the method of inversion and the principle of images

At the conclusion of Proposition LXXXI, Newton introduces Proposition LXXXII as follows.

> By the same method one may determine the attraction of a corpuscle situated within the sphere, but more expeditiously by the following theorem.

But first, what is the solution by the 'same method'?

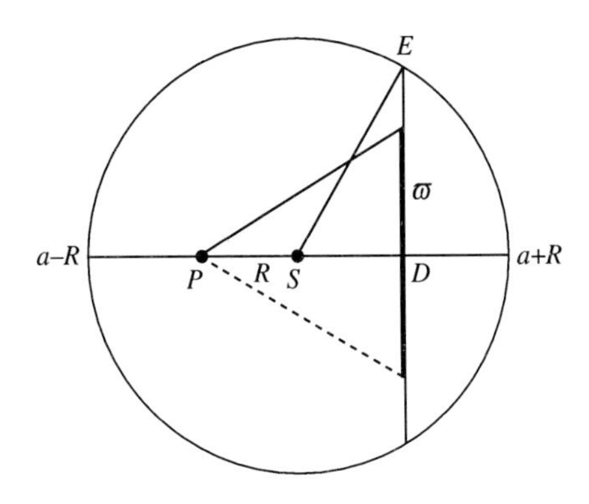

A comparison of the adjoining diagram with that on p. 287 makes it manifest that the expression,

$$F_1 = \frac{\pi}{R_1^2} \int_{R_1 - a}^{R_1 + a} V(y) y[(R_1^2 + a^2) - y^2] \, dy, \tag{1}$$

applicable to 'a corpuscle situated without the sphere' at a distance $R_1 \geqslant a$ from the centre will be replaced by

$$F_2 = \frac{\pi}{R_2^2} \int_{a - R_2}^{a + R_2} V(y) y[(R_2^2 + a^2) - y^2] \, dy, \tag{2}$$

applicable to 'a corpuscle situated within the sphere' at a distance $R_2 \leqslant a$ from the centre. Clearly equation (2) can be treated in the same manner as equation (1). But in Proposition LXXXII Newton shows how the problem can be solved 'more expeditiously'.

Proposition LXXXII. Theorem XLI

In a sphere described about the centre S with the radius SA, if there be taken SI, SA, SP continually proportional: I say, that the attraction of a corpuscle within the sphere in any place I is to its attraction without the sphere in the place P in a ratio compounded of the square root of the ratio of IS, PS, the distances from the centre, and the square root of the ratio of the centripetal forces tending to the centre in those places P and I.

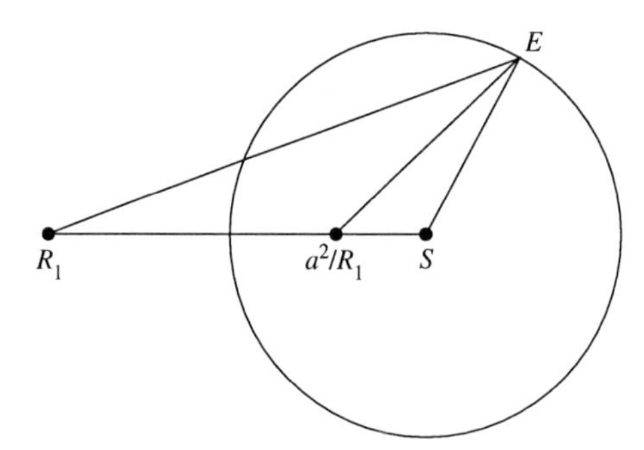

This proposition can be paraphrased as follows. Let the distances, $R_1 \geqslant a$ and $R_2 \leqslant a$, of two corpuscles situated without and within the sphere be related in the manner,

$$R_1 R_2 = a^2. \tag{3}$$

Then, for a law of centripetal attraction proportional to the nth power of the distance, the forces F_1 and F_2 are in the ratio

$$F_1 : F_2 = \sqrt{R_1^{n+1}} : \sqrt{R_2^{n+1}}. \tag{4}$$

One recognizes at once that in 1685 Newton had discovered the method of images fully 150 years before Lord Kelvin who is credited with the discovery. We append below some historical comments on the discovery of the method of images; but meantime the proof of the proposition.

Proof:

For

$$R_2 = a^2/R_1, \tag{5}$$

the limits of the integrals (1) and (2) are:

$$\left.\begin{aligned} R_1 + a &= (\sqrt{R_1} + \sqrt{R_2})\sqrt{R_1}; \qquad & a + R_2 &= (\sqrt{R_1} + \sqrt{R_2})\sqrt{R_2},\\ R_1 - a &= (\sqrt{R_1} - \sqrt{R_2})\sqrt{R_1}; \qquad & a - R_2 &= (\sqrt{R_1} - \sqrt{R_2})\sqrt{R_2}. \end{aligned}\right\} \tag{6}$$

Besides,

$$R_1^2 + a^2 = R_1(R_1 + R_2) \qquad \text{and} \qquad R_2^2 + a^2 = R_2(R_1 + R_2). \tag{7}$$

Making these substitutions in equations (1) and (2), we find:

$$F_1 = \frac{\pi}{R_1^2} \int_{(\sqrt{R_1}-\sqrt{R_2})\sqrt{R_1}}^{(\sqrt{R_1}+\sqrt{R_2})\sqrt{R_1}} V(y)y[R_1(R_1 + R_2) - y^2]\,\mathrm{d}y \tag{8}$$

and

$$F_2 = \frac{\pi}{R_2^2} \int_{(\sqrt{R_1}-\sqrt{R_2})\sqrt{R_2}}^{(\sqrt{R_1}+\sqrt{R_2})\sqrt{R_2}} V(y)y[R_2(R_1 + R_2) - y^2]\,\mathrm{d}y. \tag{9}$$

Now write

$$\left.\begin{aligned} y &= \eta\sqrt{R_1} \qquad \text{in equation (8),}\\ &\text{and}\\ y &= \eta\sqrt{R_2} \qquad \text{in equation (9).} \end{aligned}\right\} \tag{10}$$

Equations (8) and (9) take the symmetrical forms:

$$\left.\begin{aligned} F_1 &= \pi \int_{\sqrt{R_1}-\sqrt{R_2}}^{\sqrt{R_1}+\sqrt{R_2}} V(\eta\sqrt{R_1})\eta[(R_1 + R_2) - \eta^2]\,\mathrm{d}\eta,\\ F_2 &= \pi \int_{\sqrt{R_1}-\sqrt{R_2}}^{\sqrt{R_1}+\sqrt{R_2}} V(\eta\sqrt{R_2})\eta[(R_1 + R_2) - \eta^2]\,\mathrm{d}\eta. \end{aligned}\right\} \tag{11}$$

Now suppose that

$$f(y) = -\kappa(n+1)y^n \qquad \text{and} \qquad V(y) = \kappa y^{n+1}, \tag{12}$$

where κ is a constant. Equations (11) then become,

$$\left.\begin{aligned} F_1 &= \kappa\pi R_1^{(n+1)/2} \int_{\sqrt{R_1}-\sqrt{R_2}}^{\sqrt{R_1}+\sqrt{R_2}} \eta^{n+2}[(R_1 + R_2) - \eta^2]\,\mathrm{d}\eta,\\ F_2 &= \kappa\pi R_2^{(n+1)/2} \int_{\sqrt{R_1}-\sqrt{R_2}}^{\sqrt{R_1}+\sqrt{R_2}} \eta^{n+2}[(R_1 + R_2) - \eta^2]\,\mathrm{d}\eta. \end{aligned}\right\} \tag{13}$$

It is now manifest that:

$$F_1 : F_2 = \sqrt{R_1^{(n+1)}} : \sqrt{R_2^{(n+1)}}. \tag{14}$$

Q.E.D.

The foregoing proof is not really different from what Newton expresses in a page of well-knit prose. One can only marvel at the 'muscles of his intuition'.

On the discovery of the method of images

The method of images, as one learns, was discovered by Sir William Thomson (alias Lord Kelvin) in the context of electrical problems. His original investigations were published in Cambridge and Dublin Mathematical Journal, 1848, pp. 241–74; and earlier in 1844 in a brief letter to Liouville (J. de Math. Pure et Appliques, 1844, pp. 364–66). As Maxwell explains in his *Electricity and magnetism* (Oxford University Press, Third edition, 1892, p. 245, et seq.):

> In applying this method to the most elementary case of a sphere under the influence of a single electrified point, we require to expand the potential due to the electrified point in a series of solid harmonics, and to determine a second series of solid harmonics which express the potential, due to the electrification of the sphere, in the space outside.
>
> It does not appear that any of these mathematicians observed that this second series expresses the potential due to an imaginary electrified point, which has no physical existence as an electrified point, but which may be called an electrical image, because the action of the surface on external points is the same as that which would be produced by the imaginary electrified point if the spherical surface was removed.
>
> This discovery seems to have been reserved for Sir. W. Thomson, who has developed it into a method of great power for the solution of electrical problems, and at the same time capable of being presented in an elementary geometrical form.

And further:

> Let a be the radius of the sphere.
>
> Let f be the distance of the electrified point A from the centre C.
>
> Let e be the charge of this point.
>
> Then the image of the point is at B, on the same radius of the sphere at a distance a^2/f, and the charge of the image is $-ea/f$.
>
> We have shewn that this image will produce the same effect on the opposite side of the surface as the actual electrification of the surface does.

(The near identity of this illustration of Maxwell's with that on p. 294 is to be noted.) Newton's discovery of the method of images in Proposition LXXXII is in the gravitational context where one does not have 'corpuscles' of opposite 'charges'. His discovery is, therefore, all the more remarkable since the natural starting point, of surfaces

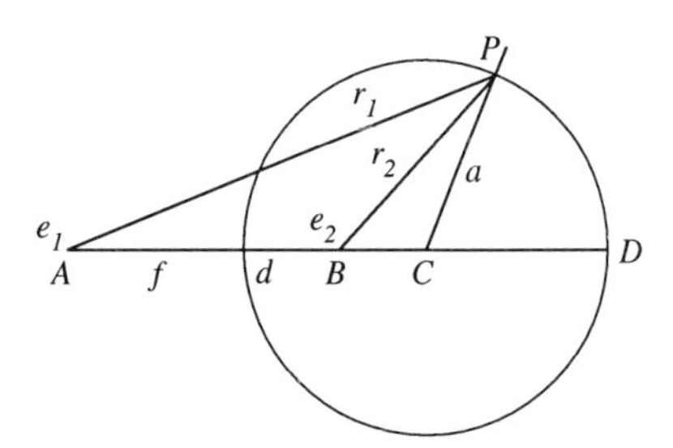

of zero potential (as for example, $e_1/r_1 + e_2/r_2 = 0$, in electrostatics) does not exist in the gravitational context. However, E. J. Routh in his *Treatise on analytical statics*, Vol. II (Cambridge University Press, Third edition, 1892, p. 84, et seq.) does introduce the method of inversion in the gravitational context, as follows.

METHOD OF INVERSION

Inversion from a point

Let O be any assumed origin, and let Q be a point moving in any given manner. If on the radius vector OQ we take a point Q' so that $OQ.OQ' = k^2$, *then Q and Q' are called inverse points.* If Q trace out a curve, Q' traces out the inverse curve; if Q trace out a surface or solid, Q' traces out the inverse surface or solid. The points Q, Q' are sometimes said to be *inverse with regard to a sphere* whose centre is O and radius k.

Let P', Q' be the inverse points of P, Q, then since the products $OP.OP'$, $OQ.OQ'$ are equal and the angles POQ, $P'OQ'$ are the same, the triangles POQ, $P'OQ'$ are similar. We therefore have

$$\frac{1}{P'Q'} = \frac{1}{PQ} \cdot \frac{OQ}{OP'} \tag{1}$$

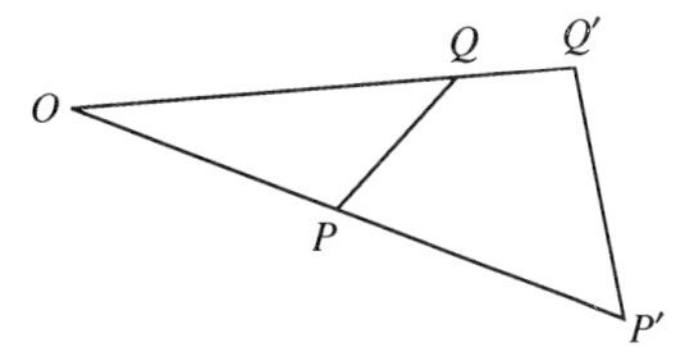

Let m, m' be the masses of two particles placed respectively at Q, Q', and let the densities be such that

$$m' = m\frac{k}{OQ}. \tag{2}$$

Multiplying equations (1) and (2) together, we see that the potential at P' of m' is equal to that at P of m, after multiplication by a quantity k/OP' which is independent of the position of Q.

And Routh goes on to prove a number of propositions on this definition; but not Newton's. Routh, like Maxwell, credits Kelvin with the discovery.

82. Propositions LXXXIII and LXXXIV

After the display of insight, imagination, and craftsmanship in the preceding propositions, Propositions LXXXIII and LXXXIV are somewhat of an anticlimax for the reader. But they are examples of Newton's reluctance to leave a subject with frayed edges.

Proposition LXXXIII. Problem XLII

> *To find the force with which a corpuscle placed in the centre of a sphere is attracted towards any segment of that sphere whatsoever.*

In the adjoining diagram the shaded portion is a segment of the sphere with centre S; $SD = \zeta$ is the perpendicular from S to the base of the segment; and $DB = a - \zeta$ is the thickness of the segment. It is evident that equation (4) of §80, under the present circumstances, takes the form

$$F = 2\pi \int_{\zeta}^{a} \mathrm{d}z\, z \int_{z}^{a} \mathrm{d}y\, f(y). \tag{1}$$

Inverting the order of the integrations, we obtain

$$F = 2\pi \int_{\zeta}^{a} \mathrm{d}y\, f(y) \int_{\zeta}^{y} \mathrm{d}z\, z = \pi \int_{\zeta}^{a} \mathrm{d}y\, f(y)[z^2]_{\zeta}^{y}, \tag{2}$$

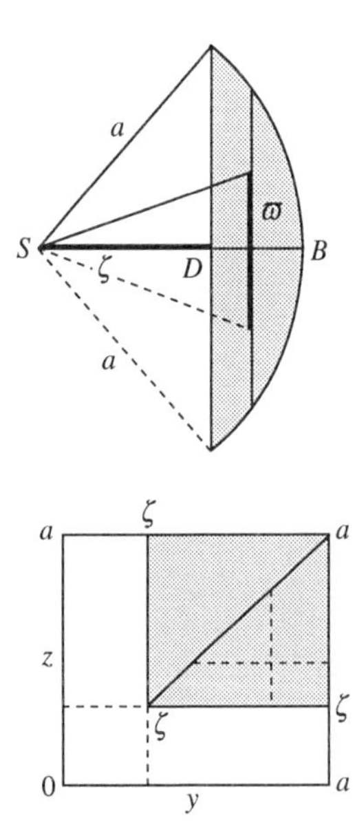

that is

$$F = \pi \int_{\zeta}^{a} f(y)(y^2 - \zeta^2)\, dy. \tag{3}$$

Q.E.I.

We shall not stop to explain that equation (3) exactly corresponds to Newton's solution.

Proposition LXXXIV. Problem XLIII

To find the force with which a corpuscle, placed without the centre of a sphere in the axis of any segment, is attracted by that segment.

The adjoining diagram, which is appropriate to the problem on hand, is self-explanatory; and the equation that provides the solution is, manifestly:

$$F = \pi \int_{R-\zeta}^{R+a} f(y)\left[y^2 - \frac{(y^2 - a^2 + R^2)^2}{4R^2} \right] dy; \tag{4}$$

or, in terms of the potential,

$$F = \frac{\pi}{4R^2}(a^2 - \zeta^2)[(2R - \zeta)^2 - a^2]\, V(R - \zeta) + \frac{\pi}{R^2} \int_{R-\zeta}^{R+a} Vy(R^2 + a^2 - y^2)\, dy. \tag{5}$$

And the solution can be completed as in §80. Q.E.I.

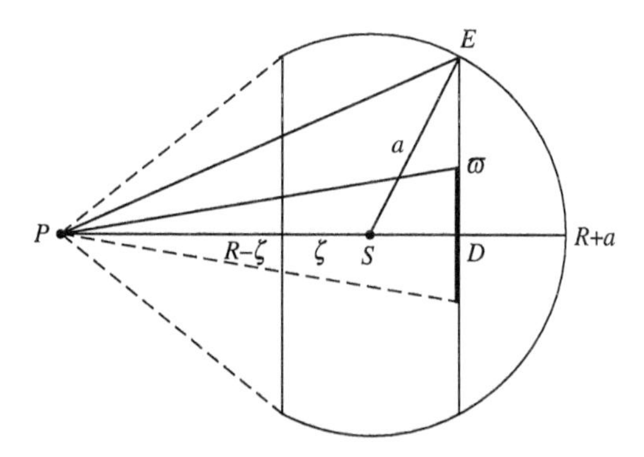

Section XII concludes with the Scholium

Scholium

The attractions of spherical bodies being now explained, it comes next in order to treat of the laws of attraction in other bodies consisting in like manner of attractive particles; but to treat of them particularly is not necessary to my design. It will be sufficient to add some general propositions relating to the forces of such bodies, and the motions thence arising, because the knowledge of these will be of some little use in philosophical inquiries.

Plate 2 Newton at age eighty-three (1726) by John Vanderbank. The artist's identity has been questioned. It may be the work of Michael Dahl. (With permission from the National Portrait Gallery.)

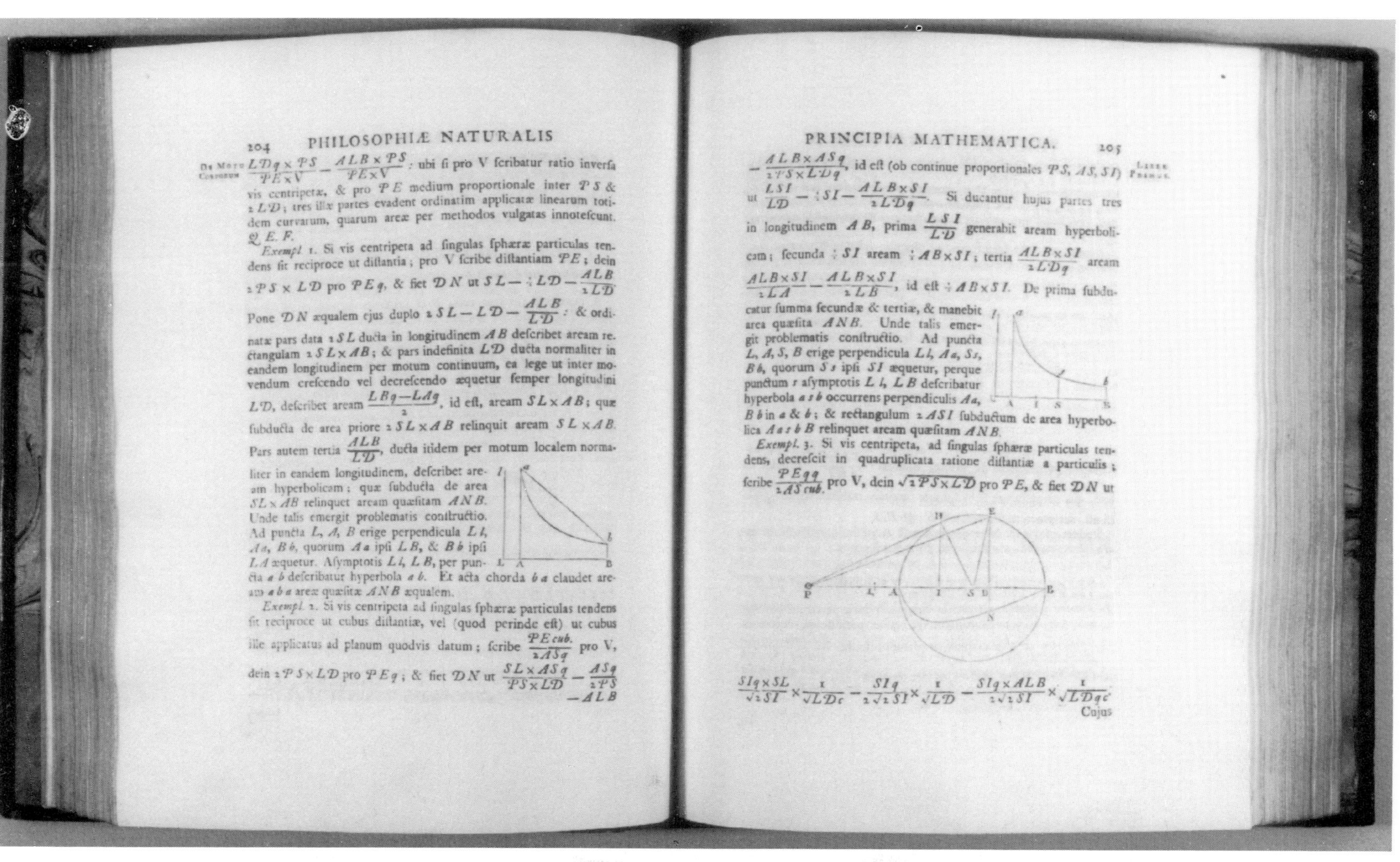

204 PHILOSOPHIÆ NATURALIS

De Motu Corporum

$\frac{LDq \times PS}{PE \times V} - \frac{ALB \times PS}{PE \times V}$: ubi ſi pro V ſcribatur ratio inverſa vis centripetæ, & pro PE medium proportionale inter PS & $2LD$; tres illæ partes evadent ordinatim applicatæ linearum totidem curvarum, quarum areæ per methodos vulgatas innoteſcunt. *Q. E. F.*

Exempl. 1. Si vis centripeta ad ſingulas ſphæræ particulas tendens ſit reciproce ut diſtantia; pro V ſcribe diſtantiam PE; dein $2PS \times LD$ pro PEq, & fiet DN ut $SL - \frac{1}{2}LD - \frac{ALB}{2LD}$.

Pone DN æqualem ejus duplo $2SL - LD - \frac{ALB}{LD}$: & ordinatæ pars data $2SL$ ducta in longitudinem AB deſcribet aream rectangulam $2SL \times AB$; & pars indefinita LD ducta normaliter in eandem longitudinem per motum continuum, ea lege ut inter movendum creſcendo vel decreſcendo æquetur ſemper longitudini LD, deſcribet aream $\frac{LBq - LAq}{2}$, id eſt, aream $SL \times AB$; quæ ſubducta de area priore $2SL \times AB$ relinquit aream $SL \times AB$. Pars autem tertia $\frac{ALB}{LD}$, ducta itidem per motum localem normaliter in eandem longitudinem, deſcribet aream hyperbolicam; quæ ſubducta de area $SL \times AB$ relinquet aream quæſitam ANB. Unde talis emergit problematis conſtructio. Ad puncta L, A, B erige perpendicula Ll, Aa, Bb, quorum Aa ipſi LB, & Bb ipſi LA æquetur. Aſymptotis Ll, LB, per puncta $a\,b$ deſcribatur hyperbola ab. Et acta chorda ba claudet aream aba areæ quæſitæ ANB æqualem.

Exempl. 2. Si vis centripeta ad ſingulas ſphæræ particulas tendens ſit reciproce ut cubus diſtantiæ, vel (quod perinde eſt) ut cubus ille applicatus ad planum quodvis datum; ſcribe $\frac{PE\,cub.}{2ASq}$ pro V, dein $2PS \times LD$ pro PEq; & fiet DN ut $\frac{SL \times ASq}{PS \times LD} - \frac{ASq}{2PS} - \frac{ALB}{}$

PRINCIPIA MATHEMATICA. 205

Liber Primus.

$-\frac{ALB \times ASq}{2PS \times LDq}$, id eſt (ob continue proportionales PS, AS, SI) ut $\frac{LSI}{LD} - \frac{1}{2}SI - \frac{ALB \times SI}{2LDq}$. Si ducantur hujus partes tres in longitudinem AB, prima $\frac{LSI}{LD}$ generabit aream hyperbolicam; ſecunda $\frac{1}{2}SI$ aream $\frac{1}{2}AB \times SI$; tertia $\frac{ALB \times SI}{2LDq}$ aream $\frac{ALB \times SI}{2LA} - \frac{ALB \times SI}{2LB}$, id eſt $\frac{1}{2}AB \times SI$. De prima ſubducatur ſumma ſecundæ & tertiæ, & manebit area quæſita ANB. Unde talis emergit problematis conſtructio. Ad puncta L, A, S, B erige perpendicula Ll, Aa, Ss, Bb, quorum Ss ipſi SI æquetur, perque punctum s aſymptotis Ll, LB deſcribatur hyperbola asb occurrens perpendiculis Aa, Bb in a & b; & rectangulum $2ASI$ ſubductum de area hyperbolica $AasbB$ relinquet aream quæſitam ANB.

Exempl. 3. Si vis centripeta, ad ſingulas ſphæræ particulas tendens, decreſcit in quadruplicata ratione diſtantiæ a particulis; ſcribe $\frac{PEqq}{2AS\,cub.}$ pro V, dein $\sqrt{2PS \times LD}$ pro PE, & fiet DN ut

$$\frac{SIq \times SL}{\sqrt{2SI}} \times \frac{1}{\sqrt{LDc}} - \frac{SIq}{2\sqrt{2SI}} \times \frac{1}{\sqrt{LD}} - \frac{SIq \times ALB}{2\sqrt{2SI}} \times \frac{1}{\sqrt{LDqc}}.$$

Cujus

Plate 3 From the original 1726 edition of the *Principia* in the Department of Special Collections of the University of Chicago Library; and published here with the permission of the Library.

83. Some personal reflections

We have quoted in Chapter 1. §5, J. W. L. Glaisher, who, referring to Newton's 'superb theorem' (without specifying which of the eleven Theorems, XXX–XLI, included in Section XII) went on to say that 'all mechanisms of the universe at once lay spread before him'. We know that Newton was fully aware of the central role of Proposition LXXVI and its Corollaries VIII and IX for the formulation of his universal law of gravitation by his reference to 'Propositions LXXV and LXXVI and their Corollaries' in Proposition VIII of Book III; and by his letter of 20 June 1686 to Halley (which we have also quoted in Chapter 1, §5). But we have only scant evidence for Newton's personal feelings on proving these propositions apart from his summing of the Propositions, LXXV–LXXVII, in the Scholium that follows, by *Quod est notatu dignum* ('which is very remarkable').

Evidence of a different sort comes from his portrait at the age of 83 that now hangs in the National Portrait Gallery, London (see Plate 2). In this portrait, we see Newton with the then new (1726) third edition of the *Principia* open at pages 204 and 205, displaying (as one can easily distinguish) the illustrations of Proposition LXXXI (see facing Plate 3). Newton, one may suppose, made this particular selection; and he must have had some reason. We can only speculate.

Proposition LXXXI could not have had for Newton any particular 'use in philosophical inquiries'. But the proposition does display Newton's familiarity with and craftsmanship in the use of the integral calculus: inversion of the order of the integrations for reducing a double integral into a single one, and integration by parts to simplify an integral. Was Newton displaying to his critics (if they can understand!) that already, in 1685 when he was writing the *Principia*, he was a virtuoso in the Art of the Calculus?

❖16❖

Attraction by non-spherical bodies

84. Introduction

As Newton stated in his concluding Scholium of Proposition LXXXIV, his next concern was the attraction by non-spherical bodies; but 'to treat them particularly was not his design'. Even so, Propositions LXXXVII and Corollaries II and III of Proposition XCI stand out.

In Proposition LXXXVII, Newton establishes a scaling law for the force exerted at an external point by an arbitrary distribution of mass specified by a density function and the law of centripetal attraction, by the elements of mass of the distribution, is inversely as the *n*th power of the distance.

And in Corollary II of Proposition XCI, Newton determines the attraction of an oblate spheroid at an external point on its minor axis; and in Corollary III, he generalizes, for spheroids, Propositions LXX and LXXIII applicable to spheres. As we have noted in Chapter 9, §49, these corollaries were added out of sequence when, at a later time, Newton was writing on the oblateness of the planets caused by their rotation in Book II, Proposition XIX.

And finally, we may note parenthetically that except in Propositions LXXXVI and LXXXVII, the considerations of this chapter are restricted to homogeneous bodies of constant density.

85. How we may discriminate between different laws of centripetal attraction

In Propositions LXXXV and LXXXVI, Newton shows how we may draw inferences of physical interest from the propositions of Chapter 15, particularly from the second and third examples of Proposition LXXXI.

Proposition LXXXV. Theorem XLII

> *If a body be attracted by another, and its attraction be vastly stronger when it is contiguous to the attracting body than when they are separated from each other by a very small interval; the forces of the particles of the attracting body decrease, in the recess of the body attracted, in more than the squared ratio of the distance of the particles.*

Let B_1 be the body that attracts B_2; and B_2 approach B_1 to come into contact with it at C. Draw a sphere with centre S that touches the inner boundary of B_1 at C with the same curvature.

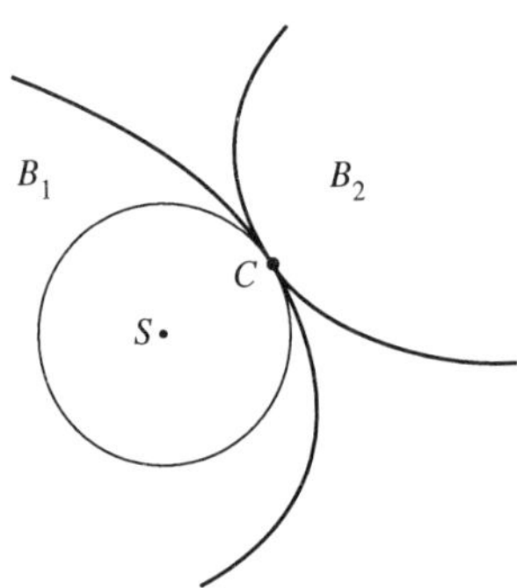

We shall first consider the case when the particles of B_1 attract the particles of B_2 with a force proportional to the inverse square of their mutual distances. The attraction of the particles of B_2 by the solid sphere will be proportional to the inverse square of their distances from the centre S. In particular, the particles of B_2 in the immediate neighbourhood of C, at contact, will be attracted proportionally to SC^{-2} which is finite. The particles further away from C will be attracted less. And as Newton explains:

> [if] we take away any parts remote from the place of contact, and add new parts anywhere at pleasure, we may change the figures of the attractive bodies at pleasure; but the parts added or taken away, being remote from the place of contact, will cause no remarkable excess of the attraction arising from the contact of the two bodies.

We conclude that the attraction of B_2 by B_1 'cannot be vastly stronger when they are contiguous than when they are separated from each other by a very small interval'. If the contrary should be the case we can infer that the attraction of the particles of B_2 by the particles of B_1 must 'decrease more than the squared ratio of the distance of the particles'.

The arguments are even stronger if the body B_1 is hollow and the body B_2, inside the hollow, comes into contact with B_1 at C (see the figure). For, drawing a sphere with centre S, inside the hollow, which touches the outer hollow side of the boundary of B_1 at C with the same curvature, the attraction on the particles of B_2 in the interior

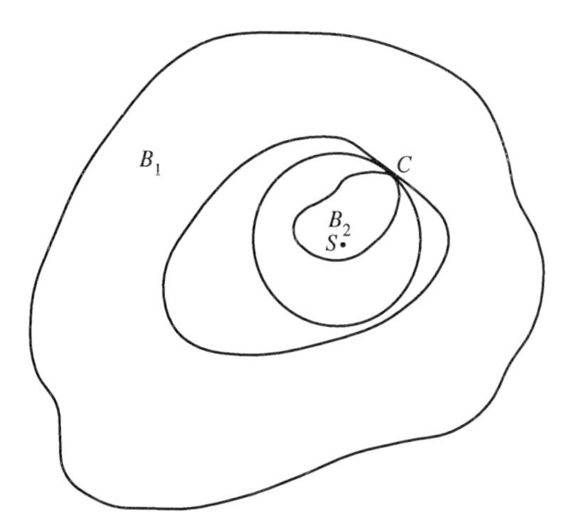

of the hollow sphere vanishes by Proposition LXX. And the argument can be completed as before.

Therefore the Proposition holds good in bodies of all figures. Q.E.D.

Proposition LXXXVI. Theorem XLIII

If the forces of the particles of which an attractive body is composed decrease, in the recession of the attractive body, as the third or more than the third power of the distance from the particles, the attraction will be vastly stronger in the point of contact than when the attracting and attracted bodies are separated from each other, though by ever so small an interval.

In Examples 2 and 3 of Proposition LXXXI, it was shown that if the centripetal force tending to the constituent particles of a homogeneous sphere varies as the inverse cube or the inverse fourth power of the distance, then the force with which the sphere attracts a corpuscle at a distance R from the centre of the sphere, tends to infinity with the behaviours (cf., equations (30) and (36) of §80, respectively)

$$-\log(R-a) \quad \text{and} \quad (R-a)^{-1}, \qquad (R \to a).$$

Arguing exactly as in Proposition LXXXV, we conclude on the basis of the foregoing behaviours that the attraction of such bodies will tend to infinity even when the attracting body is not spherical; for, as Newton explains:

> by adding to or taking from those spheres and orbs any attractive matter anywhere without the place of contact, so that the attractive bodies may receive any assigned figure, the Proposition will hold good of all bodies universally. Q.E.D.

While the proposition has been proved for the inverse-cube and the inverse-fourth power laws of centripetal attraction, Newton assumes, as intuitively obvious, that it is valid for all power laws that decrease 'more than the third power'.

It is of interest to add that by this proposition Newton implicitly explains *why bodies stick together on contact*. For, as we now know, the *van der Waals force*, which is responsible, varies as the inverse *seventh*-power of the distance.

86. The scaling law

Proposition LXXXVII. Theorem XLIV

If two bodies similar to each other, and consisting of matter equally attractive, attract separately two corpuscles proportional to those bodies, and in a like situation to them, the accelerative attractions of the corpuscles towards the entire bodies will be as the accelerative attractions of the corpuscles towards particles of the bodies proportional to the wholes, and similarly situated in them.

and

COR. I. Therefore if, as the distances of the corpuscles attracted increase, the attractive forces of the particles decrease in the ratio of any power of the distances, the accelerative attractions towards the whole bodies will be directly as the bodies, and inversely as those powers of the distances.

The attraction $\vec{F}(\vec{R})$ at an external point at $\vec{R}$, by an arbitrary distribution of mass specified by a density function $\rho(\vec{r})$ when the elements of mass are endowed with a power of centripetal attraction inversely as the nth power of the distance, is given by

$$\vec{F}(\vec{R}) = -\text{constant} \iiint_{\mathscr{V}} \rho(\vec{r}) \frac{\vec{R} - \vec{r}}{|\vec{R} - \vec{r}|^{n+1}} \, \mathrm{d}\vec{r}, \tag{1}$$

where $\mathscr{V}$ is the spatial domain in which $\rho(\vec{r})$ is non-vanishing. Replace in this equation,

$$\vec{r} \text{ by } \alpha\vec{r} \quad \text{and} \quad \vec{R} \text{ by } \alpha\vec{R}, \tag{2}$$

where α is a constant, that is, we expand (or contract) the entire system *homologously* by a factor α. The equation becomes

$$\vec{F}(\alpha\vec{R}) = -\text{constant } \alpha^{3-n} \iiint_{\mathscr{V}} \rho(\vec{r}) \frac{\vec{R} - \vec{r}}{|\vec{R} - \vec{r}|^{n+1}} \, \mathrm{d}\vec{r}. \tag{3}$$

Therefore,

$$\vec{F}(\alpha\vec{R}) = \alpha^{3-n}\vec{F}(\vec{R}). \tag{4}$$

This is the scaling law of Newton. Q.E.D.

Special cases of equation (4) which Newton particularizes are

$$n = 2, \qquad \text{when} \quad \vec{F}(\alpha\vec{R}) = \alpha\vec{F}(\vec{R}), \tag{5}$$

which is Proposition LXXII, Corollary III;

$$n = 3, \qquad \text{when} \quad \vec{F}(\alpha\vec{R}) \equiv F(\vec{R}) \tag{6}$$

and

$$n = 4, \qquad \text{when} \quad \vec{F}(\alpha\vec{R}) = \alpha^{-1}\vec{F}(\vec{R}). \tag{7}$$

It is surprising that, simple as the scaling law is, it is not to be found in any of the standard textbooks. And it is a measure of Newton's perspicacity that he should add this corollary.

> COR. II. Hence, on the other hand, from the forces with which like bodies attract corpuscles similarly situated, may be obtained the ratio of the decrease of the attractive forces of the particles as the attracted corpuscle recedes from them; if only that decrease is directly or inversely in any ratio of the distances.

87. Propositions LXXXVIII and LXXXIX

Proposition LXXXVIII. Theorem XLV

> *If the attractive forces of the equal particles of any body be as the distance of the places from the particles, the force of the whole body will tend to its centre of gravity; and will be the same with the force of a globe, consisting of similar and equal matter, and having its centre in the centre of gravity.*

In this proposition, Newton considers the one case for which the attraction by an arbitrary distribution of mass can be found. Considering first the case of n discrete mass points $m_i (i = 1, 2, \ldots, n)$ at $\vec{r}_i (i = 1, \ldots, n)$ each attracting a 'foreign' particle at $\vec{R}$ with the force per unit mass,

$$-Cm_i(\vec{R} - \vec{r}_i),$$

where C is some positive constant, we find for the attraction by all mass points,

$$\vec{F} = -C\sum_{i=1}^{n} m_i(\vec{R} - \vec{r}_i) = -CM\vec{R} + C\sum_{i=1}^{n} m_i\vec{r}_i$$

where

$$\sum_{i=1}^{m} m_i = M, \quad \text{the mass of all the particles.}$$

But by definition

$$\sum_{i=1}^{n} m_i \vec{r}_i = M\vec{G}$$

where $\vec{G}$ is the location of the centre of gravity of all the masses. Hence,

$$\vec{F} = -CM(\vec{R} - \vec{G})$$

that is, the force that the foreign particle experiences is the same as if all the mass points are at $\vec{G}$ with their total mass. Newton proves his proposition by induction. The arguments are readily generalized to apply to a continuous distribution of mass. In fact it follows directly from equation (1) of §86 for the present case $n = -1$:

$$\vec{F}(\vec{R})) = -\text{constant} \iiint_{\mathscr{V}} \rho(\vec{r})(\vec{R} - \vec{r})\, \mathrm{d}\vec{r}$$

$$= -\text{constant}\ M(\vec{R} - \vec{G}).$$

Q.E.D.

The following corollary is self-evident:

> COR. Hence the motion of the attracted body Z will be the same as if the attracting body $RSTV$ were spherical; and therefore if that attracting body be either at rest, or proceed uniformly in a right line, the body attracted will move in an ellipse having its centre in the centre of gravity of the attracting body.

Proposition LXXXIX. Theorem XLVI

If there be several bodies consisting of equal particles whose forces are as the distances of the places from each, the force compounded of all the forces by which any corpuscle is attracted will tend to the common centre of gravity of the attracting bodies; and will be the same as if those attracting bodies, preserving their common centre of gravity, should unite there, and be formed into a globe.

As a further generalization of Proposition LXXXVIII, the mass points are replaced by finite bodies with spherically symmetric distributions of density, the constituent particles of which attract an external particle—the 'foreign particle' of Proposition LXXXVIII—with a force proportional to the distances. By case 6 of Proposition LXXVII, each of these bodies will attract an external particle as though their masses are brought together at their respective centres, that is, their respective centres of gravity. The conditions of Proposition LXXXVIII are satisfied with respect to masses of the spherical bodies collected at their centres; and the proposition stated follows. And the corollary of that proposition is now replaced by

COR. Therefore the motion of the attracted body will be the same as if the attracting bodies, preserving their common centre of gravity, should unite there, and be formed into a globe. And, therefore, if the common centre of gravity of the attracting bodies be either at rest, or proceed uniformly in a right line, the attracted body will move in an ellipse having its centre in the common centre of gravity of the attracting bodies.

88. The attraction by circular discs and round solids at points along their axes

Proposition XC. Problem XLIV

If to the several points of any circle there tend equal centripetal forces, increasing or decreasing in any ratio of the distances; it is required to find the force with which a corpuscle is attracted, that is, situated anywhere in a right line which stands at right angles to the plane of the circle at its centre.

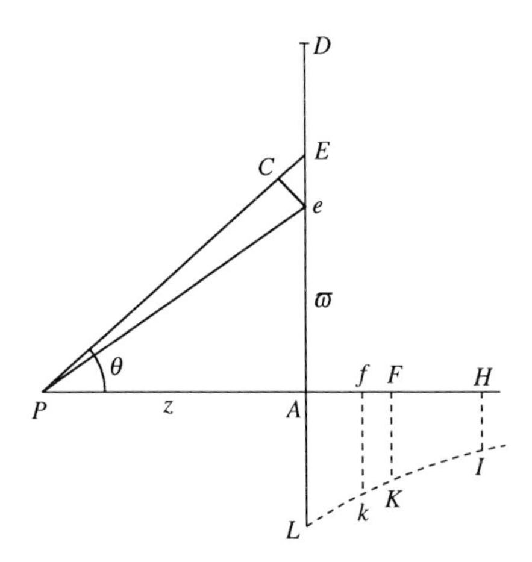

Required: The attraction per unit mass, $F(R; z)$ of a uniform circular disc, with centre A and radius $R(=AD)$ at a point P at a normal distance $z(=PA)$ from the plane of the disc, given that the law of centripetal attraction is some function $f(r)$ of the distance $r(=PE)$. In the accompanying diagram,

$$AE = \varpi, \qquad eE = \mathrm{d}\varpi, \qquad \text{and} \qquad \cos\theta = \frac{AP}{PE} = \frac{z}{\sqrt{(z^2+\varpi^2)}}. \tag{1}$$

(The lettering of the diagram to the right of LAD will be explained presently.)

First, we shall obtain the required attraction by the common procedure.

The force normal to PA will be 'destroyed by contrary attractions'. The remaining force

$F(R; z)$, per unit mass and per unit thickness, in the normal direction is manifestly,

$$F(R; z) = 2\pi \int_0^R [f(\sqrt{(z^2 + \varpi^2)})\cos\theta]\varpi \, d\varpi$$
$$= 2\pi z \int_0^R f(\sqrt{(z^2 + \varpi^2)}) \frac{\varpi \, d\varpi}{\sqrt{(z^2 + \varpi^2)}}. \tag{2}$$

Letting

$$y = \sqrt{(z^2 + \varpi^2)} \tag{3}$$

we obtain

$$F(R; z) = 2\pi z \int_z^{\sqrt{(R^2+z^2)}} f(y) \, dy. \tag{4}$$

For

$$f(y) = y^{-2} \quad \text{and} \quad y^{-n}, \tag{5}$$

we obtain, respectively,

Corollary I

$$F(R, z) = 2\pi\left[1 - \frac{z}{\sqrt{(R^2 + z^2)}}\right], \tag{6}$$

and

Corollary II

$$F(R, z) = \frac{2\pi}{n-1}\left[\frac{1}{z^{n-2}} - \frac{z}{(R^2 + z^2)^{(n-1)/2}}\right]. \tag{7}$$

Now to give Newton's proof. In the same illustration (cf. equations (1))

$$\left.\begin{aligned} &AP = z, \quad AE = \varpi, \quad eE = d\varpi; \quad \frac{CE}{eE} = \frac{AE}{PE} = \sin\theta; \\ &AD = R; \quad PF = PE = \sqrt{(z^2 + \varpi^2)}; \quad \text{and} \quad PH = PD = \sqrt{(z^2 + R^2)}. \end{aligned}\right\} \tag{8}$$

Also,

$$FK = f(\sqrt{(z^2 + \varpi^2)}) = f(PE); \quad AL = f(z), \quad \text{and} \quad HI = f(PD) = f(\sqrt{(z^2 + R^2)}). \tag{9}$$

Therefore, by equation (2), by approximating the integral by a sum (in the manner of Lemmas I–IV) and suppressing the factor 2π, we have

$$F \simeq \sum f(PE)\frac{AP}{PE}(AE)(eE) = AP \sum f(PE)\Delta(PE), \tag{10}$$

since

$$\Delta(PE) = CE = (\sin\theta)eE = \frac{AE}{PE}eE. \qquad (11)$$

On the other hand, by definition,

$$f(PE) = FK \qquad \text{and} \qquad \Delta(PE) = \text{interval}\,(fF). \qquad (12)$$

Hence

$$F \simeq AP\sum_{A}^{H} FK \times \text{interval}\,(fF) \to AP \times \text{area}\,(AHIL), \qquad (13)$$

which is Newton's expression.

Corollaries I and II are equivalent to equations (6) and (7) and derived in similar fashion.

Corollary III

In this corollary the force at a height z normal to an infinite 'plane-sheet' is obtained by letting $R \to \infty$ in equation (7). We find

$$F(\infty, z) \to \frac{1}{n-1} z^{-(n-2)}, \qquad (14)$$

again suppressing the factor 2π.

Proposition XCI. Problem XLV

To find the attraction of a corpuscle situated in the axis of a round solid, to whose several points there tend equal centripetal forces decreasing in any ratio of the distances whatsoever.

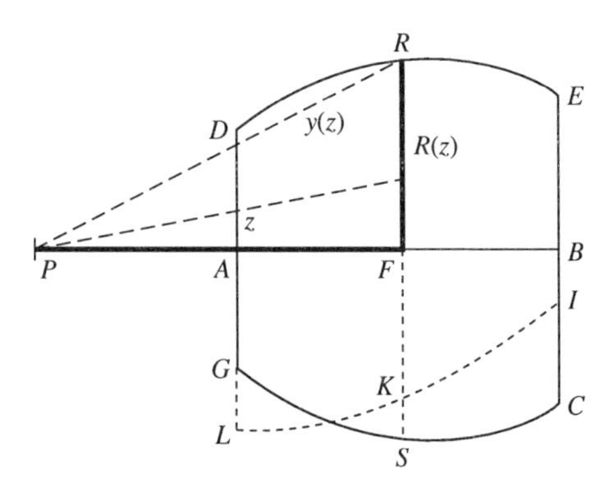

A round solid, as Newton defines it, is a body of revolution whose circular cross-sections are of variable radii. In the accompanying diagram, the axis of revolution of the round solid $GDEC$, as depicted, is $PAFB$ where P is the point at which the attraction of the solid is required. The contribution to the attraction by the circular disc FR (of thickness $\mathrm{d}z$) at the distance $PF(=z)$ from P and of radius $R(z)$, by equation (4) of Proposition XC, is

$$\mathrm{d}z\left[z\int_{PF}^{PR} f(y)\,\mathrm{d}y\right], \tag{15}$$

where we have suppressed the factor 2π. The attraction by the entire solid is therefore,

$$F=\int_{PA}^{PB}\mathrm{d}z\left[z\int_{PF}^{PR} f(y)\,\mathrm{d}y\right] \qquad (z=PF), \tag{16}$$

or, alternatively,

$$F=\int_{PA}^{PB} z\,\mathrm{d}z\left[\left[\int_{z}^{\sqrt{(R^2+z^2)}} f(y)\,\mathrm{d}y\right]\right]. \tag{17}$$

Newton expresses this result as the area $LABI$ under the curve LKI whose ordinate FK is the same as in equation (4) expressed in his notation as the limit of the finite sum (see equation (10))

$$\sum PF\Delta(PF)\sum f(PE)\Delta(PE). \tag{18}$$

Corollary I

In this corollary, we consider the special case of a right-circular cylinder when

$$R(z)=R=\text{constant}=AD=BE, \tag{19}$$

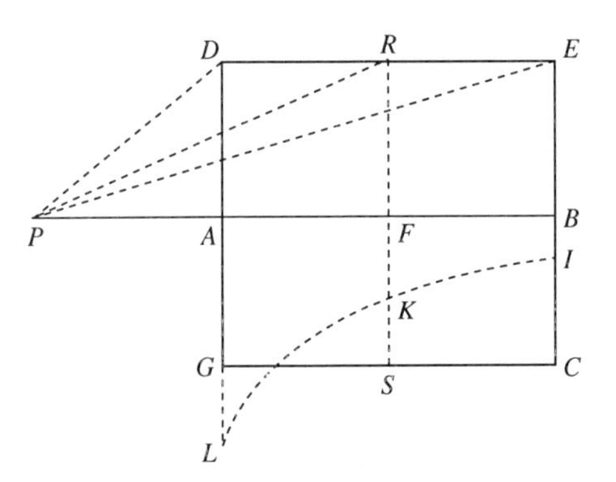

and the law of attraction is inversely as the square of the distance. Inserting the expression (6) of the preceding proposition in equation (17), we find:

$$F=\int_{PA}^{PB}\mathrm{d}z\left[1-\frac{z}{\sqrt{(z^2+R^2)}}\right]=[z-\sqrt{(z^2+R^2)}]_{PA}^{PB}, \tag{20}$$

or

$$F=PB-PA-PE+PD=AB-PE+PD. \tag{21}$$

This is Newton's result derived in essentially the same manner.

89. Corollaries II and III of Proposition XCI and Proposition XCII

In Corollary II of this proposition, Newton considers the force of attraction of an oblate spheroid at a point on the continuation of the minor axis. In terms of the accompanying diagram and its description Newton states the solution:

> the force with which the spheroid attracts the body P will be to the force with which a sphere described with the diameter AB attracts the same body as
>
> $$\frac{AS\,.\,CS^2 - PS\,.\,KMRK}{PS^2 + CS^2 - AS^2}$$
>
> is to
>
> $$\frac{AS^3}{3PS^2}.$$
>
> And by a calculation founded on the same principles may be found the forces of the segments of the spheroid.

with no more than the opening and the concluding remarks,

> Hence also is known the force by which a spheroid $AGBC$ attracts any body P situate externally in its axis AB.

and

> And by a calculation founded on the same principles may be found the forces of the segments of the spheroid.

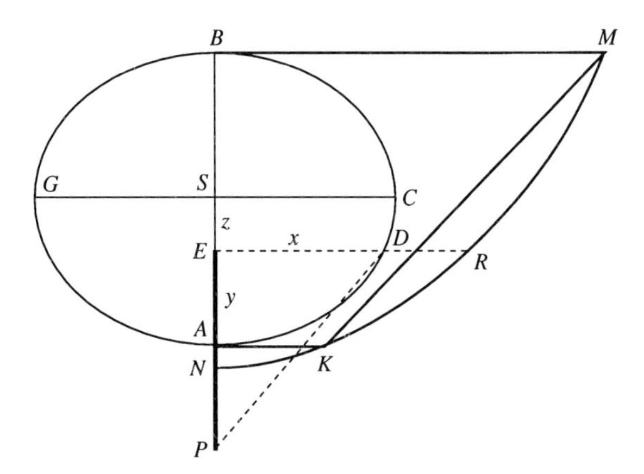

Description of the diagram:

$AGBC$: an oblate spheroid;

$GS = SC = a$: the semimajor axis;

$SA = SB = b$: the semiminor axis;

P: a point on the continuation of minor axis, BA;

$$SP = Y; \qquad ED = x; \qquad SE = z; \qquad EP = y = Y - z; \qquad PD = \sqrt{(x^2 + y^2)}.$$

By construction:

$$PD = ER; \qquad PA = AK; \qquad PB = BM;$$

Arc KRM is the locus of R as E moves from A to B.

The oblate spheroid is a solid of revolution about its minor axis and is therefore a 'round solid' in Newton's definition; and he clearly implies that the solution is to be found by making use of Corollary I of Proposition XC (equation (6) and equation (7) of this proposition). And that we shall do!

Writing

$$y = Y - z \quad \text{in place of } z \qquad \text{and} \qquad x \quad \text{in place of } \varpi, \tag{22}*$$

to be in accord with our present definitions (see the accompanying figure and its description), and combining equations (6) and (17) we obtain

$$F = \int_{Y-b}^{Y+b} \left[1 - \frac{y}{\sqrt{(x^2 + y^2)}}\right] \mathrm{d}y = 2b - \int_{Y-b}^{Y+b} \frac{y\,\mathrm{d}y}{\sqrt{(x^2 + y^2)}}, \tag{23}$$

where x must now be expressed in terms of y. Since

$$x^2 = a^2(1 - z^2/b^2). \tag{24}$$

by virtue of $ACBG$ being an ellipse,

$$x^2 + y^2 = y^2 + a^2 - \frac{a^2}{b^2}(Y - y)^2$$

$$= \frac{1}{1 - e^2}[-e^2y^2 + 2Yy - (Y^2 - b^2)] = PD^2, \tag{25}$$

where we have expressed b, in terms of the eccentricity e of the ellipse, by the defining equation

$$b^2 = a^2(1 - e^2). \tag{26}$$

Inserting for $(x^2 + y^2)$ from equation (25), the expression (23) for F becomes

$$F = 2b - \sqrt{(1 - e^2)}\,.\int_{Y-b}^{Y+b} \frac{y\,\mathrm{d}y}{\sqrt{X}} \tag{27}$$

where

$$X = -e^2y^2 + 2Yy - (Y^2 - b^2). \tag{28}$$

* We shall continue the numbering of the equations started in §88 since we are dealing with the same matters in this section.

Making use of the identity,

$$\int \frac{y\,\mathrm{d}y}{\sqrt{Z}} = \frac{1}{4\alpha\gamma - \beta^2}\left[2(2\gamma + \beta y)\sqrt{Z} - 4\beta \int \mathrm{d}y\sqrt{Z}\right], \tag{29}$$

where

$$Z = \alpha y^2 + \beta y + \gamma, \tag{30}$$

we find:

$$\begin{aligned}\int_{Y-b}^{Y+b} \frac{y\,\mathrm{d}y}{\sqrt{X}} &= \frac{1}{(1-e^2)(Y^2+e^2a^2)}\left\{-[Yy-(Y^2-b^2)]\sqrt{X}\Big|_{Y-b}^{Y+b} + 2Y\int_{Y-b}^{Y+b}\mathrm{d}y\sqrt{X}\right\}\\ &= \frac{1}{Y^2+e^2a^2}\left\{-\frac{2b}{\sqrt{(1-e^2)}}(Y^2+b^2) + \frac{2Y}{(1-e^2)}\int_{Y-b}^{Y+b}\mathrm{d}y\sqrt{X}\right\}.\end{aligned} \tag{31}$$

On substituting this equation in the expression (27) for F we obtain:

$$\begin{aligned}\tfrac{1}{2}F &= \frac{1}{Y^2+e^2a^2}\left\{b[(Y^2+e^2a^2)+Y^2+a^2(1-e^2)] - Y\int_{Y-b}^{Y+b}\mathrm{d}y\sqrt{\frac{X}{1-e^2}}\right\}\\ &= \frac{1}{Y^2+e^2a^2}\left\{a^2b + Y\left[2bY - \int_{Y-b}^{Y+b}\mathrm{d}y\sqrt{\frac{X}{1-e^2}}\right]\right\}.\end{aligned} \tag{32}$$

On the other hand, by equation (25) and the description of the figure,

$$PS^2 + CS^2 - AS^2 = Y^2 + a^2 - b^2 = Y^2 + e^2a^2; \; \sqrt{X} = PD\sqrt{(1-e^2)}. \tag{33}$$

Accordingly, an equivalent form of equation (32) is

$$\tfrac{1}{2}F = \frac{1}{PS^2 + CS^2 - AS^2}\left\{AS\,.\,CS^2 + PS\left[AB\,.\,PS - \int_{Y-b}^{Y+b} PD\,\mathrm{d}y\right]\right\}. \tag{34}$$

Since $PD = ER$ by construction,

$$\int_{Y-b}^{Y+b} PD\,\mathrm{d}y = \int_A^B ER\,\mathrm{d}y = \text{area } AKRMB, \tag{35}$$

while

$$\begin{aligned}AB\,.\,PS &= \tfrac{1}{2}AB(AP + BP) = \tfrac{1}{2}AB(AK + BM)\\ &= \text{area of trapezium } AKMB.\end{aligned} \tag{36}$$

Hence,

$$AB\,.\,PS - \int_{Y-b}^{Y+b} PD\,\mathrm{d}y = -\text{area of } KMRK; \tag{37}$$

and equation (34) becomes

$$\tfrac{1}{2}F = \frac{1}{PS^2 + CS^2 - AS^2}(AC\,.\,CS^2 - PS \times \text{area of } KMRK). \tag{38}$$

If the spheroid is replaced by a sphere of radius $b(=AS)$ its attraction at the same distance PS is

$$\tfrac{1}{2}F = \frac{AS^3}{3PS^2}. \tag{39}$$

since in our expressions for F we have uniformly suppressed the factor 2π. Q.E.D.!

It is clear that a similar expression will obtain for the attraction at points on the continuation of the major axis of a prolate spheroid.

> COR. III. If the corpuscle be placed within the spheroid and in its axis, the attraction will be as its distance from the centre.

This corollary generalizes for a spheroid Propositions LXX and LXXI applicable to a sphere: the *spherical shell* of Proposition LXX being replaced by a *homoeoid*, that is, a shell bounded by two concentric similar, and similarly situated, ellipsoids (spheroids in the context).

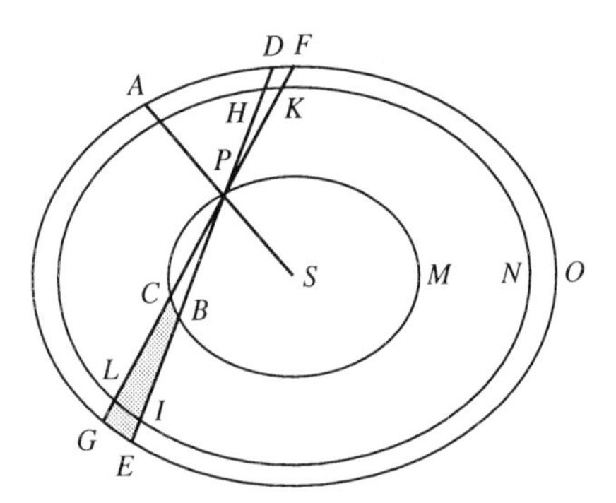

Inside the given spheroid $AGOF$ with centre S, construct two other concentric spheroids $HLINK$ and $PCBM$ similar and similarly situated, the latter passing through the point P at which the attraction of the spheroid is required. Through P draw the lines GPF and EPD intersecting the inner spheroids at the antipodal points (K, L) and (H, I). Recalling that chords (such as PC, LK, and GF; or PB, IH, and ED) intersecting concentric spheroids have the same mid-point,* we conclude that

$$DP = BE, \quad HP = BI \quad \text{and} \quad FP = CG, \quad KP = CL \tag{40}$$

and therefore

$$DH = IE \quad \text{and} \quad FK = LG. \tag{41}$$

* This fact follows from the equation, $x_1(x - x_1)/a^2 + y_1(y - y_1)/b^2 = 0$, of a chord with its mid-point at (x_1, y_1) and its invariance to the replacement of a and b by αa and αb where α is a constant.

Now let *DPF* and *EPG* represent solid cones including at their common vertex an infinitesimal solid angle while the homoeoidal shell (*AGOF*, *HLINK*) is of infinitesimal thickness. Then, the volumes of the infinitesimal frusta *DFKH* and *LIEG* will be directly as the square of the distances *PD* and *PE* by virtue of the equalities (41); their attractions at *P* will therefore be equal and opposite. And the same will be true of the pairs of frusta intercepted by the cones *DPF* and *EPG* from a neighbouring infinite sequence of spheroidal surfaces (like *AGOF* and *HLINK*) drawn in the intervals *PD* and *BE*. We conclude that the attractions of the volumes of the cone *DPF* and the frustum *CGEB* (shaded in the diagram) will be equal and opposite and cancel each other. And by summing over all similar volumes of cones and frusta, we conclude that the attractive force exerted by the entire homoeoid included between the concentric spheroids *AGOF* and *PCBM* will also cancel. And as Newton concludes:

> Therefore the body *P* is attracted by the interior spheroid *PCBM* alone, and therefore (by Cor. III, Prop. LXXII) its attraction is to the force with which the body *A* is attracted by the whole spheroid *AGOD* as the distance *PS* is to the distance *AS*. Q.E.D.

And finally in Proposition XCII, Newton frames an experiment which was to be performed a century later by Henry Cavendish.

Proposition XCII. Problem XLVI

An attracting body being given, it is required to find the ratio of the decrease of the centripetal forces tending to its several points.

The body given must be formed into a sphere, a cylinder, or some regular figure, whose law of attraction answering to any ratio of decrease may be found by Prop. LXXX, LXXXI, and XCI. Then, by experiments, the force of the attractions must be found at several distances, and the law of attraction towards the whole, made known by that means, will give the ratio of the decrease of the forces of the several parts; which was to be found.

90.

Proposition XCIII. Theorem XLVII

If a solid be plane on one side, and infinitely extended on all other sides, and consist of equal particles equally attractive, whose forces decrease, in receding from the solid, in the ratio of any power greater than the square of the distances; and a corpuscle placed towards either part of the plane is attracted by the force of the whole solid: I say, that the attractive force of the whole solid, in receding from its plane surface will decrease in the ratio of a power whose side is the distance of the corpuscle from the plane, and its index less by 3 than the index of the power of the distances.

In this proposition, we are required to find the attraction of a semi-infinite plane-parallel slab at a point either outside or inside the slab.

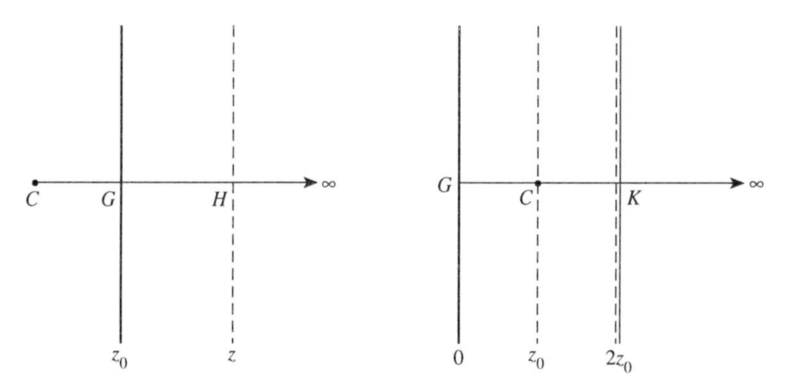

Case 1. Consider first the case when the point C at which the attraction is required is at a normal distance $z_0(=CG)$ from the bounding plane at G. Then, by Corollary III of Proposition XC (equation (14) in §88), the contribution to the attraction at C by an infinitesimal sheet of thickness, dz, at z is

$$\frac{1}{n-1} z^{-n+2}\,\mathrm{d}z, \tag{1}$$

where, as hitherto, we have suppressed the factor 2π. The attraction by the entire semi-infinite slab is, therefore,

$$F(z_0) = \frac{1}{n-1}\int_{z_0}^{\infty} z^{-n+2}\,\mathrm{d}z = \frac{1}{(n-1)(n-3)} z_0^{-n+3}. \tag{2}$$

The requirement $n > 3$ for the finiteness of the result derives from the logarithmic divergence of the integral for $n = 3$: or, as Newton explains

> the attraction of the remoter part of the infinite body is always ... infinitely greater than the attraction of the nearer part. Q.E.D.

Case 2. Consider next the case when the point C is at a depth z_0 below the bounding plane. Clearly, the attraction of the slab *above* C of thickness z_0 and the attraction of the slab of equal thickness *below* C are equal and opposite and will cancel each other. Only the contribution by the still(!) semi-infinite slab below the depth $2z_0$ will survive. Hence, the attraction of the slab at image points, equidistant from the bounding plane, are equal. Q.E.D.

Newton's discussion of case 2 is an example of the use of a 'principle of invariance'—the slab below any finite depth in a semi-infinite slab is still a semi-infinite slab—reminiscent of similar principles of invariance used extensively in modern treatments of problems in radiative transfer.

Corollary I below is a further example.

Corollary I

If the slab considered is of a finite thickness D, then the attraction of the slab at a normal distance z_0 from either bounding plane is obtained by the same principle, namely by subtracting the attractions of a semi-infinite slab at heights z_0 and $z_0 + D$ thus

$$F(z_0; D) = \frac{1}{(n-1)(n-3)} [z_0^{-n+3} - (z_0 + D)^{-n+3}]. \tag{3}$$

Corollary II

For $D \to \infty$,

$$F(z_0; D) \to \frac{1}{(n-1)(n-3)} z_0^{-n+3}. \tag{4}$$

Corollary III

In this corollary Newton argues that the same asymptotic behaviour (4) will obtain quite generally for attraction by a body of an arbitrary shape with a plane boundary so long as z_0 is very much smaller than any of the linear dimension of the body.

Consider, for example, a right circular cylinder of radius R and height $(z_1 - z_0)$. Then by Corollary II of Proposition XC (§88, equation (7))

$$F(z_0) = \frac{1}{n-1} \int_{z_0}^{z_1} \left[z^{-n+2} - \frac{z}{(R^2 + z^2)^{(n-1)/2}} \right] \mathrm{d}z, \tag{5}$$

or, after evaluating the integral, we find

$$F(z_0) = \frac{1}{(n-1)(n-3)} \{(z_0^{-n+3} - z_1^{-n+3}) - [(R^2 + z_0^2)^{(-n+3)/2} - (R^2 + z_1^2)^{(-n+3)/2}]\}. \tag{6}$$

Clearly if $(n > 3)$,

$$z_0 \ll R \qquad \text{and } z_1 \text{ and } R \text{ are comparable,} \tag{7}$$

the dominant term is

$$F(z_0) \simeq \frac{1}{(n-1)(n-3)} z_0^{-n+3}. \tag{8}$$

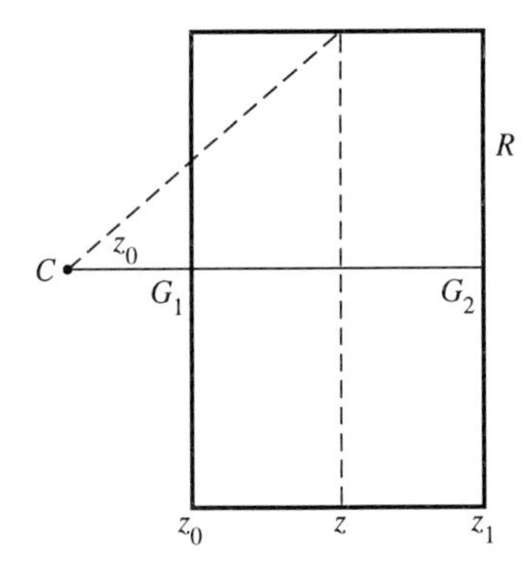

And the validity of this behaviour for more general shapes can be justified as in Propositions LXXXV and LXXXVI.

Scholium

By the propositions proved in this section, for the attraction by a semi-infinite plane-parallel slab, only the component in the z-direction normal to the plane of stratification is non-vanishing. The rectilinear motion of a particle in the z-direction can be solved by the method described in Proposition XXXIX (Chapter 8, §45, equation (4)). And the general curvilinear motion can be obtained, as Newton writes;

> (by Cor. II of the Laws) compounding that motion with a uniform motion performed in the direction of lines parallel to that plane.

There is another question one can ask in this context, namely, what is the law of force of $f(z)$ in the direction normal to the planes of stratification that will cause a body to move in the (y, z)-plane along a given curve $y(z)$ (as the semicircumference of a circle in Proposition VIII). In other words, we are required to satisfy the equation

$$\frac{\mathrm{d}^2 y}{\mathrm{d}z^2} = f(z), \tag{9}$$

for the given $y(z)$. It is a matter of simple differentiation. But 'the operation can be contracted' by expanding $y(z)$ in a convergent series—actually, a *Laurent series*—of the form

$$y = z^\alpha + \alpha z^{\alpha-1} + \tfrac{1}{2}\alpha(\alpha - 1)z^{\alpha-2} + \cdots, \tag{10}$$

where $\alpha(=m/n$ in Newton's notation) is some positive or negative constant. One can obtain two independent solutions—the 'even' and the 'odd'—by starting the series with z^α or $z^{\alpha-1}$. The even solution is given by

$$f(z) = \alpha(\alpha - 1)z^{\alpha-2} + \tfrac{1}{2}\alpha(\alpha - 1)(\alpha - 2)(\alpha - 3)z^{\alpha-4} + \cdots. \tag{11}$$

Hence, if the orbit is a parabola,

$$y = z^2, \qquad \alpha = 2 \qquad \text{and} \qquad f(z) = 2 = a \text{ constant}, \tag{12}$$

as 'Galileo hath demonstrated'. Similarly if the orbit is a hyperbola,

$$y = z^{-1}, \qquad \alpha = -1, \qquad f(z) = 2z^{-3} = 2y^3; \tag{13}$$

and the force decreases as z^{-3}.

It is of interest to verify the solution given in Proposition VIII for the semicircumference,

$$y(z) = \sqrt{(a^2 - z^2)}, \tag{14}$$

where a is the radius of the circle. By direct evaluation, we find

$$f(z) = -\frac{a^2}{(a^2 - z^2)^{3/2}} = -\frac{a^2}{y^3}, \tag{15}$$

in agreement with Proposition VIII.

By now Newton appears to have become impatient and he concludes with the statement:

> But leaving propositions of this kind, I shall go on to some others relating to motion which I have not yet touched upon.

❖17❖

A digression into Opticks

91. Introduction

In this last Section XIV of Book I of the *Principia*, Newton considers

> *The motion of very small bodies when agitated by centripetal forces tending to the several parts of any very great body,*

motivated, apparently, by the fact that

> These attractions bear a great resemblance to the reflections and refractions of light made in a given ratio of the secants, as was discovered by Snell; and consequently in a given ratio of the sines, as was exhibited by Descartes.

Besides, it gave Newton the occasion to expand on his 'query' on the nature of light though he protests that in establishing Propositions XCIV–XCVI he was

> not at all considering the nature of the rays of light, or inquiring whether they are bodies or not; but only determining the curves of bodies which are extremely like the curves of the rays.

Nevertheless, there is something strangely out of joint in the presentation of these propositions. But of this, presently.

92. Propositions XCIV–XCVI

In these three propositions, Newton considers streams of non-interacting tiny particles that have been travelling in parallel straight lines, incident on a plane-parallel slab of finite thickness. Once inside the slab, the particles experience an unspecified accelerative force in the vertical direction which ceases to be operative on their emergence from the other side. The problem is to relate the velocities and their directions at incidence and at emergence.

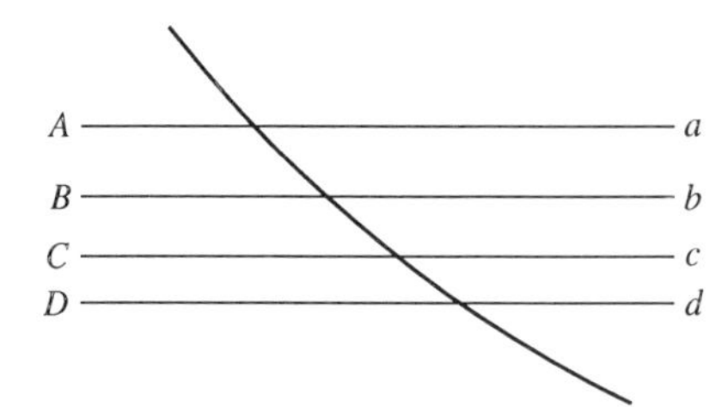

Proposition XCIV. Theorem XLVIII

> *If two similar mediums be separated from each other by a space terminated on both sides by parallel planes, and a body in its passage through that space be attracted or impelled perpendicularly towards either of those mediums, and not agitated or hindered by any other force; and the attraction be everywhere the same at equal distances from either plane, taken towards the same side of the plane: I say, that the sine of incidence upon either plane will be to the sine of emergence from the other plane in a given ratio.*

The problem as here stated is clearly a special case of the one formulated *and* solved in the Scholium concluding Section XIII and following the preceding Proposition XCIII. To quote:

> If a body is attracted perpendicularly towards a given plane, and from the law of attraction given, the motion of the body be required; the Problem will be solved by seeking (by Prop. XXXIX) the motion of the body descending in a right line towards that plane, and (by COR. II of the Laws) compounding that motion with a uniform motion performed in the direction of lines parallel to that plane.

By the procedure just outlined, the solution to the present problem is immediate (see below). Why then, does Newton, without so much as a reference to what he has stated only so recently, proceed to establish the present proposition with such great detail in the style of Proposition XXX, unless it be(?) that Newton had considered the same problem at an earlier time in the context of his optical investigations and introduced it here *ad hoc*.

We shall, however, prove the proposition following Newton's prescription. Let S_i denote the surface, $z = 0$, on which a stream of particles is incident in a direction making an angle θ_i with respect to the normal, and let S_e be the surface from which they emerge, making an angle θ_e with the outward normal. Since the particles experience no acceleration

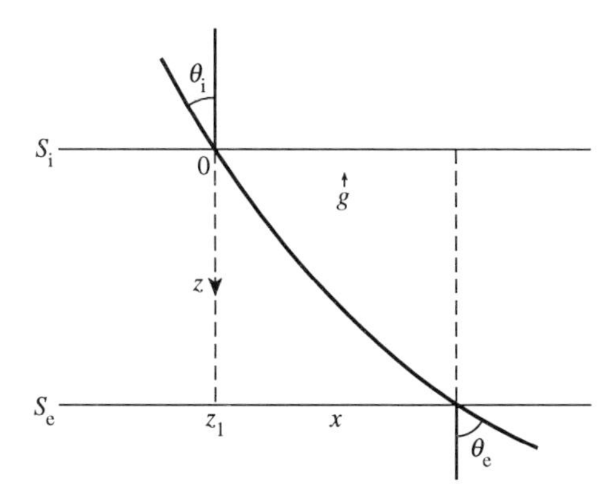

in the x-direction, either outside or inside the slab, the horizontal component of the velocity V_x in the x-direction will remain constant (by Law I); that is,

$$V_x = C_x = \text{constant}. \tag{1}$$

But in the z-direction, the velocity will change by virtue of the acceleration 'exerted according to any assigned law', $-f(z)$; thus,

$$\frac{\mathrm{d}^2 z}{\mathrm{d}t^2} = -f(z). \tag{2}$$

By the result established in Proposition XXXIX (equation (5) in the margins of Newton's illustration on p. 162 of §45 of Chapter 8):

$$\frac{1}{2}\left(\frac{\mathrm{d}z}{\mathrm{d}t}\right)^2 = \tfrac{1}{2}V_z^2 = \text{constant} - \int_0^z f(z)\,\mathrm{d}z; \tag{3}$$

or, if C_z is the initial component of the vertical velocity at incidence,

$$V_z^2 = C_z^2 - 2\int_0^z f(z)\,\mathrm{d}z. \tag{4}$$

The velocity V at any z is obtained by compounding V_z and V_x $(=C_x)$ vectorially (by Corollary II of the Laws). Thus,

$$V^2 = V_z^2 + V_x^2 = C_z^2 + C_x^2 - 2\int_0^z f(z)\,\mathrm{d}z. \tag{5}$$

If C_i denotes the velocity of the particles at incidence,

$$V^2(z) = C_i^2 - 2\int_0^z f(z)\,\mathrm{d}z. \tag{6}$$

In particular, the velocity C_e at emergence is given by

$$C_e^2 = [V(z_1)]^2 = C_i^2\left[1 - \frac{2}{C_i^2}\int_0^{z_1} f(z)\,\mathrm{d}z\right]. \tag{7}$$

Since the angles of incidence (θ_i) and emergence (θ_e) are related to C_i and C_e by

$$C_i \sin \theta_i = C_x = C_e \sin \theta_e, \tag{8}$$

we infer

$$\sin \theta_e = \gamma \sin \theta_i, \tag{9}$$

where

$$\gamma = \frac{C_i}{C_e} = \left[1 - \frac{2}{C_i^2}\int_0^{z_1} f(z)\, dz\right]^{-1/2}. \tag{10}$$

This is the case II that Newton considers. In case I he considers the simpler case when the force, like gravity, acting in the z-direction, is a constant:

$$f = g = \text{constant}. \tag{11}$$

In this case,

$$\gamma = (1 - 2gz_1/C_i^2)^{-1/2}. \quad \text{Q.E.I.} \tag{12}$$

Proposition XCV. Theorem XLIX

The same things being supposed, I say, that the velocity of the body before its incidence is to its velocity after emergence as the sine of emergence to the sine of incidence.

This is a direct consequence of equation (8):

$$\begin{aligned} C_i : C_e &= \text{velocity at incidence: velocity at emergence} \\ &= \sin \theta_e : \sin \theta_i. \end{aligned} \qquad \text{Q.E.D.}$$

Logically Proposition XCV takes precedence over Proposition XCIV; and yet...

We learn from Horace, Homer sometimes sleeps; Byron.

Proposition XCVI. Theorem L

The same things being supposed, and that the motion before incidence is swifter than afterwards: I say, that if the line of incidence be inclined continually, the body will be at last reflected, and the angle of reflection will be equal to the angle of incidence.

By equation (6), at any level z,

$$\frac{\sin \theta(z)}{\sin \theta_i} = \frac{C_i}{V(z)} = \left[1 - \frac{2}{C_i^2}\int_0^{z} f(z)\, dz\right]^{-1/2}. \tag{13}$$

The particle cannot therefore penetrate beyond the depth z_* given by (assuming that the medium extends to at least this depth)

$$1 - \frac{2}{C_\mathrm{i}^2}\int_0^{z_*} f(z)\,\mathrm{d}z = \sin^2\theta_\mathrm{i}. \tag{14}$$

For the case $f(z) = g =$ a constant, this equation gives,

$$1 - 2gz_*/C_\mathrm{i}^2 = \sin^2\theta_\mathrm{i}. \tag{15}$$

We shall quote Newton beyond this point: it provides an example of his meticulous style when he wants to explain something with clarity. (Note that the level we have distinguished by z_* is the level R in Newton's illustration below.)

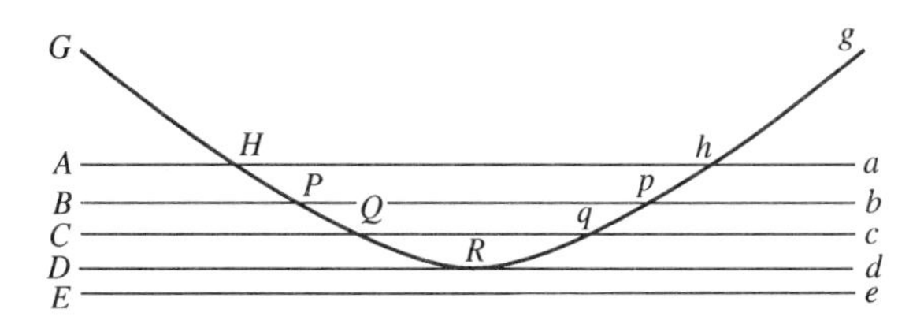

> Let the body come to this plane in the point R; and because the line of emergence coincides with that plane, it is manifest that the body can proceed no farther towards the plane Ee. But neither can it proceed in the line of emergence Rd; because it is perpetually attracted or impelled towards the medium of incidence. It will return, therefore, between the planes Cc, Dd, describing an arc of a parabola QRq, whose principal vertex (by what Galileo hath demonstrated) is in R, cutting the plane Cc in the same angle at q, that it did before at Q; then going on in the parabolic arcs qp, ph, etc., similar and equal to the former arcs QP, PH, etc., it will cut the rest of the planes in the same angles at p, h, etc., as it did before in P, H, etc., and will emerge at last with the same obliquity at h with which it first impinged on that plane at H. Conceive now the intervals of the planes Aa, Bb, Cc, Dd, Ee, etc., to be infinitely diminished, and the number infinitely increased, so that the action of attraction or impulse, exerted according to any assigned law, may become continual; and, the angle of emergence remaining all along equal to the angle of incidence, will be equal to the same also at last. Q.E.D.

93. The Scholium

We shall quote in full the Scholium which follows Propositions XCIV–XCVI: it gives an insight into the quality of Newton's thoughts on matters on which he was not fully clear himself, for example, the nature of light. Besides it does not seem to have been noted even by those (e.g. Roger Penrose in *Three hundred years of gravity*, edited by S. Hawking and W. Israel, Cambridge University Press, 1987, pp. 17–20) who have written perceptively on this subject.

Scholium

These attractions bear a great resemblance to the reflections and refractions of light made in a given ratio of the secants, as was discovered by Snell; and consequently in a given ratio of the sines, as was exhibited by Descartes. For it is now certain from the phenomena of Jupiter's satellites, confirmed by the observations of different astronomers, that light is propagated in succession, and requires about seven or eight minutes to travel from the Sun to the Earth. Moreover, the rays of light that are in our air (as lately was discovered by Grimaldi, by the admission of light into a dark room through a small hole, which I have also tried) in their passage near the angles of bodies, whether transparent or opaque (such as the circular and rectangular edges of gold, silver, and brass coins, or of knives, or broken pieces of stone or glass), are bent or inflected round those bodies as if they were attracted to them; and those rays which in their passage come nearest to the bodies are the most inflected, as if they were most attracted; which thing I myself have also carefully observed. And those which pass at greater distances are less inflected; and those at still

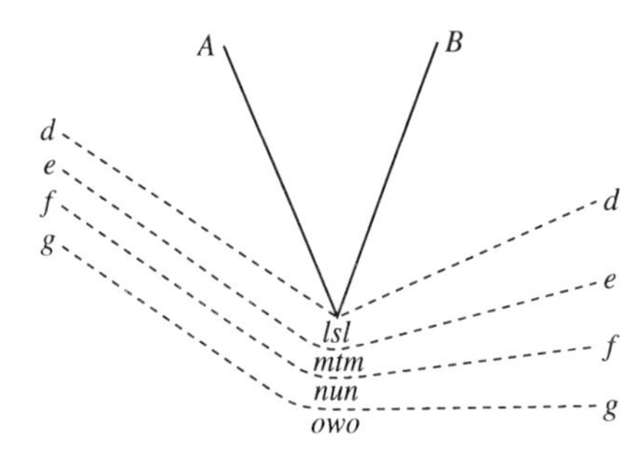

greater distances are a little inflected the contrary way, and form three fringes of colours. In the figure *s* represents the edge of a knife, or any kind of wedge *AsB*; and *gowog*, *fnunf*, *emtme*, *dlsld* are rays inflected towards the knife in the arcs *owo*, *nun*, *mtm*, *lsl*; which inflection is greater or less according to their distance from the knife. Now since this inflection of the rays is performed in the air without the knife, it follows that the rays which fall upon the knife are first inflected in the air before they touch the knife. And the case is the same of the rays falling upon glass. The refraction, therefore, is made not in the point of incidence, but gradually, by a continual inflection of the rays; which is done partly in the air before they touch the glass, partly (if I mistake not) within the

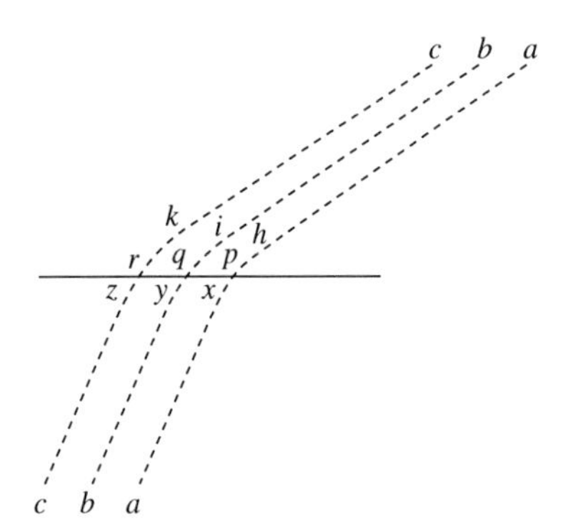

glass, after they have entered it; as is represented in the rays *ckzc*, *biyb*, *ahxa*, falling upon *r*, *q*, *p*, and inflected between *k* and *z*, *i* and *y*, *h* and *x*. Therefore because of the analogy there is between the propagation of the rays of light and the motion of bodies, I thought it not amiss to add the following propositions for optical uses; not at all considering the nature of the rays of light, or inquiring whether they are bodies or not; but only determining the curves of bodies which are extremely like the curves of the rays.

94. The ovals of Descartes

As Newton states at the end of the Scholium we have just quoted, he added Propositions XCVII and XCVIII for 'optical uses'—very far indeed from the objectives of the *Principia*. He envisages in these propositions, thin uniform 'lenses' which can be idealized as surfaces at which incident light rays are refracted according to Snell's law:

$$\gamma \sin \theta_i = \sin \theta_e, \tag{1}$$

where θ_i and θ_e are the angles of incidence and emergence with respect to the local normal at the point of incidence (and of emergence) and γ (>1) is a constant.*

Proposition XCVII, Problem XLVII

Supposing the sine of incidence upon any surface to be in a given ratio to the sine of emergence; and that the inflection of the paths of those bodies near that surface is performed in a very short space, which may be considered as a point; it is required to determine such a surface as may cause all the corpuscles issuing from any one given place to converge to another given place.

* The retention of the mnemonic device of letting the subscripts 'i' and 'e' correspond to 'incidence' and 'emergence' and the requirement $\sin \theta_e/\sin \theta_i = \gamma > 1$ can come into conflict if the light ray enters a denser from a lighter medium. When such a conflict arises (as it does in the illustration for Corollary II of Proposition XCVII) the mnemonic meaning of 'i' and 'e' will be abandoned in favour of the requirement $\gamma > 1$.

The problem is to determine the shape of the refracting surface (of the kind described), passing through a given point C, which will bring to focus at a point A all the rays 'issuing' from another point B.

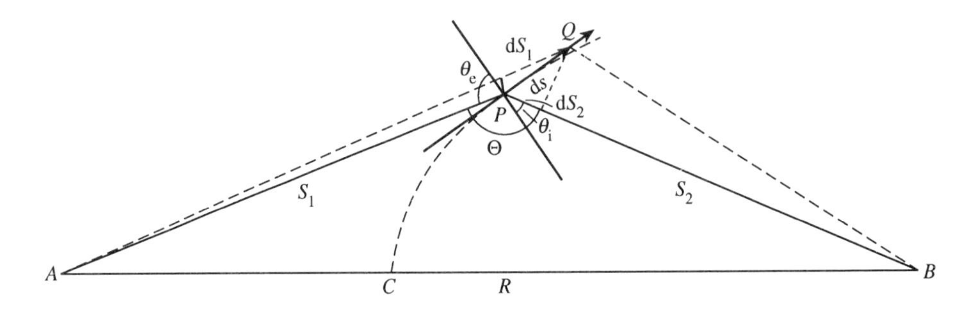

Newton's solution of the problem consists of first observing that if the curve should already be known up to a point P, one should be able to extend the solution to a little distance beyond P, if we could specify the direction of the tangent of the required curve at P; and second, if Q should be a point on the curve an infinitesimal distance beyond P and ds is the element of arc joining P and Q, then

$$dS_1 = ds(\sin\theta_e) = ds(\gamma\sin\theta_i) \qquad \text{and} \qquad dS_2 = -ds(\sin\theta_i), \tag{2}$$

where dS_1 and dS_2 are the increments in the distances S_1 and S_2 from A and B, respectively. It follows that at *every* point along the curve that is sought,

$$dS_1 : dS_2 = -\gamma : 1. \tag{3}$$

Newton's construction is to draw circles of radii

$$S_1 + \Delta S_1 \qquad \text{and} \qquad S_2 - \Delta S_2, \tag{4}$$

where ΔS_1 and ΔS_2 are infinitesimal lengths in the ratio, with centres at A and B, respectively,

$$\Delta S_1 : \Delta S_2 = \gamma : 1. \tag{5}$$

The intersection of the two circles will determine the position of the neighbouring point Q on the curve; and so *ad infinitum*. Q.E.I.

> COR. I. By causing the point A or B to go off sometimes *in infinitum*, and sometimes to move towards other parts of the point C, will be obtained all those figures which Descartes has exhibited in his *Optics* and *Geometry* relating to refractions. The invention of which Descartes having thought fit to conceal, is here laid open in this Proposition.

From equation (3) it follows that along the required curve,

$$S_1 + \gamma S_2 = \text{constant}. \tag{6}$$

Rewritten in the form,

$$n_1 r_1 + n_2 r_2 = \text{constant}, \tag{7}$$

where n_1 and n_2 are constants and r_1 and r_2 are, respectively, the distances from two fixed points, we recognize that the curve described is an oval of Descartes.

A comment

In the earlier sections of the *Principia*, whenever a problem is reduced to quadratures, as in Proposition XXXIX and XLI, Newton is accustomed to adding 'granting the quadratures of curvilinear figures' or more simply 'granting quadratures'. But, in the present instance, the problem is not reduced to quadratures but to an algorism by which the solution curve can be extended, given the derivative of the curve at any point, that is to a *first-order differential equation.* Newton is totally silent on this point!

In Appendix I we derive the appropriate differential equation and its solution, leaving in doubt whether Newton had this solution in mind when he wrote

> The invention of which Descartes having thought fit to conceal, is here laid open in this Proposition.

And in Appendix II, we reproduce in part the very first paper of Clerk Maxwell in which reference is made to the same Proposition XCVII.

Corollary II

The problem is to find the shape of the refractive surface passing through a point C which will refract all rays 'issuing' from a given point A into parallel rays along a specified direction.

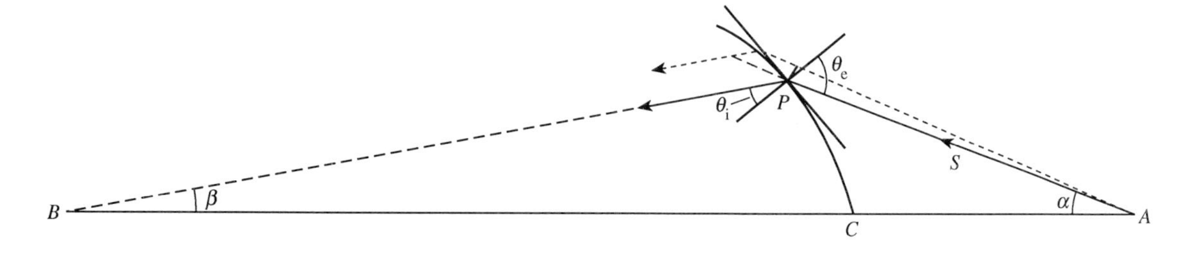

Let P be a point along the curve that is sought. Draw the tangent and its normal to the curve at P. Let θ_e and θ_i be the angles as indicated. It is required that all rays, after passing through the refractive surface, emerge parallel in a given direction. In the diagram, let the ray ASP be inclined at some chosen angle α to the direction AB. Then the refracted ray PB must intersect AB at a given angle β. It is now manifest that

$$\theta_e - \theta_i = \alpha + \beta. \tag{8}$$

Hence,

$$\begin{aligned}\sin\theta_e = \gamma\sin\theta_i &= \gamma\sin[\theta_e - (\alpha+\beta)]\\ &= \gamma[\sin\theta_e\cos(\alpha+\beta) - \cos\theta_e\sin(\alpha+\beta)];\end{aligned} \tag{9}$$

and therefore,

$$[\gamma\cos(\alpha+\beta) - 1]\sin\theta_e = \gamma\cos\theta_e\sin(\alpha+\beta), \tag{10}$$

or

$$\tan\theta_e = \frac{\gamma\sin(\alpha+\beta)}{\gamma\cos(\alpha+\beta)-1}. \tag{11}$$

Now, let ds be an element of arc including the point P. Then the increments in the lengths of the line AP and in the angle α are:

$$\mathrm{d}S = \mathrm{d}s\,(\sin\theta_e) \quad \text{and} \quad S\,\mathrm{d}\alpha = (\mathrm{d}s)\cos\theta_e. \tag{12}$$

Hence,

$$\frac{\mathrm{d}S}{S} = (\tan\theta_e)\,\mathrm{d}\alpha = \frac{\gamma\sin(\alpha+\beta)}{\gamma\cos(\alpha+\beta)-1}\,\mathrm{d}(\alpha+\beta), \tag{13}$$

since β is a constant. Integrating this equation, we obtain:

$$S = \frac{\text{constant}}{\gamma\cos(\alpha+\beta)-1}. \tag{14}$$

Since the length S at $\alpha = 0$, namely AC, is given,

$$S = AC\,\frac{\gamma\cos\beta - 1}{\gamma\cos(\alpha+\beta)-1}. \tag{15}$$

Newton stops with deducing equations (12) and determining the direction of the tangent at P; and leaving it at that!

Proposition XCVIII. Problem XLVIII

The same things supposed; if round the axis AB any attractive surface be described, as CD, regular or irregular, through which the bodies issuing from the given place A must pass; it is required to find a second attractive surface EF, which may make those bodies converge to a given place B.

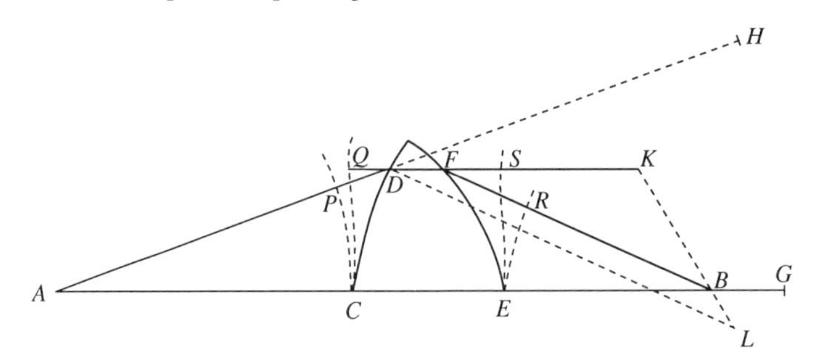

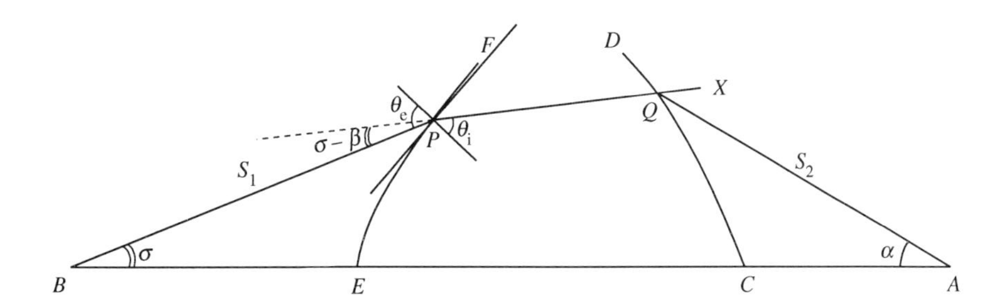

To be able to use Proposition XCVII as we have proved it, we have reversed the direction of Newton's illustration. In the context of this illustration, the problem to be solved is this: given a refractive surface EF, of some specified shape, it is required to find another refractive surface CD passing through a given point C, which will focus at A all the rays 'issuing' from B, on the axis BA, after crossing EF and being refracted by it.

Consider a particular ray BP striking EF at P. Draw the tangent and the normal to the given surface EF. The angles of incidence, θ_i, and emergence, θ_e, are related by Snell's law, that is,

$$\sin\theta_i = \frac{1}{\gamma}\sin\theta_e, \tag{16}$$

where γ denotes the refractive index of the surface EF. The direction PX of the ray, after passage through EF, being known, the inclination β with which the line PX, extended backwards, will intersect AB is clearly (see the figure)

$$\beta = \sigma - (\theta_e - \theta_i). \tag{17}$$

Accordingly, the locus of points, passing through C, that will refract *all* rays issuing from A into rays parallel to QP, by the solution for Corollary II, Proposition XCVII, equation (15), is given by

$$S_2 = AC\,\frac{\gamma\cos(\sigma - \theta_e + \theta_i) - 1}{\gamma\cos(\alpha + \sigma - \theta_e + \theta_i) - 1}. \tag{18}$$

Therefore, a circle of radius S_2 with centre A will intersect PX at Q, a point on the curve that is sought. By repeating the procedure for each ray issuing from A, we can determine the entire curve CD. Q.E.I.

Newton's construction, though on first sight looks '*ad hoc*', is, as one can verify, equivalent to the one we have given. The difference arises largely from the fact that he does not explicitly derive the solution (15). He stops after stating the relations (12) and leaves the rest to the reader's resources. It is a characteristic of Newton's style, which one often encounters in the *Principia*, that he will either omit or be silent when an equation based on the calculus has to be spelled out: he is disinclined to explain.

95. The concluding Scholium of Book I

At the end of most sections and certainly on arrival at watersheds, Newton has elaborated upon what remains to be investigated on grounds of his Natural Philosophy. One should have thought that coming to the end of Book I was such an occasion. Not so! Having digressed into *Opticks*, he cannot turn away; he continues to dwell on its problems. So it is that the present Scholium discusses Proposition XCVIII in the larger context of the construction of optical telescopes.

Scholium

> In the same manner one may go on to three or more surfaces. But of all figures the spherical is the most proper for optical uses. If the object glasses of telescopes were made of two glasses of a spherical figure, containing water between them, it is not unlikely that the errors of the refractions made in the extreme parts of the surfaces of the glasses may be accurately enough corrected by the refractions of the water. Such object glasses are to be preferred before elliptic and hyperbolic glasses, not only because they may be formed with more ease and accuracy, but because the pencils of rays situated without the axis of the glass would be more accurately refracted by them. But the different refrangibility of different rays is the real obstacle that hinders optics from being made perfect by spherical or any other figures. Unless the errors thence arising can be corrected, all the labour spent in correcting the others is quite thrown away.

Appendix I. An analytic solution for the ovals of Descartes

The differential equation that provides an analytic solution for Proposition XCVII can be readily found by methods well within the resources of the Newton of 1685. But, I have not been able to find any reference to an explicit solution for the ovals of Descartes in the literature—very limited, I must confess—that I have searched.

For convenience, the illustration for Proposition XCVII is reproduced below.

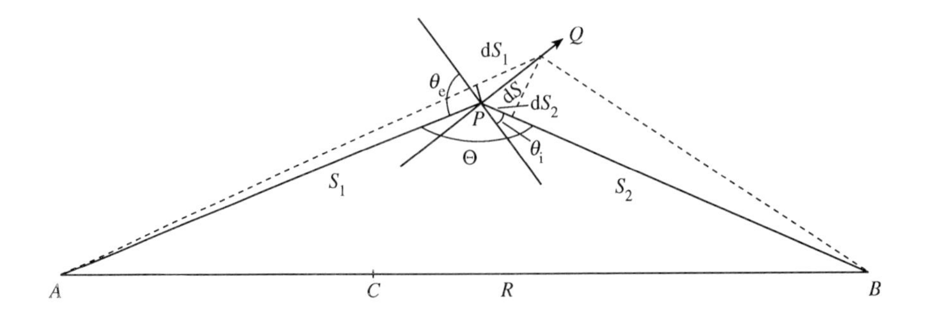

I. Basic facts and definitions:

$$S_1^2 + S_2^2 - 2S_1S_2 \cos\Theta = R^2; \tag{1}$$

$$\Theta + \theta_e - \theta_i = 180°, \qquad \text{or} \qquad \theta_e - \theta_i = 180° - \Theta = \Phi, \qquad \text{(say)} \tag{2}$$

and

$$\sin\theta_e = \gamma \sin\theta_i \qquad (\gamma > 1). \tag{3}$$

II. The expression of θ_e and θ_i in terms of Φ:

$$\sin\theta_e = \gamma \sin(\theta_e - \Phi) = \gamma(\sin\theta_e \cos\Phi - \cos\theta_e \sin\Phi), \tag{4}$$

or

$$(\gamma\cos\Phi - 1)\sin\theta_e = \gamma\cos\theta_e \sin\Phi. \tag{5}$$

Hence,

$$\tan\theta_i = +\frac{\sin\Phi}{\gamma - \cos\Phi} \qquad \text{and} \qquad \tan\theta_e = +\frac{\gamma\sin\Phi}{\gamma\cos\Phi - 1}, \tag{6}$$

or, alternatively,

$$\sin\theta_i = \frac{\sin\Phi}{\sqrt{(1 - 2\gamma\cos\Phi + \gamma^2)}}, \qquad \sin\theta_e = \frac{\gamma\sin\Phi}{\sqrt{(1 - 2\gamma\cos\Phi + \gamma^2)}}. \tag{7}$$

Letting

$$\Gamma = \sqrt{(1 - 2\gamma\cos\Phi + \gamma^2)}, \tag{8}$$

we can write,

$$\sin\theta_i = \frac{1}{\Gamma}\sin\Phi, \qquad \sin\theta_e = \frac{\gamma}{\Gamma}\sin\Phi. \tag{9}$$

III. The integral of Descartes:

$$\mathrm{d}S_1 = (\mathrm{d}s)\sin\theta_e \qquad \text{and} \qquad \mathrm{d}S_2 = -(\mathrm{d}s)\sin\theta_i; \tag{10}$$

or, by equations (9),

$$\mathrm{d}S_1 = \left(\frac{\gamma}{\Gamma}\sin\Phi\right)\mathrm{d}s \qquad \text{and} \qquad \mathrm{d}S_2 = -\left(\frac{1}{\Gamma}\sin\Phi\right)\mathrm{d}s. \tag{11}$$

Hence

$$\gamma\,\mathrm{d}S_2 + \mathrm{d}S_1 \equiv 0. \tag{12}$$

Therefore,

$$\gamma S_2 + S_1 = \text{constant} = C \qquad \text{(say)}. \tag{13}$$

which is Descartes' equation defining his ovals.

IV. The differential equation: We return to equation (1) which we have not used so far. From this equation, we obtain successively

$$\begin{aligned} S_1 S_2 \sin\Phi\, \mathrm{d}\Phi &= S_1\, \mathrm{d}S_1 + S_2\, \mathrm{d}S_2 + (S_1\, \mathrm{d}S_2 + S_2\, \mathrm{d}S_1)\cos\Phi \\ &= (S_1 + S_2\cos\Phi)\,\mathrm{d}S_1 + (S_2 + S_1\cos\Phi)\,\mathrm{d}S_2 \\ &= [\gamma(S_1 + S_2\cos\Phi) - (S_2 + S_1\cos\Phi)]\sin\Phi\,\frac{\mathrm{d}s}{\Gamma}; \end{aligned} \tag{14}$$

or

$$[S_1(\gamma - \cos\Phi) + S_2(\gamma\cos\Phi - 1)]\,\frac{\mathrm{d}s}{\Gamma S_1 S_2} = \mathrm{d}\Phi. \tag{15}$$

Rewriting equation (13) in the form

$$\gamma S_2 = C - S_1, \tag{16}$$

we find:

$$S_1(\gamma - \cos\Phi) + \frac{1}{\gamma}(C - S_1)(\gamma\cos\Phi - 1) = \frac{1}{\gamma}[\Gamma^2 S_1 + C(\gamma\cos\Phi - 1)]. \tag{17}$$

Eliminating ds from equation (15) with the aid of the equation

$$\mathrm{d}S_1 = \mathrm{d}s\,\frac{\gamma}{\Gamma}\sin\Phi, \tag{18}$$

and making use of the reduction (17), we obtain the differential equation describing the ovals of Descartes:

$$\frac{\mathrm{d}S_1}{\mathrm{d}\Phi} = \frac{S_1(C - S_1)\gamma\sin\Phi}{S_1(\gamma^2 - 2\gamma\cos\Phi + 1) + C(\gamma\cos\Phi - 1)}. \tag{19}$$

It should be noted that there is a constraint on the permissible values of C for a given γ. Thus, measuring lengths in the unit R (the distance between A and B) and distinguishing the values of S_1 and S_2 on the axis by a superscript '0' we have

$$\left.\begin{aligned} S_1^{(0)} + S_2^{(0)} &= 1 \qquad \text{(by definition)} \\ \text{and} \qquad \gamma S_2^{(0)} + S_1^{(0)} &= C \qquad \text{(by equation (13)).} \end{aligned}\right\} \tag{20}$$

From these equations we find

$$S_1^{(0)} = \frac{\gamma - C}{\gamma - 1} \qquad \text{and} \qquad S_2^{(0)} = \frac{C - 1}{\gamma - 1}; \tag{21}$$

and it follows that

$$\gamma > C > 1. \tag{22}$$

V. The reduction of equation (19) and its solution: Introducing, in place of S_1 and Φ, the variables,

$$y = S_1 - \tfrac{1}{2}C \qquad \text{and} \qquad z = \gamma^2 - 2\gamma\cos\Phi + 1, \tag{23}$$

we find, after some elementary reductions, the equation

$$(yz + Q)\frac{dy}{dz} = -\tfrac{1}{2}(y^2 - \tfrac{1}{4}C^2), \tag{24}$$

where

$$Q = \tfrac{1}{2}C(\gamma^2 - 1). \tag{25}$$

With the further change of variables,

$$\tfrac{1}{2}C\sin\theta = y, \tag{26}$$

we obtain

$$[z\sin\theta + (\gamma^2 - 1)]\frac{d\theta}{dz} = \tfrac{1}{2}\cos\theta, \tag{27}$$

or, equivalently,

$$\frac{dz}{d\theta} = 2z\tan\theta + 2(\gamma^2 - 1)\sec\theta. \tag{28}$$

The solution of this equation is,

$$z = [\text{constant} + 2(\gamma^2 - 1)\sin\theta]\sec^2\theta, \tag{29}$$

where we may recall that

$$z = \gamma^2 - 2\gamma\cos\Phi + 1, \qquad \sin\theta = \frac{2(S_1 - \tfrac{1}{2}C)}{C}, \qquad \text{and} \qquad \sec\theta = \frac{C}{2\sqrt{[S_1(C - S_1)]}}. \tag{30}$$

The foregoing solution, via the differential equation is, as one can verify, the same that one can obtain algebraically by combining equations (1) and (16).

When S_1 and S_2 are measured in units of R equation (1) becomes

$$S_1^2 + S_2^2 + 2S_1S_2\cos\Phi - 1 = 0 \qquad (\Phi = 180^\circ - \Theta). \tag{31}$$

Now, eliminating S_2 with the aid of equation (16) and simplifying, we obtain,

$$\Gamma^2S_1^2 + 2S_1C(\gamma\cos\Phi - 1) - (\gamma^2 - C^2) = 0, \tag{32}$$

where Γ^2 has the same meaning as in equation (8). Solving equation (32) for S_1 we find

$$S_1 = \frac{1}{\Gamma^2}[-C(\gamma\cos\Phi - 1) + \gamma\sqrt{(\Gamma^2 - C^2\sin^2\Phi)}], \tag{33}$$

compatible with the requirement, $S_1 > 0$. We verify that, when,

$$\Phi = 0 \qquad \text{and} \qquad \Gamma = \gamma - 1, \tag{34}$$

equation (33) gives,

$$S_1 = \frac{\gamma - C}{\gamma - 1}, \tag{35}$$

in agreement with equation (21).

The accompanying illustration exhibits the solutions for Proposition XCVII for the pairs of values ($\gamma = 1.5$, $C = 1.25$) and ($\gamma = 1.1$, $C = 1.05$). In both cases the ovals intersect AB at its mid-point. I am grateful to Dr. Andrea Malagoli for obtaining the solutions from equations (33) and (35).

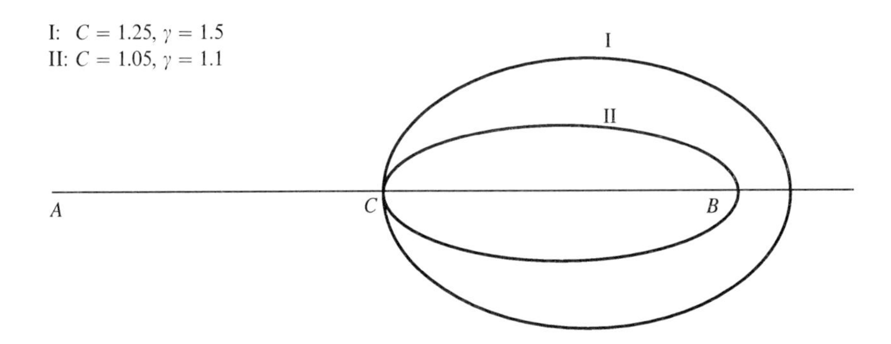

Appendix II. Maxwell on the ovals of Descartes

In 1846, Clerk Maxwell, hardly 15 at the time, wrote a charming paper on *Oval Curves*, which apart from its intrinsic interest, bears on Proposition XCVII to which reference is made. For these reasons, we are reproducing the following extracts from this paper.

> I. *On the Description of Oval Curves, and those having a plurality of Foci*; *with remarks by Professor Forbes*. Communicated by PROFESSOR FORBES.
>
> MR CLERK MAXWELL ingeniously suggests the extension of the common theory of the foci of the conic sections to curves of a higher degree of complication in the following manner:—
>
> (1) As in the ellipse and hyperbola, any point in the curve has the *sum* or *difference* of two lines drawn from two points or *foci* = a constant quantity, so the author infers, that curves to a certain degree analogous, may be described and determined by the condition that the simple distance from one focus *plus* a multiple distance from the other, may be = a constant quantity; or more generally, m times the one distance + n times the other = constant.
>
> (2) The author devised a simple mechanical means, by the wrapping of a thread round pins, for producing these curves. See Figs. 1 and 2. He then thought of extending the principle to other curves, whose property should be,

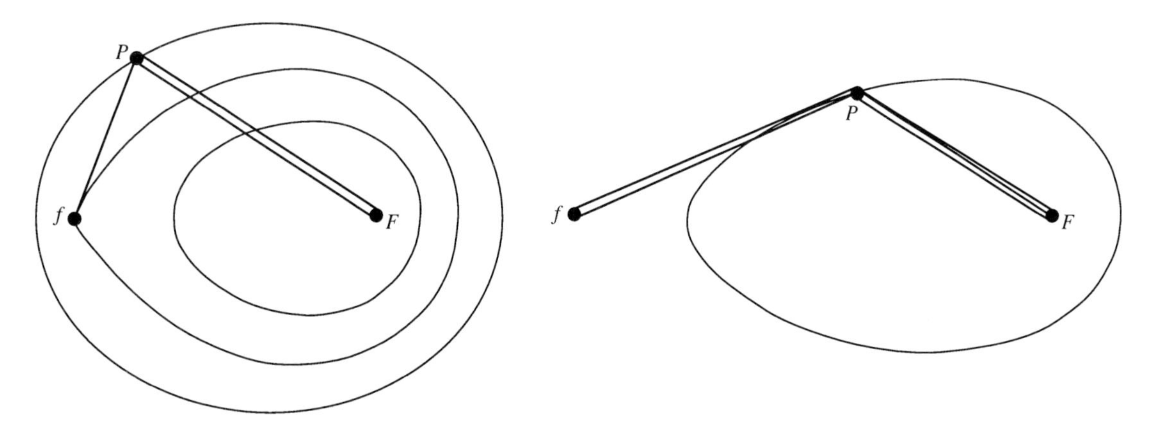

Fig. 1. Two foci. Ratios 1, 2.

Fig. 2. Two foci. Ratios 2, 3.

that the sum of the simple or multiple distances of any point of the curve from three or more points or foci, should be = a constant quantity; and this, too, he has effected mechanically, by a very simple arrangement of a string of given length passing round three or more fixed pins, and constraining a tracing point, *P*. See Fig. 3. Farther, the author regards curves of the first kind as constituting a particular class of curves of the second kind, two or more foci coinciding in one, a focus in which two strings meet being considered a double focus; when three strings meet a treble focus, &c.

. . . .

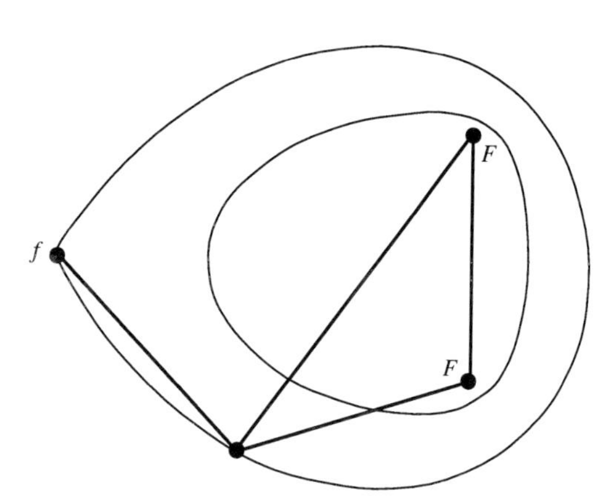

Fig. 3. Three foci. Ratios of Equality.

But the simplest analogy of all is that derived from the method of description, r and r' being the radients to any point of the curve from the two foci;

$$mr + nr' = \text{constant},$$

which in fact at once expresses on the undulatory theory of light the optical character of the surface in question, namely, that light diverging from one focus F without the medium, shall be correctly convergent at another point f within it; and in this case the ratio n/m expresses the index of refraction of the medium*.

. . . .

The ovals of Descartes were described in his *Geometry*, where he has also given a mechanical method of describing one of them†, but only in a particular case, and the method is less simple than Mr Maxwell's. The *demonstration* of the optical properties was given by Newton in the *Principia*, Book I., prop. 97, by the law of the sines; and by Huyghens in 1690, on the Theory of Undulations in his *Traité de la Lumière*. It probably has not been suspected that so easy and elegant a method exists of describing these curves by the use of a thread and pins whenever the powers of the foci are commensurable. For instance, the curve, Fig. 2, drawn with powers 3 and 2 respectively, give the proper form for a refracting surface of a glass, whose index of refraction is 1.50, in order that rays diverging from f may be refracted to F.

As to the higher classes of curves with three or more focal points, we cannot at present invest them with equally clear and curious physical properties, but the method of drawing a curve by so simple a contrivance, which shall satisfy the condition

$$mr + nr' + pr'' + \&c. = \text{constant},$$

is in itself not a little interesting; and if we regard, with Mr Maxwell, the ovals above described, as the limiting case of the others by the coalescence of two or more foci, we have a farther generalization of the same kind as that so highly recommended by Montucla‡, by which Descartes elucidated the conic sections as particular cases of his oval curves.

[From the *Proceedings of the Royal Society of Edinburgh*, Vol. II. April, 1846.]

Postscript

After this chapter was written and completed, I discovered that Proposition XCVII is included in Newton's (1670–71) Lucasian Lectures on Optics. The statement of the Proposition and the proof are essentially the same.

* This was perfectly well shewn by Huyghens in his *Traité de la Lumière*, p. 111. (1690.)

† Edit. 1683. *Geometria*, Lib. II. p. 54.

‡ *Histoire des Mathématiques*. First Edit. II. 102.

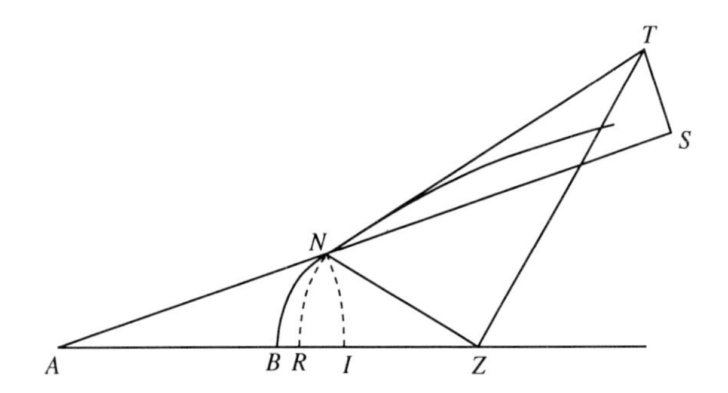

Let A be the meeting point of the incident rays, Z that of the refracted ones, and let some point B be assumed arbitrarily in the straight line AZ as the vertex of the curve. From that point B in the line BZ in the direction of the denser medium take BI of any length and BR to BI in the ratio of the sine of incidence to that of refraction. Then with centres A and Z and radii AI and ZR describe circles intersecting one another in N, and the locus of N will be the curve which effects the desired refraction.

And, Proposition XCVIII, as an isolated '*Problema*' was found in Newton's 'Waste Book'. In the accompanying handwritten copy, the demonstration is virtually the same as in the *Principia*, including the diagram and its lettering.

Problem

Given any refracting surface CD which is to refract rays diverging from the point A in any manner whatever, to find a second surface EF which will make all refracted rays DF converge on some point B.

My inference, based on internal evidence, that there is 'something strangely out of joint in the presentation of these propositions' is confirmed. It gives me encouragement to hope that my other similar inferences, in other contexts, may be equally justified.

The text and the illustrations in this postscript are taken from *The mathematical papers of Isaac Newton*, Volume III, ed. D. T. Whiteside, Cambridge University Press, 1969, pp. 496–7 and plate (facing page 530).

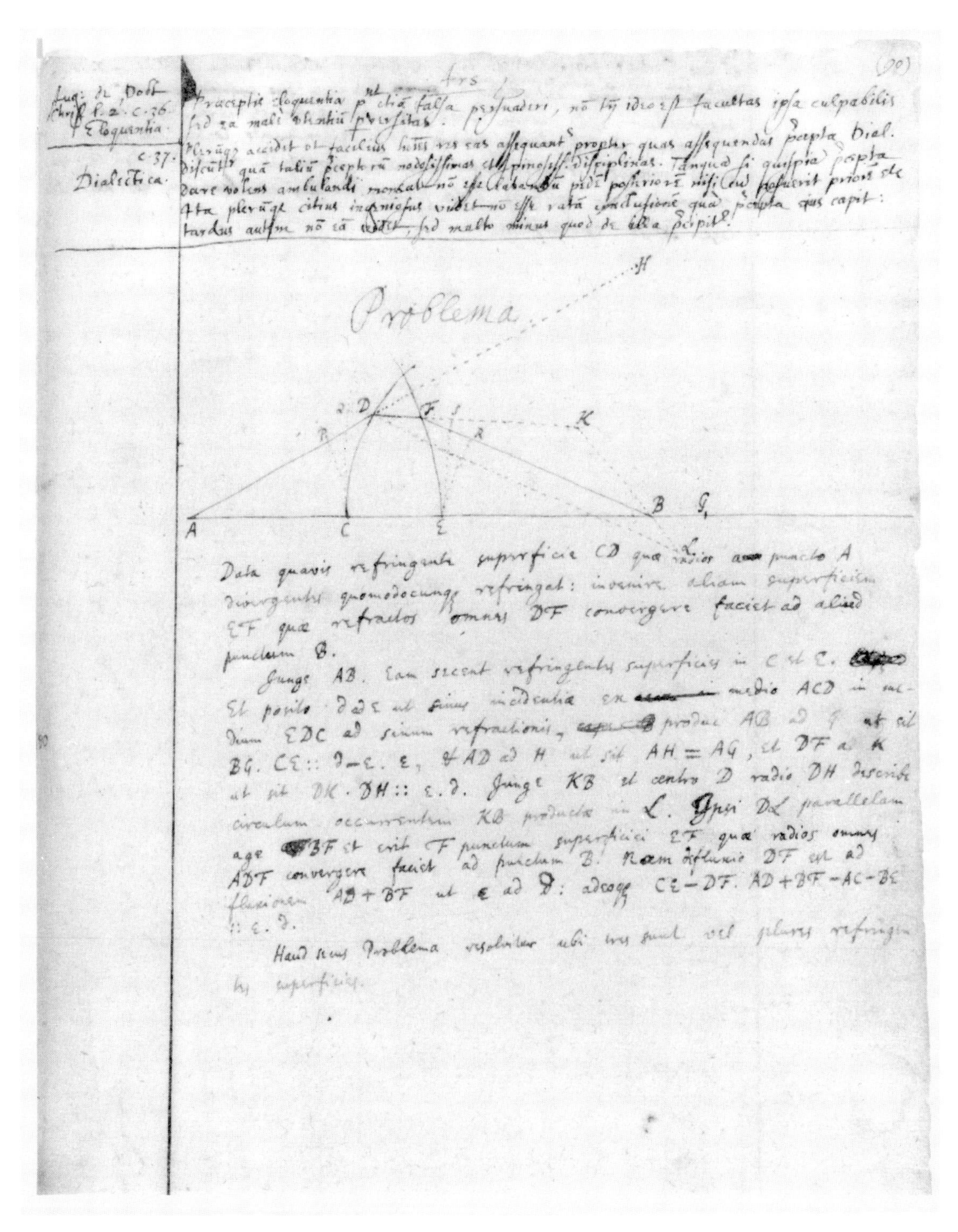

(90)

Aug: de Doct Christ l. 2. c. 36. Eloquentia.

Praeceptis eloquentiae ut etiā falsa persuadeatur, nō tamen ideo est facultas ipsa culpabilis sed ea male utentiū perversitas.

c. 37. Dialectica.

Plerumq accidit ut facilius homines res eas assequantur propter quas assequendas praecepta Dial. discuntur quā talium praeceptorū nodosissimas et spinosissimas disciplinas. Tanquā si quispiam praecepta dare volens ambulandi moneat nō esse levandum pedē posteriorē nisi cum posuerit priorē etc. [illegible] plerumq citius ingeniosus videt nō esse ratā conclusionē quā praecepta ejus capit: tardus autem nō eā videt, sed multo minus quod de illa praecipitur.

Problema

Data quavis refringente superficie CD quae radios a puncto A divergentes quomodocunq refringat: invenire aliam superficiem EF quae refractos omnes DF convergere faciet ad aliud punctum B.

Junge AB. Eam secent refringentes superficies in C et E. Et posito d ad e ut sinus incidentia ex medio ACD in medium EDC ad sinum refractionis, produc AB ad G ut sit BG. CE :: d − e. e, & AD ad H ut sit AH = AG, et DF ad K ut sit DK. DH :: e. d. Junge KB et centro D radio DH describe circulum occurrentem KB productae in L. Ipsi DL parallelam age BF et erit F punctum superficiei EF quae radios omnes ADF convergere faciet ad punctum B. Nam fluxio DF est ad fluxionem AD + BF ut e ad d: adeoq CE − DF. AD + BF − AC − BE :: e. d.

Haud secus Problema resolvitur ubi tres sunt vel plures refringentes superficies.

Plate 4 The reproduction from the Newton Archives at the Cambridge University Library, published with their permission.

Introduction to Newton's System of the world (Book III)

To pass on to Book III of the *Principia*, directly after the presentation of Book I requires an explanation. First, it may be reiterated (what was stated at the outset in the Prologue) that to present the entire *Principia* was never the intention: it is too formidable a task even to contemplate. The main objective was to present the analytical (i.e., the purely mathematical) propositions that are at the base, as Newton broadly conceived it, for his major thesis as he states it without ambiguity in Proposition VII of Book III:

> *That there is a power of gravity pertaining to all bodies, proportional to the several quantities of matter which they contain*;

and to overcome the syntactical problems associated with the often convoluted style that Newton had perforce to adopt in presenting delicate mathematical arguments in continuous prose: in fact, to bring out in sharp relief the monolithic intellectual achievement that the *Principia* is. Compatible with the stated objective, the selection of the propositions of Book I presented no difficulty: the coherence and the logical unity of the Book, emphasized in the successive Scholia, made it compulsory that at least 77 of the 98 propositions were included. The only propositions that seemed reasonable to omit were the 12 Propositions XVIII–XXIX dealing essentially with the geometry of the conic sections and the nine Propositions XLVIII–LVI dealing with the 'oscillating pendulous motions'.

In contrast, Book II does not present an equally indivisible front: the diversity of the problems treated and their base, often in experiments, are in part responsible. (Not that there are not enough propositions of great merit exhibiting Newton's virtuosity; but they are more suitable for separate consideration in miscellanea.) Besides, by Newton's own evaluation, of the 22 propositions he explicitly quotes in the first 14 propositions of Book III in which the universal law of gravitation is founded, only three are from Book II, one of which (Proposition XXIV establishing the equality of the inertial and the gravitational masses) has been included in Chapter II, §10(a). It is permissible then to start on Book III without the many digressions of Book II.

But even of Book III, our presentation will not be as complete as Book I. In Book III, Newton felt compelled to discuss the relevant astronomical observations in considerable detail. We shall not include these discussions except briefly on occasions. We shall concentrate instead on the theoretical and the mathematical propositions.

We begin our study of Book III with an introductory Prolegomenon in Newton's words: his opening magisterial enunciation of the 'Rules of Reasoning in Philosophy'; his enumeration of the six Phenomena; and the texts of 22 propositions on which he will base his Reasoning.

❖18❖

Prolegomenon

In the preceding books I have laid down the principles of philosophy; principles not philosophical but mathematical: such, namely, as we may build our reasonings upon in philosophical inquiries.

I. Newton

96.

RULES OF REASONING IN PHILOSOPHY

Rule I

We are to admit no more causes of natural things than such as are both true and sufficient to explain their appearances.

To this purpose the philosophers say that Nature does nothing in vain, and more is in vain when less will serve; for Nature is pleased with simplicity, and affects not the pomp of superfluous causes.

Rule II

Therefore to the same natural effects we must, as far as possible, assign the same causes.

As to respiration in a man and in a beast; the descent of stones in Europe and in America; the light of our culinary fire and of the Sun; the reflection of light in the Earth, and in the planets.

Rule III

The qualities of bodies, which admit neither intensification nor remission of degrees, and which are found to belong to all bodies within the reach of our experiments, are to be esteemed the universal qualities of all bodies whatsoever.

For since the qualities of bodies are only known to us by experiments, we are to hold for universal all such as universally agree with experiments; and such as are not liable to diminution can never be quite taken away. We are certainly not to relinquish the evidence of experiments for the sake of dreams and vain fictions of our own devising; nor are we to recede from the analogy of Nature, which is wont to be simple, and always consonant to itself. We no other way know . . . .

Lastly, if it universally appears, by experiments and astronomical observations, that all bodies about the Earth gravitate towards the Earth, and that in proportion to the quantity of matter which they severally contain; that the Moon likewise, according to the quantity of its matter, gravitates towards the Earth; that, on the other hand, our sea gravitates towards the Moon; and all the planets one towards another; and the comets in like manner towards the Sun; we must, in consequence of this rule, universally allow that all bodies whatsoever are endowed with a principle of mutual gravitation. For the argument from the appearances concludes with more force for the universal gravitation of all bodies than for their impenetrability; of which, among those in the celestial regions, we have no experiments, nor any manner of observation. Not that I affirm gravity to be essential to bodies: by their *vis insita* I mean nothing but their inertia. This is immutable. Their gravity is diminished as they recede from the Earth.

Rule IV

In experimental philosophy we are to look upon propositions inferred by general induction from phenomena as accurately or very nearly true, notwithstanding any contrary hypotheses that may be imagined, till such time as other phenomena occur, by which they may either be made more accurate, or liable to exceptions.

This rule we must follow, that the argument of induction may not be evaded by hypotheses.

97.

PHENOMENA

Phenomenon I

That the circumjovial planets, by radii drawn to Jupiter's centre, describe areas proportional to the times of description; and that their periodic times, the fixed stars being at rest, are as the 3/2th power of their distances from its centre.

This we know from astronomical observations.

Phenomenon II

That the circumsaturnal planets, by radii drawn to Saturn's centre, describe areas proportional to the times of description; and that their periodic times, the fixed stars being at rest, are as the 3/2th power of their distances from its centre.

Phenomenon III

That the five primary planets, Mercury, Venus, Mars, Jupiter, and Saturn, with their several orbits, encompass the Sun.

Phenomenon IV

That the fixed stars being at rest, the periodic times of the five primary planets, and (whether of the Sun about the Earth, or) of the Earth about the Sun, are as the 3/2th power of their mean distances from the Sun.

This proportion, first observed by Kepler, is now received by all astronomers; for the periodic times are the same, and the dimensions of the orbits are the same, whether the Sun revolves about the Earth, or the Earth about the Sun. And as to the measures of the periodic times, all astronomers are agreed about them. But for the dimensions of the orbits, Kepler and Boulliau, above all others, have determined them from observations with the greatest accuracy; and the mean distances corresponding to the periodic times differ but insensibly from those from which they have assigned, and for the most part fall in between them;

Phenomenon V

Then the primary planets, by radii drawn to the Earth, describe areas in no wise proportional to the times; but the areas which they describe by radii drawn to the Sun are proportional to the times of description.

Phenomenon VI

That the Moon, by a radius drawn to the Earth's centre, describes an area proportional to the time of description.

This we gather from the apparent motion of the Moon, compared with its apparent diameter. It is true that the motion of the Moon is a little disturbed by the action of the Sun: but in laying down these phenomena, I neglect those small and inconsiderable errors.

98.

PROPOSITIONS

Proposition II. Theorem II

Every body that moves in any curved line described in a plane, and by a radius drawn to a point either immovable, or moving forwards with a uniform rectilinear motion, describes about that point areas proportional to the times, is urged by a centripetal force directed to that point.

Proposition III. Theorem III

Every body, that by a radius drawn to the centre of another body, howsoever moved, describes areas about that centre proportional to the times, is urged by a force compounded of the centripetal force tending to that other body, and of all the accelerative force by which that other body is impelled.

Proposition IV. Theorem IV

The centripetal forces of bodies, which by equable motions describe different circles, tend to the centres of the same circles; and are to each other as the squares of the arcs described in equal times divided respectively by the radii of the circles.

COR. II. And since the periodic times are as the radii divided by the velocities, the centripetal forces are as the radii divided by the square of the periodic times.

COR. VI. If the periodic times are as the 3/2th powers of the radii, and therefore the velocities inversely as the square roots of the radii, the centripetal forces will be inversely as the squares of the radii; and conversely.

COR IX. From the same demonstration it likewise follows, that the arc which a body, uniformly revolving in a circle with a given centripetal force, describes in any time, is a mean proportional between the diameter of the circle, and the space which the same body falling by the same given force would describe in the same given time.

Proposition XLV. Problem XXXI

To find the motion of the apsides in orbits approaching very near to circles.

COR. I. Hence if the centripetal force be as any power of the altitude, that power may be found from the motion of the apsides; and conversely. That is, if the whole angular motion, with which the body returns to the same apse, be to the angular motion of one revolution, or 360°, as any number as m to another as n, and the altitude be called A; the force will be as the power $A^{(nn/mm)-3}$ of the altitude A; the index of which power is $(nn/mm) - 3$.

COR. II. Hence also if a body, urged by a centripetal force which is inversely as the square of the altitude, revolves in an ellipse whose focus is in the centre of the forces; and a new and foreign force should be added to or subtracted from this centripetal force, the motion of the apsides arising from that foreign force may (by the third Example) be known; and conversely.

Proposition XXXVI. Problem XXV

To determine the times of the descent of a body falling from a given place A.

Proposition LX. Theorem XXIII

If two bodies S and P, attracting each other with forces inversely proportional to the square of their distance, revolve about their common centre of gravity: I say, that the principal axis of the ellipse which either of the bodies, as P, describes by this motion about the other S, will be to the principal axis of the ellipse, which the same body P may describe in the same periodic time about the other body S fixed, as the sum of the two bodies $S + P$ *to the first of two mean proportionals between that sum and the other body S.*

Proposition XXIV. Theorem XIX

The quantities of matter in pendulous bodies, whose centres of oscillation are equally distant from the centre of suspension, are in a ratio compounded of the ratio of the weights and the squared ratio of the times of the oscillations in a vacuum.

COR. I. Therefore if the times are equal, the quantities of matter in each of the bodies are as the weights.

COR. VI. But in a non-resisting medium, the quantity of matter in the pendulous body is directly as the comparative weight and the square of the time, and inversely as the length of the pendulum. For the comparative weight is the motive force of the body in any heavy medium, as was shown above; and therefore does the same thing in such a non-resisting medium as the absolute weight does in a vacuum.

Proposition LXV. Theorem XXV

Bodies, whose forces decrease as the square of their distances from their centres, may move among themselves in ellipses; and by radii drawn to the foci may describe areas very nearly proportional to the times.

COR. II. But the perturbation will be greatest of all, if we suppose the accelerative attractions of the parts of the system towards the greatest body of all are not to each other inversely as the squares of the distances from that great body; especially if the inequality of this proportion be greater than the inequality of the proportion of the distances from the great body. For if the accelerative force, acting in parallel directions and equally, causes no perturbation in the motions of the parts of the system, it must of course, when it acts unequally, cause a perturbation somewhere, which will be greater or less as the inequality is greater or less. The excess of the greater impulses acting upon some bodies, and not acting upon others, must necessarily change their situation among themselves. And this perturbation, added to the perturbation arising from the inequality and inclination of the lines, makes the whole perturbation greater.

COR. III. Hence if the parts of this system move in ellipses or circles without any remarkable perturbation, it is manifest that, if they are at all impelled by accelerative forces tending to any other bodies, the impulse is very weak, or else is impressed very near equally and in parallel directions upon all of them.

Proposition LXIX. Theorem XXIX

In a system of several bodies A, B, C, D, etc., if any one of those bodies, as A, attract all the rest, B, C, D, etc., with accelerative forces that are inversely as the squares of the distances from the attracting body; and another body, as B, attracts also the rest, A, C, D, etc., with forces that are inversely as the squares of the distances from the attracting body; the absolute forces of the attracting bodies A and B will be to each other as those very bodies A and B to which those forces belong.

Proposition LXXIV. Theorem XXXIV

The same things supposed, I say, that a corpuscle situated without the sphere is attracted with a force inversely proportional to the square of its distance from the centre.

COR. III. If a corpuscle placed without a homogeneous sphere is attracted by a force inversely proportional to the square of its distance from the centre, and the sphere consists of attractive particles, the force of every particle will decrease as the square of the distance from each particle.

Proposition LXXV. Theorem XXXV

If to the several points of a given sphere there tend equal centripetal forces decreasing as the square of the distances from the point, I say, that another similar sphere will be attracted by it with a force inversely proportional to the square of the distance of the centres.

Proposition LXXVI. Theorem XXXVI

If spheres be however dissimilar (as to density of matter and attractive force) in the same ratio onwards from the centre to the circumference; but everywhere similar, at every given distance from the centre, on all sides round about; and the attractive force of every point decreases as the square of the distance of the body attracted: I say, that the whole force with which one of these spheres attracts the other will be inversely proportional to the square of the distance of the centres.

Proposition LXXII. Theorem XXXII

If to the several points of a sphere there tend equal centripetal forces decreasing as the square of the distances from those points; and there be given both the density of the sphere and the ratio of the diameter of the sphere to the distance of the corpuscle from its centre: I say, that the force with which the corpuscle is attracted is proportional to the semidiameter of the sphere.

Proposition LXXIII. Theorem XXXIII

If to the several points of a given sphere there tend equal centripetal forces decreasing as the square of the distances from the points, I say, that a corpuscle placed within the sphere is attracted by a force proportional to its distance from the centre.

Proposition XL. Problem IX

To find by experiment the resistance of a globe moving through a perfectly fluid compressed medium.

Proposition XXII. Theorem XVII

Let the density of any fluid be proportional to the compression, and its parts be attracted downwards by a gravitation inversely proportional to the squares of the distances from the centre: I say, that if the distances be taken in harmonic progression, the densities of the fluid at those distances will be in a geometrical progression.

Proposition I. Theorem I

The areas which revolving bodies describe by radii drawn to an immovable centre of force do lie in the same immovable planes, and are proportional to the times in which they are described.

Proposition XI. Problem VI

If a body revolves in an ellipse; it is required to find the law of the centripetal force tending to the focus of the ellipse.

Proposition XIII. Problem VIII

If a body moves in the perimeter of a parabola; it is required to find the law of the centripetal force tending to the focus of that figure.

Cor. i. From the three last Propositions it follows, that if any body P goes from the place P with any velocity in the direction of any right line PR, and at the same time is urged by the action of a centripetal force that is inversely proportional to the square of the distance of the places from the centre, the body will move in one of the conic sections, having its focus in the centre of force; and conversely. For the focus, the point of contact, and the position of the tangent, being given, a conic section may be described, which at that point shall have a given curvature. But the curvature is given from the centripetal force and velocity of the body being given; and two orbits, touching one the other, cannot be described by the same centripetal force and the same velocity.

Proposition LXVI. Theorem XXVI

If three bodies, whose forces decrease as the square of the distances, attract each other; and the accelerative attractions of any two towards the third be between themselves inversely as the squares of the distances; and the two least revolve about the greatest: I say, that the interior of the two revolving bodies will, by radii drawn to the innermost and greatest, describe round that body areas more proportional to the times, and a figure more approaching to that of an ellipse having its focus in the point of intersection of the radii, if that great body be agitated by those attractions, than it would do if that great body were not attracted at all by the lesser, but remained at rest; or than it would do if that great body were very much more or very much less attracted, or very much more or very much less agitated, by the attractions.

Proposition LXVII. Theorem XXVII

The same laws of attraction being supposed, I say, that the exterior body S does, by radii drawn to the point O, the common centre of gravity of the interior bodies P and T, describe round that centre areas more proportional to the times, and an orbit more approaching to the form of an ellipse having its focus in that centre, than it can describe round the innermost and greatest body T by radii drawn to that body.

Proposition LXX. Theorem XXX

If to every point of a spherical surface there tend equal centripetal forces decreasing as the square of the distances from those points, I say, that a corpuscle placed within that surface will not be attracted by those forces any way.

❖19❖

The universal law of gravitation

99. Introduction

After his Prolegomenon (Chapter 18), Newton proceeds to the deduction of his universal law of gravitation in the first 14 Propositions of Book III.

We shall begin by listing the 19 propositions of Book I, the three propositions of Book II, and the Rules of Reasoning that will serve as the basis of Newton's reasoning. An examination of this list is already of interest for revealing the priorities that Newton attaches to the various propositions; and in particular to Proposition XI (and XLI) in which he proved that a body will revolve in an elliptic orbit about its focus if acted upon by a centripetal force inversely as the square of the distance from the focus. It will be recalled that it was on hearing from Newton (in August, 1684) that he had proved this proposition 'some years earlier' and on receiving in the same year (November, 1684) *De Motu Corporum in Gyrum* in which the proposition is proved, that Halley persuaded Newton to embark on what was to become *Philosophie naturalis Principia Mathematica.* Nevertheless, Newton refers to Proposition XI only casually—almost as an afterthought—in the last but one of the 14 propositions that comprise this section. Newton's priorities are clearly different.

Propositions and rules (to which references are made)*

Proposition I: Prop. II, or III,
Prop. IV, Cor. VI;

Proposition II: Prop. II,
Prop. IV, Cor. VI,
Prop. XLV, Cor. I;

* The references are to Book I, except where otherwise indicated.

Proposition III:	Prop. II or III, Prop. XLV, Cor. I, Prop. XLV, Cor. II;
Proposition IV:	Prop. XXXVI, Prop. IV, Cor. IX, Prop. LX; Rules 1 and 2;
Scholium	Rules 1 and 2;
Proposition V:	Law III; Rule 2;
Scholium	Rules 1, 2 and 4;
Proposition VI:	Prop. XXIV, Cor. I & VI, (Book II); Prop. LXV, Cor. III, Prop. LXV, Cor. II; Rule 3;
Proposition VII:	Prop. LXIX, Law III, Prop. LXXIV, Cor. III;
Proposition VIII:	Prop. LXXV and Corollaries, Prop. LXXVI and Corollaries, Prop. IV, Cor. II, Prop. LXXII;
Proposition IX:	Prop. LXXIII;
Proposition X:	Prop. XL, (Book II), Prop. XXII, (Book II);
Proposition XI:	Cor. IV of Laws;
Proposition XIII:	Prop. I, Prop. XI, Prop. XIII, Cor. I, Prop. LXVI, Prop. LXVII;
Proposition XIV:	Prop. XI, Prop. I, Prop. LXX;

He attaches greater importance to the rotation of the line of apsides which will result from the slightest departure from the inverse-square law of attraction (established in Proposition XLV and its corollaries) and to the 'quiescence of the aphelion points'; to the equality of the inertial and the gravitational masses, that is, the *principle of equivalence* (Proposition XXIV of Book II); and, to the 'superb theorems' (Propositions LXX and LXXII–LXXIV). These priorities will become clearer in due course.

100. Propositions I, II and III

In Propositions I, II and III, the inference of the centripetal attractions of the planets (Jupiter, Saturn, and the Earth) towards the satellites and of the Sun towards the primary planets is based on the description of equal areas in equal times, by the satellites about the planets and the primary planets about the Sun (Propositions II and III); and the inference of their inverse-square character from the proportionality of the periods to the $\frac{3}{2}$th power of the mean distances from their respective centres (Proposition IV, Corollary VI). The elliptical character of the orbits is nowhere mentioned. The 'quiescence of the aphelion points', and the very 'sensible' rotation of the line of apsides that will result from the slightest 'aberration' from the inverse-square law of attraction (Proposition XLV and its corollaries) are for Newton (and must also be for us) the stronger arguments.

Proposition I. Theorem I

That the forces by which the circumjovial planets are continually drawn off from rectilinear motions, and retained in their proper orbits, tend to Jupiter's centre; and are inversely as the squares of the distances of the places of those planets from that centre.

The former part of this Proposition appears from Phen. I, and Prop. II or III, Book I; the latter from Phen. I, and Cor. VI, Prop. IV, of the same Book.

The same thing we are to understand of the planets which encompass Saturn, by Phen. II.

Proposition II. Theorem II

That the forces by which the primary planets are continually drawn off from rectilinear motions, and retained in their proper orbits, tend to the Sun; and are inversely as the squares of the distances of the places of those planets from the Sun's centre.

The former part of the Proposition is manifest from Phen. V, and Prop. II, Book I; the latter from Phen. IV, and Cor. VI, Prop. IV, of the same Book. But this part of the Proposition is, with great accuracy, demonstrable from the quiescence of the aphelion points; for a very small aberration from the proportion according to the inverse square of the distances would (by Cor. I, Prop. XLV, Book I) produce a motion of the apsides sensible enough in every single revolution, and in many of them enormously great.

Proposition III. Theorem III

That the force by which the Moon is retained in its orbit tends to the Earth; and is inversely as the square of the distance of its place from the Earth's centre.

That the centre of the Earth is the origin of a centripetal force acting on the Moon follows from Phenomenon VI as stated. But to infer that the centripetal force is inversely as the square of the Earth–Moon distance is somewhat more problematical in view of the slow rotation of the Moon's apogee, amounting to $3^\circ\ 3'$ in one complete revolution. There are, *prima facie*, two *ad hoc* possibilities, both of which Newton explores; either to take the rotation of the apse at its face value and determine the law of force which will be compatible with it; or to suppose that the rotation is caused by an external disturbance such as the tidal action of the Sun and determine its magnitude.

I. Considering the first possibility, we have, by Example 2 of Proposition XLV (p. 195), a body describing a revolving orbit will return to the same apse m times while the fixed orbit completes n revolutions, if the centripetal force is of the form

$$\text{C.F.} \propto r^{(n^2/m^2)-3}$$

(where we have replaced the A of Corollary I, Proposition XLV, p. 197, by r). For the case on hand

$$n:m = 360:363 = 120:121.$$

Therefore,

$$\frac{n^2}{m^2} - 3 = -\tfrac{29\,523}{14\,641} = -2\tfrac{4}{243};$$

and the required law of force is

$$r^{-2-(4/243)} = r^{-2\cdot 0165}.$$

After obtaining this result, Newton adds:

> that is to say, in the proportion of the distance somewhat greater than the inverse square, but which comes $59\frac{3}{4}$ times nearer to the proportion according to the square than to the cube. But since this increase is due to the action of the Sun (as we shall afterwards show), it is here to be neglected.

II. Considering the second possibility, that the motion of the Moon's apogee derives from the tidal action of the Sun, Newton reverses the argument of Corollary II, Proposition XLV of Book I. In that corollary (see Chapter 10, p. 198) Newton had supposed that the effect of the Sun was to superpose on the attraction of the Earth, following the inverse-square law, a force linear with the distance (in accordance with equation (13), §71, Chapter 14) so that the net centripetal force is of the form

$$\text{C.F.} \propto (r^{-2} - cr),$$

where c is some constant. Such a law of force, will produce an angle of rotation,

$$180^\circ\sqrt{\frac{1-c}{1-4c}} \simeq 180^\circ(1+\tfrac{3}{2}c)$$

in excess of 180°. With the 'realistic estimate',

$$c = \tfrac{100}{35\,745},$$

Newton had found that the Moon's apogee will move forwards by $1^\circ\,31'\,28''$ during a single revolution and had concluded: 'the apse of the Moon is twice as fast'. Therefore, to obtain a motion of the apse in agreement with the observation, Newton now sets

$$c = \tfrac{200}{35\,745} \simeq 0.005595,$$

and concludes:

> that is, as 1 to $178\frac{29}{40}$. And if we neglect so inconsiderable a force of the Sun, the remaining force, by which the Moon is retained in its orb, will be inversely as D^2. This will yet more fully appear from comparing this force with the force of gravity, as is done in the next Proposition.

Newton clearly wishes us to understand that while the centripetal forces,

$$\text{C.F.} \propto r^{-2\cdot 0165} \qquad \text{and} \qquad \text{C.F.} \propto (r^{-2} - 0.005595r)$$

provide alternative *ad hoc* assumptions for interpreting the observed motion of the Moon's apogee, a correct interpretation will be in terms of a complete self-consistent lunar theory. At the same time, one can sense Newton's unhappiness that his own attempt along the lines of Corollary VII of Proposition LXVI is insufficient. But undaunted he adds the following corollary:

> COR. If we augment the mean centripetal force by which the Moon is retained in its orb, first in the proportion of $177\frac{29}{40}$ to $178\frac{29}{40}$, and then in the proportion of the square of the semidiameter of the Earth to the mean distance of the centres of the Moon and Earth, we shall have the centripetal force of the Moon at the surface of the Earth; supposing this force, in descending to the Earth's surface, continually to increase inversely as the square of the height.

101. Proposition IV and the Moon test

In this proposition, Newton, in effect, returns to the 'Moon test' of the plague years (see Chapter 1, pp. 3–6). The sophistication of the present test, in contrast with the naiveté (by Newton's standards!) of his earlier test, is a measure of his demands on the cogency of a scientific argument.

Proposition IV. Theorem IV

> *That the Moon gravitates towards the Earth, and by the force of gravity is continually drawn off from a rectilinear motion, and retained in its orbit.*

The staple of Newton's present arguments are the following steps.

(a) The mean distance of the Moon from the Earth is

$$60 \text{ earth radii at the syzygies.} \tag{1}$$

Newton arrives at this value by quoting the authorities:

Ptolemy: 59 earth radii;
Vendelin and Huygens: 60 earth radii;
Copernicus: $60\frac{1}{3}$ earth radii;
Street: $60\frac{2}{5}$ earth radii;
Tycho: $56\frac{1}{2}$ earth radii;

and ignoring Tycho's value (which is discordant with the rest) on the grounds that Tycho's allowance for the refraction of the Sun and the Moon is 'altogether against the nature of light'.

(b) The circumference of the Earth is

$$1.232496 \times 10^8 \textit{ Paris} \text{ feet,} \tag{2}$$

as 'the *French* have found by mensuration'. It should be noted here that

$$\text{one } \textit{French} \text{ inch } (= \textit{Pouce}) = 2.7070 \text{ cm,} \tag{3}$$

in contrast to an *English* inch which is 2.5400 cm.

(c) The mean lunar day is

$$27^{\text{d}}\ 7^{\text{h}}\ 43^{\text{m}} = 39\,343 \text{ minutes.} \tag{4}$$

(d) The angular arc, $\delta\theta$, the Moon describes in its orbit, *per minute* is

$$\delta\theta = \frac{2\pi}{39\,343} \text{ radians.} \tag{5}$$

Therefore, the distance, the Moon descends towards the Earth ($= BD$ in the diagram) in *one minute* is

$$BD = [(60 \times \text{radius of earth}) \times \delta\theta] \times \tfrac{1}{2}\delta\theta; \tag{6}$$

or, using the values we have listed,

$$\begin{aligned} BD &= \frac{1}{2}\left(60 \times \frac{1.232496 \times 10^8}{2\pi}\right) \times \left(\frac{2\pi}{3.9343 \times 10^4}\right)^2 \\ &= 60\pi \frac{1.232496}{(3.9343)^2} \\ &= 15\tfrac{1}{120} \textit{ Paris} \text{ feet.} \end{aligned} \tag{7}$$ *

(This argument is the same as in the earlier Moon test.) It follows that on the *surface of the Earth* where the gravity is ($60 \times 60 = 3600$) times stronger, a body will descend in one second (which will bring the 'compensating' factor 1/3600) the *same distance*,

$$15\tfrac{1}{120} \textit{ French} \text{ feet;} \tag{8}$$

* Newton gives instead $15\frac{1}{12}$ *Paris* feet. Perhaps he wrote this value from memory (?) from an earlier computation with different parameters from those listed in this proposition.

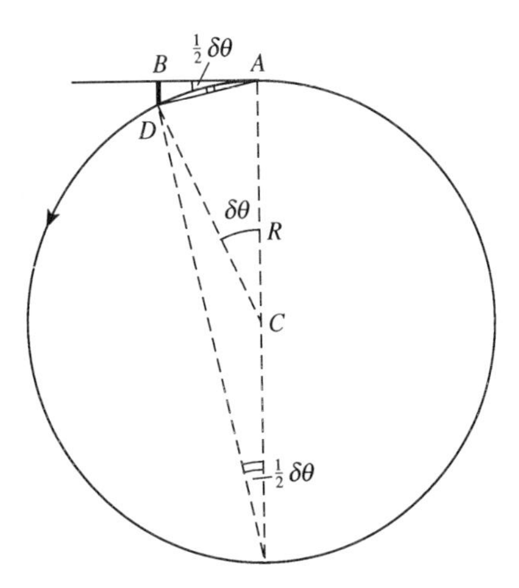

or, in centimeters (by equation (3)),

$$(15\tfrac{1}{120} \times 12) \times 2.7070 = 487.5 \text{ cm}. \tag{9}$$

It remains to compare the (predicted) descent (9) in one second, derived from astronomical observations, with the rectilinear descent of a body under the action of a constant gravity such as prevails on the Earth's surface. Under the action of a constant gravity g, a body will descend, in a time t, the distance,

$$s = \tfrac{1}{2}gt^2, \tag{10}$$

by 'Galileo's theorem'. If, in one second, a body should descend by the amount (9), then

$$g = 2s = 975.0 \text{ cm sec}^{-2} \tag{11}$$

and this is in satisfactory agreement with the known value of $g(=978 \text{ cm sec}^{-2})$. But the meaning of this agreement is not clear. Is it assumed, as in Newton's earlier test, that the attraction of the Earth on the surface does not depend on the internal variation of the density so long as it is the same 'at every given distance from the centre, on all sides round about'? But at this stage the latter assumption is not justified; and nor is it needed. Besides, how about the possible difference in the inertial and the gravitational masses of the substance of the Earth? Newton avoids these questions altogether by appealing to experiments, such as his own, in which the equality of the inertial and gravitational masses was established in Proposition XXIV of Book II (described in Chapter 2, p. 35). By this proposition, the period of oscillation, T, of a pendulum is related to its length l by

$$T^2 = \pi^2 l/g, \tag{12}$$

a relation known also to Huygens. Hence, for a *seconds pendulum*, that is, a pendulum whose period of oscillation is one second, the length is given by

$$l = g/\pi^2; \tag{13}$$

while by equation (11)

$$g = 2s. \tag{14}$$

Eliminating g from equations (13) and (14) we obtain,

$$s = \tfrac{1}{2} l \pi^2. \tag{15}$$

This is the relation that Newton uses; and on Huygens's authority, he quotes for l the value

$$l = 3 \textit{ Paris} \text{ feet and } \tfrac{7}{10} \text{ of an inch.} \tag{16}$$

Using this value of l, Newton obtains:

$$s = (15 + \tfrac{1.16}{12}) \textit{ French} \text{ feet.} \tag{17}$$

Comparing this last value with the value (8), inferred from lunar data, Newton concludes:

> And therefore the force by which the Moon is retained in its orbit becomes, at the very surface of the Earth, equal to the force of gravity which we observe in heavy bodies there. And therefore (by Rule 1 and 2) the force by which the Moon is retained in its orbit is that very same force which we commonly call gravity; for, were gravity another force different from that, then bodies descending to the Earth with the joint impulse of both forces would fall with a double velocity, and in the space of one second of time would describe $30\frac{1}{6}$ *Paris* feet; altogether against experience.

In the Scholium, which follows, Newton explains 'more diffusely'(!) the meaning of his last remark referring to his Rules 1 and 2.

Scholium

> The demonstration of this Proposition may be more diffusely explained after the following manner. Suppose several moons to revolve about the Earth, as in the system of Jupiter or Saturn; the periodic times of these moons (by the argument of induction) would observe the same law which Kepler found to obtain among the planets; and therefore their centripetal forces would be inversely as the squares of the distances from the centre of the Earth, by Prop. I, of this Book. Now if the lowest of these were very small, and were so near the Earth as almost to touch the tops of the highest mountains, the centripetal force thereof, retaining it in its orbit, would be nearly equal to the weights of any terrestrial bodies that should be found upon the tops of those mountains, as may be known by the foregoing computation. Therefore if the same little moon should be deserted by its centrifugal force that carries it through its orbit, and be disabled from going onward therein, it would descend to the Earth; and that with the same velocity, with which heavy bodies actually fall upon the tops of those very mountains, because of the equality of the forces that oblige them both to descend. And if the force by which that lowest moon would descend were

different from gravity, and if that moon were to gravitate towards the Earth, as we find terrestrial bodies do upon the tops of mountains, it would then descend with twice the velocity, as being impelled by both these forces conspiring together. Therefore since both these forces, that is, the gravity of heavy bodies, and the centripetal forces of the moons, are directed to the centre of the Earth, and are similar and equal between themselves, they will (by Rules 1 and 2) have one and the same cause. And therefore the force which retains the Moon in its orbit is that very force which we commonly call gravity; because otherwise this little moon at the top of a mountain must either be without gravity, or fall twice as swiftly as heavy bodies are wont to do.

102. The emergence of the law of gravitation

Proposition V, in effect, states that from Propositions I–IV some elements of the universal law of gravitation already emerge; but only some elements. The law fully emerges in Proposition VII after the crucial Proposition VI. Newton's statement on what does emerge is surpassingly clear; and we shall quote him *in extenso.*

Proposition V. Theorem V

That the circumjovial planets gravitate towards Jupiter; the circumsaturnal towards Saturn; the circumsolar towards the Sun; and by the forces of their gravity are drawn off from rectilinear motions, and retained in curvilinear orbits.

For the revolutions of the circumjovial planets about Jupiter, of the circumsaturnal about Saturn, and of Mercury and Venus, and the other circumsolar planets, about the Sun, are appearances of the same sort with the revolution of the Moon about the Earth; and therefore, by Rule 2, must be owing to the same sort of causes; especially since it has been demonstrated, that the forces upon which those revolutions depend tend to the centres of Jupiter, of Saturn, and of the Sun; and that those forces, in receding from Jupiter, from Saturn, and from the Sun, decrease in the same proportion, and according to the same law, as the force of gravity does in receding from the Earth.

COR. I. There is, therefore, a power of gravity tending to all the planets; for, doubtless, Venus, Mercury, and the rest, are bodies of the same sort with Jupiter and Saturn. And since all attraction (by Law III) is mutual, Jupiter will therefore gravitate towards all his own satellites, Saturn towards his, the Earth towards the Moon, and the Sun towards all the primary planets.

COR. II. The force of gravity which tends to any one planet is inversely as the square of the distance of places from that planet's centre.

COR. III. All the planets do gravitate towards one another, by Cor. I and II. And hence it is that Jupiter and Saturn, when near their conjunction, by their

> mutual attractions sensibly disturb each other's motions. So the Sun disturbs the motions of the Moon; and both Sun and Moon disturb our sea, as we shall hereafter explain.

To ensure that he is not misunderstood, Newton adds this further Scholium:

> *Scholium*
>
> The force which retains the celestial bodies in their orbits has been hitherto called centripetal force; but it being now made plain that it can be no other than a gravitating force, we shall hereafter call it gravity. For the cause of that centripetal force which retains the Moon in its orbit will extend itself to all the planets, by Rule 1, 2, and 4.

Newton's repeated insistence on the role, especially, of Rules 1 and 2 in his reasoning is significant.

103. Proposition VI: the confirmation of the equality of the inertial and the gravitational masses by astronomical data

> *Proposition VI. Theorem VI*
>
> *That all bodies gravitate towards every planet; and that the weights of bodies towards any one planet, at equal distances from the centre of the planet, are proportional to the quantities of matter which they severally contain.*

In many ways, this proposition is the centre and the core of Newton's arguments for the universality of his law of gravitation: the universal equality of the inertial and the gravitational masses. It was, as we have seen, his first concern in formulating his Laws; and towards which he carried out the first precision measurements with the pendulum (described in detail in the General Scholium at the end of Section VI of Book II). The theoretical and the conceptual base of the experiment was established in Proposition XXIV of Book II (of which an account is given in Chapter 2).

Proposition VI starts with recalling Proposition XXIV of Book II while giving a brief synopsis of his basic experiments. We shall begin with Newton's synopsis.

> It has been, now for a long time, observed by others, that all sorts of heavy bodies (allowance being made for the inequality of retardation which they suffer from a small power of resistance in the air) descend to the Earth *from equal heights* in equal times; and that equality of times we may distinguish to a great accuracy, by the help of pendulums. I tried experiments with gold, silver, lead,

> glass, sand, common salt, wood, water, and wheat. I provided two wooden boxes, round and equal: I filled the one with wood, and suspended an equal weight of gold (as exactly as I could) in the centre of oscillation of the other. The boxes, hanging by equal threads of 11 feet, made a couple of pendulums perfectly equal in weight and figure, and equally receiving the resistance of the air. And, placing the one by the other, I observed them to play together forwards and backwards, for a long time, with equal vibrations. And therefore the quantity of matter in the gold (by Cors. I and VI, Prop. XXIV, Book II) was to the quantity of matter in the wood as the action of the motive force (or *vis motrix*) upon all the gold to the action of the same upon all the wood; that is, as the weight of the one to the weight of the other: and the like happened in the other bodies. By these experiments, in bodies of the same weight, I could manifestly have discovered a difference of matter less than the thousandth part of the whole, had any such been.

After this synopsis, Newton makes the following astonishingly 'dogmatic' assertion:

> But, without all doubt, the nature of gravity towards the planets is the same as towards the Earth.

The reason is clear: his Moon test of Proposition IV was based on the length of the seconds pendulum, determined by experiments very similar to his own that established the equality of the inertial and the gravitational masses. In the reasoned explanation of his assertion which follows, Newton refers to this fact and the wider implications of his Moon test by appealing once again to Rule 2, that 'to the same natural effects, we must, as far as possible, assign the same causes'. And this is how he explains in terms, not unlike those in the Scholium, following Proposition IV.

> For, should we imagine our terrestrial bodies taken to the orbit of the Moon, and therefore, together with the Moon, deprived of all motion, to be let go, so as to fall together towards the Earth, it is certain, from what we have demonstrated before, that, in equal times, they would describe equal spaces with the Moon, and of consequence are to the Moon, in quantity of matter, as their weights to its weight. Moreover, since the satellites of Jupiter perform their revolutions in times which observe the $\frac{3}{2}$th power of the proportion of their distances from Jupiter's centre, their accelerative gravities towards Jupiter will be inversely as the squares of their distances from Jupiter's centre; that is, equal, at equal distances. And, therefore, these satellites, if supposed to fall *towards Jupiter* from equal heights, would describe equal spaces in equal times, in like manner as heavy bodies do on our Earth. And, by the same argument, if the circumsolar planets were supposed to be let fall at equal distances from the Sun, they would, in their descent towards the Sun, describe equal spaces in equal

> times. But forces which equally accelerate unequal bodies must be as those bodies: that is to say, the weights of the planets *towards the Sun* must be as their quantities of matter. Further, that the weights of Jupiter and of his satellites towards the Sun are proportional to the several quantities of their matter, appears from the exceedingly regular motions of the satellites (by Cor. III, Prop. LXV, Book I).

No matter how convincing these arguments may be, Newton makes some concrete observations which bear on the same question.

> For if some of those bodies were more strongly attracted to the Sun in proportion to their quantity of matter than others, the motions of the satellites would be disturbed by that inequality of attraction (by Cor. II, Prop. LXV, Book I). If, at equal distances from the Sun, any satellite, in proportion to the quantity of its matter, did gravitate towards the Sun with a force greater than Jupiter in proportion to his, according to any given proportion, suppose of d to e; then the distance between the centres of the Sun and of the satellite's orbit would be always greater than the distance between the centres of the Sun and of Jupiter, nearly as the square root of that proportion: as by some computations I have found.

But what is the nature of these 'computations' by which Newton established the result (underlined) that he states? Since we know from the 22 corollaries of Proposition LXVI, and Newton's insight into and familiarity with what one may expect for the variation of the elements of a Kepler orbit by a given external perturbation, it is not hard to guess what those computations must have been.

We start with equations (1) of Chapter 13, §66. It is clear that the masses $m_k (k = 1, 2, 3)$ which occur on the left-hand sides of these equations are the inertial masses while those on the right-hand sides are the gravitational masses. Since we now wish to entertain that the inertial and the gravitational masses in the astronomical contexts may differ, we shall distinguish them by $m_k^{(i)}$ and $m_k^{(g)}$ respectively. And we shall suppose that

$$m_k^{(i)} = m_k^{(g)}/\delta_k, \tag{1}$$

where $\delta_k (k = 1, 2, 3)$ are some assigned constants, presumably, close to unity. Equations (1) of Chapter 13, §66, are now replaced by

$$\left.\begin{aligned}
\frac{d^2\vec{r}_1}{dt^2} &= \frac{Gm_2\delta_1}{r_{12}^3}(\vec{r}_2 - \vec{r}_1) + \frac{Gm_3\delta_1}{r_{13}^3}(\vec{r}_3 - \vec{r}_1),\\
\frac{d^2\vec{r}_2}{dt^2} &= \frac{Gm_1\delta_2}{r_{21}^3}(\vec{r}_1 - \vec{r}_2) + \frac{Gm_3\delta_2}{r_{23}^3}(\vec{r}_3 - \vec{r}_2),\\
\frac{d^2\vec{r}_3}{dt^2} &= \frac{Gm_1\delta_3}{r_{31}^3}(\vec{r}_1 - \vec{r}_3) + \frac{Gm_2\delta_3}{r_{32}^3}(\vec{r}_2 - \vec{r}_3),
\end{aligned}\right\} \tag{2}$$

where all the masses on the right-hand side are now gravitational masses; and we have dispensed with the distinguishing superscripts (g).

Combining equation (2) in the same manner as equations (1) of Chapter 13, §66, we obtain, in place of equation (5) of §66,

$$\frac{d^2\vec{r}}{dt^2} = -\mu\frac{\vec{r}}{r^3} + Gm_3\left(\delta_2\frac{\vec{R}-\vec{r}}{\rho^3} - \delta_1\frac{\vec{R}}{R^3}\right), \tag{3}$$

where

$$\mu = G(m_1^{(g)}\delta_2 + m_2^{(g)}\delta_1). \tag{4}$$

The disturbing function is accordingly,

$$\vec{F} = GM_\odot\left[-\delta_2\frac{\vec{r}}{\rho^3} + \vec{R}\left(\frac{\delta_2}{\rho^3} - \frac{\delta_1}{R^3}\right)\right], \tag{5}$$

where, as in Chapter 13, we have replaced m_3 by the gravitational mass of the sun, $M_\odot$.

Again, by the same approximative scheme as in Chapter 13, §66(b), we find that equations (12)–(14) of that section are replaced by:

$$\left.\begin{aligned} F_r &= GM_\odot\left[-\delta_2\frac{r}{\rho^3} + R\cos(v-U)\left(\frac{\delta_2}{\rho^3} - \frac{\delta_1}{R^3}\right)\right],\\ F_\alpha &= GM_\odot\left[\qquad - R\sin(v-U)\left(\frac{\delta_2}{\rho^3} - \frac{\delta_1}{R^3}\right)\right],\\ F_h &= GM_\odot\left[\qquad - R\sin U\sin\iota\left(\frac{\delta_2}{\rho^3} - \frac{\delta_1}{R^3}\right)\right]; \end{aligned}\right\} \tag{6}$$

and further,

$$\rho = R - r\cos\psi + O(r^2/R), \tag{7}$$

and

$$\frac{\delta_2}{\rho^3} - \frac{\delta_1}{R^3} = 3r\frac{\delta_2}{R^4}\cos\psi + \frac{\delta_2-\delta_1}{R^3} + O(r^2/R^5). \tag{8}$$

Inserting equation (8) in equations (6), we now obtain, in place of equations (23)–(25) of §66(b),

$$\begin{aligned} F_r &= \frac{1}{2}\frac{GM_\odot}{R^3}[\delta_2 r(1+3\cos 2\psi) + 2(\delta_2-\delta_1)R\cos\psi],\\ F_\alpha &= \frac{1}{2}\frac{GM_\odot}{R^3}[-3r\delta_2\sin 2(v-U) - 2(\delta_2-\delta_1)R\sin(v-U)],\\ F_h &= \frac{GM_\odot}{R^3}[-3r\delta_2\sin U\sin\iota\cos\psi - (\delta_2-\delta_1)R\sin U\sin\iota]. \end{aligned} \tag{9}$$

We observe that the terms in $\delta_2 - \delta_1$ are one order *lower* in r/R than the standard tidal terms. We can, accordingly, set, already, a stringent upper limit on $\delta_2 - \delta_1$ if these terms are not to swamp the well-confirmed tidal terms: we must, in fact, require

$$\frac{R}{r}(\delta_2 - \delta_1) \ll 1. \tag{10}$$

For the Earth–Moon and Jupiter–Callisto systems:

$$\left.\begin{aligned} &\frac{\text{Earth–Sun distance}}{\text{Earth–Moon distance}} \simeq 389, \\ \text{and} \quad & \\ &\delta_2 - \delta_1 \ll 2.57 \times 10^{-3}; \end{aligned}\right\} \tag{10a}$$

$$\left.\begin{aligned} &\frac{\text{Jupiter–Sun distance}}{\text{Jupiter–Callisto distance}} \simeq 383, \\ \text{and} \quad & \\ &\delta_2 - \delta_1 \ll 2.61 \times 10^{-3}. \end{aligned}\right\} \tag{10b}$$

We can use equations (9) to obtain the variation of all the elements of the Kepler orbit as in Chapters 13 and 14. We shall however restrict ourselves to the case considered by Newton: the variation of the major axis a.

By equation (vii) Chapter 13, §66(c) and ignoring the factor $\sqrt{(1 - e^2)}$, we have,

$$\frac{\mathrm{d}\bar{a}}{\mathrm{d}t} = \frac{2\bar{a}}{rn} F_\alpha. \tag{11*}$$

To determine the effect of the term in $\delta_2 - \delta_1$ we consider the contribution by

$$f_\alpha = -\frac{GM_\odot}{R^2}(\delta_2 - \delta_1)\sin(v - U) \tag{12}$$

to F_α. Combining equations (11) and (12) we have

$$\frac{\mathrm{d}\bar{a}}{\mathrm{d}t} = -\frac{2\bar{a}_0}{rn} N^2 R(\delta_2 - \delta_1)\sin(v - U) \qquad (N^2 = GM_\odot/R^3), \tag{13}$$

or, since $m^2 = N^2/n^2$, $n\,\mathrm{d}t = \mathrm{d}v$, and

$$v - U = (1 - m)v \qquad \text{(by equation (18) of Chapter 14, §70}(b)\text{))} \tag{14}$$

we have

$$\mathrm{d}\bar{a} = -2\bar{a}_0 m^2 \frac{R}{r}(\delta_2 - \delta_1)\sin[(1 - m)v]\,\mathrm{d}v; \tag{15}$$

where $\bar{a}_0$ represents a mean value of $\bar{a}$. Integrating equation (15), we obtain:

$$\bar{a} - \bar{a}_0 = 2\bar{a}_0 \frac{m^2}{1 - m}\frac{R}{r}(\delta_2 - \delta_1)\cos\psi \tag{16}$$

* In equations (11)–(16) we are letting $\bar{a}$ and $\bar{a}_0$ denote the orbital radii when $\delta_2 \neq \delta_1$. Also, we are ignoring the factor $\sqrt{(1 - e^2)}$.

which should be contrasted with equation (47) of Chapter 14, §70(e):

$$a - a_0 = \tfrac{3}{2}a_0 \frac{m^2}{1-m} \cos 2\psi \qquad \text{(tidal effects only).} \tag{17*}$$

From (16) and (17) it follows that if the $\delta_2 - \delta_1$ term is not to swamp the tidal term, the inequality (10) must obtain.

To compare with Newton's findings, we must rewrite equation (16) by using the relation

$$m^2 = \frac{N^2}{n^2} = \frac{GM_\odot}{R^3} \frac{r^3}{G(m_1^{(i)} + m_2^{(i)})} = \frac{M_\odot}{m_1^{(i)} + m_2^{(i)}} \frac{r^3}{R^3}. \tag{18}$$

We find by ignoring the factor $(1-m)^{-1}$ in equation (16),

$$\bar{a} - \bar{a}_0 = 2\bar{a}_0 \frac{M_\odot}{m_1^{(i)} + m_2^{(i)}} \left(\frac{r}{R}\right)^2 (\delta_2 - \delta_1) \cos \psi. \tag{19}$$

In the last equation, we do not clearly need to distinguish between the inertial and the gravitational mass; and we can write

$$\bar{a} - \bar{a}_0 = 2\bar{a}_0 \frac{M_\odot}{m_1 + m_2} \left(\frac{r}{R}\right)^2 (\delta_2 - \delta_1) \cos \psi. \tag{20}$$

The corresponding form of equation (17) is

$$a - a_0 = \tfrac{3}{2}a_0 \frac{M_\odot}{m_1 + m_2} \left(\frac{r}{R}\right)^3 \cos 2\psi. \tag{21}$$

Newton's reference to 'as by some computations I have found' is precisely to the occurrence of the combination $(r/R)^2(\delta_2 - \delta_1)$ in equation (20).

The question that Newton poses is: What is the change in the satellite–Sun distance at conjunction that will compensate for the contribution arising from equation (20) with a non-vanishing, yet small, $(\delta_2 - \delta_1)$. The answer to this question follows readily from equations (20) and (21).

First, consider the change in $a - a_0$ resulting from a small change Δr in r at $\psi = 0$. By equation (21) the required change is given by

$$\Delta(a - a_0) = \tfrac{9}{2}a_0 \frac{M_\odot}{m_1 + m_2} \frac{r^2}{R^3} \Delta r; \tag{22}$$

while the change given by equation (20) is

$$\bar{a} - \bar{a}_0 = 2a_0 \frac{M_\odot}{m_1 + m_2} \left(\frac{r}{R}\right)^2 (\delta_2 - \delta_1). \tag{23}$$

* It is known that this equation is incorrect even qualitatively (see Chapter 14, §70, p. 246). However, since equations (16) and (17) are derived on the same premises, we shall continue to use equation (17) along with equation (16)—as Newton doubtless must have .

The net change resulting from the two causes is therefore,

$$\Delta(a - a_0) + \bar{a} - \bar{a}_0 = a_0 \frac{M_\odot}{m_1 + m_2}\left(\frac{r}{R}\right)^2\left[\frac{9}{2}\frac{\Delta r}{R} + 2(\delta_2 - \delta_1)\right], \tag{24}$$

where on the right-hand side we have ignored the notational difference in a_0 and $\bar{a}_0$. The condition for the net change (24) to vanish is therefore,

$$\delta_2 - \delta_1 + \frac{9}{4}\frac{\Delta r}{R} = 0. \tag{25}$$

On the other hand at the conjunction $\psi = 0$,

$$(\text{satellite–Sun distance})_{\psi=0} = \begin{aligned} &\text{planet–Sun distance } (=R) \\ &-\text{planet–satellite distance } (=r) \end{aligned} \tag{26}$$

Therefore

$$\Delta\ (\text{satellite–Sun distance}) = -\Delta r. \tag{27}$$

Hence, the solution to Newton's problem is:

$$\delta_2 - \delta_1 = 2.25\frac{\Delta\ (\text{satellite–Sun distance})_{\psi=0}}{\text{planet–Sun distance}}, \tag{28}$$

which differs from Newton's solution by having the factor 2.25 in place of 2—a difference that is of no significance in this connection. In particular if

$$\delta_2 - \delta_1 = \tfrac{1}{1000}, \qquad \text{then} \qquad \Delta\frac{|\text{satellite–Sun distance}|}{\text{planet–Sun distance}} = \frac{1}{2250}. \tag{29}$$

For the Jupiter–Callisto system, since

$$\text{Jupiter–Callisto distance} = 1883 \times 10^3 \text{ km}, \tag{30}$$

$$\tfrac{1}{2250}\ \text{Jupiter–Sun distance} = 349 \times 10^3 \text{ km}.$$

$$\simeq \tfrac{1}{6}\ \text{Jupiter–Callisto distance}. \tag{31}$$

Equations (25)–(27) are exactly the contents of Newton's statement:

> Therefore if, at equal distances from the Sun, the accelerative gravity of any satellite towards the Sun were greater or less than the accelerative gravity of Jupiter towards the Sun but by one $\frac{1}{1000}$ part of the whole gravity, the distance of the centre of the satellite's orbit from the Sun would be greater or less than the distance of Jupiter from the Sun by one $\frac{1}{2000}$ [$\frac{1}{2250}$] part of the whole distance; that is, by a fifth [sixth] part of the distance of the utmost satellite from the centre of Jupiter; an eccentricity of the orbit which would be very sensible. But the orbits of the satellites are concentric to Jupiter, and therefore the accelerative gravities of Jupiter, and of all its satellites towards the Sun, are equal among themselves.

and he continues:

> And by the same argument, the weights of Saturn and of his satellites towards the Sun, at equal distances from the Sun, are as their several quantities of matter; and the weights of the Moon and of the Earth towards the Sun are either none, or accurately proportional to the masses of matter which they contain. But some weight they have, by Cor. I and III, Prop. V. (!)

Notice the confidence (or, should one say, the audacity?) of the underlined remark.

The following further amplifications are worth quoting *in extenso.*

> But further; the weights of all the parts of every planet towards any other planet are one to another as the matter in the several parts; for if some parts did gravitate more, others less, than for the quantity of their matter, then the whole planet, according to the sort of parts with which it most abounds, would gravitate more or less than in proportion to the quantity of matter in the whole. Nor is it of any moment whether these parts are external or internal; for if, for example, we should imagine the terrestrial bodies with us to be raised to the orbit of the Moon, to be there compared with its body; if the weights of such bodies were to the weights of the external parts of the Moon as the quantities of matter in the one and in the other respectively, but to the weights of the internal parts in a greater or less proportion, then likewise the weights of those bodies would be to the weight of the whole Moon in a greater or less proportion; against what we have shown above.
>
> COR. I. Hence the weights of bodies do not depend upon their forms and textures; for if the weights could be altered with the forms, they would be greater or less, according to the variety of forms, in equal matter; altogether against experience.
>
> COR. II. Universally, all bodies about the Earth gravitate towards the Earth; and the weights of all, at equal distances from the Earth's centre, are as the quantities of matter which they severally contain. This is the quality of all bodies within the reach of our experiments; and therefore (by Rule 3) to be affirmed of all bodies whatsoever.

There can be no doubt that Newton held the *accurate proportionality of the weight* '*to the masses of matter which they contain*' as inviolable.

The rest of Corollary II deals with the 'philosophical' issue:

> If the ether, or any other body, were either altogether void of gravity, or were to gravitate less in proportion to its quantity of matter, ...

Newton denies this possibility as 'contrary to what was proved in the preceding corollary'; and in similar veins, in Corollaries III and IV:

> COR. III. All spaces are not equally full.
>
> OR. IV. If all the solid particles of all bodies are of the same density, and cannot be rarefied without pores, then a void, space, or vacuum must be granted. By bodies of the same density, I mean those whose inertias are in the proportion of their bulks.

In Corollary V, Newton makes a perceptive comparison between gravitational and magnetic forces:

> COR. V. The power of gravity is of a different nature from the power of magnetism; for the magnetic attraction is not as the matter attracted. Some bodies are attracted more by the magnet; others less; most bodies not at all. The power of magnetism in one and the same body may be increased and diminished; and is sometimes far stronger, for the quantity of matter, than the power of gravity; and in receding from the magnet decreases not as the square but almost as the cube of the distance, as nearly as I could judge from some rude observations.

It is difficult to understand Newton's motives in inserting these corollaries at this point, unless it be as defence against those who, like Huygens and Leibniz, held the notion of action at a distance as a heresy.

A final observation needs to be made. Newton's 'secretive' treatment of the effect of an inequality in the inertial and the gravitational masses on the motions in the solar system has been re-invented (albeit in a different context) in recent times and goes under the name of the 'Nordvedt effect' (!)

104. Proposition VII: the universal law of gravitation

We have now reached the climactic point of *Philosphie naturalis Principia Mathematica*. After establishing the preceding propositions, particularly, Propositions IV and VI, Newton, at long last, is ready to enunciate his law of gravitation. Let him speak.

> *Proposition VII. Theorem VII*
>
> *That there is a power of gravity pertaining to all bodies, proportional to the several quantities of matter which they contain.*
>
> That all the planets gravitate one towards another, we have proved before; as well as that the force of gravity towards every one of them, considered apart, is inversely as the square of the distance of places from the centre of the planet. And thence (by Prop. LXIX, Book I, and its Corollaries) it follows, that the gravity tending towards all the planets is proportional to the matter which they contain.

Moreover, since all the parts of any planet A gravitate towards any other planet B; and the gravity of every part is to the gravity of the whole as the matter of the part to the matter of the whole; and (by Law III) to every action corresponds an equal reaction; therefore the planet B will, on the other hand, gravitate towards all the parts of the planet A; and its gravity towards any one part will be to the gravity towards the whole as the matter of the part to the matter of the whole. Q.E.D.

This is the only proposition of the 14 in this section that Newton deigns to end with a 'Q.E.D.'—*Quad erat demonstrandum*!

Even at this high noon, Newton finds it necessary to add this defence:

COR. I. Therefore the force of gravity towards any whole planet arises from, and is compounded of, the forces of gravity towards all its parts. Magnetic and electric attractions afford us examples of this; for all attraction towards the whole arises from the attractions towards the several parts. The thing may be easily understood in gravity, if we consider a greater planet, as formed of a number of lesser planets, meeting together in one globe; for *hence it would appear* that the force of the whole must arise from the forces of the component parts. If it is objected, that, according to this law, all bodies with us must gravitate one towards another, whereas no such gravitation anywhere appears, I answer, that since the gravitation towards these bodies is to the gravitation towards the whole Earth as these bodies are to the whole Earth, the gravitation towards them must be far less than to fall under the observation of our senses.

And to avoid any possible misunderstanding, he adds this further amplification:

COR. II. The force of gravity towards the several equal particles of any body is inversely as the square of the distance of places from the particles; as appears from Cor. III, Prop. LXXIV, Book I.

105. Proposition VIII and IX: the implications of the 'superb theorems'

Proposition VIII. Theorem VIII

In two spheres gravitating each towards the other, if the matter in places on all sides round about and equidistant from the centres is similar, the weight of either sphere towards the other will be inversely as the square of the distance between their centres.

Newton's elaboration of the statement of this proposition is no more than a recollection of the role of the discovery of Propositions LXXV and LXXVI in the evolution of his thoughts. We have already quoted the text in Chapter I, §12; but it is worth quoting again in the context.

> After I had found that the force of gravity towards a whole planet did arise from and was compounded of the forces of gravity towards all its parts, and towards every one part was in the inverse proportion of the squares of the distances from the part, I was yet in doubt whether that proportion inversely as the square of the distance did accurately hold, or but nearly so, in the total force compounded of so many partial ones; for it might be that the proportion which accurately enough took place in greater distances should be wide of the truth near the surface of the planet, where the distances of the particles are unequal, and their situation dissimilar. But by the help of Prop. LXXV and LXXVI, Book I, and their Corollaries, I was at last satisfied of the truth of the Proposition, as it now lies before us.

It will be noticed that the foregoing statement *does not even* include the explicit expression of the relevance of the propositions that Corollary III of Proposition LXXV and Corollary VIII of Proposition LXXVI do:

> COR. III. Those several truths demonstrated above concerning the motion of bodies about the focus of the conic sections will take place when an attracting sphere is placed in the focus, and the bodies move without the sphere.
>
> COR. VIII. All those truths above demonstrated, relating to the motions of bodies about the foci of conic sections, will take place when an attracting sphere, of any form and condition like that above described, is placed in the focus.

Besides, it cannot be an accident that Newton refers to the 'superb theorems' only after his definitive enunciation of his law of gravitation in Proposition VII (Book III). Was Glaisher's anticipated role of the theorems on Newton's proceeding with the *Principia* in the spring of 1685 an exaggeration? My own view, as I have stated in Chapter 1, §13, is that the proof, especially of Proposition LXXVI (Book I), removed for Newton the major impediment to his acceptance, on its face value, of the positive result of his 1666 Moon test. But his revised test of Proposition IV, appealing directly to the length of the seconds pendulum, does not require any such justification. It is not surprising, therefore, that the superb theorems had lost their priority.

Corollaries I–IV

In Corollaries I, II, and III which follow, Newton breaks new ground. He shows how from the orbital elements of the planets and the satellites we can derive information on their physical elements. Newton idealizes the problem, principally to illustrate the principles involved.

Newton's assumptions are that the orbits described by the planets and the satellites are circular and further that they are objects with spherically symmetric density distributions so that Propositions LXXV and LXXVI and their corollaries (which we have quoted) are applicable. By Corollary II of Proposition IV, or, more generally by equation (27) of Chapter 13, §66,

$$\frac{2\pi}{T} = \left(\frac{\mu}{a^3}\right)^{1/2},$$

where $\mu = GM$, a is the semimajor axis, and T is the period. When the orbit is assumed to be circular, we shall let a be the mean radius of the orbit. The basic equation is

$$GM = 4\pi^2 \frac{a^3}{T^2} \qquad \text{or} \qquad M \propto a^3 T^{-2}. \tag{1}$$

And the same proportionality applies also to the weight. Next, on the assumption that the central object (be it the Sun or the planets) is spherically symmetric and of radius R, an alternative form of equation (1) is

$$\tfrac{4}{3}\pi G\bar{\rho}R^3 = 4\pi^2 a^3 T^{-2}, \tag{2}$$

where $\bar{\rho}$ is the mean density of the object. We rewrite equation (2) in the form

$$\bar{\rho} = \frac{3\pi}{G}\left(\frac{a}{R}\right)^3 \frac{1}{T^2} \qquad \text{or} \qquad \bar{\rho} \propto \left(\frac{a}{R}\right)^3 T^{-2}. \tag{3}$$

Also, under the same assumptions, the value of gravity at any distance L from the centre is

$$g_L = \frac{GM}{L^2} = \frac{4\pi^2}{L^2}\frac{a^3}{T^2}. \tag{4}$$

In particular, the value of gravity at the surface of the object is

$$g_R = \frac{GM}{R^2} = \tfrac{4}{3}\pi G\bar{\rho}R \qquad \text{or} \qquad g_R \propto R\bar{\rho}. \tag{5}$$

The equations which Newton derives in the corollaries are precisely equations (1), (3), (4), and (5). He applies these equations to the Sun–Venus, Jupiter–Callisto (or the Huygenian satellite), Saturn–Titan, and the Earth–Moon systems with the astronomical data that were available to him; and he revised his calculations, as the latter 'improved', in the successive editions of the *Principia*. And some arithmetical errors have crept into his calculations whose origins have recently been clarified.* So as not to obscure the physical and the mathematical contents of these corollaries, we shall not consider them. Instead, we shall take the astronomical data as currently available; use Newton's formulae (1), (3), and (4) to evaluate the same quantities and compare them with the currently accepted values; and lastly, compare them with values that Newton gives in the last edition of the

* Robert Gariso, *Am. J. Phys.*, 1991, **59**, 42. I am indebted to Dr. N. Swerdlow who has acted on this matter as my referee.

Data

		R(km)	a(km)	T(days)
Sun	Sun–Venus	6.96×10^5	1.082×10^8	224.7
Jupiter	Jupiter–Callisto	7.16×10^4	1.883×10^6	16.689
Saturn	Saturn–Titan	6.00×10^4	1.222×10^6	15.945
Earth	Earth–Moon	6.378×10^3	3.844×10^5	27.322

Mass $\propto a^3/T^2$

	Sun	Jupiter	Saturn	Earth
By formula	2.509×10^{19}	2.397×10^{16}	7.177×10^{15}	7.609×10^{13}
Normalized	1	1:1047	1:3496	$1:3.297 \times 10^5$
With current data	1	1:1048	1:3500	$1:3.296 \times 10^5$
Newton	1	1:1067	1:3021	$1:1.693 \times 10^5$

Mean density $\propto (a/R)^3 T^{-2}$

	Sun	Jupiter	Saturn	Earth
By formula	74.41	65.31	33.23	293.3
Normalized	100	87.8	44.7	394
With current data	100	93.1	49.9	391
Newton	100	94.5	67	400

Surface gravity: $g \propto R\bar{\rho}$

	Sun	Jupiter	Saturn	Earth
By formula (normalized)	10^4	903	385	361
With current data	10^4	958	330	358
Newton	10^4	943	529	435

Principia. Except for the mass of the Earth as a fraction of the Sun's mass in the last column of the second table, Newton's values are not outside the margin of the errors one should expect.

As physically the most interesting of the deductions, Newton draws special attention to the relative mean densities of the Sun and the planets. Thus

> The Sun, therefore, is a little denser than Jupiter, and Jupiter than Saturn, and the Earth four times denser than the Sun; for the Sun, by its great heat, is kept in a sort of rarefied state. The Moon is denser than the Earth, as shall appear afterwards.

And in the spirit of his well-known *Queries* in his *Opticks*, he continues in Corollary IV:

> COR. IV. The smaller the planets are, they are, other things being equal, of so much the greater density; for so the powers of gravity on their several surfaces come nearer to equality. They are likewise, other things being equal, of the greater density, as they are nearer to the Sun. So Jupiter is more dense than Saturn, and the Earth than Jupiter; for the planets were to be placed at different distances from the Sun, that, according to their degrees of density, they might enjoy a greater or less proportion of the Sun's heat. Our water, if it were removed as far as the orbit of Saturn, would be turned into ice, and in that of Mercury would quickly fly away in vapour; for the light of the Sun, to which its heat is proportional, is seven times denser in the orb of Mercury than with us: and by the thermometer I have found that a sevenfold heat of our summer Sun will make water boil. Nor are we to doubt that the matter of Mercury is adapted to its heat, and is therefore more dense than the matter of our Earth; since, in a denser matter, the operations of Nature require a stronger heat.

And finally, Proposition IX, which we include in this section, requires no comments.

Proposition IX. Theorem IX

That the force of gravity, considered downwards from the surface of the planets, decreases nearly in the proportion of the distances from the centre of the planets.

If the matter of the planet were of a uniform density, this Proposition would be accurately true (by Prop. LXXIII, Book I). The error, therefore, can be no greater than what may arise from the inequality of the density.

106. Propositions X–XIV

There is a distinct break in style after Proposition IX: no fresh theorems or new principles are established. The propositions that follow are more of the nature of the meditative thoughts of one who has reached the pinnacle.

> Some happy tone
> Of meditation, slipping in between
> The beauty coming and the beauty gone.
>
> Wordsworth

Proposition X. Theorem X

That the motions of the planets in the heavens may subsist an exceedingly long time.

Arguing in terms of his observations, experimental or otherwise:

> In the Scholium of Prop. XL, Book II, I have shown that a globe of water frozen into ice, and moving freely in our air, in the time that it would describe the length of its semidiameter, would lose by the resistance of the air $\frac{1}{4586}$ part of its motion; and the same proportion holds nearly in all globes, however great, and moved with whatever velocity.

and

> It is shown in the Scholium of Prop. XXII, Book II, that at the height of 200 miles above the Earth the air is more rare than it is at the surface of the Earth in the ratio of 30 to 0.0000000000003998, or as 75000000000000 to 1, nearly. And hence the planet Jupiter, revolving in a medium of the same density with that superior air, would not lose by the resistance of the medium the 1000000th part of its motion in 1000000 years.

Newton concludes:

> And therefore, the celestial regions being perfectly void of air and exhalations, the planets and comets meeting no sensible resistance in those spaces will continue their motions through them for an immense tract of time;

an argument that one readily concedes.

Next, Newton states a hypothesis, not as a tenet of his own but rather as one to see 'what may from hence follow'.

> *Hypothesis I*
>
> *That the centre of the system of the world is immovable.*
>
> This is acknowledged by all, while some contend that the Earth, others that the Sun, is fixed in that centre. Let us see what may from hence follow.

> *Proposition XI. Theorem XI*
>
> *That the common centre of gravity of the Earth, the Sun, and all the planets, is immovable.*
>
> For (by Cor. IV of the Laws) that centre either is at rest, or moves uniformly forwards in a right line; but if that centre moved, the centre of the world would move also, against the Hypothesis.

One could equally well have concluded that the hypothesis is 'against' Corollary IV of the Laws (and following Newton's example: no Q.E.D.)! But there is more serious import to the hypothesis and to Corollary II of the Laws.

If we agree to formulate the first Law in the manner of Maxwell, namely,

> The centre of mass of the system preserves in its state of rest, or of uniform motion in a straight line, except in so far as it is made to change that state by forces acting on the system from without,

and we choose not to recognize (for convenience or otherwise) any system 'from the outside' then the uniform motion in a straight line of the system we are considering becomes a matter of mere contemplation; and we may go even so far as to say that it has no 'physical content'. Apart from such semantic arguments, the matter reduces to what we wish to include in 'the system of the world' *and* pragmatically the most useful frame of reference in which to describe its changing internal state.

If the 'system of the world' (as Newton once stated) consists of the 'whole space of planetary phenomenon' then the frame in which the centre of gravity of the system is at rest is clearly the most useful as Newton argues convincingly in Proposition LXVII (in the context of the three-body problem) and in the corollary of Proposition LXVIII (in the context of many bodies like Jupiter and its satellites). We adopt the same principle in considering larger systems: the motions in a star cluster with respect to its centre of gravity; the motions in a galaxy with respect to its centre; and *ad infinitum*. It is not a question of 'reality': it is rather a question of excluding (for convenience, principally) factors that have no influence on the system we are considering. This interpretation of the hypothesis and Proposition XI are confirmed by Proposition XII which follows. (It may be noted parenthetically, that what Newton had listed as 'Hypotheses' in the first edition of the *Principia* were changed to 'Phenomena' in the second edition (ignoring the order and the content).) It would appear to the writer that the retention of *Hypothesis* in this one instance is probably a lapse. And there are lapses in the revision of Book III as we have just described in Proposition VIII.

Proposition XII. Theorem XII

That the Sun is agitated by a continual motion, but never recedes far from the common centre of gravity of all the planets.

In this proposition, Newton examines the extent to which the centre of the Sun oscillates. Considering separately the Sun–Jupiter and the Sun–Saturn systems, we find:

(Centre of Sun–Jupiter) distance:	778.3×10^6 km.
$M_\odot$/mass of Jupiter:	1046
(Centre of Sun–centre of gravity of Sun and Jupiter) distance:	7.44×10^5 km.
(Centre of Sun–Saturn) distance:	1427×10^6 km.
$M_\odot$/mass of Saturn:	3503
(Centre of Sun–centre of gravity of Sun and Saturn) distance:	4.07×10^5 km.
Radius of Sun:	6.96×10^5 km.

And as Newton concludes:

> the common centre of gravity of Jupiter and the Sun will fall upon a point a little without the surface of the Sun.

and

> the common centre of gravity of Saturn and the Sun will fall upon a point a little within the surface of the Sun.

In the same vein he continues:

> And, pursuing the principles of this computation, we should find that though the Earth and all the planets were placed one one side of the Sun, the distance of the common centre of gravity of all from the centre of the Sun would scarcely amount to one diameter of the Sun. In other cases, the distances of those centres are always less; and therefore, since that centre of gravity is continually at rest, the Sun, according to the various positions of the planets, must continually be moved every way, but will never recede far from that centre.

In the corollary, Newton returns to the theme of Proposition XI.

> Cor. Hence the common centre of gravity of the Earth, the Sun, and all the planets, is to be esteemed the centre of the world; for since the Earth, the Sun, and all the planets gravitate one towards another, and are, therefore, according to their powers of gravity, in continual agitation, as the Laws of Motion require, it is plain that their movable centres cannot be taken for the immovable centre of the world. If that body were to be placed in the centre, towards which other bodies gravitate most (according to common opinion), that privilege ought to be allowed to the Sun; but since the Sun itself is moved, a fixed point is to be chosen from which the centre of the Sun recedes least, and from which it would recede yet less if the body of the Sun were denser and greater, and therefore less apt to be moved.

From the statements underlined, it would appear to this writer that Newton adopts, with restraint, the pragmatic point of view of the concluding remarks following Proposition XI.

Proposition XIII. Theorem XIII

> *The planets move in ellipses which have their common focus in the centre of the Sun; and, by radii drawn to that centre, they describe areas proportional to the times of description.*

At long last, Newton returns to the theorem that started it all. This is what he says with some impatience:

We have discoursed above on these motions from the Phenomena. Now that we know the principles on which they depend, from those principles we deduce the motions of the heavens *a priori*. Because the weights of the planets towards the Sun are inversely as the squares of their distances from the Sun's centre, if the Sun were at rest, and the other planets did not act one upon another, their orbits would be ellipses, having the Sun in their common focus; and they would describe areas proportional to the times *of description*, by Prop. I and XI, and Cor. I, Prop. XIII, Book I. But the actions of the planets one upon another are so very small, that they may be neglected; and by Prop. LXVI, Book I, they disturb the motions of the planets around the Sun in motion, less than if those motions were performed about the Sun at rest.

I thought so once; but now I know it.

John Gay (1685–1732)

Finally, in Proposition XIV we have a brilliant summation:

Proposition XIV. Theorem XIV

The aphelions and nodes of the orbits of the planets are fixed.

The aphelions are immovable by Prop. XI, Book I; and so are the planes of the orbits, by Prop. I of the same Book. And if the planes are fixed, the nodes must be so too. It is true, that some inequalities may arise from the mutual actions of the planets and comets in their revolutions; but these will be so small, that they may be here passed by.

COR. I. The fixed stars are immovable, seeing they keep the same position to the aphelions and nodes of the planets.

COR. II. And since these stars are liable to no sensible parallax from the annual motion of the Earth, they can have no force, because of their immense distance, to produce any sensible effect in our system. Not to mention that the fixed stars, everywhere promiscuously dispersed in the heavens, by their contrary attractions destroy their mutual actions, by Prop. LXX, Book I.

And it will not be proper to omit the concluding Scholium.

Scholium

Since the planets near the Sun (viz., Mercury, Venus, the Earth, and Mars) are so small that they can act with but little force upon one another, therefore their aphelions and nodes must be fixed, except so far as they are disturbed by the actions of Jupiter and Saturn, and other higher bodies. And hence we may find, by the theory of gravity, that their aphelions move forwards a little, in respect of the fixed stars, and that as the $\frac{3}{2}$th power of their several distances from the Sun. So that if the aphelion of Mars, in the space of a hundred years, is carried forwards 33′ 20″, in respect of the fixed stars, the aphelions of the Earth, of Venus, and of Mercury, will in a hundred years be carried forwards 17′ 40″, 10′ 53″, and 4′ 16″, respectively. But these motions are so inconsiderable, that we have neglected them in this Proposition.

❖20❖

The figure of the Earth and of the planets

This admirable work [Newton's *Principia*] contains the seeds of all the great discoveries that have since been made about the system of the world.

De Laplace (1823)

107. Introduction

For Newton the central part of his law of gravitation is:

gravity towards any one part [of a body] will be to the gravity towards the whole as the matter of the part to the matter of the whole.

(Proposition VII, Book III)

and

in consequence of this rule, universally allow that all bodies whatsoever are endowed with a principle of mutual gravitation.

(Rule III)

Newton's consideration of the figure of the Earth and of the theory of the tides are towards this end: it is the gravitational attraction of the protuberant part of the Earth, caused by its own rotation or by the tidal action of the Sun and the Moon, that are the direct causes. And Newton's absorption in the lunar theory derives in part from the same concern. The very nature of these problems requires a careful examination of the relevant data. Newton is conscientious in these matters. But in our account, we shall pass them by and restrict ourselves to the theoretical aspects and the basic theorems.

108. Proposition XVIII and the historical background

Proposition XVIII. Theorem XVI

That the axes of the planets are less than the diameters drawn perpendicular to the axes.

The equal gravitation of the parts on all sides would give a spherical figure to the planets, if it was not for their diurnal revolution in a circle. By that

> circular motion it comes to pass that the parts receding from the axis endeavour to ascend about the equator; and therefore if the matter is in a fluid state, by its ascent towards the equator it will enlarge the diameters there, and by its descent towards the poles it will shorten the axis. So the diameter of Jupiter (by the concurring observations of astronomers) is found shorter between pole and pole than from east to west. And, by the same argument, if our Earth was not higher about the equator than at the poles, the seas would subside about the poles, and, rising towards the equator, would lay all things there under water.

Newton starts with the premise that 'equal gravitation of the parts on all sides would give a spherical figure to the planets' (as he has stated already in Corollary XXII, Prop. LXVI, Book I); and he infers that the rotation about the axis must make them oblate. It will, however, be noticed that the empirical evidence he gives is the *observed* ellipticity of Jupiter, and of the Earth only inferentially. The reason for this inverted order was perhaps that when Newton wrote on this matter in 1685 there was, as yet, no consensus as to the figure of the Earth. As I. Todhunter in his *A history of the mathematical theories of attraction and the figure of the Earth*, Constable, London, 1873; reprinted by Dover Publications, New York, 1962, pp. 60 and 100, has written:

> The measurement of the arc of the meridian by the French in Lapland is historically the most important of all such operations. The question as to the oblate or oblong form of the Earth was decisively settled.
>
> Two generations of the best astronomical observers formed in the school of the Cassinis had struggled in vain against the authority and the reasoning of Newton.

and

> The success of the Arctic expedition may be fairly ascribed in great measure to the skill and energy of Maupertuis: and his fame was widely celebrated. The engravings of the period represent him in the costume of a Lapland Hercules, having a fur cap over his eyes; with one hand he holds a club, and with the other he compresses a terrestrial globe (see facing page). Voltaire, then his friend, congratulated him warmly for having 'aplati les pôles et les Cassini'.

Later Maupertuis and Voltaire became involved in controversy and Voltaire wrote:

> Vouz avez confirmé dans les lieux pleins d'ennui
> Ce que Newton connut sans sortir de chez lui.

While we are digressing more than we should, it is difficult to pass by the following description of Maupertuis by Thomas Carlyle (*History of Frederick the Great*, Vol. 3, p. 60 and Vol. 6, p. 474):

Plate 5 A contemporary caricature of Maupertuis, reproduced with the permission of Roger Gaskell Rare Books.

> No reader guesses in our time what a shining celestial body the Maupertuis, who is now fallen so dim again, then was to mankind. In cultivated French society there is no such lion as M. Maupertuis since he returned from flattening the Earth in the Arctic regions. 'The Exact Sciences, what else is there to depend on?' thinks French cultivated society: 'and has not Monsieur done a feat in that line?' Monsieur, with fine ex-military manners, has a certain austere gravity, reticent loftiness, and polite dogmatism, which confirms that opinion. A studious ex-military man—was Captain of Dragoons once, but too fond of study—who is conscious to himself, or who would fain be conscious, that he is, in all points, mathematical, moral, and other, the man. A difficult man to live with in society. Comes really near the limit of what we call genius, of originality, poetic greatness in thinking, but never once can get fairly over said limit, though always struggling dreadfully to do so. Think of it! A fatal kind of man, especially if you have made a lion of him at any time. Of his envies, deep-hidden splenetic discontents and rages, with Voltaire's return for them, there will be enough to say in the ulterior stages. He wears—at least ten years hence he openly wears, though I hope it is not yet so flagrant—'a red wig with yellow bottom (*crinière jaune*);' and as Flattener of the Earth, is, with his own flattish red countenance and impregnable stony eyes, a man formidable to look upon, though intent to be amiable if you do the proper homage.

and after Maupertuis's death:

> Poor Maupertuis; a man of rugged stalwart type; honest; of an ardour, an intelligence,—not to be forgotten for La Beaumelle's pulings over them. A man of good and even of high talent; unlucky in mistaking it for the highest!

109. Proposition XIX: the method of the canals

Since Newton's account of this proposition is more than usually difficult to follow on account of the many details he has withheld—we shall spell them out presently—we shall first give a 'straightforward' account making use of formulae (e.g., Maclaurin's) with which we are familiar. But the ideas and the methods are all Newton's.

(*a*) *Newton's method of the canals*

Let

$$\epsilon = \frac{\text{equatorial radius} - \text{polar radius}}{\text{the mean radius } (R)}, \tag{1}$$

be the ellipticity and its cause measured by

$$m = \frac{\text{centrifugal acceleration at the equator}}{\text{mean gravitational acceleration on the surface}}$$

$$= \frac{\Omega^2 R}{GM/R^2} = \frac{\Omega^2 R^3}{GM}, \tag{2}$$

where Ω denotes the angular velocity of rotation at the equator.

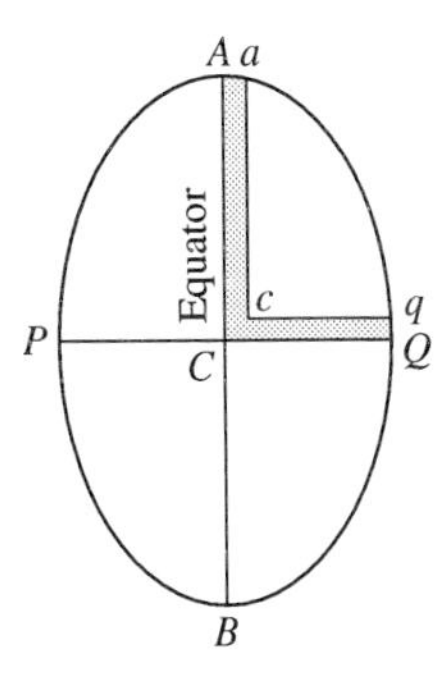

Newton imagines a hole of unit cross-section bored from a point, A, on the equator to the centre of the Earth and a similar hole bored from the pole, Q, to the centre; and he further imagined that the 'canals' so constructed were filled with water. From the fact that the water in the canal will be in equilibrium, Newton concludes that the weights of the equatorial and the polar columns of water must be equal. However, along the equator the acceleration due to gravity is 'diluted' by the centrifugal acceleration; and since both these accelerations, in a homogeneous body, vary from the centre linearly with the distance (by Propositions LXXIII and Corollary III, Proposition XCI of Book I) the 'dilution factor' remains constant and retains its value at the boundary, namely m ($\ll 1$).

If a denotes the equatorial radius, the weight of the equatorial column is given by:

$$\text{weight of equatorial column} = \tfrac{1}{2}\rho a g_{\text{eq}}(1 - m), \tag{3}$$

where g_{eq} is the acceleration due to gravity at the equator and ρ is the density. Similarly, if b denotes the polar radius

$$\text{weight of polar column} = \tfrac{1}{2}\rho b g_{\text{pole}}. \tag{4}$$

And since the two weights must be equal,

$$a g_{\text{eq}}(1 - m) = b g_{\text{pole}}. \tag{5}$$

This is Newton's equation of the canals. As Newton recognizes, the equation is valid for any spheroid, not necessarily one of small eccentricity. If, however, we assume, as is the case in the applications contemplated, that the ellipticity, as defined in equation (1), is small, we may write

$$b = a(1 - \epsilon); \tag{6}$$

and we obtain from equation (5),

$$\frac{g_{\text{pole}}}{g_{\text{eq}}} = \frac{a}{b}(1 - m) = \frac{1 - m}{1 - \epsilon} \simeq (1 + \epsilon - m). \tag{7}$$

It remains to determine the ratio of $g_{\text{pole}}/g_{\text{eq}}$.

From Maclaurin's well-known formula for the gravitational attraction of an oblate spheroid of finite eccentricity, e, we find on suppressing the factor, $G\rho$:

$$g_{\text{pole}} = 4\pi a \frac{\sqrt{(1 - e^2)}}{e^3}[e - (\sin^{-1} e)\sqrt{(1 - e^2)}]. \tag{8}$$

and

$$g_{\text{eq}} = 2\pi a \frac{\sqrt{(1 - e^2)}}{e^3}[\sin^{-1} e - e\sqrt{(1 - e^2)}]. \tag{9}$$

(We may parenthetically note that Newton must have known the first of these equations: it follows from Corollary II, Proposition XCI, Book I, as we shall show presently.) From equations (8) and (9) it follows that, for small eccentricity,

$$g_{\text{pole}} \simeq \tfrac{4}{3}\pi a(1 - \tfrac{1}{10}e^2) \simeq \tfrac{4}{3}\pi a(1 - \tfrac{1}{5}\epsilon), \tag{10}$$

and

$$g_{\text{eq}} \simeq \tfrac{4}{3}\pi a(1 - \tfrac{1}{5}e^2) \simeq \tfrac{4}{3}\pi a(1 - \tfrac{2}{5}\epsilon) \tag{11}$$

since

$$e^2 = 2\epsilon \qquad \text{for small } \epsilon. \tag{12}$$

From equations (7), (10), and (11) it follows

$$\frac{g_{\text{pole}}}{g_{\text{eq}}} = 1 + \tfrac{1}{5}\epsilon = 1 + \epsilon - m. \tag{13}$$

Therefore,

$$\tfrac{4}{5}\epsilon = m \qquad \text{or} \qquad \epsilon = \tfrac{5}{4}m. \tag{14}$$

This is Newton's result.

It is so elegant
So intelligent

T. S. Eliot

(b) What Newton withheld

It is manifest from equation (7) that what is required to obtain the ratio of ϵ to m is a knowledge of $g_{\text{pole}}/g_{\text{eq}}$, to $O(\epsilon)$. It will become abundantly clear that Newton was well aware of the relations

$$g_{\text{eq}} = \tfrac{4}{3}\pi a(1 - \tfrac{2}{5}\epsilon) \qquad \text{and} \qquad g_{\text{pole}} = \tfrac{4}{3}\pi a(1 - \tfrac{1}{5}\epsilon). \tag{15}$$

Newton does refer to Corollary II, Proposition XCI, Book I in which the attraction of an oblate spheroid at an external point along the minor axis is determined. But it is a long way from the conundrum-like fashion in which the attraction is expressed in the corollary to its value at the pole, on the axis, for $\epsilon \ll 1$. (See the remarks following equation (12) in Chapter 9, §49.) Besides, the corollary is of no avail for obtaining the corresponding value of the gravity at the equator. With our present familiarity with Newton's methods we can guess with some surety what his procedure must have been.

(i) $g_{\text{pole}} = \frac{4}{3}\pi a(1 - \frac{1}{5}\epsilon)$:

In Chapter 16, §89, equations (27) and (28), we showed that the integral which resolves Newton's conundrum is:

$$F = 2b - \sqrt{(1 - e^2)} \int_{R-b}^{R+b} \frac{y\,\mathrm{d}y}{\sqrt{[-e^2y^2 + 2Ry - (R^2 - b^2)]}}. \tag{16}$$

At the pole, where $R = b$, equation (16) reduces to

$$F = 2b - \sqrt{(1 - e^2)} \int_{0}^{2b} \frac{y\,\mathrm{d}y}{\sqrt{(-e^2y^2 + 2by)}}. \tag{17}$$

The evaluation of this integral—'child's play' for Newton—yields:

$$F = 2\frac{b}{e^3}[e - (\sin^{-1} e)\sqrt{(1 - e^2)}] = 2a\frac{\sqrt{(1 - e^2)}}{e^3}[e - (\sin^{-1} e)\sqrt{(1 - e^2)}], \tag{18}$$

a formula one normally attributes to Maclaurin. From equation (18), one readily finds:

$$g_{\text{pole}} = \tfrac{4}{3}\pi a(1 - \tfrac{1}{5}\epsilon). \tag{19}$$

(ii) $g_{\text{eq}} = \frac{4}{3}\pi a(1 - \frac{2}{5}\epsilon)$:

Corollary II of Proposition XCI, Book I, does not apply for attraction at an external point along the major axis of an oblate spheroid—the cross-sections normal to this axis are not circular but elliptical; and Proposition XC of Book I is of no avail. Newton circumvents the problem by an igenious argument—a further example of Newton's analytical and physical insights.

Consider a sphere of radius a. By contracting the scale along the y-axis uniformly by a factor $(1 - \epsilon)$, we obtain an oblate spheroid. And by expanding the scale along the x-axis by a factor $(1 + \epsilon)$, we obtain a prolate spheroid. Since the volumes (to order ϵ), diminished in one case and enhanced in the other case, are equal and similarly disposed with respect to the sphere, Newton concludes,

$$g^{(\text{ob})}_{\text{eq at }B} = \tfrac{1}{2}[g^{(\text{prolate})}_{\text{pole at }A} + g^{(\text{sphere})}_{\text{at }B}]; \tag{20}$$

or, as he writes:

$$g^{(\text{ob})}_{\text{eq at }B} = \sqrt{[g^{(\text{prolate})}_{\text{pole at }A} \cdot g^{(\text{sphere})}_{\text{at }B}]}. \tag{21}$$

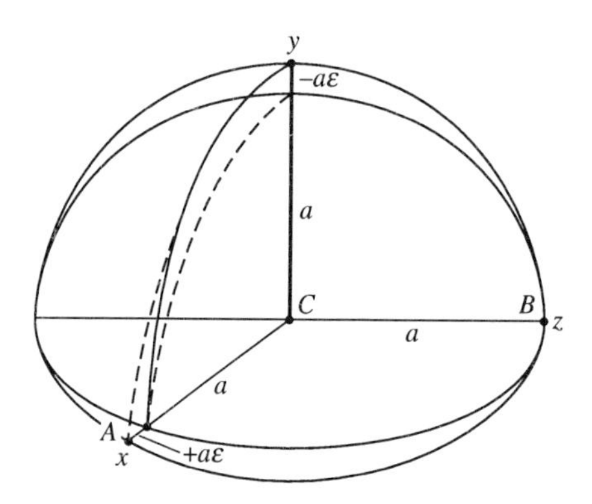

But what about the attraction of a prolate spheroid at an external point along its major axis? Since the cross-sections normal to this axis are circular, Corollary II of Proposition XCI has an obvious adaptation! And we find by the same procedure that led to equation (16) that in its place we now have

$$F = 2a - \int_{-a}^{+a} \frac{a - x}{\sqrt{[(a - x)^2 + z^2]}} \, dx, \tag{22}$$

where

$$\frac{x^2}{a^2} + \frac{z^2}{c^2} = 1 \qquad \text{and} \qquad c^2 = a^2(1 - e^2). \tag{23}$$

Letting

$$y = a - x, \tag{24}$$

we find, successively,

$$\begin{aligned}(a - x)^2 + z^2 &= y^2 + c^2\left(1 - \frac{x^2}{a^2}\right)\\ &= y^2 + c^2 - \frac{c^2}{a^2}(a - y)^2\\ &= y^2 + c^2 - \frac{c^2}{a^2}(a^2 - 2ay + y^2)\\ &= y^2\left(1 - \frac{c^2}{a^2}\right) + 2\frac{c^2}{a}y = e^2y^2 + 2a(1 - e^2)y.\end{aligned} \tag{25}$$

Therefore,

$$F = 2a - \int_0^{2a} \frac{y \, dy}{\sqrt{[2a(1 - e^2)y + e^2y^2]}}. \tag{26}$$

The same 'child's play' now gives

$$F = 2a\frac{1 - e^2}{e^2}\left[\frac{1}{2e}\log\frac{1 + e}{1 - e} - 1\right]; \tag{27}$$

and Newton must have known this formula. For small ϵ,

$$F \simeq 2a\,\frac{1-e^2}{e^2}\left[\tfrac{1}{3}e^2 + \tfrac{1}{5}e^4 + \tfrac{1}{7}e^6\right]$$

$$\simeq \frac{2a}{3}(1-e^2)(1+\tfrac{3}{5}e^2+\cdots) \simeq \frac{2a}{3}(1-\tfrac{2}{5}e^2), \tag{28}$$

or, restoring the factor 2π,

$$g^{(\text{prolate})}_{\text{at } A} = \tfrac{4}{3}\pi a(1-\tfrac{4}{5}\epsilon); \tag{29}$$

and, by equation (21),

$$g^{(\text{ob})}_{\text{eq at } B} = \tfrac{4}{3}\pi a(1-\tfrac{2}{5}\epsilon). \quad \text{(Q.E.D.!)} \tag{30}$$

(c) *Newton's determination of $g^{(\mathrm{ob})}_{\mathrm{pole}}/g^{(\mathrm{ob})}_{\mathrm{eq}}$ and ϵ/m*

The successive steps by which Newton determines $g^{(\text{ob})}_{\text{pole}}/g^{(\text{ob})}_{\text{eq}}$ are listed below.

1. $$g^{(\text{oblate})}_{\text{eq}} = \tfrac{1}{2}\left[g^{(\text{prolate})}_{\text{pole at } A} + g^{(\text{sphere})}_{(1)}\right] = \tfrac{2}{3}\pi a(1-\tfrac{4}{5}\epsilon+1) = \tfrac{4}{3}\pi a(1-\tfrac{2}{5}\epsilon).$$

2. $$\frac{g^{(\text{oblate})}_{\text{pole}}}{g^{(\text{sphere})}_{(1-\epsilon)}} = \frac{1-\frac{1}{5}\epsilon}{1-\epsilon} = 1+\tfrac{4}{5}\epsilon.$$

3. $$\frac{g^{(\text{sphere})}_{(1)}}{g^{(\text{oblate})}_{\text{eq}}} = \frac{1}{1-\frac{2}{5}\epsilon} = 1+\tfrac{2}{5}\epsilon.$$

4. $$\frac{g^{(\text{sphere})}_{(1-\epsilon)}}{g^{(\text{sphere})}_{(1)}} = 1-\epsilon.$$

$2\times 3\times 4$:

$$\frac{g^{(\text{oblate})}_{\text{pole}}}{g^{(\text{oblate})}_{\text{eq}}} = 1+\tfrac{1}{5}\epsilon.$$

Newton could well have omitted this baffling(?) sequence of steps had he not chosen to withhold what he knew; or, was he being solicitous to his readers?

With the foregoing explanations, the following annotated version of Newton's text should be easy to follow though it cannot diminish our wonder.

Therefore if *APBQ* represent the figure of the Earth, now no longer spherical, but generated by the rotation of an ellipse about its lesser axis *PQ*; and *ACQqca* a canal full of water, reaching from the pole *Qq* to the centre *Cc*, and thence rising to the equator *Aa*; the weight of the water in the leg of the canal *ACca* will be to the weight of water in the other leg *QCca* as 289 to 288, because the centrifugal force arising from the circular motion sustains and takes off one of

$$\frac{\text{centrifugal acceleration at equator}}{\text{gravitational acceleration at equator}} = \frac{\Omega^2 a}{GM/a^2} = 1/289 = m$$

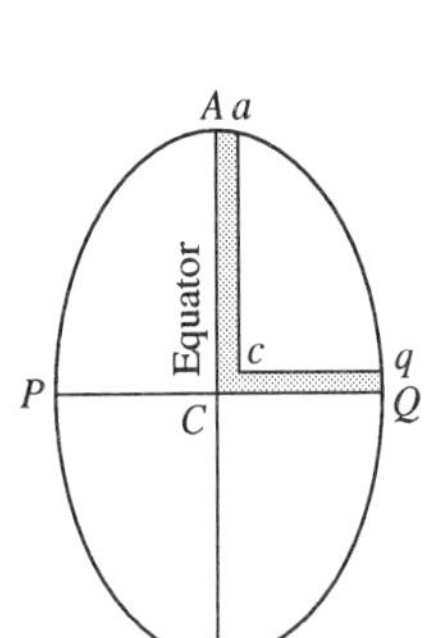

the 289 parts of the weight (in the one leg), and the weight of 288 in the other sustains the rest. But by computation (from COR. II, Prop XCI, Book I) I find, that, if the matter of the earth was all uniform, and without any motion, and its axis *PQ* were to the diameter *AB* as 100 to 101, the force of gravity in the place *Q* towards the Earth would be to the force of gravity in the same place *Q* towards a sphere described about the centre *C* with the radius *PC*, or *QC*, as 126 to 125. And, by the same argument, the force of gravity in the place *A* towards the spheroid generated by the rotation of the ellipse *APBQ* about the axis *AB* is to the force of gravity in the same place *A*, towards the sphere described about the centre *C* with the radius *AC*, as 125 to 126. But the force of gravity in the place *A* towards the Earth is a mean proportional between the forces of gravity towards the spheroid and this sphere; because the sphere, by having its diameter *PQ* diminished in the proportion of 101 to 100, is transformed into the figure of the Earth; and this figure, by having a third diameter perpendicular to the two

$$a:b = 101:100; \qquad \epsilon = 1/100$$

$$\frac{g^{(\text{ob})}_{\text{pole}}}{g^{\text{sphere}}_{(1-\epsilon)}} = \frac{1-\frac{1}{5}\epsilon}{1-\epsilon} = 1 + \tfrac{4}{5}\epsilon : \tfrac{126}{125} \quad (1)$$

$$g^{(\text{ob})}_{\text{equator}} = \left[g^{(\text{prolate})}_{\text{pole}} \cdot g^{\text{sphere}}_{(1)}\right]^{1/2} = \tfrac{4}{3}\pi a(1 - \tfrac{2}{5}\epsilon).$$

diameters AB and PQ diminished in the same proportion, is converted into the said spheroid; and the force of gravity in A, in either case, is diminished nearly in the same proportion. Therefore the force of gravity in A towards the sphere described about the centre C with the radius AC, is to the force of gravity in A towards the Earth as 126 is to $125\frac{1}{2}$. And the force of gravity in the place Q towards the sphere described about the centre C with the radius QC, is to the force of gravity in the place A towards the sphere described about the centre C with the radius AC, in the proportion of the diameters (by Prop. LXXII, Book I), that is, as 100 to 101. If, therefore, we compound those three proportions 126 to 125, 126 to $125^{1/2}$, and 100 to 101, into one, the force of gravity in the place Q towards the Earth will be to the force of gravity in the place A towards the earth as $126 \cdot 126 \cdot 100$ to $125 \cdot 125^{1/2} \cdot 101$; or as 501 to 500.

$$\frac{g_{(1)}^{\text{sphere}}}{g_{\text{equator}}^{(\text{ob})}} = \frac{1}{1 - \frac{2}{5}\epsilon} = 1 + \tfrac{2}{5}\epsilon : \frac{126}{125\frac{1}{2}} \qquad (2)$$

$$\frac{g_{(1-\epsilon)}^{\text{sphere}}}{g_{1}^{\text{sphere}}} = 1 - \epsilon : \frac{100}{101} \qquad (3)$$

$$(1) \times (2) \times (3) = \frac{g_{\text{pole}}^{(\text{ob})}}{g_{\text{equator}}^{(\text{ob})}} = 1 + \tfrac{1}{5}\epsilon : \frac{501}{500}$$

Now since (by COR. III, Prop. XCI, Book I) the force of gravity in either leg of the canal $ACca$, or $QCcq$, is as the distance of the places from the centre of the Earth, if those legs are conceived to be divided by transverse, parallel, and equidistant surfaces, into parts proportional to the wholes, the weights of any number of parts in the one leg $ACca$ will be to the weights of the same number of parts in the other leg as their magnitudes and the accelerative forces of their gravity conjointly, that is, as 101 to 100, and 500 to 501, or as 505 to 501. And therefore if the centrifugal force of every part in the leg $ACca$, arising from the diurnal motion, was to the weight of the same part as 4 to 505, so that from the weight of every part, conceived to be divided

$$\epsilon = 1:100$$
$$a:b = 101:100 = 505:500$$

into 505 parts, the centrifugal force might take off four of those parts, the weights would remain equal in each leg, and therefore the fluid would rest in an equilibrium. But the centrifugal force of every part is to the weight of the same part as 1 to 289; that is, the centrifugal force, which should be $\frac{4}{505}$ parts of the weight, is only $\frac{1}{289}$ part thereof. And, therefore, I say, by the rule of proportion, that if the centrifugal force $\frac{4}{505}$ make the height of the water in the leg *ACca* to exceed the height of the water in the leg *QCcq* by $\frac{1}{100}$ part of its whole height, the centrifugal force $\frac{1}{289}$ will make the excess of the height in the leg *ACca* only $\frac{1}{289}$ part of the height of the water in the other leg *QCcq*; and therefore the diameter of the Earth at the equator is to its diameter from pole to pole as 230 to 229. And since the mean semidiameter of the Earth, according to Picard's mensuration, is 19 615 800 *Paris* feet, or 3923.16 miles (reckoning 5000 feet to a mile), the Earth will be higher at the equator than at the poles by 85 472 feet, or $17^{1/10}$ miles. And its height at the equator will be about 19 658 600 feet, and at the poles 19 573 000 feet.

$$g_{\text{pole}} : g_{\text{eq}} = 501 : 500$$

$$(1 - m)\, 505/500 = 501/500$$

$$m = 4/505 \simeq \tfrac{4}{5}\epsilon$$

$$m = 1/289$$

$$\epsilon = 1/230$$

Additional explanation:

$$\tfrac{1}{2} a g_{\text{eq}}(1 - m) = \tfrac{1}{2} b g_{\text{pole}};$$

therefore,

$$\frac{g_{\text{polar}}}{g_{\text{equator}}} = \frac{a}{b}(1 - m) : \frac{501}{500} = (1 - m)\frac{505}{500}.$$

(*d*) *Application to the figure of Jupiter*

In the remaining part of this proposition, Newton returns to the figure of Jupiter with which he began in Proposition XVIII.

First, he expresses the relation $\epsilon/m = \frac{5}{4}$ in a form in which it can be directly applied to any object of known angular velocity of rotation and mean density $\bar{\rho}$. Since

$$m = \frac{\Omega^2 a}{GM/a^2} = \frac{\Omega^2 a}{\frac{4}{3}G\pi\bar{\rho}a^3/a^2} = \frac{3\Omega^2}{4\pi G\bar{\rho}}, \tag{31}$$

an alternative expression for ϵ is

$$\epsilon = \frac{15\Omega^2}{16\pi G\bar{\rho}}. \tag{32}$$

Accordingly,

$$\epsilon_{\text{Jupiter}} = \epsilon_{\text{Earth}}\left(\frac{\Omega_{\text{Jupiter}}}{\Omega_{\text{Earth}}}\right)^2 \frac{\bar{\rho}_{\text{Earth}}}{\bar{\rho}_{\text{Jupiter}}}. \tag{33}$$

With the current data,

	Rotation period	Mean density (g cm^{-3})
Jupiter	$9^{\text{h}}\ 50^{\text{m}}$	1·314
Earth	$23^{\text{h}}\ 56^{\text{m}}$	5·518

we obtain:

$$\epsilon_{\text{Jupiter}} = \frac{1}{230}\left(\frac{23^{\text{h}}\ 56^{\text{m}}}{9^{\text{h}}\ 50^{\text{m}}}\right)^2 \frac{5{\cdot}518}{1{\cdot}314} \simeq \frac{1}{9{\cdot}25}, \tag{34*}$$

where we have substituted ϵ for the Earth as deduced in §(c). In place of equation (34), Newton gives:

$$\epsilon_{\text{Jupiter}} \simeq \frac{1}{230} \times \left(\frac{24}{10}\right)^2 \simeq \frac{29}{5} \times \frac{400}{94\frac{1}{2}} \simeq \frac{1}{9\frac{1}{3}}, \tag{35}$$

where $400/94\frac{1}{2}$ is the ratio of the mean densities of Jupiter and the Earth as he had derived in Corollary III, Proposition VIII. Newton expresses the result (35) in the form

$$a : b = 10\tfrac{1}{3} : 9\tfrac{1}{3}; \tag{36}$$

and he compares it with the range of values, $12:11\text{–}14\frac{1}{2}:13\frac{1}{2}$, derived from observations. After some comments in the manner of his *Queries*, Newton concludes:

So that the theory agrees with phenomena.

* If we had used for ϵ_{Earth} the currently accepted value 1/289 (derived by Newton!), we should have got $\epsilon_{\text{Jupiter}} \simeq 1/11{\cdot}6$.

110. The variation of gravity over an oblate spheroid

Proposition XX. Problem IV

To find and compare together the weights of bodies in the different regions of our Earth.

Having shown in Proposition XIX that a slowly rotating homogeneous body is a spheroid of small eccentricity, Newton considers in this proposition the variation of gravity over the surface of a spheroid on the 'hypothesis of the [Earth] being a spheroid'.

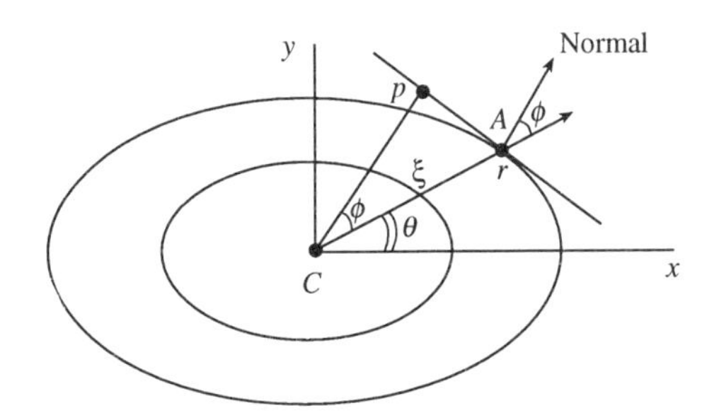

Newton solves this problem by a generalization of his method of the canals. Towards this end, he conceives of a canal CA, of unit cross-section, bored radially outwards from the centre C of the spheroid to its boundary at A; and of the canal being filled with water of density σ (say). In Corollary III of Proposition XCI, Book I, he has already shown that in the interior of a spheroid the attraction towards the centre varies linearly as the distance ξ from the centre, that is,

$$\text{C.F.} = C\xi \tag{1}$$

where C is some constant. Therefore,

$$\text{the weight of water in the canal} = C\sigma \int_0^r \xi \, d\xi = \tfrac{1}{2}C\sigma r^2, \tag{2}$$

where r is the length AC of the canal. Since the value of gravity at A is

$$g = Cr \tag{3}$$

it follows that

$$\text{the weight of water in the canal} = \tfrac{1}{2}g\sigma r. \tag{4}$$

For the mutual equilibrium of all the canals, similarly bored, the necessary condition is that the resolved components of the weights of the canals normal to the boundary are equal;* that is, (see the figure)

$$gr\cos\phi = K \tag{5}$$

* Huygens is attributed to having formulated this principle; and it cannot be doubted that Newton knew of it either independently or through Huygens.

where K is a constant; or, alternatively,

$$gp = K, \tag{6}$$

where p is the perpendicular distance of C from the tangent at A. From equation (6) it follows,

$$g - g_{\text{eq}} = K\left(\frac{1}{p} - \frac{1}{a}\right), \tag{7}$$

where a denotes the semimajor axis of the spheroid.

From the equation of the tangent at (x_1, y_1); namely,

$$\frac{xx_1}{a^2} + \frac{yy_1}{b^2} = 1, \tag{8}$$

we have, by a familiar formula,

$$\begin{aligned}\frac{1}{p^2} &= \frac{x^2}{a^4} + \frac{y^2}{b^4} = \frac{x^2}{a^4} + \frac{1}{b^2}\left(1 - \frac{x^2}{a^2}\right)\\ &= \frac{1}{b^2}\left[1 - \frac{x^2}{a^4}(a^2 - b^2)\right] = \frac{1}{a^2b^2}(a^2 - e^2x^2),\end{aligned} \tag{9}$$

or

$$\frac{1}{p} = \frac{1}{ab}(a^2 - e^2x^2)^{1/2}. \tag{10}$$

Therefore,

$$\begin{aligned}\frac{1}{p} - \frac{1}{a} &= \frac{1}{a}\left[\frac{a}{b}\left(1 - \frac{e^2x^2}{a^2}\right)^{1/2} - 1\right]\\ &= \frac{1}{a}\left[\frac{1}{(1-e^2)^{1/2}}\left(1 - \frac{e^2x^2}{a^2}\right)^{1/2} - 1\right]\\ &\simeq \frac{e^2}{2a}\left(1 - \frac{x^2}{a^2}\right) = \frac{\epsilon}{a}\sin^2\theta,\end{aligned} \tag{11}$$

setting $x = a\cos\theta$.

We conclude that, for $\epsilon \ll 1$, the variation of gravity over the surface is given by

$$g - g_{\text{eq}} \simeq K\frac{\epsilon}{a}\sin^2\theta, \tag{12}$$

in agreement with Newton's result:

> the increase of weight in passing from the equator to the pole is ... as the square of the sine of the latitude.

To this result, Newton adds:

> the arcs of degrees of latitude in the meridian increase nearly in the same proportion,

meaning the variation of the curvature. The curvature of an orbit is implicit in Proposition IX, Book I, (Chapter 5, §24, equation (xiv); see also, Chapter 6, §31, and the 'digression' following); but here it is explicit.

Writing the equation of an ellipse in the form,

$$y = \frac{b}{a}(a^2 - x^2)^{1/2}, \tag{13}$$

we find

$$\frac{dy}{dx} = -\frac{bx}{a(a^2 - x^2)^{1/2}}; \qquad \frac{d^2y}{dx^2} = -\frac{ab}{(a^2 - x^2)^{3/2}}. \tag{14}$$

The expression for the curvature follows from its definition; thus.

$$\rho = \left| \frac{[1 + (dy/dx)^2]^{3/2}}{d^2y/dx^2} \right| = \frac{1}{ab}(a^2 - e^2x^2)^{3/2}. \tag{15}$$

Therefore, by equation (10)

$$\rho - \rho_{eq} = a^2b^2\left(\frac{1}{p^3} - \frac{1}{a^3}\right) = \frac{1}{ab}(a^2 - e^2x^2)^{3/2} - \frac{b^2}{a}$$

$$= \frac{a^2}{b}[(1 - e^2x^2/a^2)^{3/2} - (1 - e^2)^{3/2}]$$

$$\simeq \frac{3}{2}\frac{a^2e^2}{b}\sin^2\theta = 3\frac{a^2}{b}\epsilon\sin^2\theta; \tag{16}$$

or, in the same approximation,

$$\rho - \rho_{eq} \simeq 3a\epsilon\sin^2\theta, \tag{17}$$

even as Newton stated.

Newton compares the predicted variation of gravity according to equation (12) with the observations over the surface of the Earth available in his time. Considering the lack of consensus at that time on the figure of the Earth, we shall content ourselves with quoting his concluding remark as a summation of his feelings (?).

> M. Richer repeated his observations, made in the island of Cayenne, every week for ten months together, and compared the lengths of the pendulum which he had there noted in the iron rods with the lengths thereof which he observed in France. This diligence and care seems to have been wanting to the other observers. If this gentleman's observations are to be depended on, the Earth is higher under the equator than at the poles, and that by an excess of about 17 miles; as appeared above by the theory.

A personal reflection

As I have stated in the Prologue, I have made no attempt to read any of the extant Newtoniana: I have preferred to read and assimilate the *Principia*, to the extent I can, on my own. But I have on occassions sought the views of the masters of earlier times; and, in the context of this chapter, *A history of the mathematical theories of attraction and the figure of the Earth* (Constable, London, 1873; reprinted by Dover Publications, New York, 1962) by I. Todhunter. In his first chapter on Newton (pp. 1–27), in discussing Proposition XX, Todhunter writes:

> his investigation of the very important proposition now before us is thus rendered obscure and imperfect.

And he quotes statements of other writers with similar overtones. I do not share their views. Their principal error, I believe, is to read the written text of *Principia* (in its different editions) and assume that it represents the sum of Newton's knowledge. And they do not seem to be cognizant of Newton's withholding a great deal—a fact made abundantly clear in the earlier chapters. And even in this Proposition XX, the quotation following equation (12), in which this result is stated, is preceded only by 'this theorem arises'!

My impression is very strong that Newton becomes impatient and 'secretive' when he comes to the denouement of his technically more difficult propositions and corollaries; he increasingly seems to have adopted the attitude:

> Learn to understand and you shall hear it. But in other terms—no. If you would not rise to us, we cannot stoop to you.
>
> (John Ruskin; in *Sesame and Lilies*, 1865)

❖21❖

On the theory of tides

But among all the great men who have philosophized about this remarkable effect, I am more astonished at Kepler than at any other. Despite his open and acute mind, and though he has at his fingertips the motions atributed to the earth, he has nevertheless lent his ear and his assent to the moon's dominion over the waters, to occult properties, and to such puerilities.

Galileo Galilee* (1629)

Thus I have explained the causes of the motions of the Moon and of the sea.

Isaac Newton† (1687)

Concerning the Cause of the tides given by M. Newton, I am by no means satisfied by it, nor by all other Theories that he builds upon his principle of Attraction, which to me seems absurd, . . . And I have often wondered how he could have given himself all the trouble of making such a number of investigations and difficult calculations that have no other foundation than this very principle.

Christian Huygens‡ (1690)

111. Introduction

The cultural climate in which Newton initiated the current theory of the oceanic tides is manifest from the quotations that head his chapter.

The basic kinematical elements of the theory outlined by Newton are the following:

* Fourth Dialogue in his *System of the World*: in Stillman Drake's translation, p. 462 (University of California Press, 1970). It has been remarked that in judging Galileo's statement on Kepler, one must remember that the '"moon's dominion over the waters" is an ancient superstition and not a scientific anticipation of Newton's general gravitational law' (Drake, *loc. cit.*, p. 491). Superstition or not, Sagredo's follow up comment,

> It is my guess that what has happened to these more reflective men is what is happening at present to me; namely, inability to understand the interrelation of the three periods, annual, monthly, and diurnal, and how their causes may seem to depend upon the sun and the moon without either of these having anything to do with the water itself.

continued to be valid, in spite of all the valiant and confident efforts of Galileo and Descartes, till Newton resolved the problem.

† At the conclusion of Proposition XXIV (see p. 411).

‡ In a letter to Leibniz.

(i) the tidal action of one gravitating body on another is to effect equal bulges at antipodal points of the attracting body;
(ii) the Moon and the Sun effect such bulges in the ocean (assumed to cover the Earth) in the instantaneous directions of the Moon and of the Sun, respectively, the effect of the latter being about one-half (or more precisely 0.44*) of the former;
(iii) the bulges caused by the Moon and the Sun are carried, independently, by friction with the changing positions of the Moon and of the Sun by the orbital motion of the Moon about the Earth and of the Earth about the Sun, both coupled with the rotation of the Earth.

The known periodicities of the tides (which troubled Sagredo) follow from the premises in self-evident fashion (see illustration): the *high tide* occurs at (b) when a particular location on the Earth is carried past the bulge; at new and full moon (c) the Sun and the Moon act together to give *spring tides* which are greater than the *neap tides* at first and last quarter.

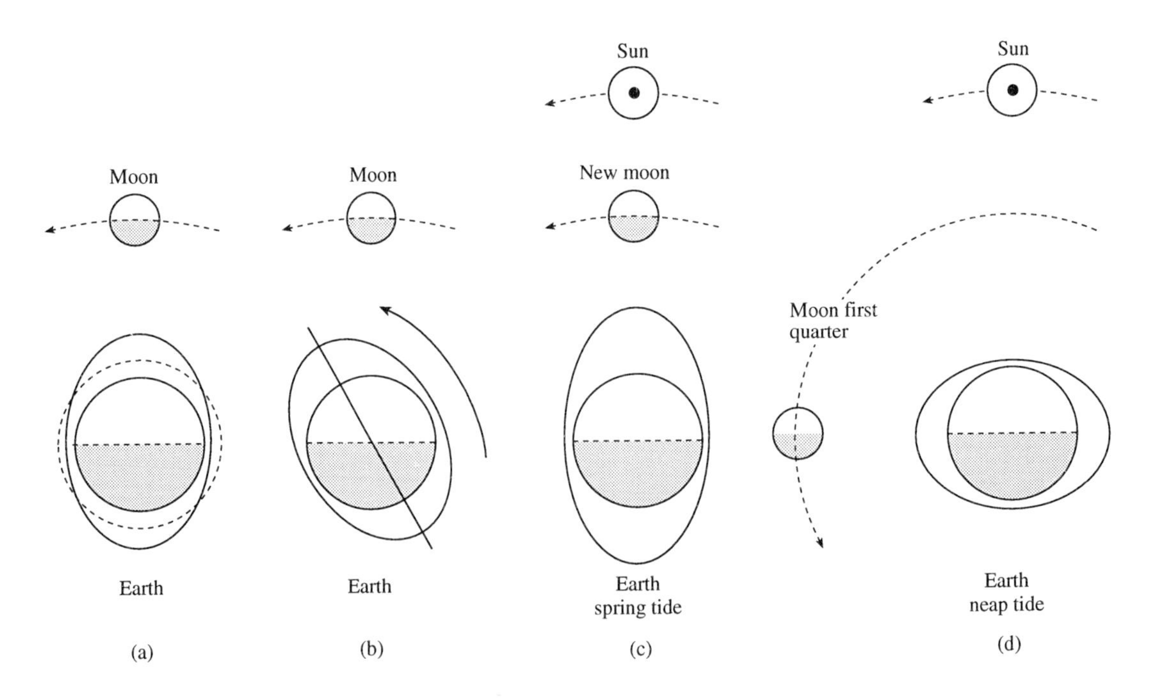

In the Appendix (§115) we present a current version of the elementary equilibrium theory of the tides to facilitate the understanding of Newton's propositions.

* Newton did not have this precise estimate since the ratio of the mass of the Earth to that of the Moon was unknown at his time. But he estimated a value of 0.3 from his theory of the tides in Proposition XXXVIII (see p. 414).

112. A recapitulation.

The principal elements of the theory have already been set out in Corollaries XVIII-XXII of Proposition LXVI, Book I (§73 of Chapter 14). It may be useful for the reader at this point to review those corollaries, particularly the summary following Corollary XVIII (p. 260), Corollary XIX, equations (1) and (2), and Corollary XX, equations (7)–(9).

As we have pointed out in §73, following Corollary XXI (p. 264), Newton's more detailed considerations are blemished by not realizing that the only tide-generating force is the tangential component and that the normal component is balanced by the pressure; and assuming, instead, that the total tidal force acts on the fluid and causes its motions.* The corrections that are required on this account (which, it will be noted, do not affect the essential content of the theory) are indicated in the annotated version of Proposition XXIV in §113.

The tidal force of a distant body acting on the boundary of a spherical body

The required force has, in effect, been evaluated in Chapter 13, §66 (equations (23) and (24)) and Chapter 14, §70 (equations (12) and (13)). We shall obtain it once again, in the present context, more in the manner of Newton.

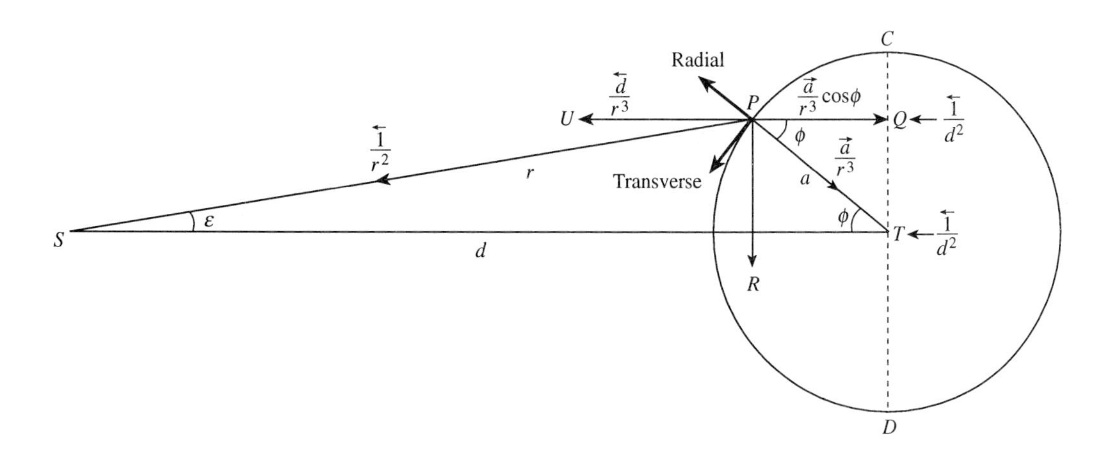

* I am somewhat puzzled that the underlying principle, 'the surface of the sea is such that at every point *the direction of the plumb-line is perpendicular to the surface*', formulated by Huygens in his *Discours de la Cause de la Pesanteur*, published in 1690, was overlooked by Newton and, more particularly, by Roger Cotes when the first edition of the *Principia* was being revised. For this and other reasons, I am not altogether impressed by the perspicacity of Cotes in his reading of the *Principia* apart from correcting typographical misprints and numerical errors, but not all of them.

Description of figure* and definitions:

S is the distant body;
T is the centre of a spherical body of radius a;
P is a point on the sphere at which the tidal force due to S is required;
$r =$ distance PS;
$d =$ distance ST;
$\epsilon = \angle \widehat{TS,SP}$ considered small; and
$\phi = \angle \widehat{ST,TP}$.

From these definitions it follows:

$$\sin\epsilon = \frac{a}{r}\sin\phi \qquad \text{and} \qquad \frac{r}{d} = 1 - \frac{a}{d}\cos\phi + O(a^2/d^2).$$

We shall neglect quantities of $O(a^2/d^2)$ or, equivalently, $O(\epsilon^2)$.†

The attraction $1/r^2$ in the direction PS

$$= \frac{d}{r^3} \text{ in the direction } PU - \frac{a}{r^3}\cos\phi \text{ in the direction } PQ$$

$$\simeq \frac{d}{r^3}\left(1 - \frac{a}{d}\cos\phi\right) \simeq \frac{d}{r^3}\frac{r}{d} = \frac{1}{r^2} \text{ in the direction } PU$$

$$\simeq \frac{1}{r^2} \text{ in the direction } PS \text{ to } O(\epsilon^2). \tag{1}$$

Therefore, the force relative to the centre of mass of T, that is, the tidal force in the direction PU is, by Proposition LXVI, Corollary 14 (p. 257),

$$\frac{d}{r^3} - \frac{1}{d^2} = \frac{d}{r^3}\left(1 - \frac{r^3}{d^3}\right) \simeq \frac{3a}{r^3}\cos\phi \simeq \frac{3a}{d^3}\cos\phi = 3\frac{x}{d^3}. \tag{2}$$

We conclude:

$$\text{the radial tidal force} = \frac{3a}{d^3}\cos^2\phi - \frac{a}{r^3}$$

$$\simeq \frac{a}{d^3}(3\cos^2\phi - 1) = \frac{1}{2}\frac{a}{d^3}(1 + 3\cos 2\phi); \tag{3}$$

and

$$\text{the transverse tidal force} = \frac{3a}{d^3}\cos\phi\sin\phi = \frac{3}{2}\frac{a}{d^3}\sin 2\phi. \tag{4}$$

* In the figure, $\vec{1}$ stands for a unit vector in the associated direction indicated. Thus, $\vec{1}/r^2$ and $\vec{1}/d^2$ stand for $\vec{r}/r^3$ and $\vec{d}/d^3$, respectively.

† In the sequel we shall replace $\vec{r}$ by $\vec{d}$ whenever it is consistent with the basic requirement $a \ll |\vec{r}|$ and therefore also $a \ll |\vec{d}|$.

It should be noted that the factor that has been suppressed in the foregoing expressions is $GM_{\odot}$. In particular the tidal force acting in the direction PU is

$$3GM_{\odot}x/d^3 \tag{5}$$

which on the surface of the Earth becomes

$$3GM_{\odot}a/d^3. \tag{6}$$

113. Proposition XXIV: an annotated version*

In this section we shall present an annotated version of Proposition XXIV correcting for the error mentioned in §112 and be in accord with the true dependences $\sin 2\lambda \sin 2\delta_{☾}$ and $\sin 2\lambda \sin 2\delta_{\odot}$ on latitude and declination of the diurnal lunar and solar tides (cf. equation (15) of the Appendix, §115). The parts that are crossed out are to be replaced by the entries in the margin.

Proposition XXIV. Theorem XIX

That the flux and reflux of the sea arise from the actions of the Sun and Moon.

By Cor. XIX and XX, Prop. LXVI, Book I, it appears that the waters of the sea ought twice to rise and twice to fall every day, as well lunar as solar; and that the greatest height of the waters in the open and deep seas ought to follow the approach of the luminaries to the meridian of the place by a less interval than six hours;[1] as happens in all that eastern tract of the Atlantic and Ethiopic seas between France and the Cape of Good Hope; and on the coasts of Chile and Peru in the South Sea; in all which shores the flood falls out about the second, third, or fourth hour, unless where the motion propagated from the deep ocean is by the shallowness of the channels, through which it passes to some particular places, retarded to the fifth, sixth, or seventh hour, and even later. The hours I reckon from the approach of each luminary to the meridian

* In preparing this version, I have benefited by an earlier reading by J. Proudman in *Isaac Newton* (ed. Greenstreet, G. Bell & Sons Ltd., London 1927), pp. 90–93.

1 Corollary XIX (equation (2)) on p. 261.

of the place, as well under as above the horizon; and by the hours of the lunar day I understand the 24th parts of that time which the Moon, by its apparent diurnal motion, employs to come about again to the meridian of the place which it left the day before.[2] [The force of the Sun or Moon in raising the sea is greatest in the approach of the luminary to the meridian of the place; but the force impressed upon the sea at that time continues a little while after the impression, and is afterwards increased by a new though less force still acting upon it. This makes the sea rise higher and higher, till, this new force becoming too weak to raise it any more, the sea rises to its greatest height. And this will come to pass, perhaps, in one or two hours, but more frequently near the shores in about three hours, or even more, where the sea is shallow.]

In the main erroneous

Explanation of spring and neap tides

The two luminaries excite two motions, which will not appear distinctly, but between them will arise one mixed motion compounded out of both. In the conjunction or opposition of the luminaries their forces will be conjoined, and bring on the greatest flood and ebb. In the quadratures the Sun will raise the waters which the Moon depresses, and depress the waters which the Moon raises, and from the difference of their forces the smallest of all tides will follow.[2] And because (as experience tells us) the force of the Moon is greater than that of the Sun, the greatest height of the waters will happen about the ~~third lunar hour~~. Out of the syzygies and quadratures, the greatest tide, which by the single force of the Moon ought to fall out at ~~the third~~

[time of the lunar tide]

[2] Corollary XIX, p. 261, equations (2) and (3).

~~lunar hour~~, and by the single force of the Sun at ~~the third solar hour~~, by the compounded forces of both must fall out in an intermediate time that approaches nearer to ~~the third hour~~ of the Moon than to that of the Sun. And, therefore, while the Moon is passing from the syzygies to the quadratures, [during which time the third hour of the Sun precedes the third hour of the Moon] the greatest height of the waters will also precede ~~the third hour of~~ the Moon, and that, by the greatest interval, a little after the octants of the moon; and, by like intervals, the greatest tide will follow ~~the third lunar hour~~, while the Moon is passing from the quadratures to the syzygies.[2] Thus it happens in the open sea; for in the mouths of rivers the greater tides come later to their height.

[one time]
[at another time]
[that]
[Partially erroneous]
[that due to]
[that due to the Moon]

But the effects of the luminaries depend upon their distances from the Earth; for when they are less distant, their effects are greater, and when more distant, their effects are less, and that as the cube of their apparent diameter.[3] Therefore it is that the Sun, in the winter time, being then in its perigee, has a greater effect, and makes the tides in the syzygies somewhat greater, and those in the quadratures somewhat less than in the summer season; and every month the Moon, while in the perigee, raises greater tides than at the distance of fifteen days before or after, when it is in its apogee. From this it comes to pass that two highest tides do not follow one the other in two immediately succeeding syzygies.

Semi-annual solar tides

Diurnal lunar tide

Dependence on declination

The effect of either luminary doth likewise depend upon its declination[4] or distance

[3] Equations (2)–(4) of §112.
[4] See Equation (15) of Appendix (§115) for the dependence on sin 2δ.

from the equator; for if the luminary was placed at the pole, it would constantly attract all the parts of the waters without any intensification or remission of its action, and could cause no reciprocation of motion. And, therefore, as the luminaries decline from the equator towards either pole they will, by degrees, lose their force, and on this account will excite lesser tides in the solstitial than in the equinoctial syzygies. But in the solstitial quadratures they will raise greater tides than in the quadratures about the equinoxes; because the force of the Moon, then situated in the equator, most exceeds the force of the Sun. Therefore the greatest tides occur in those syzygies, and the least in those quadratures, which happen about the time of both equinoxes; and the greatest tide in the syzygies is always succeeded by the least tide in the quadratures, as we find by experience. But, because the Sun is less distant from the Earth in winter than in summer, it comes to pass that the greatest and least tides more frequently appear before than after the vernal equinox, and more frequently after than before the autumnal.

Dependence on latitude

Moreover, the effects of the luminaries depend upon the latitudes[5] of places. Let *ApEP* represent the Earth covered with deep waters; *C* its centre; *P*, *p* its poles; *AE* the equator; *F* any place without the equator; *Ff* the parallel of the place; *Dd* the correspondent parallel on the other side of the equator; *L* the place of the Moon ~~three~~ [six] hours before; *H* the place of the Earth directly under it; *h* the opposite place; *K*, *k* the places at 90 degrees distance; *CH*, *Ch*,

[5] See equation (15) of Appendix (§115) for the dependence on sin 2λ.

the greatest heights of the sea from the centre of the Earth; and *CK*, *Ck* its least heights: and if with the axes *Hh*, *Kk*, an ellipse is described, and by the revolution of that ellipse about its longer axis *Hh* a spheroid *HPKhpk* is formed, this spheroid will nearly represent the figure of the sea; and *CF*, *Cf*, *CD*, *Cd*, will represent the heights of the sea in the places *Ff*, *Dd*. But further; in the said revolution of the ellipse any point *N* describes the circle *NM* cutting the parallels *Ff*, *Dd*, in any places *RT*, and the equator *AE* in *S*; *CN* will represent the height of the sea in all those places, *R*, *S*, *T*, situated in this circle. Therefore, in the

not always

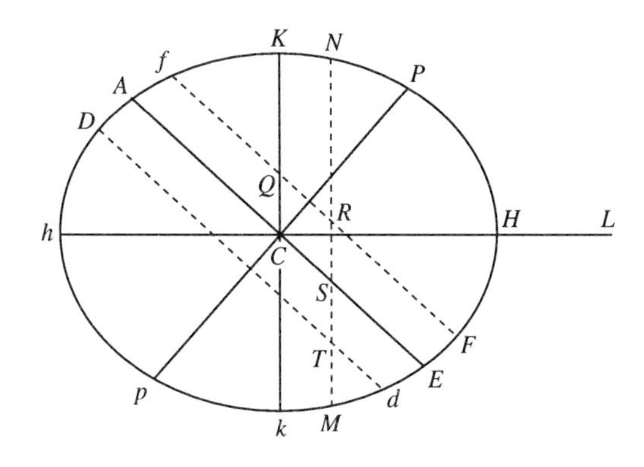

diurnal revolution of any place *F*, the greatest flood will be in *F*, at the ~~third~~ [sixth] hour after the appulse of the Moon to the meridian above the horizon; and afterwards the greatest ebb in *Q*, at the ~~third~~ [sixth] hour after the setting of the Moon; and then the greatest flood in *f*, at the ~~third~~ [sixth] hour after the appulse of the Moon to the meridian under the horizon; and, lastly, the greatest ebb in *Q*, at the ~~third~~ [sixth] hour after the rising of the Moon; and the latter flood in *f* will be less than the preceding flood in *F*. For the whole sea is divided into two hemispherical floods, one in the hemisphere *KHk* on the north side, the other in the opposite hemisphere *Khk*, which we may therefore call the northern and the southern floods. These floods, being always opposite the one to the other, come by turns to the meridians of all places, after an interval of twelve lunar hours. And as the northern countries partake more of the northern flood, and the southern countries more of the southern flood, thence arise tides, alternately greater and less in all places without the equator, in which the luminaries rise and set. But the greatest tide will happen when the Moon declines towards the vertex of the place, about the ~~third~~ [sixth] hour after the appulse of the Moon to the meridian above the horizon; and when the Moon changes its declination *to the other side of the equator*, that which was the greater tide will be changed into a lesser. And the greatest difference of the floods will fall out about the times of the solstices; especially if the ascending node of the Moon is about the first of Aries. So it is found by experience that [if] the morning tides in winter exceed those of the evening, [then] the evening tides in summer exceed those of the morning; at Plymouth by the height of one foot, but at Bristol by the height of fifteen inches, according to the observations of Colepress and Sturmy.

But the motions which we have been describing suffer some alteration from that force of reciprocation, which the waters, being once moved, retain a little while *by their inertia*. Whence it comes to pass that the tides may continue for some time, though the actions of the luminaries should cease. This power of retaining the impressed motion lessens the difference of the alternate tides, and makes those tides which immediately succeed after the syzygies greater, and those which follow next after the quadratures less. And hence it is that the alternate tides at Plymouth and Bristol do not differ much more from each other than by the height of a foot or fifteen inches, and that the greatest tides at those ports are not the first but the third after the syzygies. And, besides, all the motions are retarded in their passage through shallow channels, so that the greatest tides of all, in some straits and mouths of rivers, are the fourth or even the fifth after the syzygies.

Further, it may happen that the tide may be propagated from the ocean through different channels towards the same port, and may pass quicker through some channels than through others; in which case the same tide, divided into two or more succeeding one another, may compound new motions of different kinds. Let us suppose two equal tides flowing towards the same port from different places, one preceding the other by six hours; and suppose the first tide to happen at the third hour of the approach of the Moon to the meridian of the port. If the Moon at the time of the approach to the meridian was in the equator, every six hours alternately there would arise equal floods, which, meeting with as many equal ebbs, would so balance each other that for that day the water would stagnate and be quiet. If the Moon then declined from the equator, the tides in the ocean would be alternately greater and less, as was said; and from thence two greater and two less tides would be alternately propagated towards that port. But the two greater floods would make the greatest height of the waters to fall out in the middle time between both; and the greater and less floods would make the waters to rise to a mean height in the middle time between them, and in the middle time between the two less floods the waters would rise to their least height. Thus in the space of twenty-four hours the waters would come, not twice, as commonly, but once only to their greatest, and once only to their least height; and their greatest height, if the Moon declined towards the elevated pole, would happen at the sixth or thirtieth hour after the approach of the Moon to the meridian; and when the Moon changed its declination, this flood would be changed into an ebb. An example of this Dr. Halley has given us, from the observations of seamen in the port of Batshaw, in the kingdom of Tunquin, in the latitude of 20° 15′ north. In that port, on the day which follows after the passage of the Moon over the equator, the waters stagnate: when the Moon declines to the north, they begin to flow and ebb, not twice, as in other ports, but once only every day; and the flood happens at the setting, and the greatest ebb at the rising of the Moon. This tide increases with the declination of the Moon till the seventh or eighth day; then for the seven or eight days following it decreases at the same rate as it had increased, and ceases when the Moon changes its declination, crossing over the equator to the south. After which the flood is immediately changed into an ebb; and thenceforth the ebb happens at the setting and the flood at the rising of the Moon; till the Moon, again passing the equator, changes its declination. There are two inlets to this port and the neighbouring channels, one from the seas of China, between the continent and the island of Leuconia; the other from the Indian sea, between the continent and the island of Borneo. But whether there be really two tides propagated through the said channels, one from the Indian sea in the space of twelve hours, and one from the sea of China in the space of six hours, which therefore happening at the third and ninth lunar hours, by being compounded together, produce those motions; or whether there be any other circumstances in the state of those seas, I leave to be determined by observations on the neighbouring shores.

Of this last paragraph, J. Proudman has commented:

> It is impossible not to contrast the soundness of this remarkable paragraph with the absurdity of the popular explanation of the double high water at Southampton, more than two centuries later.

And Newton concludes the proposition as follows:

> Thus I have explained the causes of the motions of the Moon and of the sea. Now it is fit to subjoin something concerning the amount of those motions.

114. Propositions XXV, XXXVI, and XXXVII

In these propositions, Newton recapitulates the relevant results of Corollaries I–IV of Proposition LXVI of Book I, in the context of the influence of the Sun on the sea. But we can see the 'lion's paw' in the corollary of Proposition XXXVI.

Proposition XXV. Problem VI

To find the forces with which the Sun disturbs the motions of the Moon.

It has been shown in Chapter 14, §70, equations (16)–(18), that the perturbing centripetal force of the Sun on the orbit of the Moon (assumed circular) is given by

$$F_r = \tfrac{1}{2}m_{☾}^2(1 + 3\cos 2\phi)n^2 \times PT, \tag{1}$$

where

$$m_{☾}^2 = \frac{N^2}{n^2} = \left(\frac{\text{orbital period of the Moon about the earth}}{\text{orbital period of the Earth about the Sun}}\right)^2, \tag{2}$$

and

$$n^2 = \frac{GM_\oplus}{R_☾^3} = \frac{g_☾}{R_☾}, \tag{3}$$

where $g_☾$ is the value of gravity at the Moon's orbit and $R_☾$ the radius of the orbit. We have further replaced ψ by ϕ to be in accordance with our present notation. We shall write the expression for F_r alternatively in the form

$$F_r = \tfrac{1}{2}m_☾^2(1 + 3\cos 2\phi)g_☾\frac{PT}{R_☾}. \tag{4}$$

With the known periods of the Earth and of the Moon (cf. Chapter 14, equation (42))

$$m_☾^2 = (0.0748)^2 = 0.005595. \tag{5}$$

or, as Newton writes,

$$m_☾^2 = 1:178\tfrac{29}{40}. \tag{6}$$

Equation (4) expresses F_r in terms of $g_☾$. It is convenient in the present context to write it in terms of $g_\oplus$, the gravity at the surface of the Earth. Allowing for the fact that the Earth–Moon distance is 60 times the radius of the Earth and that the radius of the Moon's orbit about the Earth–Moon–Sun centre of gravity is (60/60.5) times the Earth–Moon distance,

$$g_\oplus = g_☾ \times (60)^2 \times (60/60\tfrac{1}{2}). \tag{7}$$

Inserting this value in equation (4), we obtain

$$F_r = \tfrac{1}{2}m_\oplus^2(1 + 3\cos 2\phi)\,g_\oplus\frac{PT}{R_☾}, \tag{8}$$

where

$$m_\oplus^2 = \frac{m_☾^2}{(60)^2(60/60\frac{1}{2})} = 1:6.3809 \times 10^5. \quad \text{Q.E.I.} \tag{9}$$

Proposition XXXVI. Problem XVII

To find the force of the Sun to move the sea.

The tidal attraction, $T_\oplus$, of the Sun on the sea is obtained by setting

$$PT = a = \text{the radius of the Earth} \tag{10}$$

in equation (8), and we obtain

$$T_\oplus = \tfrac{1}{2}m_s^2(1 + 3\cos 2\phi)g_\oplus \tag{11}$$

where

$$m_s^2 = m_\oplus^2(a/R_☾) = m_\oplus^2/60\tfrac{1}{2} = 1:3.8605 \times 10^7. \tag{12}$$

It can be readily verified that equation (11) agrees with equation (3) of §112.

At quadratures ($\phi = \pi/2$ and $3\pi/2$) and at syzygies ($\phi = 0$ and π)

$$T_\oplus\,(\text{quadratures}) = -m_s^2 g_\oplus \qquad \text{and} \qquad T_\oplus\,(\text{syzygies}) = 2m_s^2 g_\oplus. \tag{13}$$

And finally by equation (2) of §112,

$$T_{\oplus} \text{ (in the direction of } PU) = -3m_s^2 g_{\oplus} \cos\phi = -\frac{g_{\oplus}}{1.2868 \times 10^7}\cos\phi. \tag{14}$$

Equations (11)–(14) are all given by Newton.

Corollary. To obtain the height of the solar diurnal tide at syzygies, Newton adopts a procedure that is reminiscent of his evaluation in Chapter 20, §109(b), of the attractive force at the equator of an oblate spheroid in terms of the attractive force at the end of the long axis of a prolate spheroid.

In the accompanying illustration, the sections of the spheroids (denoted by the parallel rulings) normal to the major axis of a tidally formed prolate spheroid are shown. It will be observed that the distribution of the transverse components of the centrifugal force, $\Omega^2 x$, and of the tidal force, $-3GM_{\odot}x/d^3$, directed towards the Sun are similar even if the forces are directed in the opposite directions. It follows that the ellipticities, ϵ_Ω and ϵ_T, consequent to the respective perturbations, must be proportional; and therefore, by equation (11)

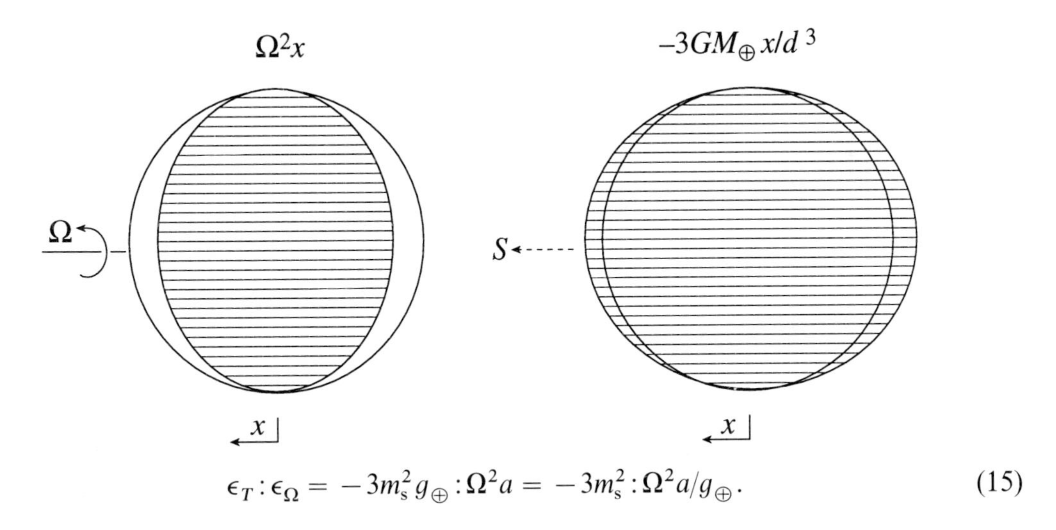

$$\epsilon_T : \epsilon_\Omega = -3m_s^2 g_{\oplus} : \Omega^2 a = -3m_s^2 : \Omega^2 a/g_{\oplus}. \tag{15}$$

But $\Omega^2 a/g_{\oplus}$ is what we have denoted by m in Chapter 20, §109, equation (2). Hence

$$\epsilon_T = -3\epsilon_\Omega m_s^2/m. \tag{16}$$

The height H of the tides at syzygies is therefore

$$H = a|\epsilon_T| = 3a\epsilon_\Omega m_s^2/m. \tag{17}$$

where a (as hitherto) denotes the radius of the Earth. This is Newton's formula.

For the Earth,

$$a = 6378 \times 10^5 \text{ cm}, \qquad m = 1/289 \qquad \text{and} \qquad m_s^2 = 1/3.8605 \times 10^7. \tag{18}$$

And for a homogeneous Earth it has been shown in Chapter 20, §109, equation (14), that

$$\epsilon_\Omega = \tfrac{5}{4}m = \tfrac{1}{230}. \tag{19}$$

Inserting the foregoing values in equation (17) we obtain,

$$H = \frac{6378 \times 10^5}{230} \times \frac{3}{3.8605 \times 10^7} \times 289 = 62.28 \text{ cm}, \tag{20}$$

or, converting to *French* inches,

$$H = 62.28/2.7070 = 23.0 \textit{ French} \text{ inches}, \tag{21}$$

in agreement with Newton's value, 85472/44527 *French* feet.

Proposition XXXVII. Problem XVIII

To find the force of the Moon to move the sea.

> The force of the Moon to move the sea is to be deduced from its ratio to the force of the Sun, and this ratio is to be determined from the ratio of the motions of the sea, which are the effects of those forces. Before the mouth of the river Avon, three miles below Bristol, the height of the ascent of the water in the vernal and autumnal syzygies of the luminaries (by the observations of Samuel Sturmy) amounts to about 45 feet, but in the quadratures to 25 only. The former of those heights arises from the sum of the aforesaid forces, the latter from their difference. If, therefore, S and L are supposed to represent respectively the forces of the Sun and Moon while they are in the equator, as well as in their mean distances from the Earth, we shall have $L + S$ to $L - S$ as 45 to 25, or as 9 to 5.

After similar discussions, Newton concludes:

> therefore, till we procure something that is more certain, we shall use the proportion of 9 to 5.

Newton thus accepts the provisional value

$$L/S = 3\tfrac{1}{2}. \tag{22}$$

From our present knowledge of the Moon's mass the true ratio is (cf. §115)

$$L/S = 2.23. \tag{23}^*$$

Because of the larger value of L/S that Newton adopts and of the many pitfalls in deducing reliable information from a knowledge of the tides derived from experience, we shall not consider the remaining specific corollaries—there are ten of them—of this proposition.

* He later revises this value, in this same proposition, to 4.4815.

115. Appendix: the equilibrium theory of the tides

In the accompanying illustration, O denotes the centre of the Earth and M that of a distant body (the Moon or the Sun). Let P be a point on the earth near its surface at a distance r from its centre. Let $OM = D$ and $\vartheta = \angle \widehat{POM}$ be the geocentric zenith distance of P.

The gravitational potential of the attraction of M at P is

$$-\frac{GM}{\sqrt{(D^2 - 2rD\cos\vartheta + r^2)}}. \tag{1}$$

To obtain the acceleration relative to the centre of mass of the Earth, we must subtract from (1) the potential of the uniform field of force, GM/D^2, that M exerts over the whole mass of the Earth, namely,

$$-\frac{GM}{D^2} r\cos\vartheta. \tag{2}$$

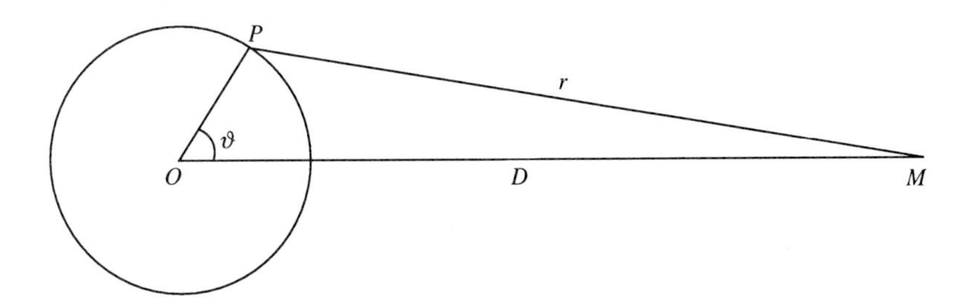

The potential of the tidal force is therefore,

$$T = -\frac{GM}{\sqrt{(D^2 - 2rD\cos\vartheta + r^2)}} + \frac{GM}{D^2} rP_1(\cos\vartheta). \tag{3}$$

Expanding in powers of r/D and retaining only the dominant term, we have

$$T = -\frac{GMr^2}{D^3} P_2(\cos\vartheta) = -\tfrac{3}{2}GM\frac{r^2}{D^3}(\cos^2\vartheta - \tfrac{1}{3}). \tag{4}$$

In the equilibrium theory of the tides, it is assumed that the free surface of the seas takes, at each instant, the form that it will have if the disturbing body were to retain unchanged its position relative to the rotating Earth. In other words, it is assumed that the free surface is a *level surface* under the combined action of gravity, rotation of the Earth, and the disturbing force, that is, it is a surface of constant total potential,

$$\Psi = \tfrac{1}{2}\Omega^2\varpi^2 - T, \tag{5}$$

where Ω is the angular velocity of rotation and ϖ is the distance from its axis; and the equation of the level surface is

$$\Psi - \tfrac{1}{2}\Omega^2\varpi^2 + T = \text{constant}. \tag{6}$$

If $\bar{\zeta}$ denotes the elevation of water above the undisturbed surface, S_0 (say), the equation determining $\bar{\zeta}$, in a first approximation, is given by

$$[\Psi - \tfrac{1}{2}\Omega^2\varpi^2]_{S_0} + \left[\frac{\partial}{\partial z}(\Psi - \tfrac{1}{2}\Omega^2\varpi^2)\right]_{S_0} \bar{\zeta} + T = \text{constant}, \tag{7}$$

where $\partial/\partial z$ denotes differentiation in the direction of the outward normal. Ignoring the first term in the left-hand side, which is a constant, we can write

$$\bar{\zeta} = -T/g + \text{constant}, \tag{8}$$

where

$$g = \left[\frac{\partial}{\partial z}(\Psi - \tfrac{1}{2}\Omega^2\varpi^2)\right]_{S_0} \tag{9}$$

denotes the effective gravity on the undisturbed surface.

In the theory of tides, it is customary not to allow for the variations in g derived from the ellipticity of the Earth and assume

$$g \simeq GM_{\oplus}/a^2. \tag{10}$$

We thus obtain from equations (4), (8), and (10),

$$\bar{\zeta} = \frac{GMa^2}{D^3}\frac{a^2}{GM_{\oplus}}\frac{1}{2}(3\cos^2\vartheta - 1) + C, \tag{11}$$

or

$$\bar{\zeta} = \frac{a^4}{2M_{\oplus}}\frac{M}{D^3}(3\cos^2\vartheta - 1) + C. \tag{12}$$

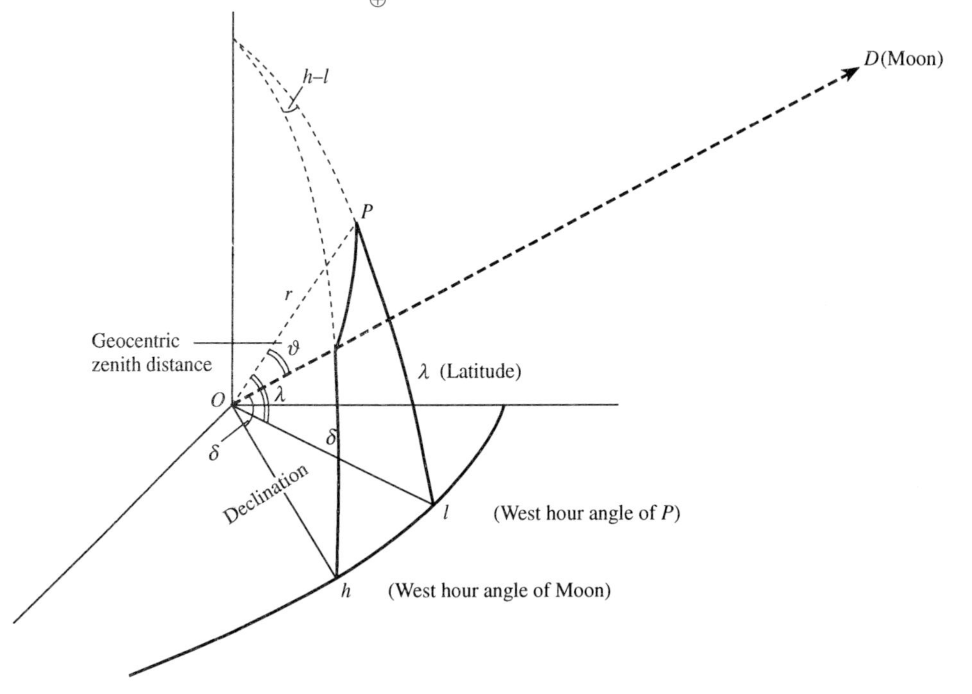

To bring out explicitly the dependence of $\bar{\zeta}$ on the latitude λ, the declination δ, and the difference in the hour angles of P and M, we make use of the trigonometric relation,

$$\cos\vartheta = \sin\lambda\sin\delta + \cos\lambda\cos\delta\cos(h - l), \tag{13}$$

and find:

$$\begin{aligned} 3\cos^2\vartheta - 1 = (1 - 3\sin^2\lambda)(\tfrac{3}{2}\cos^2\delta - 1) \\ + \tfrac{3}{2}\cos^2\lambda\cos^2\delta\cos 2(h - l) \\ + \tfrac{3}{2}\sin 2\lambda\sin 2\delta\cos(h - l). \end{aligned} \tag{14}$$

Inserting this expression in equation (12) and combining the actions of the Moon and the Sun, we obtain:

$$\begin{aligned} \frac{2M_\oplus}{a^4}\bar{\zeta} = {} & (1 - 3\sin^2\lambda)\left[\underbrace{\frac{M_☾}{D_☾^3}(\tfrac{3}{2}\cos^2\delta_☾ - 1)}_{\text{fortnightly declinational lunar tide}} + \underbrace{\frac{M_\odot}{D_\odot^3}(\tfrac{3}{2}\cos^2\delta_\odot - 1)}_{\text{solar declinational semi-annual tide}}\right] \quad \text{long-period tides} \\ & + \tfrac{3}{2}\sin 2\lambda\left[\underbrace{\frac{M_☾}{D_☾^3}\sin 2\delta_☾\cos(h_☾ - l)}_{\text{diurnal lunar tide}} + \underbrace{\frac{M_\odot}{D_\odot^3}\sin 2\delta_\odot\cos(h_\odot - l)}_{\text{diurnal solar tide}}\right] \\ & + \tfrac{3}{2}\cos^2\lambda\left[\underbrace{\frac{M_☾}{D_☾^3}\cos^2\delta_☾\cos 2(h_☾ - l)}_{\text{semi diurnal lunar tide}} + \underbrace{\frac{M_\odot}{D_\odot^3}\cos^2\delta_\odot\cos 2(h_\odot - l)}_{\text{semi diurnal solar tide}}\right] \end{aligned} \tag{15}$$

where

$$\frac{M_\odot}{D_\odot^3} : \frac{M_☾}{D_☾^3} = \frac{1.99 \times 10^{30}}{7.35 \times 10^{22}}\left(\frac{3.815}{1496}\right)^3 = 0.449 = 1:2.23. \tag{16}$$

The different periodic dependences on latitudes, declinations, and hour angles of the Moon and of the Sun identify the different tides. In particular the spring and the neap tides occur when

$$h_☾ - h_\odot = 0 \quad \text{or} \quad \pi \qquad \text{and} \qquad \pi/2 \quad \text{or} \quad 3\pi/2, \qquad \text{respectively.} \tag{17}$$

Also, it may be noted that the place where Newton went wrong is on the dependence of the diurnal tides on $\sin 2\lambda \sin 2\delta$.

For the purposes of a *dynamical* theory of the tides, one must allow for the rotation of the Earth in an essentially non-trivial way; the correct theory was developed by Laplace a century later.

❖22❖

The lunar theory

By these computations of the lunar motions I was desirous of showing that by the theory of gravity the motions of the moon could be calculated from their physical causes. The calculus of this motion is difficult . . .

Isaac Newton
In the Scholium following Proposition XXXV, Book III

116. Introduction

The basic elements of the lunar theory have already been set out in the first seventeen corollaries of Proposition LXVI of Book I, albeit with only the barest indications of how they were established. In the fourteen Propositions, XVII, XXI–XXIII, and XXVI–XXXV in Book III (now to be considered), the theory is elaborated with derivations and applications to the perturbations of the orbit of the Moon about the Earth, caused by the tidal action of the Sun. Considering the wealth of information contained in these propositions, we shall first abstract their substance.

1. Propositions XVII and XXI–XXIII recall and summarize the relevant corollaries of Proposition LXVI of Book I as a preliminary to what is to come.
2. Propositions XXVI, XVII, and XXIX provide additional details to the 'advance' account given in §69 and in the earlier parts of §70, including the important relation (41).
3. Propositions XXX–XXXIII deal with the problem of the regression of the nodes in a highly original manner with many perceptive details.
4. Propositions XXXIV and XXXV deal with the variable inclination of the orbital plane in a like manner.

The rotation of the line of the apsides, having already been considered in great detail in Propositions XLV (Corollary II) and LXVI (Corollaries VII and VIII) is dismissed with only passing references in Proposition XXII and in the Scholium.

In summary, one may in truth say that there is hardly anything in any modern textbook on celestial mechanics (e.g., J. M. A. Danby, *Fundamentals of ceslestial mechanics*, (Willmann-Bell, Inc., Richmond, VA., U.S.A., 1989) Chapter 11, §119) and the *entire* Chapter 12 on 'The motion of the Moon' that one cannot find in the propositions that we have enumerated, and indeed with deeper understanding.

117. Propositions XVII and XXI

Propositions XVI and XVIII, while recalling the relevance of two of the corollaries (XX and XXII) of Proposition LXVI of Book I, add some new facets.

Proposition XVII. Theorem XV

That the diurnal motions of the planets are uniform, and that the libration of the Moon arises from its diurnal motion.

Newton first points out that the uniform rotation of the planets is to be understood in terms of the ideas outlined in the last Corollary XXII of Proposition LXVI of Book I. The reference is clearly to the underlined parts of the corollary quoted in full in Chapter 14 (p. 263). As Newton explains:

> because the lunar day, arising from its uniform revolution about its axis, is menstrual, *that is, equal to the time of its periodic revolution in its orbit*, hence the same face of the Moon will be always nearly turned to the upper focus of its orbit; but, as the situation of that focus requires, will deviate a little to one side and to the other from the Earth in the lower focus; and this is the libration in longitude; for the libration in latitude arises from the Moon's latitude, and the inclination of its axis to the plane of the ecliptic.

At this point, Newton refers to his having sent to N. Mercator, in some letters, his theory of the libration in longitude of the Moon and who later 'explained it more fully' in his *Astronomy* published in 1676.

From the solution to Kepler's equation given in Proposition XXXI of Book I (Chapter 7, p. 140) it is clear that Newton must have traced the longitudinal libration of the Moon to the difference, $u - \varphi$, between the mean and the true anomalies. In the notation of Chapter 13, §65, equations (20)–(22), u and φ are defined by

$$r^2 \frac{\mathrm{d}\varphi}{\mathrm{d}t} = \text{the constant of areas,} \qquad h = [a\mu(1 - e^2)]^{1/2} \tag{1}$$

and

$$n = \left(\frac{t}{a^3}\right)^{1/2} = (1 - e \cos u) \frac{\mathrm{d}u}{\mathrm{d}t}. \tag{2}$$

Newton concludes that in similar fashion the utmost satellite of Saturn must show the same face to Saturn; and 'so also the utmost satellite of Jupiter'.

Proposition XXI. Theorem XVII

That the equinoctial points go backwards, and that the axis of the Earth, by a nutation in every annual revolution, twice vibrates towards the ecliptic, and as often returns to its former position.

The regression of the nodes (in conjunction with the variation of inclination of the orbital plane) is considered in Corollary XX (and also in Corollary XI) of Proposition LXVI of Book I.

The reference to nutation at this point is curious; it anticipates the solution to the related problem of the precession of the equinoxes in a later Proposition XXXIX. But Newton must have known, already, at this time that the nutation $\dot{\theta}$ (in the notation of Chapter 23, §128, equations (50)–(52)) must be related to the average precession of the equinoxes, by the proportionality

$$\dot{\theta} = \langle \dot{\psi} \rangle_{\mathrm{AV}} \tan\theta \sin 2(vt - \psi) \simeq -50'' \tan\theta \sin 2(vt - \psi). \tag{3}$$

Using the empirically known value of $\langle \dot{\psi} \rangle_{\mathrm{AV}}$, Newton's conclusion 'that motion of nutation must be very small' follows.

118. Propositions XXII and XXIII

Proposition XXII. Theorem XVIII

That all the motions of the Moon, and all the inequalities of those motions, follow from the principles which we have laid down.

Starting with the statement:

> That the greater planets, while they are carried about the Sun, may in the meantime carry other lesser planets, revolving about them, and that those lesser planets must move in ellipses which have their foci in the centres of the greater, appears from Prop. LXV, Book I. But then their motions will be in several ways disturbed by the action of the Sun, and they will suffer such inequalities as are observed in our Moon.

Newton enumerates the 'inequalities'—meaning the perturbations in the elements of the orbit—established in the various corollaries of Proposition LXVI, Book I. We shall list them more expansively than does Newton.

1. *On the velocity along the orbit*: Corollary III, §70, equation (35).*

$$V_1 > V_0; \qquad V_1 : V_0 \simeq \frac{1+x}{1-x}\left(1 + \frac{3}{2}\frac{m^2}{1-m}\right).$$

2. *On the constant of areas*: Corollary II, §70, equation (43).

$$h_1 = \langle h \rangle_{\mathrm{av}}\left(1 + \frac{3}{4}\frac{m^2}{1-m}\right); \qquad h_0 = \langle h \rangle_{\mathrm{av}}\left(1 - \frac{3}{4}\frac{m^2}{1-m}\right).$$

* It may be recalled that the subscripts '0' and '1' refer, respectively, to quadratures and syzygies.

3. *On the curvature of the orbit*: Corollary IV, §70, equation (39).

$$\rho_1 > \rho_0; \qquad \rho_1 : \rho_0 \simeq 1 + 3m^2\left(1 + \frac{1}{1-m}\right).$$

4. *On the prolateness of the orbit*: Corollary IV, §70, equation (42).

$$1 - x : 1 + x = 69/70.$$

5. *On the variation of the eccentricity*: Corollary IX, §72, equation (13).

$$(e - e_0)_1 = 2m^2 \cos\gamma; \qquad (e - e_0)_0 = -3m^2 \sin\gamma.$$

6. *On the rotation of the line of the apsides*: Corollaries VII and VIII, §71, equations (6) and (11).

$$e\left(\frac{\mathrm{d}\varpi}{\mathrm{d}t}\right)_0 = +m^2 n \frac{r}{a}(1 - e^2)^{1/2} \cos\varphi,$$
$$e\left(\frac{\mathrm{d}\varpi}{\mathrm{d}t}\right)_1 = -2m^2 n \frac{r}{a}(1 - e^2)^{1/2} \cos\varphi;$$

but

$$\langle\varpi\rangle_{\mathrm{AV}} = \tfrac{3}{4}m^2\sqrt{(1 - e^2)}\,.\,nt.$$

7. *On the regression of the nodes*: Corollary XI, §72, equation (23).*

$$\left(\frac{\mathrm{d}\Omega}{\mathrm{d}t}\right)_1 = -3m^2 n \sin^2 v; \qquad \left(\frac{\mathrm{d}\Omega}{\mathrm{d}t}\right)_0 = 0.$$

8. *On the variation of the inclination*: Corollary X, §72, equation (20).

$$\iota - \iota_0 = \tfrac{3}{8}m^2 \sin\iota\left(\cos 2v + \frac{1}{m}\cos 2U - \frac{1}{1-m}\cos 2\psi\right).$$

(For a fuller statement, see the quotation following this equation on p. 253.)

9. *On the mean motion*: Corollary VI, §70, equation (52).

$$n = \left(\frac{\mu}{a^3}\right)^{1/2}\left[1 - \tfrac{1}{4}m^2 r^3(1 + 3\cos 2\psi)\right];$$

and

$$n \text{ at perihelion} < n \text{ at aphelion}.$$

* There is a misprint carried on from the first edition to the third edition and in *all* the translations: 'syzygies' and 'quadratures' should be interchanged—a further example of Cotes's ineptitude to correct the *Principia* where it matters. (For more on this, see the concluding section of the next chapter.)

And Newton concludes,

> But there are yet other inequalities not observed by former astronomers, by which the motions of the Moon are so disturbed that to this day we have not been able to bring them under any certain rule.

referring in particular to the dependencies of the disturbing function on N^2 $(=GM_{\odot}/R_{\odot}^3)$ and of the inequalities themselves on (N^2/n) in Corollaries XIV and XVI, respectively (see Chapter 14, p. 257).

In the main, however, this proposition, anticipates what is yet to come.

Proposition XXIII. Problem V

To derive the unequal motions of the satellites of Jupiter and Saturn from the motions of our Moon.

Continuing his reference to the dependence of the variation of the elements on N^2/n in the preceding proposition, Newton points out how we can infer the inequalities to be expected in other planet–satellite systems from what we know for one of them. Thus, from the known mean annual regression of the nodes of 19.2862° in the Earth–Moon system, we can infer the mean regression to be expected in the Jupiter–Callisto system. Thus,

$$\frac{\overline{\Delta\Omega}_{♃}}{\overline{\Delta\Omega}_{\oplus}} = \left(\frac{\text{period}_{\oplus}}{\text{period}_{♃}}\right)^2 \frac{\text{period}_{\text{callisto}}}{\text{period}_{☾}} = \frac{1}{(11{\cdot}862)^2}\,\frac{16{\cdot}689}{27{\cdot}32} \times 19{\cdot}286^{\circ} \quad \text{per year}$$
$$= 8{\cdot}373^{\circ} \quad (=8^{\circ}\,22') \quad \text{per hundred years,} \qquad (1)$$

in agreement with the value, 8° 24′, Newton gives.

Newton further points out that the advance of the line of the apsides can be similarly predicted. But in this case the observed motion in the Earth–Moon system is ‘twice as swift’ as that predicted by the first-order theory ‘on account of a cause which I cannot stop here to explain’.

119.

Proposition XXVI. Problem VII

To find the hourly increment of the area which the Moon, by a radius drawn to the Earth, describes in a circular orbit.

Newton’s solution to the problem posed in this proposition has in the main been paraphrased in Chapter 14, §69 with the only difference that, for the sake of simplicity, Newton treats *ab initio* the asymptotic case when the Sun is allowed to recede to ‘infinity’ and the figure on p. 238 of Chapter 14 is replaced by the accompanying illustration. (The lettering is the same.)

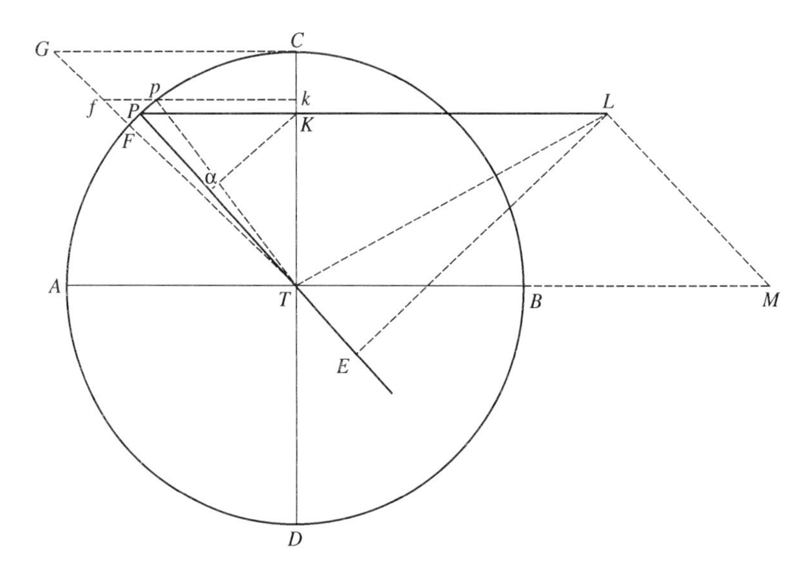

Besides, the distinction between expressing the time-dependent variations in terms of the *nodical month* (as in equation (29) §70) and in the *synodic month* is now explicitly made, the importance of which will become apparent in Propositions XXX–XXXV.

First, we may note for convenience, in comparing Newton's numerical values with their analytical equivalents, that

$$m^2 = (0{\cdot}0748025)^2 = 0{\cdot}005595414 \simeq \frac{1000}{178718} \tag{1}$$

and

$$\tfrac{3}{2}m^2 = 0{\cdot}00839323 \simeq \frac{100}{11915}. \tag{2}$$

Thus, equation (43), §70 is now replaced by

$$\frac{h_1}{h_0} = \frac{1 + \frac{3}{4}m^2}{1 - \frac{3}{4}m^2} = \frac{11965}{11865}. \tag{3}$$

Since

$$m_{\text{synodic}} = m_{\text{nodical}} \frac{29{\cdot}531}{27{\cdot}322} = 1{\cdot}08085 m_{\text{nodical}}, \tag{4}$$

equation (3) expressed in terms of the nodical month becomes

$$\left(\frac{h_1}{h_0}\right)_{\text{synodic month}} = \frac{11131}{11023}. \tag{5}$$

Therefore, equation (29), §70,

$$\begin{aligned} \tfrac{1}{2}\,\mathrm{d}h &= -\tfrac{3}{4}m^2\langle nr^2\rangle_{\text{AV}} \sin 2\psi . n\,\mathrm{d}t \\ &= -\frac{1}{238{\cdot}3}\langle nr^2\rangle_{\text{AV}} \sin 2\psi (n_{\text{nodical}})\,\mathrm{d}t, \end{aligned} \tag{6}$$

which, when expressed in terms of nodical months, after multiplication by 1·08085, becomes

$$\tfrac{1}{2}\,\mathrm{d}h = -\frac{1}{220{\cdot}5}\langle nr^2\rangle_{\mathrm{AV}}\sin 2\psi(n_{\mathrm{synodic}})\,\mathrm{d}t. \tag{7}$$

Newton gives 1/219·46 in place of 1/220·5.

The next Proposition XXVII is self-explanatory.

Proposition XXVII. Problem VIII

From the hourly motion of the Moon to find its distance from the Earth.

The area which the Moon, by a radius drawn to the Earth, describes in every moment of time, is as the hourly motion of the Moon and the square of the distance of the Moon from the Earth conjointly. And therefore the distance of the Moon from the Earth varies directly as the square root of the area and inversely as the square root of the hourly motion, taken jointly. Q.E.I.

COR. I. Hence the apparent diameter of the Moon is given; for it is inversely as the distance of the Moon from the Earth. Let astronomers try how accurately this rule agrees with the phenomena.

COR. II. Hence also the orbit of the Moon may be more exactly defined from the phenomena than hitherto could be done.

120.

Proposition XXVIII. Problem IX

To find the diameters of the orbit, in which, without eccentricity, the Moon would move.

This proposition is devoted to the derivation of the important equation (41) of §70 of Chapter 14. Most of the substance of this proposition has already been included in §§70(b) and (c). But one detail which was passed over (rather casually, it must be admitted) is spelled out here. The detail that was omitted was a rationale for the assumption (21), §70(c). Newton derived this relation on the assumption that the orbit is a prolate ellipse—an assumption 'permissible to inventors in researches as difficult as these', as Laplace commented.

We shall give Newton's derivation of the relation

$$r = 1 - x\cos 2\psi. \tag{1}$$

On the assumption that the orbit is a prolate ellipse, we can write,

$$\frac{1}{r^2} = \frac{\cos^2\psi}{a^2} + \frac{\sin^2\psi}{b^2} \qquad (a^2 < b^2). \tag{2}$$

We obtain successively,

$$\frac{1}{r^2} = \frac{1}{2a^2b^2}[a^2(1 - \cos 2\psi) + b^2(1 + \cos 2\psi)]$$

$$= \frac{1}{2a^2b^2}[(a^2 + b^2) + (b^2 - a^2)\cos 2\psi] \tag{3}$$

or

$$\frac{1}{r} = \frac{\sqrt{(a^2 + b^2)}}{ab\sqrt{2}}\left(1 + \frac{b^2 - a^2}{a^2 + b^2}\cos 2\psi\right)^{1/2}. \tag{4}$$

For $a - b \ll 1$, justifiable in the context,

$$r \simeq \frac{ab\sqrt{2}}{\sqrt{(a^2 + b^2)}}\left(1 - \frac{1}{2}\frac{b^2 - a^2}{a^2 + b^2}\cos 2\psi\right). \tag{5}$$

Now, writing

$$\left.\begin{array}{lll} a = 1 - x, & b = 1 + x, & \\ a^2 + b^2 = 2(1 + x^2) & \text{and} & b^2 - a^2 = 4x, \end{array}\right\} \tag{6}$$

we obtain

$$r = 1 - x\cos 2\psi + O(x^2). \tag{7}$$

From this point on the analysis proceeds as in §70(b) and (c) with the end result

$$x = \tfrac{3}{2}m^2\frac{1 + \dfrac{1}{1 - m}}{4(1 - m)^2 - 1} = m^2(1 + \tfrac{19}{6}m) + O(m^4) \tag{8}$$

which, as we have noted in §70(c), is correct to $O(m^3)$.* Voilà!

121.

Proposition XXIX. Problem X

To find the variation of the Moon.

The perturbed elliptical orbit, described about the centre, will clearly not describe equal areas in equal times; and as a consequence there will be a gain or loss in longitude (i.e., a change in the angle ψ) relative to the unperturbed uniformly described circular orbit. The resulting inequality is the *variation*. Newton solves this problem in a manner that manifests once again his passing inventiveness.

* See, for example, E. W. Brown, *An introductory treatise on the lunar theory* (Cambridge University Press, 1896), p. 110.

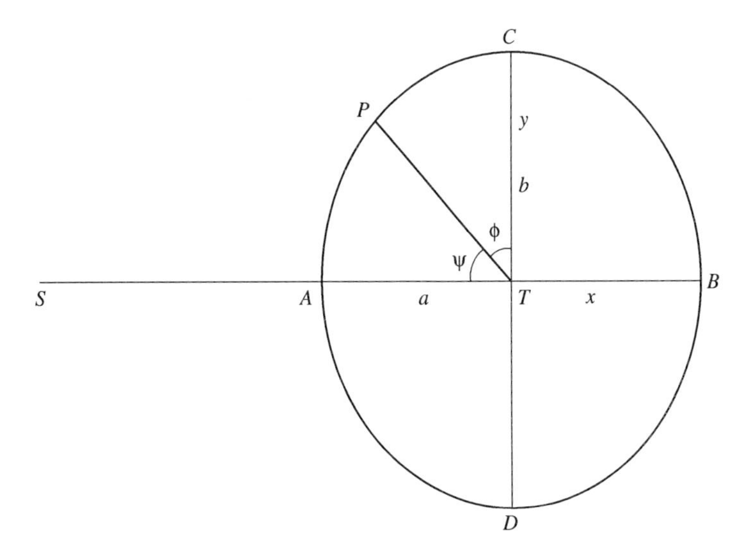

Let the equations governing the elliptical orbit be

$$x = r\sin\phi, \qquad y = r\cos\phi \qquad \left(\phi = \frac{\pi}{2} - \psi\right), \tag{1}$$

and

$$\frac{x^2}{a^2} + \frac{y^2}{b^2} = 1 \qquad (a < b), \tag{2}$$

or, equivalently,

$$\frac{1}{r^2} = \frac{\sin^2\phi}{a^2} + \frac{\cos^2\phi}{b^2}. \tag{3}$$

Newton first shows that this elliptical orbit *can* be considered as describing equal areas in equal times with a constant of areas h by the *postulated* equations,

$$r^2\frac{\mathrm{d}\phi}{\mathrm{d}t} = C, \tag{4}$$

and

$$\tan\phi = \frac{a}{b}\tan ht, \qquad h = \frac{C}{ab}. \tag{5}$$

For, by differentiating equation (5) with respect to time we obtain,

$$\sec^2\phi\,\frac{\mathrm{d}\phi}{\mathrm{d}t} = \frac{a}{b}h\sec^2 ht. \tag{6}$$

The left-hand side of this equation can be successively transformed with the aid of equations (3), (4), and (5) in the manner,

$$\frac{C}{r^2}\sec^2\phi = hab\sec^2\phi\left(\frac{\sin^2\phi}{a^2}+\frac{\cos^2\phi}{b^2}\right) = \frac{a}{b}h\left(1+\frac{b^2}{a^2}\tan^2\phi\right), \tag{7}$$

while the right-hand side becomes

$$\frac{a}{b}h(1+\tan^2 ht) = \frac{a}{b}h\left(1+\frac{b^2}{a^2}\tan^2\phi\right). \tag{8}$$

Q.E.D!

Returning to the problem on hand, when the orbit does not describe equal areas in equal times and

$$r^2\frac{\mathrm{d}\phi}{\mathrm{d}t} = C \quad \text{(no longer a constant)}, \tag{9}$$

we assume that

$$\tan\phi = \lambda\tan ht, \tag{10}$$

where λ is a constant to be determined. Proceeding as before, we find:

$$\begin{aligned}\lambda h\sec^2 ht &= \sec^2\phi\frac{\mathrm{d}\phi}{\mathrm{d}t} = \frac{C}{r^2}\sec^2\phi \\ &= C\sec^2\phi\left(\frac{\sin^2\phi}{a^2}+\frac{\cos^2\phi}{b^2}\right) = C\left(\frac{1}{b^2}+\frac{1}{a^2}\tan^2\phi\right) \\ &= C\left(\frac{1}{b^2}+\frac{\lambda^2}{a^2}\tan^2 ht\right);\end{aligned} \tag{11}$$

and therefore,

$$C = \frac{\lambda h}{(\lambda^2/a^2)\sin^2 ht + (1/b^2)\cos^2 ht}. \tag{12}$$

We now determine λ by evaluating the constant of areas C at quadratures and at syzygies. At quadratures,

$$\phi = ht = 0 \qquad \text{and} \qquad C = b^2\lambda h = h_0; \tag{13}$$

while at syzygies,

$$\phi = ht = \pi/2 \qquad \text{and} \qquad C = a^2 h/\lambda = h_1. \tag{14}$$

Combining these last two equations, we find

$$\frac{h_0}{h_1} = \frac{b^2\lambda^2}{a^2} \qquad \text{or} \qquad \lambda = \frac{a}{b}\sqrt{\frac{h_0}{h_1}}; \tag{15}$$

or, substituting for h_1/h_0 from equation (3), §119, and remembering that $a/b = (1 - x)/(1 + x)$, we obtain the required solution

$$\lambda = \frac{a}{b}\left(1 - \frac{3}{4}\frac{m^2}{1 - m}\right) = \frac{1 - x}{1 + x}\left(1 - \frac{3}{4}\frac{m^2}{1 - m}\right). \tag{16}$$

Now substituting for x from equation (8), §120, and for m its value, 0·074803, we find

$$\lambda = 0{\cdot}98126. \tag{17}$$

Equation (10) relating λ and the constant of areas h is therefore

$$\tan\phi = 0{\cdot}98126 \tan ht, \tag{18}$$

in agreement with Newton's coefficient

$$\frac{68{\cdot}6877}{70} = 0{\cdot}98125. \tag{19}$$

At the octants,

$$ht = \pi/4, \qquad \tan\phi = 0{\cdot}98126, \qquad \phi = 44{\cdot}458^\circ = 44^\circ\,27'\,29'', \tag{20}$$

and

$$45^\circ - \phi = 32'\,31'' \quad \text{(Newton gives } 32'\,32''\text{)}. \tag{21}$$

This calculation assumes that the constant of areas is expressed in terms of nodical months; it is more relevant in the context to express it in terms of synodic months. In the latter case, the value $32'\,31''$ should be multiplied by 1.08085 and we shall have instead,

$$45^\circ - \phi = 35'\,9'', \tag{22}$$

again in agreement with Newton's value, $35'\ 10''$.

More generally, by writing equation (18) in the form

$$\tan\phi = (1 - \varepsilon)\tan ht \quad (\varepsilon = 0{\cdot}01874), \tag{23}$$

and solving to the first order in ε by writing $\phi = ht + \delta$ and considering δ as small, we find

$$\left.\begin{aligned} \delta &= -\tfrac{1}{2}\varepsilon \sin(2ht) = -0{\cdot}00937 \sin 2\psi \quad \text{radians,} \\ &= -32{\cdot}21'' \sin 2\psi \quad (h \text{ measured in sidereal time}), \\ &= -34{\cdot}82'' \sin 2\psi \quad (h \text{ measured in synodic time}), \end{aligned}\right\} \tag{24}$$

an equation that is implicit in Newton.

Newton concludes with the statement:

> And this is its magnitude in the mean distance of the Sun from the Earth, neglecting the differences which may arise from the curvature of the great orbit,

and the stronger action of the Sun upon the Moon when horned and new, than when gibbous and full. In other distances of the Sun from the Earth, the greatest variation is in a ratio compounded, directly of the square of the ratio of the time of the synodic revolution of the Moon (the time of the year being given), and inversely as the cube of the ratio of the distance of the Sun from the Earth. And, therefore, in the apogee of the Sun, the greatest variation is 33′ 14″, and in its perigee 37′ 11″, if the eccentricity of the Sun is to the transverse semidiameter of the great orbit as $16\frac{15}{16}$ to 1000.

Hitherto we have investigated the variation in an orbit not eccentric, in which, to wit, the Moon in its octants is always in its mean distance from the Earth. If the Moon, on account of its eccentricity, is more or less removed from the Earth than if placed in this orbit, the variation may be something greater, or something less, than according to this rule. But I leave the excess or defect to the determination of astronomers from the phenomena.

122. The variation in the ascending node, Ω: Proposition XXX

It was shown in Book I, Proposition LXVI, Corollary XI (Chapter 14, §72(c), equations (10)) that the variation in the angle Ω of an initially circular orbit, by the tidal action of the Sun, is given by

$$\frac{d\Omega}{dt} = -3m^2 n \sin v \sin U \cos \psi. \qquad (1)$$

In Proposition XXX, Newton provides an *ab initio* derivation of this equation.

Proposition XXX. Problem XI

To find the hourly motion of the nodes of the Moon in a circular orbit.

The following is an insubstantial rearrangement of Newton's arguments.

As has been shown in Proposition XXVI (or alternatively in Chapter 14, §70, equations (10)) $3PK$ is proportional to the disturbing force acting in the direction of the Sun. It is clear that only the component of this force in the direction of the ascending node in the plane of the ecliptic can affect the direction of Ω. Referring to the adjoining figure (which is the same as the figure in Chapter 13, §66, (b), with the correspondence $E \to T$ and $M \to P$ in the figure on p. 239), Newton first observes that the required component of $3PK$ is

$$\begin{aligned} -3PK(\sin \angle \overrightarrow{SE}, \overrightarrow{Ev}) \sin \iota (\sin \angle \overrightarrow{vE}, \overrightarrow{EM}) &= -3TP\frac{PK}{TP} \sin U \sin \iota \sin v \\ &= -3TP \cos \psi \sin U \sin \iota \sin v. \qquad (2) \end{aligned}$$

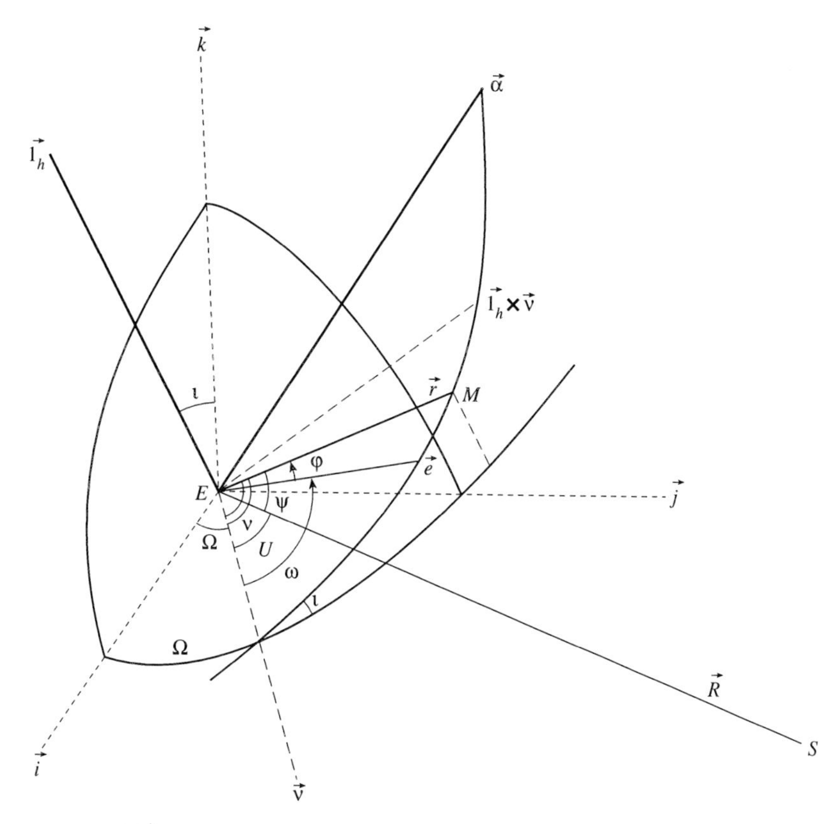

The force normal to $\overrightarrow{vE}$ along the ecliptic is, therefore,

$$-3N^2 r \sin\iota \sin\nu \sin U \cos\psi = F_{\|} \quad \text{(say)}. \tag{3}$$

Making use of the elementary relations (see Chapter 4, §20)

$$\left.\begin{aligned} F_{\|} &= v \sin\iota \frac{\mathrm{d}\Omega}{\mathrm{d}t} \\ \text{and} \qquad \frac{r}{h} &= \frac{r}{r^2\,\mathrm{d}\varphi/\mathrm{d}t} = \frac{1}{r\,\mathrm{d}\varphi/\mathrm{d}t} = \frac{1}{v_s} \end{aligned}\right\} \tag{4}$$

where v_s denotes the tangential velocity along the ecliptic, we obtain

$$\frac{\mathrm{d}\Omega}{\mathrm{d}t} = -3N^2 \frac{r^2}{h} \sin\nu \sin U \cos\psi. \tag{5}$$

Remembering that $m^2 = N^2/n^2$ and $h/r^2 = n$, for a circular orbit, we have

$$\frac{\mathrm{d}\Omega}{\mathrm{d}t} = -3m^2 n \sin\nu \sin U \cos\psi. \tag{6}$$

Since (as in the figure below)

$$\sin v = \frac{PH}{PT}, \qquad \sin U = \frac{AZ}{AT} \qquad \text{and} \qquad \cos\psi = \frac{IT}{PT}, \tag{7}$$

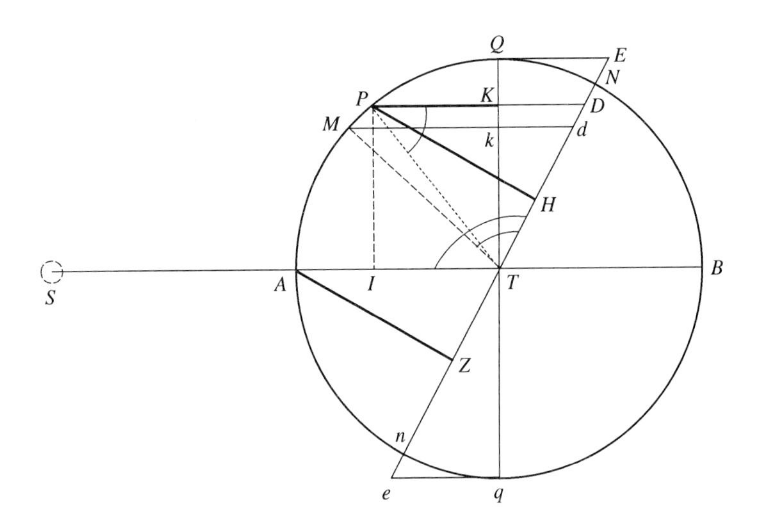

we have the result as expressed by Newton:

> the velocity of the nodes is as $IT \cdot PH \cdot AZ$, or as the product of the sines of the three angles TPI, PTN, and STN.

Rewriting equation (6) in the form,

$$\mathrm{d}\Omega = -\tfrac{3}{4}m^2 n(1 + \cos 2\psi - \cos 2v - \cos 2U)\,\mathrm{d}t, \tag{8}$$

we conclude that

$$\langle \mathrm{d}\Omega \rangle_{\mathrm{AV.}} = -\tfrac{3}{4}m^2 \quad \text{per nodical month.} \tag{9}$$

With the numerical values

$$m = 0{\cdot}0748026, \qquad m^2 = 0{\cdot}00559544, \tag{10}$$

and

$$2\pi(3m^2/4) = 90{\cdot}646' \quad \text{per nodical month,} \tag{11}$$

we find

$$\left.\begin{aligned} \langle \mathrm{d}\Omega \rangle_{\mathrm{AV}} &= -\left(\frac{90{\cdot}646'}{60m}\right)^{\circ} = -20{\cdot}1967^{\circ} \quad \text{per year} \\ &= -8{\cdot}2942'' \quad \text{per hour.} \end{aligned}\right\} \tag{12}$$

We can therefore write

$$\mathrm{d}\Omega = -33{\cdot}1768''(= -33''\ 10^{\mathrm{th}}\ 36^{\mathrm{IV}})\cos\psi \sin v \sin U \quad \text{per hour;} \tag{13}$$

or, as Newton expresses;

> hourly motion will be to $33^{s}\ 10^{th}\ 33^{IV}\ 12^{V}$ as the product of the sines of the three angles *TPI*, *PTN*, and *STN* (or of the distances of the Moon from the quadrature, of the Moon from the node, and of the node from the Sun) to the cube of the radius;

and he concludes with the statement of surpassing clarity:

> And as often as the sine of any angle is changed from positive to negative, and from negative to positive, so often must the regressive be changed into a progressive, and the progressive into a regressive motion. Whence it comes to pass that the nodes are progressive as often as the Moon happens to be placed between either quadrature, and the node nearest to that quadrature. In other cases they are regressive, and by the excess of the regress above the progress, they are monthly transferred backwards.

Corollary I

In this corollary, Newton provides an alternative from of equation (6) which he will use to advantage in Proposition XXXI.

Writing (see figure on p. 432)

$$\cos\psi \sin v \sin U = \cos\psi \frac{PH}{PT} \sin U, \tag{14}$$

where $PT = AT = r$, the radius of the unperturbed orbit, we observe that

$$\sin U = \frac{AZ}{AT} = \frac{PH}{PD} \qquad \text{or} \qquad PH = PD \sin U. \tag{15}$$

Therefore,

$$\cos\psi \sin v \sin U = \frac{1}{r} \cos\psi\, PD \sin^2 U. \tag{16}$$

Since*

$$\cos\psi = \frac{Kk}{PM}, \tag{17}$$

* *M* is a neighbouring point of *P* and *MkD* is drawn parallel to *PKD*.

we can write,

$$\cos\psi \sin v \sin U = \frac{1}{rPM} Kk.PD \sin^2 U$$

$$= \frac{\text{area of } PDdM}{2 \text{ area of sector } TMP} \sin^2 U \tag{18}$$

Equation (6) has, therefore, the alternative from

$$\frac{d\Omega}{dt} = -33{\cdot}1768'' \frac{\text{area of } PDdM}{2 \text{ area of sector } TPM} \frac{AZ^2}{AT^2} \text{ per hour.} \tag{19}$$

Q.E.D.

Corollary II

At syzygies,

$$\text{area of } PDdM = 2 \text{ area of sector } TMP$$

and

$$\left(\frac{d\Omega}{dt}\right)_{\text{syzygies}} = -33{\cdot}1768'' \frac{AZ^2}{AT^2} = -33{\cdot}1768'' \sin^2 U \quad \text{per hour.} \tag{20}$$

Therefore, averaged over all initial values of U,

$$\left\langle \left(\frac{d\Omega}{dt}\right)_{\text{syzygies}} \right\rangle_{\text{AV}} = -16{\cdot}588'' \quad \text{per hour.} \tag{21}$$

Q.E.D.

123.

Proposition XXXI. Problem XII

To find the hourly motion of the nodes of the Moon in an elliptic orbit.

In this proposition, Newton does *not* consider the generalization of Proposition XXX for the case of an initially unperturbed elliptic orbit, as one might have supposed (see below for the treatment of this case). He considers, instead, a second approximation to the problem considered in Proposition XXX by what amounts to an iterative procedure.

As has been shown in Proposition XXVIII (§120), an initially circular orbit becomes a prolate ellipse, with a ratio of the semiaxes,

$$\frac{b}{a} = \frac{1-x}{1+x} \qquad \text{where} \quad x = m^2(1 + \tfrac{19}{6}m), \tag{1}$$

(by equation (8) of §120), described about its centre, Newton addresses himself to the question of the regression of the nodes of *this* prolate ellipse.

Newton first considers the perturbation of the prolate elliptical orbit by the same disturbing force $3PK$ acting in the direction of the Sun as in Proposition XXX (as we may) and finds the same expression (§122, equation (5)),

$$\frac{\mathrm{d}\Omega}{\mathrm{d}t} = -\frac{1}{v_{\mathrm{s}}} 3N^2 r \sin v \sin U \cos \psi$$

$$= -3N^2 \frac{r^2}{h} \sin v \sin U \cos \psi, \tag{2}$$

where it should be noted that, in contrast to equation (5) (§122) of Proposition XXX, neither r nor h is now a constant. Newton then transforms this equation as in Corollary I of Proposition XXX. For simplicity, we shall first consider equation (2) more directly.

Since,

$$N = mn \qquad \text{and} \qquad n = \langle h \rangle_{\mathrm{av}}/ab, \tag{3}$$

equation (2) can be written alternatively in the form

$$\frac{\mathrm{d}\Omega}{\mathrm{d}t} = -3m^2 n \frac{r^2}{abh/\langle h \rangle_{\mathrm{av}}} \sin v \sin U \cos \psi. \tag{4}$$

Now substituting in the last equation, the known solutions (§120, equation (7) and §70(d), equation (32))

$$r = (1 - x \cos 2\psi) \qquad \text{and} \qquad \frac{h}{\langle h \rangle_{\mathrm{av}}} = 1 + \frac{3}{4}\frac{m^2}{1-m} \cos 2\psi, \tag{5}$$

we find to $O(x^2)$,

$$\frac{\mathrm{d}\Omega}{\mathrm{d}t} = -3m^2 n(1 - 2x \cos 2\psi)\left(1 - \frac{3}{4}\frac{m^2}{1-m}\cos 2\psi\right)\sin v \sin U \cos \psi, \tag{6}$$

or by equation (1), to the same order,

$$\frac{\mathrm{d}\Omega}{\mathrm{d}t} = -3m^2 n\{1 - [\tfrac{3}{4}m^2(1+m) + 2m^2(1 + \tfrac{19}{6}m)]\cos 2\psi\} \sin v \sin U \cos \psi. \tag{7}$$

Simplifying the quantity in braces in this last equation, we can write

$$\frac{\mathrm{d}\Omega}{\mathrm{d}t} = -\tfrac{3}{4}m^2 n(1 - q\cos 2\psi)(1 + \cos 2\psi - \cos 2v - \cos 2U), \tag{8}$$

where

$$q = \tfrac{11}{4}m^2(1 + \tfrac{85}{33}m) = 0{\cdot}018352. \tag{9}$$

Rewriting equation (9) in the form

$$\frac{d\Omega}{dt} = -\tfrac{3}{4}m^2n\{(1 + \cos 2\psi - \cos 2v - \cos 2U) - q\cos 2\psi - \tfrac{1}{2}q[1 + \cos 4\psi - \cos 2(\psi + v) - \cos 2(\psi - v) - \cos 2(\psi + U) - \cos 2(\psi - U)]\}, \tag{10}$$

and remembering that $v - U = \psi$, we can also write

$$\frac{d\Omega}{dt} = -\tfrac{3}{4}m^2n\{(1 + \cos 2\psi - \cos 2v - \cos 2U) - \tfrac{1}{2}q[2\cos 2\psi + 1 + \cos 4\psi - \cos 2(\psi + v) - \cos 2U - \cos 2v - \cos 2(\psi - U)]\}. \tag{11}$$

We now observe that the first term in curly brackets in the foregoing equation attains the average value 1 when the orbit has described the semicircle $(0, \pi)$ while the second term attains the average value $-\frac{1}{2}q$ when the orbit has described an octant $(0, \pi/2)$. Hence, when the orbit has described the semicircle, the average value of $\dot{\Omega}$ is

$$\left(\frac{d\Omega}{dt}\right)_{\text{syzygies}} = -3m^2n(1 - q)\sin^2 U \quad \text{per hour}, \tag{12}$$

or

$$\left\langle\left(\frac{d\Omega}{dt}\right)_{\text{syzygies}}\right\rangle_{\text{AV}} = -16{\cdot}5884'' \times 0{\cdot}98165 = -16{\cdot}284'' \quad \text{per hour}; \tag{13}$$

$$= -16''\ 17^{\text{th}}\ 4^{\text{IV}} \quad \text{per hour}, \tag{14}$$

in agreement with Newton's value, $16''\ 16^{\text{th}}\ 38^{\text{IV}}$.

(*a*) *Newton's procedure*

Newton first transforms equation (2) in the same manner as in Proposition XXX, Corollary 1. Proceeding as in §122, we now find (see figure on p. 438):

$$\begin{aligned}\sin\theta \sin v \sin U &= \sin\theta \frac{PH}{r}\sin U \\ &= \sin\theta \frac{PD}{r}\sin^2 U \quad (\sin PH = PD \sin U) \\ &= \frac{\sin\theta}{rPL}(PD.PL)\sin^2 U \\ &= \frac{\sin\theta}{PL/MQ}\,\frac{PD.PL}{rMQ}\sin^2 U \\ &= \frac{\sin\theta}{PL/MQ}\,\frac{\text{area of } PDdM}{2\ \text{area of sector } TPM}\sin^2 U.\end{aligned} \tag{15}$$

Noting that

$$PL = PM\cos\chi \qquad \text{and} \qquad MQ = PM\sin(\theta + \chi), \tag{16}$$

where χ denotes the angle which the tangent at P makes with the x-direction, we obtain

$$\sin\theta\sin v\sin U = \frac{\sin\theta\sin(\theta+\chi)}{\cos\chi}\,\frac{\text{area of } PDdM}{2\text{ area of sector } TPM}\sin^2 U$$

$$= \sin^2\theta\,(1 + \cot\theta\tan\chi)\,\frac{\text{area of } PDdM}{2\text{ area of sector } TPM}\sin^2 U. \tag{17}$$

By making use of the known relation*

$$\tan\chi = \frac{b^2}{a^2}\cot\theta, \tag{18}$$

equation (17) reduces equation (4) to the form,

$$\frac{\mathrm{d}\Omega}{\mathrm{d}t} = -3m^2 n\,\frac{r^2}{abh/\langle h\rangle_{\mathrm{av}}}\left(\sin^2\theta + \frac{b^2}{a^2}\cos^2\theta\right)\frac{\text{area of } PDdM}{2\text{ area of sector } TPM}\sin^2 U. \tag{19}$$

At syzygies, where

$$r^2 = b^2, \qquad \theta = \pi/2, \qquad \text{and} \qquad \text{area of } PDdM = 2\text{ area of sector } TPM, \tag{20}$$

$$\left(\frac{\mathrm{d}\Omega}{\mathrm{d}t}\right)_{\text{syzygies}} = -33{\cdot}1768''\left[\frac{\langle h\rangle_{\mathrm{av}}}{h}\right]_{\text{syzygies}}\frac{b}{a}\sin^2 U. \tag{21}$$

* This relation follows from a known property of the ellipse: that the line,

$$y = -x\tan\chi + c, \tag{i}$$

touches the ellipse at x and y for c given by

$$c = \sqrt{(a^2\tan^2\chi + b^2)}, \tag{ii}$$

that is,

$$y + x\tan\chi = \sqrt{(a^2\tan^2\chi + b^2)}. \tag{iii}$$

Inserting the relations,

$$y = r\sin\theta, \qquad x = r\cos\theta, \qquad \text{where } \frac{1}{r^2} = \frac{\cos^2\theta}{a^2} + \frac{\sin^2\theta}{b^2}, \tag{iv}$$

in (iii), we obtain

$$(\sin\theta + \cos\theta\tan\chi)^2 = \left(\frac{\cos^2\theta}{a^2} + \frac{\sin^2\theta}{b^2}\right)(a^2\tan^2\chi + b^2). \tag{v}$$

Expanding and simplifying, we find

$$2\sin\theta\cos\theta\tan\chi = \frac{a^2}{b^2}\sin^2\theta\tan^2\chi + \frac{b^2}{a^2}\cos^2\theta, \tag{vi}$$

or

$$\frac{a^2}{b^2}\tan^2\chi + \frac{b^2}{a^2}\cot^2\theta - 2\cot\theta\tan\chi = 0. \tag{vii}$$

It follows that

$$\frac{a}{b}\tan\chi = \frac{b}{a}\cot\theta \qquad \text{or} \qquad \tan\chi\tan\theta = \frac{b^2}{a^2}. \tag{viii}$$

This last relation is not commonly found in standard books on conic sections, but it is derived in George Salmon, *A treatise on conic sections*, Longmans, Green & Co., London, 1879, p. 167.

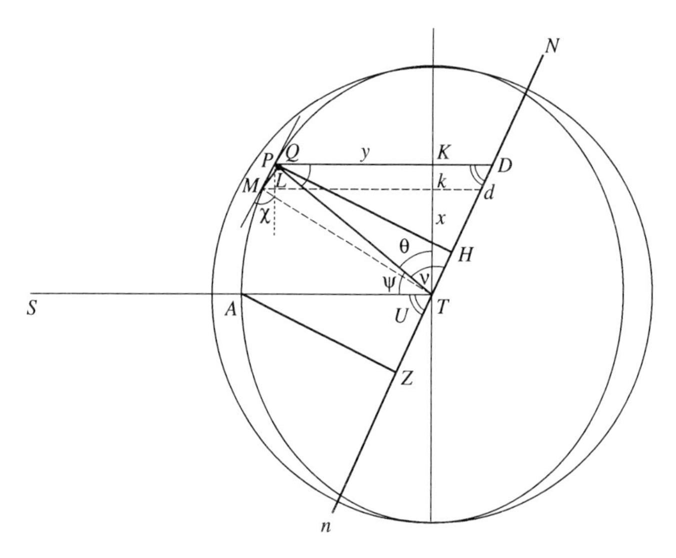

Since (see Chapter 14, §70(e), equation (42))

$$\frac{b}{a} = \frac{1-x}{1+x} = 0{\cdot}98571 \quad (\simeq \tfrac{69}{70}), \tag{22}$$

we can write,

$$\begin{aligned}\left(\frac{\mathrm{d}\Omega}{\mathrm{d}t}\right)_{\text{syzygies}} &= -33{\cdot}1768''(1 - 0{\cdot}01429)\left[\frac{\langle h\rangle_{\text{av}}}{h}\right]_{\text{syzygies}} \sin^2 U \\ &= -32.703''(=32''\ 42^{\text{th}}\ 22^{\text{IV}})\left[\frac{\langle h\rangle_{\text{av}}}{h}\right]_{\text{syzygies}} \sin^2 U,\end{aligned} \tag{23}$$

which agrees at this point with Newton's coefficient, $32^{\text{s}}\ 42^{\text{th}}\ 7^{\text{IV}}$.

Next, Newton inserts for $[\langle h\rangle_{\text{av}}/h]_{\text{syzygies}}$ the value

$$[\langle h\rangle_{\text{av}}/h]_{\text{syzygies}} = 1 - \frac{3}{4}\frac{m^2}{1-m} = (1 - 0{\cdot}0045356). \tag{24}$$

Hence

$$\begin{aligned}\left(\frac{\mathrm{d}\Omega}{\mathrm{d}t}\right)_{\text{syzygies}} &= -32{\cdot}703'' \times 0{\cdot}99546 \sin^2 U \\ &= -32{\cdot}5545'' \sin^2 U = -32''\ 33^{\text{th}}\ 16^{\text{IV}} \sin^2 U\end{aligned} \tag{25}$$

in agreement with Newton's value $-32^{\text{s}}\ 33^{\text{th}}\ 15^{\text{IV}}$.

The required agreement follows also from the fact that

$$\begin{aligned}(1 - 0{\cdot}01429)(1 - 0{\cdot}004536) &\simeq (1 - 0{\cdot}01429 - 0{\cdot}004536) \\ &= (1 - 0{\cdot}01883) \simeq (1 - q),\end{aligned} \tag{26}$$

to the accuracy expected.

The reader is left marvelling how Newton is able to explain all the requisite analysis (including the relation (18)!) in four pages of prose. One can understand Newton's increasing impatience to set out all the necessary details.

(b) Newton's transformation of the equation for $\mathrm{d}\Omega/\mathrm{d}t$ *for a Kepler ellipse*

To illustrate the elegance and generality of Newton's transformation of the equation governing the regression of the nodes, we shall consider the case of the unperturbed orbit describing a Kepler ellipse. (We may assume that Newton had carried out this transformation.)

We start with the equation (3) of Chapter 13, §66(c),

$$\frac{\mathrm{d}\Omega}{\mathrm{d}t} = -3m^2 n \frac{r^2}{ab} \cos\psi \sin v \sin U, \tag{27}$$

where we have replaced $a^2\sqrt{(1-e^2)}$ by ab. It is perhaps worth noticing that by making use of relation

$$(\pi ab)/\pi = h(\text{period}/2\pi) = h/n, \tag{28}$$

equation (27) takes the same form,

$$\frac{\mathrm{d}\Omega}{\mathrm{d}t} = -3N^2 \frac{r^2}{h} \sin\varphi \sin v \sin U \qquad \left(\varphi = \frac{\pi}{2} - \psi\right), \tag{29}$$

as equation (5) of §122 (Proposition XXX)—an example of the keenness of Newton's insight.

Making the same sequence of transformations as in equation (15), we now obtain

$$\frac{\mathrm{d}\Omega}{\mathrm{d}t} = -3m^2 n \frac{r^2}{ab} \frac{\sin\varphi}{PL/MQ} \frac{\text{area of } PDdM}{2 \text{ area of sector } TPM} \sin^2 U. \tag{30}$$

If χ denotes the angle which the tangent at P makes with PT (i.e., the direction of $\vec{r}$)

$$PL = PM\cos(\chi - \varphi) \qquad \text{and} \qquad MQ = PM \sin\chi, \tag{31}$$

and

$$PL/MQ = \cos(\chi - \varphi)/\sin\chi = (1 + \cot\varphi\cot\chi)\sin\varphi. \tag{32}$$

Equation (30) now reduces to

$$\frac{\mathrm{d}\Omega}{\mathrm{d}t} = -3m^2 n \frac{r^2}{ab} \frac{1}{(1 + \cot\varphi\cot\chi)} \frac{\text{area of } PDdM}{2 \text{ area of sector } TPM} \sin^2 U. \tag{33}$$

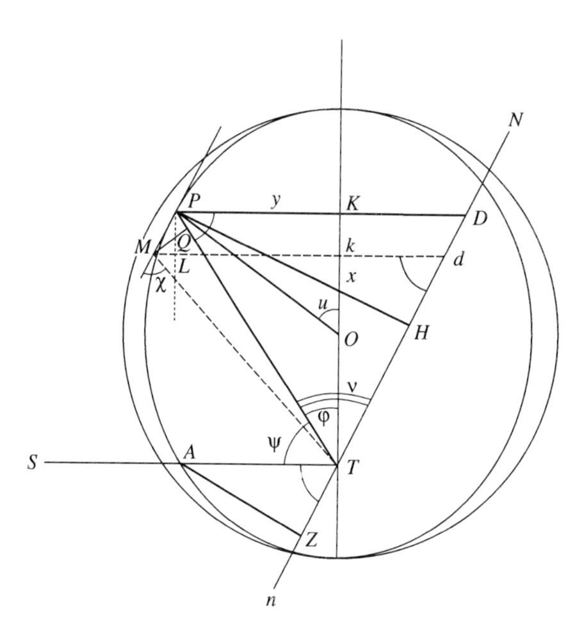

By making use of the known properties of conic sections and of equations (20)–(24), of Chapter 13, §65(e), we find*

$$\cot\varphi\cot\chi = \frac{e}{1-e^2}\frac{r}{a}\cos\varphi = \frac{e\cos\varphi}{1+e\cos\varphi}; \tag{34}$$

and equation (33) reduces to

$$\frac{d\Omega}{dt} = -3m^2n\frac{r^2}{ab}\frac{1}{1+e\cos\varphi/(1+e\cos\varphi)}\frac{\text{area of } PDdM}{2\text{ area of sector } TPM}\sin^2 U. \tag{35}$$

* In the lettering of the adjoining figure, QR is the tangent at P to the ellipse with foci at S and S', and SQ is drawn perpendicular to PR. Then,

$$\sin\chi = SQ/r, \qquad SQ^2 = p^2, \qquad \text{and} \qquad b^2 = p^2\left(\frac{2a}{r}-1\right); \tag{i}$$

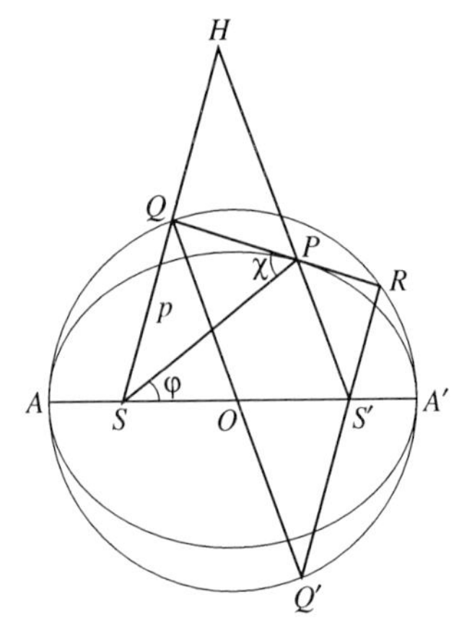

(*Footnote continues opposite*)

At syzygies,

$$\left.\begin{aligned}&\varphi = \pi/2, \qquad r = l = b^2/a, \qquad \text{and} \qquad \text{area of } PDdM = 2 \text{ area of sector } TPM;\\ &\text{and}\\ &\qquad \frac{d\Omega}{dt} = -3m^2 n\left(\frac{b}{a}\right)^3 \sin^2 U,\end{aligned}\right\} \quad (36)$$

a simple generalization of equation (20) of §122.

124. Propositions XXXII and XXXIII

In these two propositions Newton continues his consideration of the regression of the nodes along lines that are both novel and original.

Proposition XXXII. Problem XIII

To find the mean motion of the nodes of the Moon.

In Propositions XXX and XXXI it has been shown that

$$\left(\frac{d\Omega}{dt}\right)_{\text{syzygies}} = -2R \sin^2 U \quad \text{per sidereal year,} \tag{1}$$

where

$$R = 20{\cdot}1967^\circ \tag{2}$$

for a circular orbit, and

$$R = 20{\cdot}1967^\circ(1 - q) = 19{\cdot}826^\circ (= 19^\circ\ 49'\ 34'') \tag{3}$$

for the elliptical orbit described about the centre.

One is generally wont to consider R as the mean annual regression of the nodes. But as Newton points out:

---

and it follows:

$$\begin{aligned}\cot^2\chi &= \operatorname{cosec}^2\chi - 1 = \frac{r^2}{p^2} - 1 = \frac{r^2}{b^2}\left(\frac{2a}{r} - 1\right) - 1\\ &= \frac{r^2}{a^2(1-e^2)}\left(\frac{2a}{r} - 1\right) - 1 = \frac{1}{1-e^2}\left[e^2 - \left(\frac{r}{a} - 1\right)^2\right]\\ &= \frac{1}{1-e^2}(e^2 - e^2\cos^2 u) \quad \text{[by equation (23), Chapter 13, §65(e)]}\end{aligned} \tag{ii}$$

Therefore

$$\cot\chi = \frac{e}{\sqrt{(1-e^2)}}\sin u = \frac{e}{1-e^2}\frac{r}{a}\sin\varphi \quad \text{[by Chapter 13, §65(e), equations (20)–(24)]} \tag{iii}$$

and

$$\cot\varphi\cot\chi = \frac{e}{1-e^2}\frac{r}{a}\cos\varphi \tag{iv}$$

which is the relation to be established.

> Thus it would be if the node was after every hour drawn back again to its former place, that so, after a complete revolution, the Sun at the year's end would be found again in the same node which it had left when the year began. But, because of the motion of the node in the meantime, the Sun must needs meet the node sooner; and now it remains that we compute the abbreviation of the time.

'Since'—to continue with the quotation—from an assigned value of U 'the Sun, in the course of the year travels 360 degrees, and the node in the same time would be carried' $2R \sin^2 U$, the mean motion for an increment of $\mathrm{d}U$ in U is

$$-\frac{2R \sin^2 U}{360^\circ + 2R \sin^2 U}\,\mathrm{d}U = -\frac{Q \sin^2 U}{1 + Q \sin^2 U}\,\mathrm{d}U, \tag{4}$$

where

$$Q = R/180. \tag{5}$$

Therefore, the mean regression coefficient, as the Sun continues on its onward course in its annual orbit, is

$$\bar{Q} = \int_0^{\pi/2} \frac{Q \sin^2 U}{1 + Q \sin^2 U}\,\mathrm{d}U. \tag{6}$$

On evaluating this integral, we find:

$$\bar{Q} = \tfrac{1}{2}\pi\left[1 - \frac{1}{\sqrt{(Q+1)}}\right]. \tag{7}$$

For Newton's preferred value of $R = 19{\cdot}826^\circ$,

$$Q = 0{\cdot}11014, \tag{8}$$

and

$$\bar{Q} = \tfrac{1}{2}\pi(1 - 0{\cdot}949094) = \tfrac{1}{2}\pi \times 0{\cdot}050904 = 0{\cdot}07996. \tag{9}$$

This value for $\bar{Q}$ should be contrasted with

$$\bar{Q} = \tfrac{60}{793} = 0{\cdot}0756 \tag{10}$$

which Newton had 'found by the method of infinite series, nearly'.

To return to the original problem, we conclude that during a sidereal year, 'the Sun is carried' by the amount,

$$R(1 - \bar{Q}) = 19{\cdot}826^\circ \times 0{\cdot}92004 = 18{\cdot}2407^\circ = 18^\circ\ 14'\ 15''. \tag{11}$$

Hence, the mean regression coefficient is

$$-\frac{18{\cdot}2407 \times 360}{360 - 18{\cdot}2407} = -\frac{18{\cdot}2407 \times 360}{341{\cdot}7593} = -19{\cdot}214^\circ = -19^\circ\ 12'\ 50''. \tag{12}$$

which is to be compared with Newton's value, $19^\circ\ 18'\ 1''$, derived with his approximate value for $\bar{Q}$, but in sufficiently good agreement with the value $19^\circ\ 21'\ 21''\ 50'''$ 'by astronomical tables'.

Proposition XXXIII. Problem XIV

To find the true motion of the nodes of the Moon.

To determine the true motion of the nodes one must integrate the equation governing $d\Omega/dt$, that is, equation (8), §122, for the circular orbit, and equation (11) of §123, for the elliptical orbit.

We have already shown in Chapter 14 that equation (24), §72(c), for an initially circular orbit, can be directly integrated with the substitutions

$$\psi = (1 - m)v, \qquad U = mv, \qquad \text{and} \qquad dv = n\,dt, \tag{13}$$

to give, retaining only the dominant one among the periodic terms,

$$\Omega = -\tfrac{3}{4}m^2nt + \tfrac{3}{8}m \sin 2U. \tag{14}$$

The first term in equation (15) gives a secular decrease in Ω given by the 'mean' regression coefficient R, and the second term represents a semi-annual periodic variation with an amplitude

$$\tfrac{3}{8}m = 1{\cdot}6072^\circ. \tag{15}$$

For a semi-synodic-year the corresponding amplitude is

$$1{\cdot}6072^\circ/1{\cdot}08085 = 1{\cdot}4887^\circ = 1^\circ\,29'\,20''. \tag{16}$$

By the same substitutions (4) we similarly obtain from equation (11) of §123,

$$\Omega = (-\tfrac{3}{4}m^2nt + \tfrac{3}{8}m \sin 2U)(1 - \tfrac{1}{2}q). \tag{17}$$

The secular regression coefficient is now the mean of the two values (2) and (3), that is,

$$\left.\begin{aligned} R &= \tfrac{1}{2}(20{\cdot}197^\circ + 19{\cdot}826^\circ) = 20{\cdot}012^\circ = 20^\circ\,0'\,39'' \quad \text{per year} \\ &= 16{\cdot}436'' = 16''\,26^{\text{th}}\,10^{\text{IV}} \quad \text{per hour.} \end{aligned}\right\} \tag{18}$$

It is not clear whether Newton's value $16''\,19^{\text{th}}\,26^{\text{IV}}$ (different from $16''\,16^{\text{th}}\,37^{\text{IV}}$ and $16''\,35''\,17^{\text{th}}$, but not explained) is to be considered as discrepant from $16^\circ\,26^{\text{th}}\,10^{\text{IV}}$. (It is known that the numerical values given in these propositions were subject to many revisions—authorized and unauthorized!)

As for the amplitude of the semi-annual periodic variation for the iterated elliptical orbit, we have, in place of (15),

$$\tfrac{3}{8}m(1 - \tfrac{1}{2}q)/1{\cdot}08085 = 1{\cdot}4733^\circ = 1^\circ\,28'\,24'', \tag{19}$$

again a little different from the value $1^\circ\,30'$, that Newton gives.

125. The variation of the inclination

In Propositions XXXIV and XXXV Newton turns his attention to the variation of the inclination—a topic he has already elaborated on in Proposition LXVI, Corollary X, Book

I (see Chapter 14, §72(b)). He now provides an *ab initio* derivation of the basic equation (17), §72(b), namely,

$$\frac{d\iota}{dt} = -3m^2 n \frac{r^2}{ab} \sin\iota \cos v \sin U \cos\psi. \tag{1}$$

Proposition XXXIV. Problem XV

To find the hourly variation of the inclination of the Moon's orbit to the plane of the ecliptic.

Arguing as in Proposition XXX (§122), Newton first observes that the component of the force $3PK$, normal to the orbital plane, that affects the inclination is (see the figure on p. 431)

$$-3PK(\cos\angle\overrightarrow{EM},\overrightarrow{Ev})\sin\iota\,(\sin\angle\overrightarrow{Ev},\overrightarrow{ES}) = -3TP\frac{PK}{TP}\cos v \sin\iota \sin U$$
$$= -3TP\cos\psi\cos v\sin\iota\sin U. \tag{2}$$

The force that affects the inclination is therefore

$$-3N^2 r \sin\iota\cos v\sin U\cos\psi = F_\perp \quad \text{(say)}. \tag{3}$$

By making use of the elementary relations (see Chapter 4, §20)

$$\frac{d\iota}{dt} = \frac{v_s}{\rho} \qquad \text{and} \qquad F_\perp = \frac{v_s^2}{\rho} = \frac{h}{r}\frac{v_s}{\rho}, \tag{4}$$

where ρ denotes the radius curvature of the orbit and v_s the tangential velocity along it at P, we obtain the formula

$$\frac{d\iota}{dt} = -3N^2\frac{r^2}{h}\sin\iota\cos v\sin U\cos\psi. \tag{5}$$

With the further substitutions,

$$n = h/ab \qquad \text{and} \qquad N = mn, \tag{6}$$

we recover formula (1). And by restricting ourselves to a circular orbit, we obtain,

$$\frac{d\iota}{dt} = -3m^2 n \sin\iota\cos v\sin U\cos\psi. \tag{7}$$

Alternatively, by comparison with equation (13), §122, we can also write

$$\frac{d\iota}{dt} = -33''\,10'''\,36^{\text{IV}}\sin\iota\cos v\sin U\cos\psi \quad \text{per hour}. \tag{8}$$

Now, referring to the accompanying figure (which is the same as the figure on p. 432, in the context of Proposition XXX, with the additions that the neighbouring lines KP and kM are extended to meet TF, normal to nN, at D and d, respectively), we observe that

$$\cos\psi = \frac{IT}{PT}, \qquad \sin U = \frac{AZ}{AT}, \qquad \text{and} \qquad \cos v = \frac{TG}{PT}; \tag{9}$$

and since $AT = PT =$ the radius of the circular orbit, we can rewrite equation (8) in the form

$$\frac{\mathrm{d}\iota}{\mathrm{d}t} = -33''\,10'''\,36^{\mathrm{IV}}\,\frac{IT}{AT}\,\frac{AZ}{AT}\,\frac{TG}{AT}\sin\iota \quad \text{per hour.} \tag{10}$$

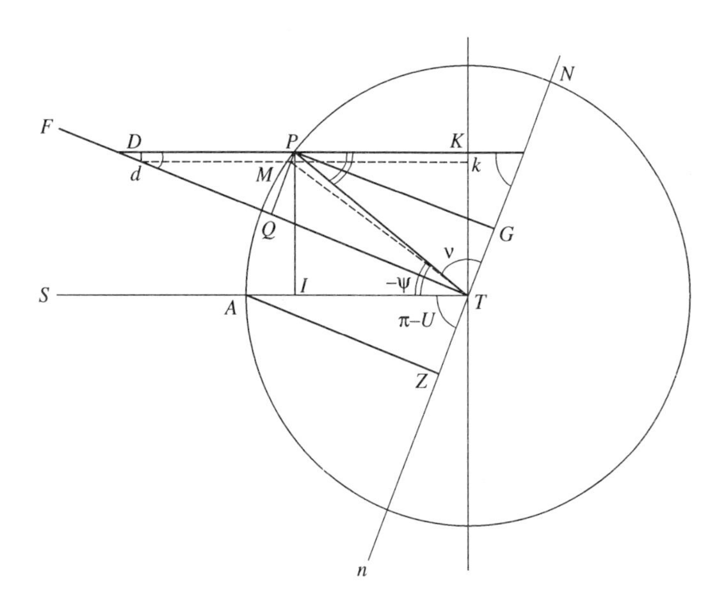

This is Newton's result which he expresses in the manner:

> the hourly variation of the inclination will be to the angle $33''\,10'''\,33^{\mathrm{IV}}$ as $IT.AZ.TG.\dfrac{Pp}{PG}$ to AT^3. Q.E.I.

where as he notes later,

> Pp is to PG as the sine of the aforesaid inclination to the radius.

Corollaries I–III

In these corollaries, Newton transforms equation (10) in the manner of equation (13), §122 for $\mathrm{d}\Omega$.

Referring once again to the figure, we now observe that

$$\cos\psi = \frac{Kk}{PM}, \qquad \cos\nu = \frac{TG}{PT}, \qquad \text{and} \qquad TG = PQ = PD\cos U, \tag{11}$$

we can therefore rewrite equation (7) in the form

$$\frac{\mathrm{d}\iota}{\mathrm{d}t} = -3m^2 n\sin\iota\,\frac{Kk.PD}{PT.PM}\cos U\sin U, \tag{12}$$

or, alternatively in either of the forms:

$$\frac{d\iota}{dt} = -3m^2n \frac{\text{area of } PDdM}{2 \text{ area of sector } TPM} \sin \iota \sin U \cos U. \tag{13}$$

or,

$$\frac{d\iota}{dt} = -33'' \, 10''' \, 36^{IV} \frac{\text{area of } PDdM}{\text{area of sector } TPM} \sin \iota \frac{AZ.ZT}{2AT^2}. \tag{14}$$

This last result is stated in different forms in Corollaries I–III.

Corollary IV

When the node is at quadrature,

$$U = \pi/2 \qquad \text{and} \qquad \psi = v - \pi/2 \tag{15}$$

and equation (8) gives,

$$\frac{d\iota}{dt} = +33{\cdot}1768'' \sin \iota \cos \psi \sin \psi \quad \text{per hour.} \tag{16}$$

In the time the Moon describes its orbit from the syzygies to the quadratures, namely, in

$$\tfrac{1}{4} \text{ period} = \frac{\pi}{2n} = 177{\cdot}167 \quad \text{hours,} \tag{17}$$

the variation in the inclination is

$$\Delta\iota \, (\text{for } \tfrac{1}{4} \text{ period}) = \frac{1}{\pi}(177{\cdot}167 \times 33{\cdot}1768'')\sin \iota \cos \psi \sin \psi \quad \text{degrees.} \tag{18}$$

For an inclination $\iota = 5^\circ \, 1'$* which Newton assumes,

$$\sin \iota = 0{\cdot}08745, \tag{19}$$

equation (18) gives:

$$\Delta\iota \, (\text{for } \tfrac{1}{4} \text{ period}) = 163{\cdot}612''(=2' \, 43'' \, 37''')\sin \psi \cos \psi, \tag{20}$$

with the coefficient agreeing with 163″ (=2′ 43″) that Newton gives.

More generally than equation (20), we can write

$$\Delta\iota \, (\text{for } \tfrac{1}{4} \text{ period}) = -0{\cdot}681725(\sin 2v - \sin 2\psi + \sin 2U). \tag{21}$$

Proposition XXXV. Problem XVI

To a given time to find the inclination of the Moon's orbit to the plane of the ecliptic.

The problem which Newton addresses himself in this proposition is to determine the dependence of the variation in the inclination on the parameters of the orbit, ι, ψ, and U; in other words, to integrate equation (7). Newton obtains the solution by a combination of physical reasoning and mathematical approximation. But we shall verify that his

* The currently accepted value is 5° 9′.

findings are in surprisingly close accord with the results obtained from a direct integration of the equation—a splendid tour de force.

Writing equation (7) in the form (see Chapter 14, §72(b), equation (19)),

$$\frac{d\iota}{dt} = -\tfrac{3}{4}m^2 \sin \iota(\sin 2v - \sin 2\psi + \sin 2U)n \tag{22}$$

and replacing U, ψ, and v, as in equation (13) of §124, by

$$U = mv, \qquad \psi = (1 - m)v, \qquad \text{and} \qquad dv = n\,dt, \tag{23}$$

we can integrate the equation directly and obtain

$$\iota - \iota_0 = \tfrac{3}{8}m^2 \sin \iota \left(\cos 2v - \frac{\cos 2\psi}{1 - m} + \frac{1}{m}\cos 2U\right), \tag{24}$$

or, to a sufficient accuracy,

$$\iota - \iota_0 = \tfrac{3}{8}m[\cos 2U + m(\cos 2v - \cos 2\psi)]\sin \iota. \tag{25}$$

Alternatively, we may also write,

$$\iota - \iota_0 = [\tfrac{3}{8}m \cos 2U - \tfrac{3}{4}m^2 \sin(2\psi + U)\sin U]\sin \iota. \tag{26}$$

Inserting in equation (26) the numerical values,

$$\tfrac{3}{8}m = 1{\cdot}6072^\circ \qquad \text{and} \qquad \tfrac{3}{8}m^2 = 7{\cdot}2139', \tag{27}$$

and adopting for the inclination ι the value $5^\circ\ 1'$ for which

$$\sin \iota = 0{\cdot}08745, \tag{28}$$

we obtain,

$$\iota - \iota_0 = 8{\cdot}4326' \cos 2U - 1{\cdot}2616' \sin(2\psi + U)\sin U. \tag{29}$$

From this equation we find: for,

$$\left.\begin{array}{lll}
U = 0: & \iota - \iota_0 = +8{\cdot}4326', & \text{(i)}\\
U = \pi/4: & \iota - \iota_0 = -0{\cdot}6303'\,(\sin 2\psi + \cos 2\psi), & \text{(ii)}\\
U = \pi/2: & \iota - \iota_0 = -8{\cdot}4326' - 1{\cdot}2616' \cos 2\psi; & \text{(iii)}\\
\psi = 0: & \iota - \iota_0 = +8{\cdot}4326' \cos 2U - 1{\cdot}2616' \sin^2 U, & \text{(iv)}\\
\psi = \pi/4: & \iota - \iota_0 = +8{\cdot}4326' \cos 2U - 1{\cdot}2616' \sin U \cos U, & \text{(v)}\\
\psi = \pi/2: & \iota - \iota_0 = +8{\cdot}4326' \cos 2U + 1.2616' \sin^2 U. & \text{(vi)}
\end{array}\right\} \tag{30}$$

From the foregoing equations (i) and (ii) it follows that the total maximum variation, 'abstracting from the situation of the Moon in its orbit' (as Newton says below), is $16{\cdot}865'$ ($=16'\ 52''\ 1'''$) in close enough agreement to Newton's finding:

> And this [$16'\ 23\tfrac{1}{2}''$] is the greatest variation of the inclination, abstracting from the situation of the Moon in its orbit.

Similarly, it follows from equations (iv) and (vi) that the difference in inclination between quadratures and syzygies of the Moon is $2.523 \sin^2 U$ or $2.523'$ ($=2' 31''$) for $U = \pi/2$. Except for the replacement of $2' 31''$ by $2' 43''$ this result is again in agreement with what Newton finds:

> if the nodes are in the syzygies, the inclination suffers no change from the various positions of the Moon. But if the nodes are in the quadratures, the inclination is less when the Moon is in the syzygies than when it is in the quadratures by a difference of $2' 43''$.

And finally, from equations (i) and (iv) it follows that

$$\left.\begin{array}{l} \text{for} \quad \psi = 0, \quad \text{when} \quad U = 0 \quad \text{and} \quad U = \pi/2, \\ \text{we have, respectively,} \\ \iota - \iota_0 = 8{\cdot}4326' \quad \text{and} \quad \iota - \iota_0 = -8{\cdot}4326' - 1{\cdot}2616' = -9{\cdot}6942'; \\ \text{and their difference is} \\ 17{\cdot}1268' \simeq 17'\ 8''; \end{array}\right\} \tag{31}$$

and from equations (i) and (iii) it follows that

$$\left.\begin{array}{l} \text{for} \quad \psi = \pi/2, \quad \text{when} \quad U = 0 \quad \text{and} \quad U = \pi/2, \\ \text{we have, respectively,} \\ \iota - \iota_0 = 8{\cdot}4326' \quad \text{and} \quad \iota - \iota_0 = -8{\cdot}4326' + 1{\cdot}2616' = -7{\cdot}171'; \\ \text{and their difference is} \\ 15{\cdot}6036' \simeq 15'\ 36''. \end{array}\right\} \tag{32}$$

Hence, for an unperturbed inclination of $5° 1'$, the inclination at syzygies of the Moon will have become $5° 19'$ and it will decrease to $5° 4'$ at quadratures. All these conclusions, except for small differences in the numerical values, are the same as Newton's:

> [The inclination] becomes $15' 2''$, when the Moon is in the quadratures; and, increased by the same, becomes $17' 45''$ when the Moon is in the syzygies. If, therefore, the Moon be in the syzygies, the whole variation in the passage of the nodes from the quadratures to the syzygies will be $17' 45''$; and, therefore, if the inclination be $5° 17' 20''$, when the nodes are in the syzygies, it will be $4° 59' 35''$ when the nodes are in the quadratures and the Moon in the syzygies.

And as he triumphantly concludes:

> The truth of all this is confirmed by observations.

126. Scholium

In this Scholium, concluding the sequence of propositions on the lunar theory, Newton returns to Corollaries VI and VII of Proposition LXVI of Book I to consider, for the first time, the problem of the *annual equation* and to reconsider the elusive problem of the motion of the apogee (i.e., of the rotation of the apse.)

(*a*) *The annual equation*

In Corollary VI of Proposition LXVI of Book I (see Chapter 14, §70, equation (51)) Newton had shown that the perturbed period of the orbit of the Moon is given by

$$P = 2\pi\left(\frac{a_☾}{\mu}\right)^{3/2}\left[1 + \frac{1}{4}m^2\left(\frac{a_\odot}{R_\odot}\right)^3(1 + 3\cos 2\psi)\right], \qquad (1)^*$$

where $a_☾$ and $a_\odot$ denote the semimajor axes of the orbits of the Moon and of the Earth—the lunar orbit always considered circular.

First, assuming that the Earth's orbit around the Sun is also circular and further averaging over ψ, we obtain for the mean period $\bar{P}$ the formula,

$$\bar{P} = P_☾(1 + \tfrac{1}{4}m^2) \qquad (2)$$

where $P_☾$ denotes the unperturbed orbital period of the Moon. The change in the period caused by the tidal action of the Sun is therefore

$$\Delta\bar{P} = \tfrac{1}{4}P_☾ m^2; \qquad (3)$$

or, inserting numerical values, we find:

$$\Delta\bar{P} = \tfrac{1}{4} \times 29{\cdot}530 \times 24 \times 0{\cdot}00559544 = 0{\cdot}9914 \quad \text{hours.} \qquad (4)$$

It is known that the value by 'astronomical tables' is close to one hour (but somewhat in excess).

Returning to equation (1), we shall find that the variation in the Earth–Sun distance, $R_\odot$, causes a quarterly change in the mean period of the Moon. To show this, we replace $(a_\odot/R_\odot)$ in equation (1), to the first order in the eccentricity $e_\odot$ of the Earth's orbit, by

$$\frac{a_\odot}{R_\odot} = 1 + e_\odot \cos\varphi. \qquad (5)$$

To the same order in $e_\odot$, equation (1) becomes

$$P = P_☾[1 + \tfrac{1}{4}m^2(1 + 3e_\odot\cos\varphi)(1 + 3\cos 2\psi)]; \qquad (6)$$

or, averaging over ψ, we have

$$\bar{P} = P_☾[1 + \tfrac{1}{4}m^2(1 + 3e_\odot\cos\varphi)]. \qquad (7)$$

Therefore,

$$\bar{P}_{,\varphi} = -\tfrac{3}{4}P_☾ m^2 e_\odot \sin\varphi. \qquad (8)$$

* In writing this equation, the coefficient 'N^2' that occurs in the expressions for the disturbing functions (Chapter 13, §66(b), equations (23)–(26)) has been replaced, as well we may, by

$$\frac{GM_\odot}{a_\odot^3}\left(\frac{a_\odot}{R_\odot}\right)^3 = N^2\left(\frac{a_\odot}{R_\odot}\right)^3,$$

where N^2 has now the conventional meaning for elliptic orbits.

Since the orbital period of the Earth is $P_{\odot} = P_{☾}/m$, we can write

$$\bar{P}_{\odot,\varphi} = \tfrac{1}{4}P_{\odot}(-3me_{\odot} \sin \varphi). \tag{9}$$

The coefficient of $\frac{1}{4}P_{\odot}$ in this formula is the *annual equation*:

$$\text{annual equation} = -3me_{\odot} \sin \varphi. \tag{10}$$

For the currently accepted value, 0·016709, for $e_{\odot}$, we find

$$\text{annual equation} = -12{\cdot}92' \sin \varphi = -(12'\,56'')\sin \varphi; \tag{11}$$

and for the eccentricity, $16\frac{7}{8}/1000$, assumed by Newton,

$$\text{annual equation} = -13{\cdot}02' \sin \varphi = -(13'\,1'')\sin \varphi. \tag{12}$$

But Newton actually gives $-11'\,49''$ for the same coefficient. The origin of this error must be in the numerical computation.* But the underling cause is explained with admirable clarity:

> By the same theory I moreover found that the annual equation of the mean motion of the Moon arises from the varying dilatation which the orbit of the Moon suffers from the action of the Sun according to Cor. VI, Prop. LXVI, Book I. The force of this action is greater in the perigean Sun, and dilates the Moon's orbit; in the apogean Sun it is less, and permits the orbit to be again contracted. The Moon moves slower in the dilated and faster in the contracted orbit; and the annual equation, by which this inequality is regulated, vanishes in the apogee and perigee of the Sun. In the mean distance of the Sun from the Earth it rises to about 11′ 50″; in other distances of the Sun it is proportional to the equation of the Sun's centre, and is added to the mean motion of the moon, while the Earth is passing from its aphelion to its perihelion, and subtracted while the Earth is in the opposite semicircle. Taking for the radius of the great orbit 1000, and $16\frac{7}{8}$ for the Earth's eccentricity, this equation, when of the greatest magnitude, by the theory of gravity comes out 11′ 49″.

(*b*) *The motion of the apogee and the 'Portsmouth equation'*

In the preceeding propositions we have seen how Newton successfully solved the perturbation equations for the orbit, the period, and the annual equation (all requiring a second approximation) and the first-order variational equations for the regression of the nodes and the eccentricity and the inclination of the orbit; and obtained satisfactory

* From the fact that the present extended version of the Scholium appears for the first time in the second edition of the *Principia* one can trace the error to the inadequate attention (or perspicacity?) of Roger Cotes.

agreements with astronomical data. But the motion of the apogee remained discrepant though his solution of the variational equation is entirely correct—it is in fact the same solution that is still reproduced in current standard textbooks on celestial mechanics (e.g. J.M.A. Danby *Fundamentals of celestial mechanics*, (Willmann-Bell, Inc., Richmond, VA, 1989), p. 377, equation (12.5.3)). But the discrepancy remained for Newton a sore from his first discovery in Corollary II of Proposition XLV, Book I, wherein he found that the motion of the Moon 'is twice as fast'. He returns to it again in Corollary VII of Proposition LXVI and again in Proposition III of Book III when formulating his law of gravitation. And in this, his final Scholium on lunar perturbations, he returns to it once again and reasons cogently that allowing properly for the eccentricity of the lunar orbit, caused by the tidal action of the Sun, might resolve the issue. His surmise was entirely justified; for as we now know, allowing for the eccentricity modifies the mean motion, $3m^2/4$, given by the first-order theory, to

$$\tfrac{3}{4}m^2(1 + \tfrac{78}{8}m) = \tfrac{3}{4}m^2(1 + 0{\cdot}7293) \quad \text{for } m = 0{\cdot}0748, \tag{13}$$

in other words accounting for the greater part of the discrepancy. We shall not attempt here to describe Newton's train of reasoning but rather consider the result of his efforts in this direction found among his unpublished manuscripts. While it is beyond my capacity (or indeed my interests) to attempt a conventional 'historical' analysis, I shall restrict myself to the following extract from F. Tisserand's classic *Mécanique céleste* (Tome III, Chapitre III, pp. 44–45).

Referring to the then recently published book *A catalogue of the Portsmouth collection of books and papers written by or belonging to Sir Isaac Newton*, (Cambridge, 1888), Tisserand writes*

> The most interesting [among the manuscripts] concerns the motion of the lunar apogee. Newton first establishes two lemmas which deal with the apogee, allowing for the eccentricity of the lunar orbit in an elliptic orbit of small eccentricity, such as results from a perturbing force acting either in the radial or in the perpendicular direction. These two lemmas has been prepared for publication and were undoubtedly meant to appear in a new edition of the *Principia*. Newton then applies these two lemmas to find the hourly motion of the perigee and arrives at a result which can be represented by the formula
>
> $$\frac{\mathrm{d}\varpi}{\mathrm{d}t} = \tfrac{3}{4}m^2n[1 + \tfrac{11}{2}\cos(2v' - 2\varpi)]. \tag{14}$$
>
> According to the Preface, the derivation of this formula is not entirely satisfactory and the corrections carried out in the manuscript show that Newton was not very sure of the coefficient $\tfrac{11}{2}$.

* I am indebted to my colleague Professor Raghavan Narasimhan for this translation from the French.

We shall show later that the exact formula, restricted to its first few terms is

$$\frac{d\varpi}{dt} = \tfrac{3}{4}m^2n[1 + 5\cos(2v' - 2\varpi)], \tag{15*}$$

which, apart from the replacement of $\frac{11}{2}$ by 5, is the same as the formula (14). The preface adds that Newton deduces, entirely correctly, from formula (14) that the mean annual motion of the apogee is about 38° 51′ 51″ whereas what is given by astronomical tables is 40° 41′ 5″.

A curious aspect of this extract from Tisserand is that the exact solution of equation (14) or, for that matter, equation (15) comes nowhere near resolving the discrepancy.

Consider the more general equation

$$\frac{d\varpi}{dv} = \tfrac{3}{4}m^2[1 + \alpha\cos(2mv - 2\varpi)], \tag{16}$$

where α is a constant, unspecified for the present, and v' in equations (14) and (15) has been replaced by mv as follows from the equation,

$$v' = v - \psi = v - (1 - m)v = mv. \tag{17}$$

With the substitution,

$$\varpi = \tfrac{3}{4}m^2v + \phi, \tag{18}$$

equation (16) becomes,

$$\frac{d\phi}{dv} = \tfrac{3}{4}m^2\alpha\cos(2qv - 2\phi), \tag{19}$$

where

$$q = m(1 - \tfrac{3}{4}m). \tag{20}$$

By the further substitution,

$$\phi = -\tfrac{1}{2}y + qv, \tag{21}$$

equation (19) can be brought to the standard form

$$\frac{dy}{dv} = a\cos y + b, \tag{22}$$

where

$$a = -\tfrac{3}{2}m^2\alpha \qquad \text{and} \qquad b = 2q. \tag{23}$$

The solution of equation (22) is

$$\tan^{-1}\left[\left(\frac{b + a}{b - a}\right)^{1/2}\cot\frac{y}{2}\right] + \tfrac{1}{2}v(b^2 - a^2)^{1/2} = C, \tag{24}$$

* I have been unable to find equation (15) (or anything like it) in Tisserand's Volume III of *Mécanique Céleste*; perhaps it is buried among the mass of formulae that is beyond my unearthing!

where C is a constant of integration. For a and b given in equation (23), the foregoing solution can be brought to the form

$$\left(\frac{1-\varepsilon}{1+\varepsilon}\right)^{1/2} \cot\frac{y}{2} = \tan[C - qv(1-\varepsilon^2)^{1/2}], \tag{25}$$

where

$$\varepsilon = \frac{3}{4}\frac{\alpha m}{(1-3m/4)}. \tag{26}$$

The requirement that

$$\phi = 0 \quad \text{when} \quad \alpha = 0, \tag{27}$$

determines

$$C = \tfrac{1}{2}\pi; \tag{28}$$

and the solution reduces to:

$$\tan\tfrac{1}{2}y = \left(\frac{1-\varepsilon}{1+\varepsilon}\right)^{1/2} \tan[qv(1-\varepsilon^2)^{1/2}]. \tag{29}$$

Finally, by combining equations (21) and (29), we have

$$\phi = qv - \tan^{-1}\left\{\left(\frac{1-\varepsilon}{1+\varepsilon}\right)^{1/2} \tan[qv(1-\varepsilon^2)^{1/2}]\right\}. \tag{30}$$

For $v = \pi$ and $\alpha = \frac{11}{2}$, we find from a numerical evaluation of equation (30) that

$$\phi = 0{\cdot}2218 - 0{\cdot}1504 = 0{\cdot}0714 \tag{31}$$

while

$$\tfrac{3}{4}m^2\pi = 0{\cdot}01318. \tag{32}$$

We conclude that α is much too large by a factor of six to account for a discrepancy of a factor 2.

An estimate of α that will account for the sought-after factor 2 can be made by solving equation (30) to $O(\varepsilon)$, assuming that ε is small (as is indeed necessary, as we shall presently verify).

Letting,

$$qv = \tfrac{1}{2}y + \phi, \tag{33}$$

we readily find that to $O(\varepsilon)$ the solution of equation (29) is

$$\phi = \tfrac{1}{2}\varepsilon\sin(2\sin qv); \tag{34}$$

or, consistently in our scheme of approximation,

$$\delta = \varepsilon qv = \tfrac{3}{4}m^2\alpha v. \tag{35}$$

In other words, in this approximation,

$$\varpi = \tfrac{3}{4}m^2(1+\alpha)v. \tag{36}$$

We conclude that

$$\alpha \leqslant 1 \tag{37}$$

for a resolution of the discrepancy. By noting that for

$$\alpha = 1, \qquad \varepsilon = 0{\cdot}0594, \tag{38}$$

we verify the consistency of the adopted scheme of approximation. The question that remains is: Did Newton contemplate a value of α as small as 1?

Newton's unfinished manuscripts and thoughts aside, it is manifest that his compelling conclusion, that a satisfactory solution for the motion of the apogee of the Moon must go much beyond the first-order variational equations, cleared the way for his successors to proceed to higher approximations along the same trail.

> When you have eliminated the impossible, whatever remains, *however improbable*, must be the truth.
>
> Sir Arthur Conan Doyle
> *The Sign of Four*

Newton himself was reluctant to proceed to higher approximations (the very thought 'made his head ache'), though that he *had* contemplated higher approximation is clear from Corollary XII of Proposition LXVI.

I am not personally convinced that in this context 'genius' was needed to proceed systematically to higher approximations. Given the right idea,

> Any well trained professional could supply what is wanted.
>
> G.H. Hardy

❖ 23 ❖

The precession of the equinoxes

127. Introduction

Lemmas I, II, and III, culminating in Proposition XXXIX 'to find the precession of the equinoxes' is one of those sections, interspersed throughout the *Principia*, in which new ideas are formulated in the context of new problems that are constantly arising.

In Lemmas I, II, and III, Newton introduces, for the first time, notions that are relevant to the dynamics of rigid bodies: the *moment of inertia* (which determines the 'efficacy to wheel [a body] with circular motion about a centre'); the *moment of momentum* (which measures the 'motions of the whole [body] about its axis of rotation') and the *torque*—'the power to wheel about'—exerted by an external force on a non-spherical body; and, indeed, the principles underlying gyroscopic motion. All these notions are defined in the context of a dynamical theory of the precession of the equinoxes which derives from the oblateness of the Earth and the inclination of its axis of rotation to the plane of the ecliptic.

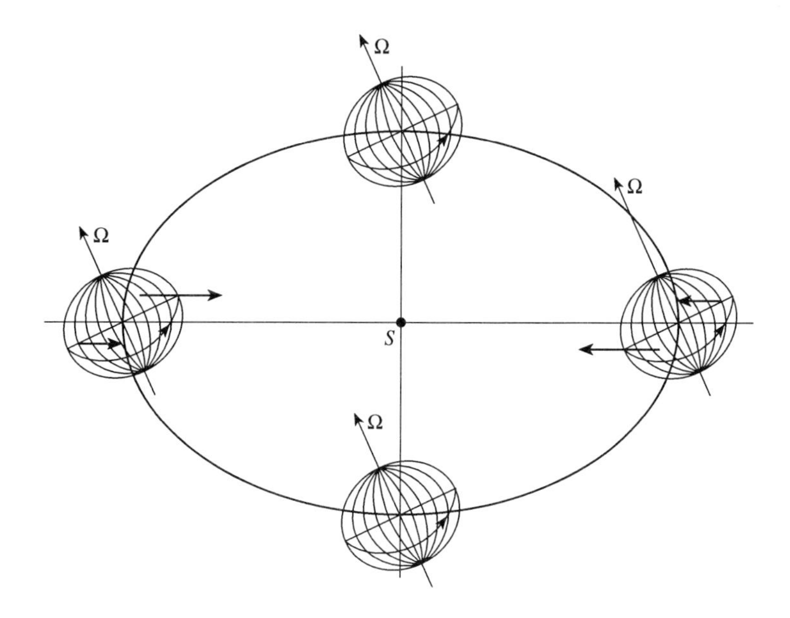

In the treatment of the problem of the precession of the equinoxes in Proposition XXXIX, we have a dazzling display of the depth of Newton's incisive physical insight in the manner in which he identifies the various contributory elements which will come as a surprise even to a modern reader (see §129 below). But on the analytical side, Newton does not pursue the problem far enough with his normal standards of rigour; and the constant of proportionality he obtains is too small by a factor 0·43. This *under-estimate* combined with his *over-estimate* by a factor 2 ($\simeq 4{\cdot}5/2{\cdot}2$) of the tidal influence of the Moon relative to the Sun (and allowing also for his *over-estimate* of the ellipticity of the Earth by a factor 1·20 ($=289/230$) led him to the optimistic conclusion,

> the force of the Moon to move the sea was to the force of the Sun nearly as 4·4815 to I; and the force of the Moon to move the equinoxes is to that of the Sun in the same proportion. Whence the annual precession of the equinoxes proceeding from the force of the Moon comes out $40'' \ 52''' \ 52^{\text{iv}}$, and the total annual precession arising from the united forces of both will be $50'' \ 00''' \ 12^{\text{iv}}$, the amount of which motion agrees with the phenomena; for the precession of the equinoxes, by astronomical observations, is about $50''$ yearly

which was illusory. (We shall comment on this and related matters in the concluding §131.)

For a full appreciation of Newton's startling approach to the problem of the precession of the equinoxes and of the source of his *under-estimate* of the constant of proportionality in his final result, we shall first present an *ab initio* treatment of the same problem that will be accessible to a modern reader.

128. On the precession of the equinoxes: a current treatment*

We begin with a list of definitions from which the notation we are adopting will be manifest.

Frame of reference:	(x, y, z);
Unit vectors along x, y, and z:	$(\vec{i}, \vec{j}, \vec{k})$, $\vec{r} = x\vec{i} + y\vec{j} + z\vec{k}$;
Force:	$\vec{F} = X\vec{i} + Y\vec{j} + Z\vec{k}$;
Equation of motion:	$\vec{F} = m\ddot{\vec{r}}$;
Moment of force:	$\vec{M} = \vec{r} \times \vec{F}$;
Angular velocity:	$\vec{\omega} = \omega_x\vec{i} + \omega_y\vec{j} + \omega_z\vec{k}$;
Velocity associated with $\vec{\omega}$:	$\vec{v} = \vec{\omega} \times \vec{r}$:
Angular momentum of an element of mass, dm:	$\mathrm{d}\vec{h} = \mathrm{d}m\vec{r} \times \vec{v}$.

* The treatment is based on that given in J. B. Scarborough, *The gyroscope: theory and applications*, Interscience Publishers, 1958, Chapter X, pp. 231–243.

Combining the definitions of $\vec{v}$ and $d\vec{h}$, we have

$$d\vec{h} = dm\vec{r} \times \vec{v} = dm\vec{r} \times (\vec{\omega} \times \vec{r}) = dm[r^2\vec{\omega} - \vec{r}(\vec{r}\cdot\vec{\omega})]. \tag{1}$$

Hence the angular momentum of a rigid body about the centre of rotation is given by

$$\vec{H} = \int_V d\vec{h} = \vec{\omega}\int_V r^2\,dm - \int_V \vec{r}(\vec{r}\cdot\vec{\omega})\,dm, \tag{2}$$

where the integration is effected over the volume V occupied by the body. On reducing the expression for $\vec{H}$, we have

$$\begin{aligned}\vec{H} &= \int_V dm[(x^2+y^2+z^2)(\omega_x\vec{i}+\omega_y\vec{j}+\omega_z\vec{k}) - (x\omega_x + y\omega_y + z\omega_z)(x\vec{i}+y\vec{j}+z\vec{k})]\\ &= \int_V dm\{[\omega_x(x^2+y^2+z^2) - (\omega_x x^2 + \omega_y xy + \omega_z xz)]\vec{i}\\ &\quad + [\omega_y(x^2+y^2+z^2) - (\omega_x xy + \omega_y y^2 + \omega_z zy)]\vec{j}\\ &\quad + [\omega_z(x^2+y^2+z^2) - (\omega_x xz + \omega_y yz + \omega_z z^2)]\vec{k}\}.\end{aligned} \tag{3}$$

The components of $\vec{H}$ are therefore,

$$\left.\begin{aligned} H_x &= +A\omega_x - F\omega_y - E\omega_z\\ H_y &= -F\omega_x + B\omega_y - D\omega_z,\\ H_z &= -E\omega_x - D\omega_y + C\omega_z,\end{aligned}\right\} \tag{4}$$

where

$$\left.\begin{aligned} A &= \int_V dm(y^2+z^2), & B &= \int_V dm(z^2+x^2), & C &= \int_V dm(x^2+y^2)\\ D &= \int_V dmyz, & E &= \int_V dmzx, & F &= \int_V dmxy,\end{aligned}\right\} \tag{5}$$

define the moment of inertia tensor. If the coordinate frame is chosen along the principal axes of the moment of inertia tensor, then

$$D = E = F = 0 \tag{6}$$

and

$$A = \int_V dm(y^2+z^2), \qquad B = \int_V dm(z^2+x^2), \qquad \text{and} \qquad C = \int_V dm(x^2+y^2); \tag{7}$$

and the components of H are,

$$H_x = A\omega_x, \qquad H_y = B\omega_y, \qquad \text{and} \qquad H_z = C\omega_z. \tag{8}$$

In particular, if the body is axisymmetric about the axis of rotation in the direction $\vec{k}$, then

$$A = B \neq C. \tag{9}$$

And, finally, the equation of motion governing $\vec{H}$ is

$$\frac{d\vec{H}}{dt} = \frac{d}{dt}\int_V dm\vec{r} \times \vec{v} = \int_V dm\left(\frac{d\vec{r}}{dt} \times \vec{v} + \vec{r} \times \frac{d\vec{v}}{dt}\right)$$
$$= \int_V \vec{r} \times d\vec{F} = \int_V d\vec{m} = \vec{M}. \tag{10}$$

(*a*) *Euler's equations*

We now consider a rigid body rotating with an angular velocity $\vec{\omega}$ about a centre fixed in the body. Let $OXYZ$ be a coordinate frame attached rigidly to the body and sharing its motion. To describe the motion of the body, we choose another frame $OX_1Y_1Z_1$ fixed in space but with the same centre.

Let $(\vec{i}, \vec{j}, \vec{k})$ denote unit vectors along OX, OY, and OZ; and let $\vec{\omega}$ be the angular velocity of the frame $OXYZ$ (i.e., of the centre of inertia of the body) and $\vec{H}$ the angular momentum of the body in the fixed frame $OX_1Y_1Z_1$. If $(\omega_x, \omega_y, \omega_z)$ and (H_x, H_y, H_z) denote the components of $\vec{\omega}$ and $\vec{H}$ resolved along the axes OX, OY, and OZ fixed in the body, then the definitions,

$$\vec{\omega} = \omega_x\vec{i} + \omega_y\vec{j} + \omega_z\vec{k}, \qquad \text{and} \qquad \vec{H} = H_x\vec{i} + H_y\vec{j} + H_z\vec{k}, \tag{11}$$

hold. Since $\vec{i}$, $\vec{j}$, and $\vec{k}$ are no longer time-independent (as viewed in the frame $OX_1Y_1Z_1$, fixed in space) we must allow for it when obtaining the equation of motion of $\vec{H}$. Thus

$$\frac{d\vec{H}}{dt} = \vec{i}\frac{dH_x}{dt} + \vec{j}\frac{dH_y}{dt} + \vec{k}\frac{dH_z}{dt} + H_x\frac{d\vec{i}}{dt} + H_y\frac{d\vec{j}}{dt} + H_z\frac{d\vec{k}}{dt}. \tag{12}$$

To evaluate the time derivatives of $\vec{i}$, $\vec{j}$, and $\vec{k}$ we consider them as position vectors, $\vec{r}$ drawn from O. And, since for any vector $\vec{r}$,

$$\frac{d\vec{r}}{dt} = \vec{\omega} \times \vec{r} = \begin{vmatrix} \vec{i} & \vec{j} & \vec{k} \\ \omega_x & \omega_y & \omega_z \\ r_x & r_y & r_z \end{vmatrix}, \tag{13}$$

we can replace $\vec{r}$ in this equation by $\vec{i}$, $\vec{j}$, or $\vec{k}$. Therefore, in particular,

$$\frac{d\vec{i}}{dt} = \begin{vmatrix} \vec{i} & \vec{j} & \vec{k} \\ \omega_x & \omega_y & \omega_z \\ 1 & 0 & 0 \end{vmatrix} = \omega_z\vec{j} - \omega_y\vec{k}, \tag{14}$$

and the expressions for $d\vec{j}/dt$ and $d\vec{k}/dt$ can be obtained by cylically permuting $(\vec{i}, \vec{j}, \vec{k})$ and $(\omega_x, \omega_y, \omega_z)$. Inserting the resulting expressions in equation (12), we obtain:

$$\begin{aligned}\frac{d\vec{H}}{dt} &= \vec{i}\dot{H}_x + \vec{j}\dot{H}_y + \vec{k}\dot{H}_z + H_x(\omega_z\vec{j} - \omega_y\vec{k}) + H_y(\omega_x\vec{k} - \omega_z\vec{i}) + H_z(\omega_y\vec{i} - \omega_x\vec{j}) \\ &= \vec{i}(\dot{H}_x + \omega_y H_z - \omega_z H_y) + \vec{j}(\dot{H}_y + \omega_z H_x - H_z\omega_x) + \vec{k}(\dot{H}_z + \omega_x H_y - \omega_y H_x) \\ &= \vec{i}M_x + \vec{j}M_y + \vec{k}M_z.\end{aligned} \tag{15}$$

If the unit vectors $\vec{i}$, $\vec{j}$, and $\vec{k}$ are chosen along the principal axes of the inertia tensor, equation (8) holds and equation (15) gives

$$\left.\begin{aligned} A\dot{\omega}_x + (C - B)\omega_z\omega_y &= M_x, \\ B\dot{\omega}_y + (A - C)\omega_z\omega_x &= M_y, \\ C\dot{\omega}_z + (B - A)\omega_x\omega_y &= M_z, \end{aligned}\right\} \tag{16}$$

and, if the body is axisymmetric about the rotation axis in the direction $\vec{k}$, when $A = B$, equations (16) reduce to:

$$\left.\begin{aligned} A\dot{\omega}_x + (C - A)\omega_y\omega_z &= M_x, \\ A\dot{\omega}_y + (A - C)\omega_z\omega_x &= M_y, \\ C\dot{\omega}_z &= M_z. \end{aligned}\right\} \tag{17}$$

(*b*) *Euler's angles*

The definition of Euler's angles, θ, ϕ, and ψ, and the equations of motion governing them are manifest from the accompanying illustrations. We have:

$$\left.\begin{aligned} \omega_x &= \dot{\theta}\sin\phi - \dot{\psi}\sin\theta\cos\phi, \\ \omega_y &= \dot{\theta}\cos\phi + \dot{\psi}\sin\theta\sin\phi, \\ \omega_z &= \dot{\phi} + \dot{\psi}\cos\theta, \end{aligned}\right\} \tag{18}$$

where $\dot{\psi}$ and $\dot{\theta}$ determine the rates of precession and nutation, respectively.

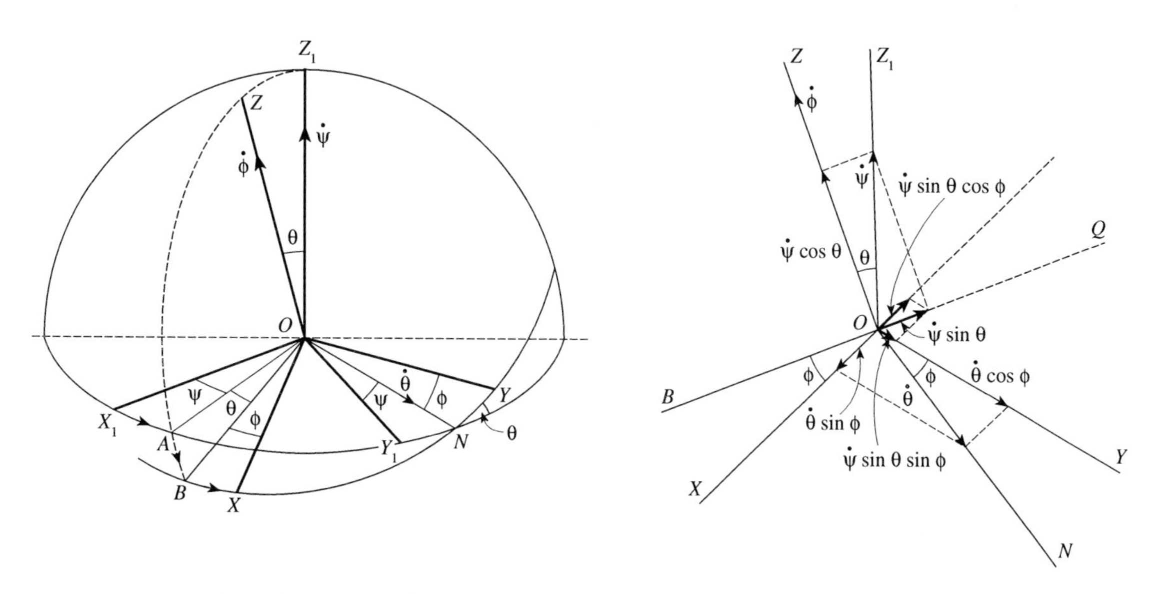

(*c*) *The equations governing precession and nutation*

Before we can apply equations (17) and (18) to the problem of precession and nutation, we must determine the moment of the force, derived from the tidal effects of the Sun and the Moon, or more generally from a distant body S (say). We shall determine the required moment $\vec{M}$ in two steps: first in the frame, $OXYZ$, rigidly attached to the centre of the Earth; and then transform it to the frame in which the centre of the Earth describes its orbit about S (the Sun or the Moon) in the plane of the ecliptic.

The configuration to be considered in the frame $OXYZ$ is illustrated in the adjoining diagram where:

S is the position of the distant body;
O the centre of the Earth; and
P is the location of an element of mass of the Earth at

$$\left.\begin{aligned}&\vec{\rho} = \vec{i}x + \vec{j}y + \vec{k}z,\\ &\vec{R} = \overrightarrow{OS}, \qquad \vec{r} = \overrightarrow{PS}, \qquad \text{and} \qquad \theta = \angle(\overrightarrow{OS}, \vec{j}).\end{aligned}\right\} \tag{19}$$

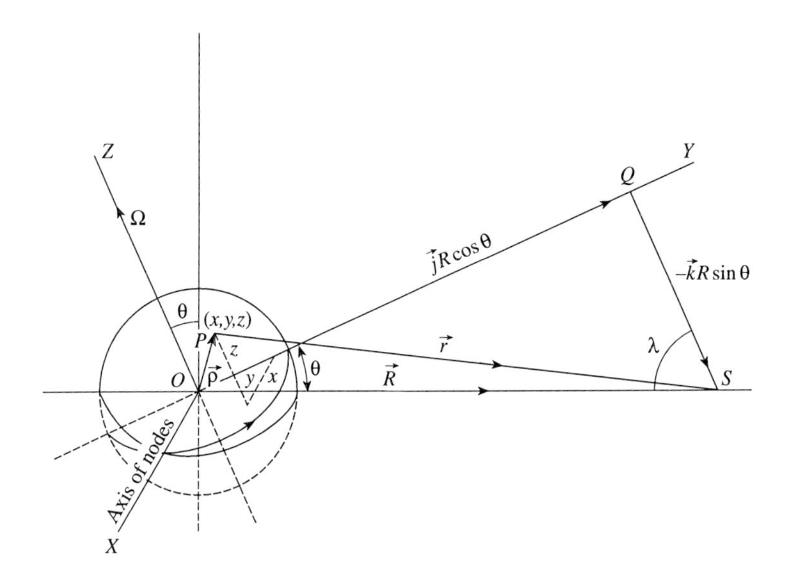

With these definitions,

$$\vec{R} = \overrightarrow{OQ} + \overrightarrow{QS} = \vec{j}R\cos\theta - \vec{k}R\sin\theta, \tag{20}$$

$$\vec{r} = \vec{R} - \vec{\rho} = -\vec{i}x + \vec{j}(R\cos\theta - y) - \vec{k}(R\sin\theta + z), \tag{21}$$

and the force acting on dm is

$$\mathrm{d}\vec{F} = GM\,\mathrm{d}m\,\frac{\vec{r}}{r^3} \quad (r = |\vec{r}|) \tag{22}$$

where M is the mass of S. The moment of this force $\mathrm{d}\vec{F}$ is

$$\mathrm{d}\vec{M} = \vec{\rho} \times \mathrm{d}\vec{F} = \frac{GM\,\mathrm{d}m}{r^3}\begin{vmatrix} \vec{i} & \vec{j} & \vec{k} \\ x & y & z \\ -x & (R\cos\theta - y) & -(R\sin\theta + z) \end{vmatrix}$$

$$= \frac{GM\,\mathrm{d}m}{r^3}[-\vec{i}R(y\sin\theta + z\cos\theta) + \vec{j}Rx\sin\theta + \vec{k}Rx\cos\theta]. \tag{23}$$

Therefore,

$$\mathrm{d}M_x = -\frac{GM\,\mathrm{d}m}{r^3}R(y\sin\theta + z\cos\theta), \tag{24}$$

$$\mathrm{d}M_y = +\frac{GM\,\mathrm{d}m}{r^3}Rx\sin\theta \quad \text{and} \quad \mathrm{d}M_z = +\frac{GM\,\mathrm{d}m}{r^3}Rx\cos\theta. \tag{25}$$

Next we have to express r in terms of $\vec{\rho}$ and $\vec{R}$ with the aid of the relation,

$$\left.\begin{aligned} r^2 &= R^2 + \rho^2 - 2\vec{R}\cdot\vec{\rho} \\ &= R^2 + \rho^2 - 2R(\vec{j}\cos\theta - \vec{k}\sin\theta)\cdot(\vec{i}x + \vec{j}y + \vec{k}z) \\ &= R^2 + \rho^2 - 2R(y\cos\theta - z\sin\theta). \end{aligned}\right\} \tag{26}$$

or

$$r^2 = R^2\left[1 + \left(\frac{\rho}{R}\right)^2 - \frac{2}{R}(y\cos\theta - z\sin\theta)\right]. \tag{27}$$

Neglecting the term $(\rho/R)^2$ consistently with the circumstances of the problem, we have

$$r^2 \simeq R^2\left[1 - \frac{2}{R}(y\cos\theta - z\sin\theta)\right], \tag{28}$$

and it follows:

$$\frac{1}{r^3} \simeq \frac{1}{R^3}\left[1 + \frac{3}{R}(y\cos\theta - z\sin\theta)\right]. \tag{29}$$

Inserting this last expression in equation (24), we have

$$M_x = -\frac{GM}{R^2}\left\{\int_V (y\sin\theta + z\cos\theta)\left[1 + \frac{3}{R}(y\cos\theta - z\sin\theta)\right]\mathrm{d}m\right.$$

$$= -3\frac{GM}{R^3}\int_V \mathrm{d}m(y^2 - z^2)\sin\theta\cos\theta; \tag{30}$$

or, by equations (5) and (9),

$$M_x = -3\frac{GM}{R^3}(C - A)\sin\theta\cos\theta. \tag{31}$$

Since it is manifest that

$$M_y = M_z = 0, \tag{32}$$

we can write vectorially,

$$\vec{M} = \frac{3GM}{R^5}(C - A)(\vec{R}\cdot\vec{k})(\vec{R} \times \vec{k}). \tag{33}$$

We must now transform $\vec{M}$ to the rest frame $OX_1Y_1Z_1$ in which the Earth describes its orbit about S in the ecliptic plane OX_1Y_1, and the X_1-axis is chosen in the direction of the vernal equinox. Let the equatorial plane of the Earth, in the frame $OXYZ$, rigidly attached to the Earth, intersect the ecliptic along OX with an inclination θ while $\angle(\overrightarrow{OX}_1, \overrightarrow{OX}) = \psi$. Draw OM orthogonal to OX and lying in the ecliptic plane. And, finally, let

$$\left.\begin{aligned} &\vec{i}_1 \text{ and } \vec{j}_1 \text{ be unit vectors along } \overrightarrow{OX}_1 \text{ and } \overrightarrow{OY}_1, \\ \text{and} \quad &\vec{j}' \text{ is a unit vector along } \overrightarrow{OM}. \end{aligned}\right\} \tag{34}$$

The disposition of the various vectors and the relations among them are exhibited in the accompanying illustration. And we observe:

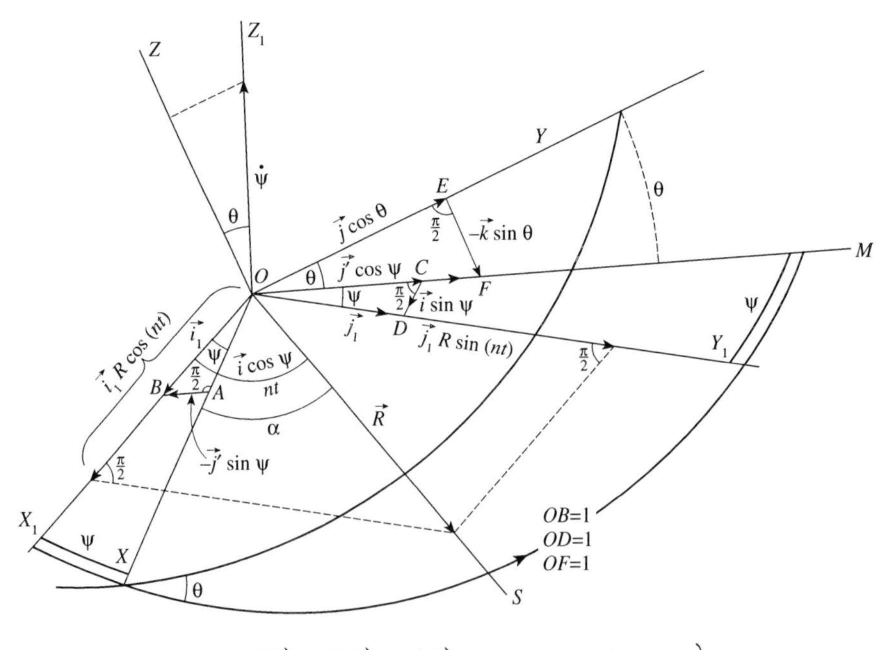

$$\left.\begin{aligned} &\vec{i}_1 = \overrightarrow{OB} = \overrightarrow{OA} + \overrightarrow{AB} = \vec{i}\cos\psi - \vec{j}'\sin\psi \\ \text{and} \quad &\vec{j}' = \overrightarrow{OF} = \overrightarrow{OE} + \overrightarrow{EF} = \vec{j}\cos\theta - \vec{k}\sin\theta. \end{aligned}\right\} \tag{35}$$

Therefore

$$\vec{i}_1 = \vec{i}\cos\psi - \vec{j}\sin\psi\cos\theta + \vec{k}\sin\psi\sin\theta \tag{36}$$

Similarly,

$$\begin{aligned}\vec{j}_1 &= \overrightarrow{OD} = \overrightarrow{OC} + \overrightarrow{CD} = \vec{j}'\cos\psi + \vec{i}\sin\psi\\ &= \vec{i}\sin\psi + \vec{j}\cos\theta\cos\psi - \vec{k}\cos\psi\sin\theta.\end{aligned} \tag{37}$$

We shall assume that the Earth describes its orbit about S in a circular orbit with an angular velocity v $(= 2\pi/\text{period})$. The position vector, $\vec{R}$, of S at time t, after vernal equinox, is given by

$$\vec{R} = R(\vec{i}_1\cos vt + \vec{j}_1\sin vt). \tag{38}$$

Substituting for $\vec{i}_1$ and $\vec{j}_1$ from equations (36) and (37) we find:

$$\begin{aligned}\vec{R} &= R[(\vec{i}\cos\psi - \vec{j}\sin\psi\cos\theta + \vec{k}\sin\psi\sin\theta)\cos vt\\ &\quad + (\vec{i}\sin\psi + \vec{j}\cos\psi\cos\theta - \vec{k}\cos\psi\sin\theta)\sin vt]\\ &= R[\vec{i}\cos(vt-\psi) + \vec{j}\cos\theta\sin(vt-\psi) - \vec{k}\sin\theta\sin(vt-\psi)].\end{aligned} \tag{39}$$

Returning to equation (33), we now find:

$$\left.\begin{aligned}&\vec{R}\cdot\vec{k} = -R\sin\theta\sin(vt-\psi)\\ \text{and}\quad &\\ &\vec{R}\times\vec{k} = +R[\vec{i}\cos\theta\sin(vt-\psi) - \vec{j}\cos(vt-\psi)].\end{aligned}\right\} \tag{40}$$

Accordingly,

$$\vec{M} = -\frac{3GM}{R^3}(C-A)\sin\theta\sin(vt-\psi)[\vec{i}\cos\theta\sin(vt-\psi) - \vec{j}\cos(vt-\psi)] \tag{41}$$

Therefore,

$$M_x = -\frac{3GM}{2R^3}(C-A)\sin\theta\cos\theta[1-\cos 2(vt-\psi)] \tag{42}$$

$$M_y = +\frac{3GM}{2R^3}(C-A)\sin\theta\sin 2(vt-\psi), \qquad \text{and} \qquad M_z = 0. \tag{43}$$

Finally, as is manifest from the illustration,

$$\omega_x = \dot{\theta} \qquad \text{and} \qquad \omega_y = \dot{\psi}\sin\theta. \tag{44}$$

The equations for $\dot{\theta}$ and $\dot{\psi}$ now follow from equations (17). Since $M_z = 0$ and there is no external moment about the axis of rotation $\vec{k}$, it follows from the defining equation (see the illustration on the right on p. 459)

$$H_z = C(\dot{\phi} + \omega_z), \tag{45}$$

and the last of the equations (17), that

$$C\frac{\mathrm{d}}{\mathrm{d}t}(\dot{\phi} + \omega_z) = 0. \tag{46}$$

We conclude that

$$\dot{\phi} + \omega_z = \text{constant} = \Omega. \tag{47}$$

where Ω denotes the angular velocity of daily rotation of the Earth. The two remaining equations of (17) now give

$$A\dot{\omega}_x - A\omega_y\omega_z + C\omega_y\Omega = -\frac{3GM}{2R^3}(C-A)\sin\theta\cos\theta[1-\cos 2(vt-\psi)], \tag{48}$$

$$A\dot{\omega}_y + A\omega_x\omega_z - C\omega_x\Omega = +\frac{3GM}{2R^3}(C-A)\sin\theta\sin 2(vt-\psi). \tag{49}$$

Since the rotation of the frame $OXYZ$ is exceedingly slow, we may neglect the first two terms on the left-hand sides of equations (48) and (49) and obtain:

$$\dot{\psi} = -\frac{3GM}{2\Omega R^3}\left(\frac{C-A}{C}\right)\cos\theta[1-\cos 2(vt-\psi)], \tag{50}$$

and

$$\dot{\theta} = -\frac{3GM}{2\Omega R^3}\left(\frac{C-A}{C}\right)\sin\theta\sin 2(vt-\psi). \tag{51}$$

The mean rate of precession, $\langle\dot{\psi}\rangle_{\text{AV}}$, in which we are primarily interested, is given by

$$\langle\dot{\psi}\rangle_{\text{AV}} = -\frac{3}{2}\frac{GM}{R^3\Omega}\frac{C-A}{C}\cos\theta. \tag{52}$$

If the Earth is considered as an oblate spheroid with semiaxes a and b ($<a$), then the moments of inertia C and A are:

$$C = \tfrac{2}{5}Ma^2 \qquad \text{and} \qquad A = \tfrac{1}{5}M(a^2+b^2). \tag{53}$$

Therefore,

$$\frac{1}{C}(C-A) = \frac{1}{2a^2}(a^2-b^2) = \frac{a+b}{2a}\left(1-\frac{b}{a}\right). \tag{54}$$

If the ellipticity

$$\varepsilon = (1-b/a) \ll 1, \tag{55}$$

then to the first order in ε,

$$(C-A)/C = \varepsilon; \tag{56}$$

and equation (52) gives

$$\langle\dot{\psi}\rangle_{\text{AV}} = -\frac{3}{2}\varepsilon\frac{GM}{R^3\Omega}\cos\theta. \tag{57}$$

Finally, we may note that the angular velocity of the daily rotation of the Earth is

$$\Omega = 2\pi \times 366\tfrac{1}{4} \quad \text{radians per year;} \tag{58}$$

and the inclination of the Earth's rotation axis to the ecliptic is

$$\theta = 23{\cdot}45^\circ = \cos^{-1} 0{\cdot}91741. \tag{59}$$

(d) The solar contribution to the precession

The solar contribution to the precession is obtained by identifying S with the Sun, in which case

$$\frac{GM}{R^3} = \frac{GM_\odot}{R_\odot^3} = \left(\frac{2\pi}{T}\right)^2, \tag{60}$$

where, by Kepler's third law, T ($= 365\frac{1}{4}$ days) is the period of revolution of the Earth about the Sun. Inserting the values (58)–(60) in equation (58) and setting, as one normally does,

$$\varepsilon = 1/305{\cdot}6 \tag{61}$$

we obtain:

$$\langle\dot{\psi}\rangle_{\mathrm{AV};\odot} = -\frac{3(4\pi^2) \times 0{\cdot}91741}{2(2\pi \times 366\frac{1}{4}) \times 305{\cdot}6} = -7{\cdot}725 \times 10^{-5} \quad \text{radians per year}, \tag{62}$$

or

$$\langle\dot{\psi}\rangle_{\mathrm{AV};\odot} = -15{\cdot}94'' \quad \text{per year}. \tag{63}$$

It should be pointed out that the choice $\varepsilon = 1/305{\cdot}6$ is *not* based on a theory of the constitution of the Earth's interior but is an empirically determined value that will predict for the combined lunisolar precession the observed value $-50''$ per year (see §(e) below). If on the other hand, we choose

$$\varepsilon = 1/230, \tag{64}$$

as predicted by Newton for a homogeneous Earth (see Chapter 20, §109(c)) then we should have obtained, in place of (63), the value

$$\langle\dot{\psi}\rangle_{\mathrm{AV};\odot} = -15{\cdot}94''\frac{305{\cdot}6}{230} = -21{\cdot}18'' \quad \text{per year}. \tag{65}$$

(e) The lunar contribution to the precession and the lunisolar precession

If we now identify S with the Moon, then

$$\frac{GM}{R^3} = \frac{GM_☾}{R_☾^3} = \frac{G(M_\oplus + M_☾)}{R_☾^3}\,\frac{M_☾}{M_\oplus + M_☾} = \left(\frac{2\pi}{T_☾}\right)^2 \frac{1}{82{\cdot}5}, \tag{66}$$

where

$$T_☾ = \frac{27{\cdot}32}{365\frac{1}{4}}\ \text{year}, \tag{67}$$

is the period of revolution of the Moon about the Earth and

$$82{\cdot}5 = 1 + M_\oplus / M_☾. \tag{68}$$

Inserting the values (58), (59), (61), (66) and (67) in equation (58), we obtain, (cf. equation (62))

$$\langle\dot{\psi}\rangle_{AV;☾} = -\frac{3(4\pi^2) \times 0{\cdot}91741}{2(2\pi \times 366\frac{1}{4})82{\cdot}5}\left(\frac{365\frac{1}{4}}{27{\cdot}32}\right)^2 \frac{1}{305{\cdot}6}$$

$$= -1{\cdot}674 \times 10^{-4} \quad \text{radians per year,} \tag{69}$$

or

$$\langle\dot{\psi}\rangle_{AV;☾} = -34{\cdot}5'' \quad \text{per year.} \tag{70}$$

Equations (62) and (69) together give for the total lunisolar precession the value,

$$\langle\dot{\psi}\rangle_{AV;\,total} = -50{\cdot}4'' \quad \text{per year,} \tag{71}$$

in close agreement with the observed value of $\sim -50''$ per year. Of course, the value of $\varepsilon = 1/305{\cdot}6$ was chosen precisely to obtain this agreement!

At the rate of precession (71), the time required for a complete revolution of the vernal equinox about the Sun is $\sim$25,700 years.

The ratio of the tidal effects of the Moon and the Sun follows from

$$\frac{\langle\dot{\psi}\rangle_{AV;☾}}{\langle\dot{\psi}\rangle_{AV;\odot}} = \frac{34{\cdot}5''}{15{\cdot}93''} = 2{\cdot}164. \tag{72}$$

This value should be contrasted with the value 2.23 derived in Chapter 21, §115, equation (16).

Finally, had we chosen for ε the value $\frac{1}{230}$, Newton had determined for a homogenous Earth, we should have obtained,

$$\langle\dot{\psi}\rangle_{AV;☾} = -45{\cdot}83'' \quad \text{per year,} \tag{73}$$

and

$$\langle\dot{\psi}\rangle_{AV;\,total} = -67{\cdot}0'' \quad \text{per year.} \tag{74}$$

We shall return to the values (65), (73), and (74) in §130.

And now to Newton's treatment of the same problem in Lemmas I–III and Proposition XXXIX.

129. Moment of momentum, moment of inertia, and circulation

In Lemmas I and II, Newton introduces the concepts of the moment of momentum and of the moment of inertia, in the context of a rigid body rotating with an angular velocity Ω about a fixed axis passing through its centre of mass. A casual reader may not readily recognize that ideas that are familiar to him are defined here for the first time by Newton's style of explaining them in the novel context of the precession of the equinoxes. For this reason, we shall rearrange his arguments to bring out his meaning.

Lemma I

If APEp represent the Earth uniformly dense, marked with the centre C, the poles P, p, and the equator AE; and if about the centre C, with the radius CP, we suppose the sphere Pape to be described, and QR to denote the plane on which a right line, drawn from the centre of the Sun to the centre of the Earth, stands at right angles; and further suppose that the several particles of the whole exterior Earth PapAPepE, without the height of the said sphere, endeavour to recede towards this side and that side from the plane QR, every particle by a force proportional to its distance from that plane: I say, in the first place, that the whole force and efficacy of all the particles that are situated in AE, the circle of the equator, and disposed uniformly without the globe, encompassing the same after the manner of a ring, to wheel the Earth about its centre, is to the whole force and efficacy of as many particles in that point A of the equator which is at the greatest distance from the plane QR, to wheel the Earth about its centre with a like circular motion as is 1 *to* 2. *And that circular motion will be performed about an axis lying in the common section of the equator and the plane QR.*

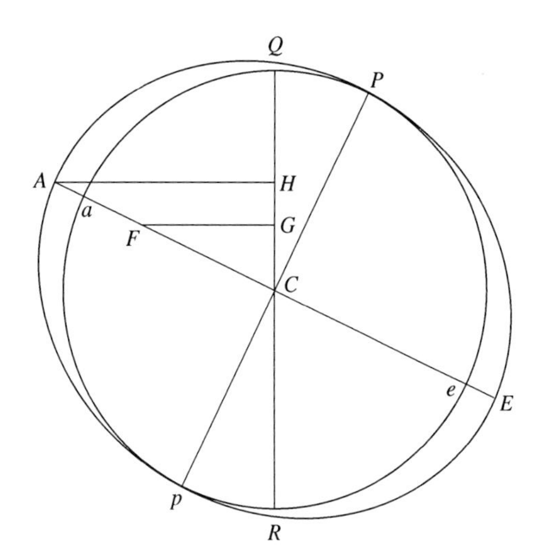

The Lemma in effect states that the velocity $\vec{v}$ of an element of mass, dm at a point $\vec{r}$ in a rigid body rotating with an angular velocity $\vec{\Omega}$ is

$$\vec{v} = \vec{\Omega} \times \vec{r} \tag{1}$$

while its moment of momentum (or, angular momentum) is given by

$$\mathrm{d}\vec{H} = \mathrm{d}m\vec{r} \times (\vec{\Omega} \times \vec{r}). \tag{2}$$

The angular momentum, $\vec{H}$, of the whole body is therefore

$$\vec{H} = \int_V \mathrm{d}m\vec{r} \times (\vec{\Omega} \times \vec{r}). \tag{3}$$

where the integral is extended over the volume V occupied by the body.

Newton arrives at the definition (2) as arising from a 'force' proportional to the distance $\vec{r}$ and calls the integral (3) the 'whole force' that 'endeavours ... to wheel the body about its centre'.

The definitions (1) and (2) are explained in Lemma I in the context of a plane circular disc of radius a rotating with an angular velocity Ω in a direction normal to the disc. Then, an element of mass, $\mathrm{d}M(r)$, at a point P on the disc at $\vec{r}$ has a velocity $\vec{v}$ of magnitude

$$v = \Omega r \sin\theta \tag{4}$$

normal to the plane $CPQ\Omega$. By equation (2), its moment of momentum,

$$\mathrm{d}H(r) = \mathrm{d}M(r)(\Omega r^2 \sin\theta) \tag{5}$$

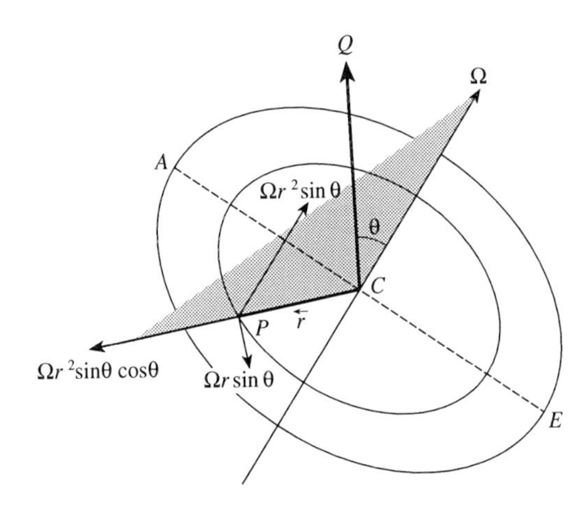

lies *in* the plane $CPQ\Omega$ and parallel to the direction $\vec{\Omega}$. The component of this moment of momentum normal to the direction CQ is

$$\mathrm{d}H_r(\vec{r};\ \perp^r CQ) = \mathrm{d}M(r)[\Omega r^2 \sin\theta\cos\theta]. \tag{6}$$

If the same mass $\mathrm{d}M(r)$ is placed at the circumference of the disc, its moment of momentum will be

$$\mathrm{d}H_a(a;\ \perp^r CQ) = \mathrm{d}M(r)[\Omega a^2 \sin\theta\cos\theta]. \tag{7}$$

The ratio of the expressions (6) and (7) is independent of θ and is given by

$$\mathrm{d}H_r(\vec{r}\colon \perp^r CQ)\colon \mathrm{d}H_a(a;\ \perp^r CQ) = \mathrm{d}M(r)\Omega r^2\colon \mathrm{d}M(r)\Omega a^2. \tag{8}$$

Therefore, the ratio of the moment of momentum of the entire disc to the moment of momentum of all the mass distributed over the circumference is

$$\int_0^a \mathrm{d}H_r(\vec{r};\ \perp^r CQ)\colon H_a(a;\ \perp^r CQ) = \Omega\int_0^a \mathrm{d}M(r)r^2\colon Ma^2\Omega, \tag{9}$$

where M denotes the mass of the disc. If the disc is of uniform density, the mass, $M(r)$, interior to r is

$$M(r) = Mr^2/a^2, \tag{10}$$

and

$$\Omega \int_0^a \mathrm{d}M(r) r^2 = \frac{\Omega a^2}{M} \int_0^a \mathrm{d}M(r) M(r) = \tfrac{1}{2} M \Omega a^2. \tag{11}$$

From equation (9) it now follows:

$$\int_0^a \mathrm{d}H_r(\vec{r};\, \perp^r CQ) \colon H_a(a;\, \perp^r CQ) = 1:2. \tag{12}$$

Q.E.D.

And Newton concludes:

> And because the action of those particles is exerted in the direction of lines perpendicularly receding from the plane QR, and that equally from each side of this plane, they will wheel about the circumference of the circle of the equator, together with the adherent body of the Earth, round an axis which lies as well in the plane QR as in that of the equator.

It will be noticed that Newton proves more than that the moment of inertia of a uniform circular disc about its normal is

$$I = \tfrac{1}{2} M a^2. \tag{13}$$

He proves, in addition, the invariance of the ratio (9) to the angle of inclination θ of the reference direction CQ; and this fact is essential for his identification of $\tfrac{1}{2} M\Omega a^2$ as the force that 'wheels the disc'.

Lemma II

> *The same things still supposed, I say, in the second place, that the total force or power of all the particles situated everywhere about the sphere to turn the Earth about the said axis is to the whole force of the like number of particles, uniformly disposed round the whole circumference of the equator AE in the fashion of a ring, to turn the whole Earth about with the like circular motion as is 2 to 5.*

In this lemma, Newton evaluates the angular momentum of a rotating homogeneous spherical body of mass M and radius a about the axis of rotation along the Z-axis (say). By the definition of the quantity in question,

$$M_z = 4\pi \int_0^a \int_0^\pi \rho [\vec{r} \times (\vec{\Omega} \times \vec{r})]_z r^2 \sin\theta \,\mathrm{d}\theta \,\mathrm{d}r. \tag{14}$$

Since,

$$[\vec{r} \times (\vec{\Omega} \times \vec{r})]_z = [\vec{\Omega} r^2 - \vec{r}(\vec{r} \cdot \vec{\Omega})]_z = \Omega(r^2 - z^2)$$
$$= \Omega(x^2 + y^2) = \Omega r^2 \cos^2 \theta, \tag{15}$$

$$M_z = 4\pi\Omega\rho \int_0^a \int_0^\pi r^4 \sin\theta \cos^2\theta \, d\theta \, dr = \frac{8\pi}{15} \Omega a^5 \rho, \tag{16}$$

or

$$M_z(\text{sphere}) = \tfrac{2}{5} M\Omega a^2 \qquad (M = \tfrac{4}{3}\pi\rho a^3). \tag{17}$$

On the other hand, if the mass M is distributed uniformly over the circumference of the equator, as a spherical shell, then its angular momentum will be

$$M_z(\text{circumferential ring}) = M\Omega a^2. \tag{18}$$

Hence

$$M_z(\text{sphere}) : M_z(\text{circumferential ring}) = 2:5, \tag{19}$$

as stated.

Newton's proof of (19) appears more complicated for two reasons: first, his apparent(?) reluctance to spell out, in detail, his evaluation of the double integral (14) by his 'method of fluxions'; and second, his wanting to establish at the same time, the invariance of the ratio 2:5 to the reference direction QCP, as in Lemma I (which in the present context is manifest from spherical symmetry—an argument which Newton could very well have used).

Lemma III

The same things still supposed, I say, in the third place, that the motion of the whole Earth about the axis above named arising from the motions of all the particles, will be to the motion of the aforesaid ring about the same axis in a ratio compounded of the ratio of the matter in the Earth to the matter in the ring; and the ratio of three squares of the quadrantal arc of any circle to two squares of its diameter, that is, in the ratio of the matter to the matter, and of the number 925275 *to the number* 1000000.

This lemma is strangely worded and conceals the significant notion of *circulation* that it defines.

By 'motion', in the present context, Newton means *circulation* as the line integral of the velocity along a closed curve, as commonly defined. In particular, the circulation along a circle of radius a normal to the axis of rotation (by equation (4)) is

$$C_z(r) = 2\pi\Omega r \sin\theta. \tag{20}$$

Therefore, the circulation of the motions in the whole sphere is

$$C_z = 2\pi\Omega \int_0^a \int_0^\pi r^3 \sin^2\theta \, \mathrm{d}\theta \, \mathrm{d}r = \tfrac{1}{4}\pi^2 a^4 \Omega, \tag{21}$$

or

$$C_z = \frac{3\pi}{16} Va\Omega, \tag{22}$$

where V denotes the volume $(=\frac{4}{3}\pi a^3)$ of the sphere.

At the same time, the circulation, ΔC, of a thin circular ring of radius a and thickness $\mathrm{d}a$ about the equator is

$$\Delta C = 2 \int_0^\pi (\Omega a \sin\varphi) a \, \mathrm{d}a \, \mathrm{d}\varphi = 4\Omega a^2 \, \mathrm{d}a, \tag{23}$$

or

$$\Delta C = \frac{2}{\pi} a\Omega\Delta V, \quad \text{where } \Delta V = 2\pi a \, \mathrm{d}a. \tag{24}$$

Hence

$$\frac{C}{\Delta C} = \frac{3\pi^2}{32} \frac{V}{\Delta V} = 0{\cdot}925275 \frac{V}{\Delta V}. \tag{25}$$

or, if the matter is of uniform density,

$$\frac{\Delta C}{C} = \frac{32}{3\pi^2} \frac{\Delta M}{M} = 1{\cdot}08076 \frac{\Delta M}{M}. \tag{26}$$

It has, to the writer's knowledge, never been recognized that Newton defined the notion of circulation some two hundred years before Lord Kelvin to whom it is commonly credited. And more significantly, in the Hypothesis II which follows he conjectured its conservation.

Hypothesis II

If the other parts of the Earth were taken away, and the remaining ring was carried alone about the Sun in the orbit of the Earth by the annual motion, while by the diurnal motion it was in the meantime revolved about its own axis inclined to the plane of the ecliptic by an angle of $23\frac{1}{2}$ degrees, the motion of the equinoctial points would be the same, whether the ring were fluid, or whether it consisted of a hard and rigid matter.

130. Proposition XXXIX: to find the precession of the equinoxes

Newton starts with the startling assertion that the solar contribution to the average rate, $\langle\dot{\psi}\rangle_{AV;\odot}$, of the precession of the equinoxes must be proportional to:

$$\langle\dot{\psi}\rangle_{AV;\odot} \propto -20^\circ\ 11'\ 46'' \times \left(\frac{23^h\ 56^m}{27^d\ 7^h\ 43^m \times 24}\right), \tag{1}$$

where

$$-20^\circ\ 11'\ 46'' = \text{the annual rate of regression of the nodes of the lunar orbit;} \tag{2}$$

$$23^h\ 56^m = \text{the sidereal day} = 2\pi/\Omega; \tag{3}$$

and

$$27^d\ 7^h\ 43^m \times 24 = \text{the orbital period (in hours) of the Moon about the Earth} = T_{☾}. \tag{4}$$

Newton argues as follows:

> Since, therefore, the nodes of the Moon in [the] orbit would be yearly transferred 20° 11′ 46″ backwards, and, if there were more moons, the motion of the nodes of every one (by Cor. XVI, Prop. LXVI, Book I) would be as its periodic time, if upon the surface of the Earth a moon was revolved in the time of a sidereal day, the annual motion of the nodes of this moon would be to 20° 11′ 46″ as $23^h\ 56^m$, the sidereal day, is to $27^d\ 7^h\ 43^m$, the periodic time of our Moon, that is, as 1436 is to 39343. And the same thing would happen to the nodes of a ring of moons encompassing the Earth, whether these moons did not mutually touch each the other, or whether they were molten, and formed into a continued ring, or whether that ring should become rigid and inflexible.

The argument is clearly based on his conjectured conservation of circulation ('Hypothesis' II).

The proportionality (1) becomes recognizable if we substitute for the regression of the nodes the expression,

$$\langle\dot{\psi}\rangle_{AV} = -\tfrac{3}{4}m^2 n. \tag{5}$$

derived in Proposition LXVI, Corollary XI, of Book I (Chapter 14, §72, equation (27)) and for the sidereal day and the orbital period as defined in equations (3) and (4). We obtain:

$$\langle\dot{\psi}\rangle_{AV;\odot} \propto -\tfrac{3}{4}m^2 n\left(\frac{2\pi}{\Omega T_{☾}}\right). \tag{6}$$

Since by Chapter 14, §70, equations (11), (16), and (18),

$$m^2 = \frac{N^2}{n^2}, \qquad N^2 = \left(\frac{2\pi}{T_\odot}\right)^2 = \frac{GM_\odot}{R_\odot^3}, \qquad \text{and} \qquad n^2 = \left(\frac{2\pi}{T_{☾}}\right)^2. \tag{7}$$

the proportionality (6) becomes,

$$\langle\dot{\psi}\rangle_{\text{AV};\odot} \propto -\frac{3\,GM_\odot}{4\,\Omega R_\odot^3} \tag{8}$$

which is, indeed, the principal factor in equation (57), §128.

To obtain the remaining factors in the expression for $\langle\dot{\psi}\rangle_{\text{AV};\odot}$, Newton approximates (idealizes?) the spheriodal shape of the Earth by replacing the bulges outside the inscribed sphere of radius b (the semiminor axis) by an equivalent circular disc having the same mass ΔM:

$$\Delta M = \tfrac{4}{3}\pi\rho(a^2 b - b^3) = \tfrac{4}{3}\pi\rho a^2 b\,\frac{a+b}{a}\left(1 - \frac{b}{a}\right)$$

$$= M\,\frac{a+b}{a}\,\varepsilon, \tag{9}$$

where

$$1 - b/a = \varepsilon = \text{the ellipticity}. \tag{10}$$

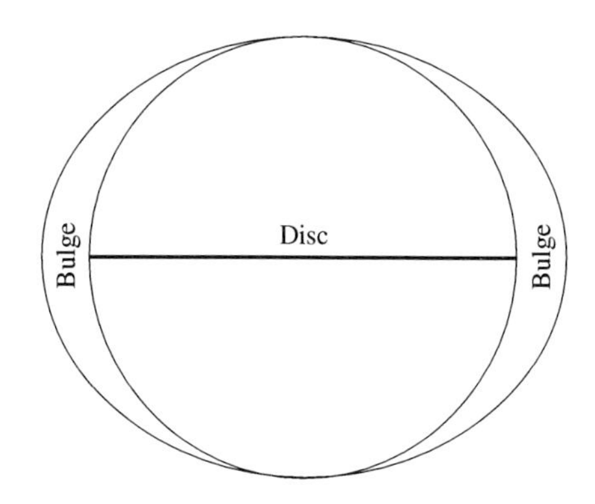

Therefore, to the first order in ε,

$$\Delta M/M = 2\varepsilon. \tag{11}$$

Besides, we must require that the ratio of the circulation, ΔC, of the disc to the circulation, $C + \Delta C$ of the spheroidal Earth is consistent with the masses of the disc and of the spheroidal Earth as required by the conservation of circulation. By Lemma III and equation (26) of §129, the requirement is

$$\frac{\Delta C}{C+\Delta C} = \frac{1}{1+\Delta C/C}\,\frac{\Delta C}{C} = \frac{1{\cdot}08076(\Delta M/M)}{1+1{\cdot}08076(\Delta M/M)}; \tag{12}$$

or, by equation (11)

$$\frac{\Delta C}{C+\Delta C} = \frac{1{\cdot}08076}{1+2{\cdot}16152\varepsilon}(2\varepsilon). \tag{13}$$

And this factor should be multiplied by

$$\frac{2}{5} = \frac{\text{moment of inertia of a circumferential ring of mass } M}{\text{moment of inertia of a sphere of equal mass } M} \tag{14}$$

to 'make up [for] that exterior part of the Earth' (i.e., the bulges) distributed over the surface of the Earth. As Newton explains:

> the total force or power of all the particles to wheel about the Earth round any diameter of the equator, and therefore to move the equinoctial points, would become less than before in the proportion of 2 to 5.

With these additional factors, equation (8) gives:

$$\langle\dot{\psi}\rangle_{\mathrm{AV};\odot} \propto -\frac{3}{5}\frac{GM_\odot}{R_\odot^3\Omega}\frac{1{\cdot}08076}{1 + 2{\cdot}16152\varepsilon}\varepsilon. \tag{15}$$

To obtain the complete expression for $\langle\dot{\psi}\rangle_{\mathrm{AV};\odot}$ we multiply (15) by

$$\cos\theta = \cos 23\tfrac{1}{2}^\circ = 0{\cdot}91706 \tag{16}$$

> because the plane of the equator is inclined to that of the ecliptic, this motion is to be diminished in the ratio of the sine 91706 (which is the cosine of $23\frac{1}{2}$ degrees) to the radius 100000.

Thus, Newton's final expression for the solar contribution to the precession of the equinoxes is

$$\langle\dot{\psi}\rangle_{\mathrm{AV};\odot} = -\frac{3}{5}\frac{GM_\odot}{R_\odot^3\Omega}\frac{1{\cdot}08076}{1 + 2{\cdot}16152\varepsilon}\varepsilon\cos\theta. \tag{17}$$

To quote Littlewood in a similar context: the reader is 'left in helpless wonder'.

A comparison of equation (17) and equation (57) of §128 shows that Newton's formula differs from the 'exact' formula by a factor

$$\frac{2}{5} \times \frac{1{\cdot}08076}{1 + 2{\cdot}16152\varepsilon} = 0{\cdot}4283 \quad \text{for } \varepsilon = \frac{1}{230}, \tag{18}$$

the value he had derived earlier in Proposition XIX. Therefore, by Newton's formula with his chosen value for ε, we should obtain (cf. §128, equation (65))

$$\langle\dot{\psi}\rangle_{\mathrm{AV};\odot} = -21{\cdot}17'' \times 0{\cdot}4283 = -9{\cdot}05'' \quad \text{per year}, \tag{19}$$

which is the value that Newton gives—an under-estimate by a factor $\frac{3}{5}$ compared to the 'true' value $-15{\cdot}94''$ per year (by §128, equation (63)).

Since Newton assumes that the ratio of the lunar to the solar tidal influences is 4·4815, his value for the lunar contribution to the precession is

$$\langle\dot{\psi}\rangle_{\mathrm{AV};☾} = -9{\cdot}05'' \times 4{\cdot}4815 = -40{\cdot}56'' \quad \text{per year}. \tag{20}$$

Adding the contributions (19) and (20), we obtain for the total lunisolar precession, the value

$$\langle\dot{\psi}\rangle_{\mathrm{AV;total}} \simeq -50'' \quad \text{per year,} \tag{21}$$

'which agrees with the phenomena'—a fortuitous coincidence. Had Newton used for the ratio of the lunar to the solar tidal influences the current value 2·164 (by §128, equation (72)), he would have found:

$$\langle\dot{\psi}\rangle_{\mathrm{AV;☾}} = -9{\cdot}05'' \times 2{\cdot}164 \simeq -19{\cdot}58'' \quad \text{per year,} \tag{22}$$

and

$$\langle\dot{\psi}\rangle_{\mathrm{AV;total}} = -9{\cdot}05'' - 19{\cdot}58'' = -28{\cdot}63'' \quad \text{per year.} \tag{23}$$

Besides, Newton's adopted value of $\varepsilon = \frac{1}{230}$ is an over-estimate compared to the true value $\frac{1}{289}(\simeq 0{\cdot}0034)$. Altogther, I am convinced that Newton himself could not have really believed that he had quantitatively accounted for the precession of the equinoxes.

131. A personal reflection

I find it hard to believe that Newton's concluding remark in Proposition XXXIX, that he had accounted for the precession of the equinoxes 'agreeing with the phenomena', has invariably—at least, to my knowledge—been accepted with apparent approval, when it should have been obvious, even to a casual reader, that the agreement must be fortuitous: Newton's theory is based on an idealized model of the spheroidal Earth in which the bulges are replaced by an equivalent thin circular disc in the equatorial plane, *albeit* by a treatment of surpassing insight; the choice of $\varepsilon = \frac{1}{230}$ for the ellipticity of the Earth is substantially different from its true value $\frac{1}{289}$ that Newton himself had derived in Proposition XIX (Book III); and finally Newton's choice of 4·4815 for the ratio of the tidal effects of the Moon and of the Sun is at best unreliable (by a factor ~ 2 as we now know).

The fact that there is *no* substantive change in the basic arguments (and conclusions) of Proposition XXXIX (Book III) from the first to the second edition of the *Principia* makes me suspicious of the supposed 'critical acumen' of Roger Cotes 'in dealing with problems of mathematical physics' and of his 'deep involvement' in the theory of tides and of lunar motion. As far as I can gather, Cotes never expressed any qualms about Newton's conclusion of Proposition XXXIX. Considering, in juxtaposition, that Cotes did not realize that, in Proposition XXIV (Book III) on the theory of tides, Newton's including the normal component of the tidal force along with the transverse component is contrary to what Huygens states, explicitly enough in his 1690 *Discours de la cause de la pesanteur*, that 'the surface of the sea is such that at every point the *direction of the plumbline is perpendicular to the surface*', I am not convinced of Cotes's 'acumen' as has been trumpeted.

It seems to me that Newton's assessment of Cotes's perspicacity cannot have been enhanced by his writing, as late as August 1709, that he had at last verified the truth of Corollary II of Proposition XCI (Book I) (see p. 313) only after consulting Newton's 1704 *Tractus de Quadratura Curvarum*, raising the question 'Did Cotes really understand Proposition XXIV (Book III)?' But even at that late date, did Cotes realize that Newton could very well have included in Proposition XLI (Book I) the example of the inverse-square law of centripetal attraction and disposed of, once and for all, the doubts of Bernoulli regarding the completeness of Proposition XI? Newton could very well have asked himself: 'Is Cotes really adequate to understand the essential *Principia*?' I believe that it was Newton's growing disillusionment with Cotes's adequacy that led him to write more than once in the vein:

> I would not have you be at the trouble of examining all the Demonstrations in the *Principia*. Its impossible to print the book wthout some faults & if you print by the copy sent you, correcting only such faults as occurr in reading over the sheets to correct them as they are printed off, you will have labour more than it's fit to give you.

24

On comets

And now we have described the system of the Sun, the Earth, Moon, and planets, it remains that we add something about the comets.

Isaac Newton
(At the conclusion of Proposition XXXIX, Book III)

132. Introduction

The seven Lemmas, IV–X, and the three Propositions, XL–XLII, devoted to comets and occupying 43 pages (439–481) in the original 1713 second edition of the *Principia* are quite exceptional: basic matters are treated in the normal tenor of the *Principia* in Lemmas V–X and in the three propositions in barely ten pages; the remaining 33 pages are concerned with the results of numerical computations of cometary orbits and the physics of comets. And yet, Lemma V, in just two pages, lays the foundation of the theory of interpolation in terms of finite-difference formulae of which James Stirling has written:*

> Many celebrated mathematicians after Newton have discussed the question of describing a curve of the parabolic order through given points. But all their solutions are the same as those given above, which hardly differ from those of Newton in the 5th Lemma of the third book of the *Principia*, and in the *Methodus differentialis* edited by Mr. Jones. Newton describes a parabolic curve through given points; others have considered the determination of the curve from definite data; but however it may be expressed and however it may be carried out, it is the same problem, and certainly the discovery of the mathematical forms for the values of the interpolated ordinate is exceedingly ingenious and worthy of the celebrity of its author; but after the mathematical forms are given the investigation of the problem is easy, nothing more being required than the solution of simple equations.

And in Proposition XLI,

> Newton describes a simplified graphical solution which he calls 'approximate' (*quam proxime*), but his lemmas contain as complete and exact a solution of

* In fact the formulae commonly known under the names of Stirling and Bessel are due to Newton.

> this *longe difficillium* problem as may be desired, and this on seven pages of most sublime geometrical reasoning quite as simple as powerful.
>
> A. N. Kriloff
>
> (Monthly Notices of the Royal Astronomical Society, Vol. 85, p. 656, 1925)

In the account which follows, we shall, consistent with our purpose, restrict ourselves to the 10 pages dealing with the basic matters.

133. Lemma IV and Proposition XL

In Lemma IV, Newton is concerned with the place of comets in the solar system.

Lemma IV

The coments are more remote than the Moon, and are in the regions of the planets.

Newton argues that the comets are further away than the Moon because they exhibit no diurnal parallax. He further shows how, from the variations in the direct and the retrograde motions of the comets, their distances can be ascertained. Newton's explanation (as in all similar explanation of physical matters) has surpassing clarity.

> Let ♈QA, ♈QB, ♈QC be three observed longitudes of the comet about the time of its first appearing, and ♈QF its last observed longitude before its disappearing. Draw the right line ABC, whose parts AB, BC, intercepted between the right lines QA and QB, QB and QC, may be one to the other as the two times between the three first observations. Produce AC to G, so that AG may be to AB as the time between the first and last observations is to the time between the first and second; and join QG. Now if the comet did move uniformly in a right line, and the Earth either stood still, or was likewise carried forwards in a right line by a uniform motion, the angle ♈QG would be the longitude of the comet at the time of the last observation. Therefore, the angle FQG, which is the difference of the longitude, proceeds from the inequality of the motions of the comet and the Earth; and if the Earth and comet move contrary ways, this angle is added to the angle ♈QG, and accelerates the apparent motion of the comet; but if the comet moves the same way with the

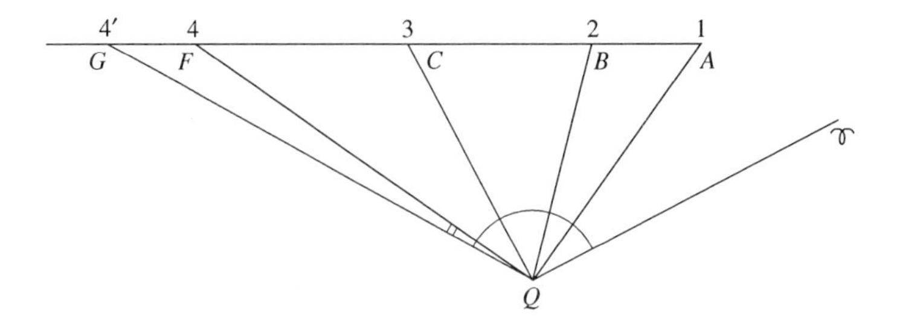

Earth, if is subtracted, and either retards the motion of the comet, or perhaps renders it retrograde, as we have just now explained. This angle, therefore, proceeding chiefly from the motion of the Earth, is justly to be esteemed the parallax of the comet, there being neglected thereby some little increment or decrement that may arise from the unequal motion of the comet in its orbit. From this parallax we thus deduce the distance of the comet. Let S represent the Sun, acT the great orbit, a the Earth's place in the first observation, c the place of the Earth in the third observation, T the place of the Earth in the last observation, and $T\Upsilon$ a right line drawn to the beginning of Aries. Set off the angle ΥTV equal to the angle ΥQF, that is, equal to the longitude of the comet at the time when the earth is in T; join ac, and produce it to g, so that ag may be to ac as AG is to AC; and g will be the place at which the Earth would have arrived in the time of the last observation, if it had continued to move uniformly in the right line ac. Therefore, if we draw $g\Upsilon$ parallel to $T\Upsilon$, and make the angle ΥgV equal to the angle ΥQG, this angle ΥgV will be equal to the longitude of the comet seen from the place g, and the angle TVg will be the parallax which arises from the Earth's being transferred from the place g into the place T; and therefore V will be the place of the comet in the plane of the ecliptic. And this place V is commonly lower than the orbit of Jupiter.

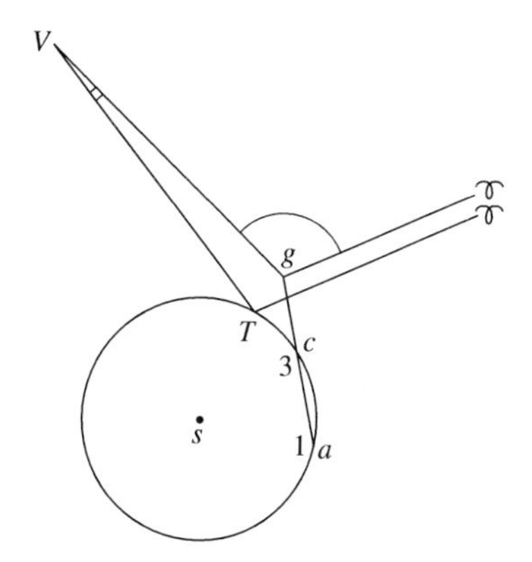

And he continues:

The near approach of the comets is further confirmed from the light of their heads; for the light of a celestial body, illuminated by the Sun, and receding to remote parts, diminishes as the fourth power of the distance; namely, as the square, on account of the increase of the distance from the Sun, and as another square, on account of the decrease of the apparent diameter. Therefore, if both the quantity of light and the apparent diameter of a comet are given, its distance will be given also, by taking the distance of the comet to the distance of a planet directly as their diameters and inversely as the square root of their lights.

and concludes:

> Therefore the comets shine by the Sun's light, which they reflect. I am out in my judgment, if they are not a sort of planets revolving in orbits returning into themselves with a continual motion;

And Proposition XL which follows is an effective summary.

Proposition XL. Theorem XX

That the comets move in some of the conic sections, having their foci in the centre of the Sun, and by radii drawn to the Sun describe areas proportional to the times.

This Proposition appears from COR. I, Prop. XIII, Book I, compared with Prop. VIII, XII, and XIII, Book III.

COR. I. Hence if comets revolve in orbits returning into themselves, the orbits will be ellipses; and their periodic times will be to the periodic times of the planets as the $\frac{3}{2}$th power of their principal axes. And therefore the comets, which for the most part of their course are more remote than the planets, and upon that account describe orbits with greater axes, will require a longer time to finish their revolutions. Thus if the axis of a comet's orbit was four times greater than the axis of the orbit of Saturn, the time of the revolution of the comet would be to the time of the revolution of Saturn, that is, to 30 years, as $4\sqrt{4}$ (or 8) is to I, and would therefore be 240 years.

COR. II. But their orbits will be so near to parabolas, that parabolas may be used for them without sensible error.

COR. III. And, therefore, by Cor. VII, Prop. XVI, Book I, the velocity of every comet will always be to the velocity of any planet, supposed to be revolved at the same distance in a circle about the Sun, nearly as the square root of double the distance of the planet from the centre of the Sun to the distance of the comet from the Sun's centre. Let us suppose the radius of the great orbit, or the greatest semidiameter of the ellipse which the Earth describes, to consist of 100000000 parts; and then the Earth by its mean diurnal motion will describe 1720212*

* The constant that is defined here is

$$\kappa = \frac{2\pi}{\text{sidereal year}\,[1 + (m_{\oplus} + m_{☾})/M_{\odot}]^{1/2}}.$$

With the values,

$$\text{sidereal year} = 365{\cdot}25636 \quad \text{mean solar days,}$$

and

$$M_{\odot}/(m_{\oplus} + m_{☾}) = 329100,$$

we find

$$\kappa = 0{\cdot}017202098,$$

to be contrasted with Newton's value,

$$\kappa = 0{\cdot}01720212.$$

The constant κ is commonly called the *Gaussian constant of gravitation*—not even hyphenated!

of those parts, and $71675\frac{1}{2}$ by its hourly motion. And therefore the comet, at the same mean distance of the Earth from the Sun, with a velocity which is to the velocity of the Earth as $\sqrt{2}$ to I, would by its diurnal motion describe 2432747 parts, and $101364\frac{1}{2}$ parts by its hourly motion. But at greater or less distances both the diurnal and hourly motion will be to this diurnal and hourly motion inversely as the square root of the distances, and is therefore given.

COR. IV. Therefore if the latus rectum of the parabola is four times the radius of the great orbit, and the square of that radius is supposed to consist of 100000000 parts, the area which the comet will daily describe by a radius drawn to the Sun will be $1216373\frac{1}{2}$ parts, and the hourly area will be $50682\frac{1}{4}$ parts. But, if the latus rectum is greater or less in any ratio, the diurnal and hourly area will be less or greater inversely as the square root of that ratio.

134. Lemma V: Newton's theory of interpolation

It has always been recognized that in Lemma V of Book III Newton provided the complete solution of the problem of interpolation in the calculus of finite differences—a problem to which he referred in a letter he wrote in 1676 to Henry Oldenburg, the then Secretary of the Royal Society:

> When simple series are not obtainable with sufficient ease I have another method not yet published by which the problem is easily dealt with. It is based upon a convenient, ready, and general solution of the problem. *To describe a geometrical curve which shall pass through any given points...*
>
> Although the problem may seen to be intractable at first sight it is never-the-less quite the contrary. Perhaps indeed it is one of the prettiest problems that I can ever hope to solve.

The solution to this problem, starkly enunciated in this lemma gives no hint as to how it was obtained. But the solution with sufficient detail is given in *Methodus differentialis*—a collection of a number of shorter works of Newton, published by William Jones in 1711 under the title *Analysis per quantitatum series, fluxiones ac differentias; cum enumeratione linearum tertii ordinis*. We shall first summarize Newton's treatment in *Methodus differentialis* following the excellent account given by Duncan C. Fraser in his article 'Newton and Interpolation' in *Isaac Newton*, 1642–1727 (G. Bell and Sons. Ltd., London, 1927), pp. 45–54.

(a) Newton's treatment in *Methodus differentialis*

Newton employs divided differences according to the following scheme (not using his notation).*

Value of independent variable	Value of function	First divided differences	Second divided differences	Third divided differences	Fourth divided differences
a	Ua				
		$\dfrac{Ua - Ub}{a - b} = \grave{\Delta}(ab)$			
b	Ub		$\dfrac{\grave{\Delta}(ab) - \grave{\Delta}(bc)}{a - c} = \grave{\Delta}^2(abc)$		
		$\dfrac{Ub - Uc}{b - c} = \grave{\Delta}(bc)$		$\dfrac{\grave{\Delta}^2(abc) - \grave{\Delta}^2(bcd)}{a - d} = \grave{\Delta}^3(abcd)$	
c	Uc		$\dfrac{\grave{\Delta}(bc) - \grave{\Delta}(cd)}{b - d} = \grave{\Delta}^2(bcd)$		$\dfrac{\grave{\Delta}^3(abcd) - \grave{\Delta}^3(bcde)}{a - e} = \grave{\Delta}^4(abcde)$
		$\dfrac{Uc - Ud}{c - d} = \grave{\Delta}(cd)$		$\dfrac{\grave{\Delta}^2(bcd) - \grave{\Delta}^2(cde)}{b - e} = \grave{\Delta}^3(bcde)$	
d	Ud		$\dfrac{\grave{\Delta}(cd) - \grave{\Delta}(de)}{d - e} = \grave{\Delta}^2(cde)$		
		$\dfrac{Ud - Ue}{d - e} = \grave{\Delta}(de)$			
e	Ue				

* It will be noticed that Newton's table of differences is left-handed and not right-handed as in current practice. But since his conventions regarding measuring x and taking differences are also contrary to current practice, the form of this final results are unaffected.

(*i*) *Proposition* 1

In Proposition 1, it is shown how to obtain expressions for the divided differences. On the assumption that the ordinate at a point can be represented by the expression,

$$F(A + x) = a + bx + cx^2 + dx^3 + ex^4, \tag{1}$$

the following table of *divided differences* is constructed in terms of the values of the ordinates at

$$A + p, \quad A + q, \quad A + r, \quad A + s, \quad \text{and} \quad A + t. \tag{2}$$

that is, when x take the values,

$$p, \quad q, \quad r, \quad s, \quad \text{and} \quad t. \tag{3}$$

It should be noted that no particular rule is laid on the ordering of the sequence, p, q, r, s, and t: ascending, descending, or otherwise.

Abscissa	Ordinates
$A + p$	$a + bp + cp^2 + dp^3 + ep^4 = \alpha$
$A + q$	$a + bq + cq^2 + dq^3 + eq^4 = \beta$
$A + r$	$a + br + cr^2 + dr^3 + er^4 = \gamma$
$A + s$	$a + bs + cs^2 + ds^3 + es^4 = \delta$
$A + t$	$a + bt + ct^2 + dt^3 + et^4 = \varepsilon$

Divisor	Difference	Quotients
$p - q$	$\alpha - \beta$	$b + c(p + q) + d(p^2 + pq + q^2) + e(p^3 + p^2q + pq^2 + q^3) = \zeta$
$q - r$	$\beta - \gamma$	$b + c(q + r) + d(q^2 + qr + r^2) + e(q^3 + q^2r + qr^2 + r^3) = \eta$
$r - s$	$\gamma - \delta$	$b + c(r + s) + d(r^2 + rs + s^2) + e(r^3 + r^2s + rs^2 + s^2) = \theta$
$s - t$	$\delta - \varepsilon$	$b + c(s + t) + d(s^2 + st + t^2) + e(s^3 + s^2t + st^2 + t^3) = \kappa$
$p - r$	$\zeta - \eta$	$c + d(p + q + r) + e(p^2 + pq + q^2 + pr + qr + r^2) = \lambda$
$q - s$	$\eta - \theta$	$c + d(q + r + s) + e(q^2 + qr + r^2 + qs + rs + s^2) = \mu$
$r - t$	$\theta - \kappa$	$c + d(r + s + t) + e(r^2 + rs + s^2 + rt + st + t^2) = \nu$
$p - s$	$\lambda - \mu$	$d + e(p + q + r + s) = \xi$
$q - t$	$\mu - \nu$	$d + e(q + r + s + t) = \pi$
$p - t$	$\xi - \pi$	$e = \sigma$

(ii) Proposition 2

In Proposition 2, it is pointed out how the values of e, d, c, b, and a can be successively determined from the values of σ, ξ, λ, and ζ. Thus

$$\left.\begin{aligned} e &= \sigma, \\ d &= \xi - e(p+q+r+s), \\ c &= \lambda - d(p+q+r) - e(p^2+pq+q^2+pr+qr+r^2), \\ b &= \zeta - c(p+q) - d(p^2+pq+q^2) - e(p^3+p^2q+pq^2+q^3). \end{aligned}\right\} \tag{4}$$

While Newton gives detailed instructions how each of the successive coefficients are to be determined, he does not (or desists to?) give the resulting formula for $F(A+x)$. But following his instructions verbatim we find:

$$\begin{aligned} \sigma &= -\frac{\pi}{p-t} + \frac{\xi}{p-t} \\ &= -\frac{\pi}{p-t} + \frac{1}{p-t}\left(\frac{\lambda}{p-s} - \frac{\mu}{p-s}\right) \\ &= -\frac{\pi}{p-t} - \frac{\mu}{(p-t)(p-s)} + \frac{1}{(p-t)(p-s)}\left(\frac{\zeta}{p-r} - \frac{\eta}{p-r}\right) \\ &= -\frac{\pi}{p-t} - \frac{\mu}{(p-t)(p-s)} - \frac{\eta}{(p-t)(p-s)(p-r)} \\ &\quad + \frac{1}{(p-t)(p-r)(p-s)}\left(\frac{\alpha}{p-q} - \frac{\beta}{p-q}\right), \end{aligned} \tag{5}$$

or

$$\begin{aligned} F(A+p) = F(A+q) &+ \eta(p-q) + \mu(p-q)(p-r) + \pi(p-q)(p-r)(p-s) \\ &+ \sigma(p-q)(p-r)(p-s)(p-t). \end{aligned} \tag{6}$$

By extending the difference table to include $A+x$, as shown below,

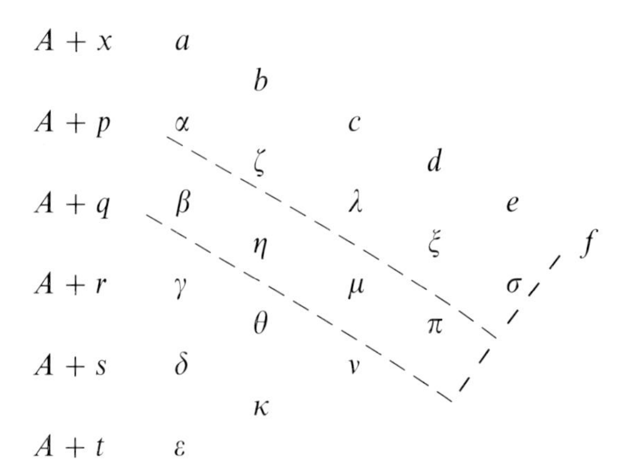

we find in similar fashion,

$$F(A+x) = \alpha + \zeta(x-p) + \lambda(x-p)(x-q) + \xi(x-p)(x-q)(x-r) \\ + \sigma(x-p)(x-q)(x-r)(x-s) + f(x-p)(x-q)(x-r)(x-s)(x-t), \tag{7}$$

a form of the solution that can be extended indefinitely to include higher-order differences.

Now letting,

$$F(A+x) = U_x, \qquad F(A+p) = U_p,$$

$\zeta = \alpha - \beta =$ the first (divided) difference corresponding to the entries at $A + p$ and $A + q = \grave{\Delta}(pq)$,

$\lambda = \zeta - \eta =$ the second (divided) difference corresponding to the entries at $A + p$, $A + q$, and

$$A + r = \grave{\Delta}^2(pqr), \tag{8}$$

etc.,

we can write

$$U_x = U_p + (x-p)\grave{\Delta}(pq) + (x-p)(x-q)\grave{\Delta}^2(pqr) \\ + (x-p)(x-q)(x-r)\grave{\Delta}^3(pqrs) + (x-p)(x-q)(x-r)(x-s)\grave{\Delta}^4(pqrst) \\ + \text{remainder term}. \tag{9}$$

For entries at equidistant intervals, we may set

$$p = 0, \qquad q = 1, \qquad r = 2, \qquad s = 3, \qquad \text{and} \qquad t = 4; \tag{10}$$

and the ordinate at x (preceding 0) will be given by

$$U_x = U_0 + x\grave{\Delta}(0, 1) + x(x-1)\grave{\Delta}^2(0, 1, 2) + x(x-1)(x-2)\grave{\Delta}^3(0, 1, 2, 3) \\ + x(x-1)(x-2)(x-3)\grave{\Delta}^4(0, 1, 2, 3, 4) + \cdots. \tag{11}$$

In equation (11), the differences are still divided differences related to ordinary differences by

$$\grave{\Delta}(0, 1) = \Delta U_0, \qquad \grave{\Delta}^2(0, 1, 2) = \tfrac{1}{2}\Delta^2 U_0, \qquad \grave{\Delta}^3(0, 1, 2, 3) = \frac{1}{3!}\Delta^3 U_0, \qquad \text{etc.} \tag{12}$$

Therefore, expressed in terms of ordinary differences, equation (11) takes the form

$$U_x = U_0 + x\Delta U_0 + \tfrac{1}{2}x(x-1)\Delta^2 U_0 + \frac{1}{3!}x(x-1)(x-2)\Delta^3 U_0 \\ + \frac{1}{4!}x(x-1)(x-2)(x-3)\Delta^4 U_0 + \cdots, \tag{13}$$

the familiar 'binomial formula'. It has been obtained following Newton's instructions; but he does not give it.

*(iii) A modern version of Proposition 2**

Starting with the definitions,

$$
\left.\begin{aligned}
[x_0x_1] &= \frac{f(x_0)-f(x_1)}{x_0-x_1} = \frac{f(x_1)-f(x_0)}{x_1-x_0} = [x_1x_0], \\
[x_1x_2] &= \frac{f(x_1)-f(x_2)}{x_1-x_2}, \\
[x_0x_1x_2] &= \frac{[x_0x_1]-[x_1x_2]}{x_0-x_2} = \frac{[x_0x_1]-[x_2x_1]}{x_0-x_2}, \\
&\cdots\cdots\cdots\cdots\cdots\cdots\cdots\cdots\cdots\cdots \\
[x_0x_1\cdots x_n] &= \frac{[x_0x_1\cdots x_{n-1}]-[x_1x_2\cdots x_n]}{x_0-x_n}.
\end{aligned}\right\} \qquad (14)
$$

we form the table of divided differences:

$$
\begin{array}{llllll}
x_0 & f(x_0) & & & & \\
 & & [x_0x_1] & & & \\
x_1 & f(x_1) & & [x_0x_1x_2] & & \\
 & & [x_1x_2] & & [x_0x_1x_2x_3] & \\
x_2 & f(x_2) & & [x_1x_2x_3] & & [x_0x_1x_2x_3x_4] \\
 & & [x_2x_3] & & [x_1x_2x_3x_4] & \cdot \\
x_3 & f(x_3) & & [x_2x_3x_4] & \cdot & \cdot \\
 & & [x_3x_4] & \cdot & \cdot & \cdot \\
x_4 & f(x_4) & \cdot & \cdot & \cdot & \\
\cdot & \cdot & \cdot & \cdot & [x_{n-3}x_{n-2}x_{n-1}x_n] & \\
\cdot & \cdot & \cdot & [x_{n-2}x_{n-1}x_n] & & \\
\cdot & \cdot & [x_{n-1}x_n] & & & \\
x_n & f(x_n) & & & &
\end{array} \qquad (15)
$$

Replacing x_0 by x in equations (14) and reordering them, we have

* The account which follows in §§(iii) and (iv) is partially based on *The calculus of finite differences* by L. M. Milne-Thomson (Macmillan & Co., Ltd., London, 1951), Chapters 1–3.

$$\left.\begin{aligned}
[xx_1x_2\ldots x_n] &= -\frac{[x_1x_2\ldots x_n]}{x-x_n}+\frac{[xx_1\ldots x_{n-1}]}{x-x_n},\\
[xx_1\ldots x_{n-1}] &= -\frac{[x_1x_2\ldots x_{n-1}]}{x-x_{n-1}}+\frac{[xx_1\ldots x_{n-2}]}{x-x_{n-1}},\\
[xx_1\ldots x_{n-2}] &= -\frac{[x_1x_2\ldots x_{n-2}]}{x-x_{n-2}}+\frac{[xx_1\ldots x_{n-3}]}{x-x_{n-2}},\\
&\cdots\cdots\cdots\cdots\cdots\cdots\\
[xx_1x_2] &= -\frac{[x_1x_2]}{x-x_2}+\frac{[xx_1]}{x-x_2},\\
[xx_1] &= -\frac{f(x_1)}{x-x_1}+\frac{f(x)}{x-x_1}.
\end{aligned}\right\} \tag{16}$$

By repeatedly substituting the second member on the right-hand side of each identity with its value given by the succeeding identity—the same procedure adopted in equation (5)—we find:

$$\begin{aligned}
[xx_1x_2\ldots x_n] = &-\frac{[x_1x_2\ldots x_n]}{x-x_n}-\frac{[x_1x_2\ldots x_{n-1}]}{(x-x_n)(x-x_{n-1})}\\
&-\frac{[x_1x_2\ldots x_{n-2}]}{(x-x_n)(x-x_{n-1})(x-x_{n-2})}-\cdots-\frac{[x_1x_2]}{(x-x_n)(x-x_{n-1})\ldots(x-x_2)}\\
&-\frac{f(x_1)}{(x-x_n)\ldots(x-x_2)(x-x_1)}+\frac{f(x)}{(x-x_n)\ldots(x-x_1)},
\end{aligned} \tag{17}$$

or

$$\begin{aligned}
f(x) = f(x_1) &+ (x-x_1)[x_1x_2]+(x-x_1)(x-x_2)[x_1x_2x_3]+\cdots\\
&+(x-x_1)(x-x_2)\ldots(x-x_{s-1})[x_1x_2\ldots x_s]+\cdots\\
&+(x-x_1)(x-x_2)\ldots(x-x_{n-1})[x_1x_2\ldots x_n]\\
&+(x-x_1)(x-x_2)\ldots(x-x_n)[xx_1x_2\ldots x_n],
\end{aligned} \tag{18}$$

or

$$f(x) = f(x_1)+\sum_{s=1}^{n-1}(x-x_1)(x-x_2)\ldots(x-x_s)[x_1x_2\ldots x_{s+1}]+R_n(x), \tag{19}$$

where

$$R_n(x) = (x-x_1)(x-x_2)\ldots(x-x_n)[xx_1x_2\ldots x_n] \tag{20}$$

is the 'remainder'.

Equation (18) is none other than Newton's (forward) interpolation formula.

An explicit formula for $[xx_1x_2\ldots x_n]$ can also be obtained. From

$$[xx_1] = \frac{f(x)}{x-x_1}+\frac{f(x_1)}{x_1-x}, \tag{21}$$

we deduce,

$$[xx_1x_2] = \frac{[xx_1] - [x_1x_2]}{x - x_2}$$
$$= \frac{1}{x - x_2}\left\{\frac{f(x) - f(x_1)}{x - x_1} - \frac{f(x_1) - f(x_2)}{x_1 - x_2}\right\}$$
$$= \frac{f(x)}{(x - x_1)(x - x_2)} + \frac{f(x_1)}{(x_1 - x)(x_1 - x_2)} + \frac{f(x_2)}{(x_2 - x)(x_2 - x_1)}; \quad (22)$$

and by induction it follows:

$$[xx_1x_2\ldots x_n] = \frac{f(x)}{(x - x_1)(x - x_2)\ldots(x - x_n)} + \frac{f(x_1)}{(x_1 - x)(x_1 - x_2)\ldots(x_1 - x_n)} + \cdots$$
$$+ \frac{f(x_n)}{(x_n - x)(x_n - x_1)\ldots(x_n - x_{n-1})}. \quad (23)$$

It is now manifest that interchanging any two of the arguments leaves the divided difference unchanged. Therefore, $[xx_1x_2x_3\ldots x_n]$ *is a symmetric function of the arguments*, an extremely important property as will appear in §(iv) below.

(iv) A modern version of Proposition 3

Newton's statement of Proposition 3, in terms of a particular example, is so brief that it conceals the basic ideas. For this reason, we shall first present a modern version.

First some definitions. If Δx is an increment in a variable x, the corresponding increment in a function $f(x)$ is defined by

$$\Delta f(x) = f(x + \Delta x) - f(x) \quad (24)$$

and is called its *first difference*. The most important case (to which we shall restrict ourselves in the following) is when Δx is a constant equal to ω (say). Then,

$$\Delta f(x) = f(x + \omega) - f(x). \quad (25)$$

Second and higher-order differences can be obtained by repeated application of the operator Δ. Thus

$$\Delta^2 f(x) = \Delta[\Delta f(x)] = \Delta[f(x + \omega) - f(x)]$$
$$= f(x + 2\omega) - f(x + \omega) - [f(x + \omega) - f(x)]$$
$$= f(x + 2\omega) - 2f(x + \omega) + f(x), \quad (26)$$

and more generally by the inductive relation,

$$\Delta^n f(x) = \Delta[\Delta^{n-1} f(x)]. \tag{27}$$

We now express the divided differences with the aid of Nörlund's *quotient operator* $\underset{\omega}{\Delta}$ defined by its effect on a function $f(x)$:

$$\underset{\omega}{\Delta} f(x) = \frac{1}{\omega}[f(x+\omega) - f(x)] = \frac{1}{\omega}\Delta f(x). \tag{28}$$

Higher-order differences can be obtained by repeated applications of $\underset{\omega}{\Delta}$.

Two identities which follow from the definition of $\underset{\omega}{\Delta}$ are

$$f(x+\omega) = f(x) + \omega \underset{\omega}{\Delta} f(x) = (1 + \omega \underset{\omega}{\Delta}) f(x) \tag{29}$$

and

$$\Delta^n f(x) = \omega^n \underset{\omega}{\Delta}{}^n f(x). \tag{30}$$

In terms of the operator

$$E^\omega = 1 + \omega \underset{\omega}{\Delta} \tag{31}$$

we can write equations (29) and (30) in the forms:

$$f(x+\omega) = E^\omega f(x) \tag{32}$$

and

$$\Delta^n f(x) = (E^\omega - 1)^n f(x). \tag{33}$$

It follows from equation (32) that by repeated application of E^ω to $f(x)$, we obtain

$$f(x + s\omega) = E^{\omega s} f(x). \tag{34}$$

Next, by expanding $(E^\omega - 1)^n$, on the right-hand side of equation (33), by the binomial theorem, we obtain

$$\Delta^n f(x) = \sum_{m=0}^{n} (-1)^m C_m^n E^{\omega(n-m)} f(x), \tag{35}$$

where C_m^n denotes the binomial coefficient,

$$C_m^n = \frac{n!}{(n-m)!\, m!}; \tag{36}$$

or, by equation (34),

$$\Delta^n f(x) = \sum_{m=0}^{n} (-1)^m C_m^n f[x + (n-m)\omega]. \tag{37}$$

Now, returning to equation (23) and writing

$$x_s = x + s\omega \quad (s = 1, 2, \ldots), \tag{38}$$

consistently with the ordinates being separated by the constant interval ω, we obtain for the divided difference $[xx_1x_2 \ldots x_n]$ the expression,

$$[xx_1x_2\ldots x_n] = \frac{f(x+n\omega)}{n!\,\omega^n} - \frac{f(x+(n-1)\omega)}{(n-1)!\,1!\,\omega^n} + \frac{f(x+(n-2)\omega)}{(n-2)!\,2!\,\omega^n} - \cdots, \tag{39}$$

By comparison with equation (37) it is now manifest that

$$[xx_1x_2\ldots x_n] = \frac{\omega^{-n}}{n!}\,\Delta^n f(x). \tag{40}$$

Consider now the following difference table:

Argument	Function				
a	$f(a)$				
		$\Delta f(a)$			
$a+\omega$	$f(a+\omega)$		$\Delta^2 f(a)$		
		$\Delta f(a+\omega)$		$\Delta^3 f(a)$	
$a+2\omega$	$f(a+2\omega)$		$\Delta^2 f(a+\omega)$		.
		$\Delta f(a+2\omega)$		$\Delta^3 f(a+\omega)$	
$a+3\omega$	$f(a+3\omega)$		$\Delta^2 f(a+2\omega)$		.
		.		.	
.	.		.		.

and make the substitutions (cf. equations (38) and (40))

$$\left.\begin{aligned} x_s &= a+(s-1)\omega \quad (s=1,2,\ldots,n) \\ \text{and}\qquad [x_1\ldots x_{s+1}] &= \frac{\omega^{-s}}{s!}\,\Delta^s f(a), \end{aligned}\right\} \tag{41}$$

in Newton's general interpolation formula (18). We obtain:

$$\begin{aligned} f(x) = f(a) + (x-a)\omega^{-1}\Delta f(a) + \frac{(x-a)(x-a-\omega)}{2!}\,\omega^{-2}\Delta^2 f(a) + \cdots \\ + \frac{(x-a)(x-a-\omega)\cdots(x-a-n\omega+2\omega)}{(n-1)!}\,\omega^{-n+1}\Delta^{n-1}f(a) + R_n(x). \end{aligned} \tag{42}$$

In contrast to the *forward-difference formula* (42) we can obtain a *backward-difference formula* by considering the scheme

$$
\begin{array}{cccccc}
\cdot & \cdot & & \cdot & & \cdot \\
 & & \cdot & & \cdot & \\
a-3\omega & f(a-3\omega) & & \Delta^2 f(a-4\omega) & & \cdot \\
 & & \Delta f(a-3\omega) & & \Delta^3 f(a-4\omega) & \\
a-2\omega & f(a-2\omega) & & \Delta^2 f(a-3\omega) & & \cdot \\
 & & \Delta f(a-2\omega) & & \Delta^3 f(a-3\omega) & \\
a-\omega & f(a-\omega) & & \Delta^2 f(a-2\omega) & & \\
 & & \Delta f(a-\omega) & & & \\
a & f(a); & & & &
\end{array}
$$

and making the substitutions,

$$
\left.\begin{aligned}
x_s &= a-(s-1)\omega, \quad (s=1,2,\ldots,n), \\
&\text{and} \\
[x_1x_2\ldots x_{s+1}] &= \frac{\omega^{-s}}{s!}\,\Delta^s f(a-s\omega),
\end{aligned}\right\} \tag{43}
$$

in the same general formula (18), we now obtain:

$$
f(x) = f(a) + (x-a)\omega^{-1}\Delta f(a-\omega) + \frac{(x-a)(x-a+\omega)}{2!}\,\omega^{-2}\Delta^2 f(a-2\omega) + \cdots. \tag{44}
$$

We finally turn to *central-difference formulae* that one can derive directly from Newton's interpolation formula (18) by making use of the symmetry of the divided difference $[x_1x_2x_3\ldots x_n]$ in its arguments—a fact to which particular attention was drawn in §(iii) following equation (23). These formulae have commonly been coupled with the names of Gauss, Stirling, and Bessel. But they are given in so clear algebraic forms in Proposition 3 (see §(v) below) that anyone with some discernment can so recognize them, and E. T. Whittaker and G. Robinson in their *The calculus of observations* (Blackie and Son Ltd., London and Glasgow, second edition 1926, pp. 21–24) do, in fact, hyphenate the names of Gauss, Stirling, and Bessel with Newton's!

In the difference table below, in which the schemes following equations (40) and (42) are combined, we have reordered the ordinates by the arguments $x_1, x_2, \ldots$, etc., in the first two columns on the left, labelled F and B, to signify 'forward' and 'backward', respectively.

B	F						
				.		.	
x_6	x_7	$a-3\omega$	$f(a-3\omega)$		$\Delta^2 f(a-4\omega)$		.
				$\Delta f(a-3\omega)$		$\Delta^3 f(a-4\omega)$	
x_4	x_5	$a-2\omega$	$f(a-2\omega)$		$\Delta^2 f(a-3\omega)$		$\Delta^4 f(a-4\omega)$
				$\Delta f(a-2\omega)$		$\Delta^3 f(a-3\omega)$	
x_2	x_3	$a-\omega$	$f(a-\omega)$		$\Delta^2 f(a-2\omega)$		$\Delta^4 f(a-3\omega)$
				$\Delta f(a-\omega)$		$\Delta^3 f(a-2\omega)$	
x_1	x_1	a	$f(a)$		$\Delta^2 f(a-\omega)$		$\Delta^4 f(a-2\omega)$
				$\Delta f(a)$		$\Delta^3 f(a-\omega)$	
x_3	x_2	$a+\omega$	$f(a+\omega)$		$\Delta^2 f(a)$		$\Delta^4 f(a-\omega)$
				$\Delta f(a+\omega)$		$\Delta^3 f(a)$	
x_5	x_4	$a+2\omega$	$f(a+2\omega)$		$\Delta^2 f(a+\omega)$		$\Delta^4 f(a)$
				$\Delta f(a+2\omega)$		$\Delta^3 f(a+\omega)$	
x_7	x_6	$a+3\omega$	$f(a+3\omega)$		$\Delta^2 f(a+2\omega)$		$\Delta^4 f(a+\omega)$
				.		.	
.	.	.			.		.

For the reordering of the arguments for the forward case, it is manifest from equation (40) that

$$\left.\begin{aligned}[x_1x_2] &= \frac{1}{\omega}\Delta f(a), \qquad [x_1x_2x_3] = \frac{\omega^{-2}}{2!}\Delta^2 f(a-\omega),\\ [x_1x_2x_3x_4] &= \frac{\omega^{-3}}{3!}\Delta^3 f(a-\omega), \qquad [x_1x_2x_3x_4x_5] = \frac{\omega^{-4}}{4!}\Delta^4 f(a-2\omega),\\ [x_1x_2x_3x_4x_5x_6] &= \frac{\omega^{-5}}{5!}\Delta^5 f(a-2\omega).\end{aligned}\right\} \tag{45}$$

More generally, by letting,

$$x_1 = a, \qquad x_{2s} = a + s\omega, \qquad \text{and} \qquad x_{2s+1} = a - s\omega. \tag{46}$$

we have

$$\left.\begin{aligned}[x_1x_2\ldots x_{2s+1}] &= \frac{\omega^{-2s}}{(2s)!}\Delta^{2s} f(a-s\omega),\\ [x_1x_2\ldots x_{2s+2}] &= \frac{\omega^{-2s-1}}{(2s+1)!}\Delta^{2s+1} f(a-s\omega).\end{aligned}\right\} \tag{47}$$

Inserting these expressions in formula (18), we obtain:

$$f(x) = f(a) + \omega^{-1}(x-a)\Delta f(a) + \frac{\omega^{-2}}{2!}(x-a)(x-a-\omega)\Delta^2 f(a-\omega)$$
$$+ \frac{\omega^{-3}}{3!}(x-a)(x-a-\omega)(x-a+\omega)\Delta^3 f(a-\omega)$$
$$+ \frac{\omega^{4}}{4!}(x-a)(x-a-\omega)(x-a+\omega)(x-a-2\omega)\Delta^4 f(a-2\omega)$$
$$+ \frac{\omega^{5}}{5!}(x-a)(x-a-\omega)(x-a+\omega)(x-a-2\omega)(x-a+2\omega)\Delta^5 f(a-2\omega) + \cdots. \tag{48}$$

It is convenient to replace

$$x \quad \text{by} \quad a + x\omega \tag{49}$$

in equation (48) when it takes the simpler form:

$$f(a+x\omega) = f(a) + x\Delta f(a) + \frac{1}{2!}x(x-1)\Delta^2 f(a-\omega)$$
$$+ \frac{1}{3!}(x+1)x(x-1)\Delta^3 f(a-\omega) + \frac{1}{4!}(x+1)x(x-1)(x-2)\Delta^4 f(a-2\omega)$$
$$+ \frac{1}{5!}(x+2)(x+1)x(x-1)(x-2)\Delta^5 f(a-2\omega) + \cdots. \tag{50}$$

This is the '*Newton–Gauss*' *forward formula.* To obtain the corresponding *backward formula* we have only to write $f(a - x\omega)$ in the form $f[a + x(-\omega)]$ and change the sign of ω in equation (50). This procedure corresponds to the reordering of the arguments according to the column B that is, by reversing the order and letting (cf. equation (46))

$$x_1 = a, \qquad x_{2s} = a - s\omega, \qquad \text{and} \qquad x_{2s+1} = a + s\omega. \tag{51}$$

In place of equation (50) we then have

$$f(a-x\omega) = f(a) - x\Delta f(a-\omega) + \frac{1}{2!}x(x-1)\Delta^2 f(a-\omega)$$
$$- \frac{1}{3!}(x+1)x(x-1)\Delta^3 f(a-2\omega) + \frac{1}{4!}(x+1)x(x-1)(x-2)\Delta^4 f(a-2\omega)$$
$$- \frac{1}{5!}(x+2)(x+1)x(x-1)(x-2)\Delta^5 f(a-3\omega) + \cdots. \tag{52}$$

Returning to formula (50),

$$f(a + x\omega) = f(a) + x\Delta f(a) + \tfrac{1}{2}x(x-1)\Delta^2 f(a-\omega) + \tfrac{1}{6}(x+1)x(x-1)\Delta^3 f(a-\omega) + \tfrac{1}{24}(x+1)x(x-1)(x-2)\Delta^4 f(a-2\omega) + \cdots, \tag{53}$$

we first rearrange the terms as follows:

$$\begin{aligned} f(a + x\omega) = f(a) &+ x[\Delta f(a) - \tfrac{1}{2}\Delta^2 f(a-\omega)] + \tfrac{1}{2}x^2\Delta^2 f(a-\omega) \\ &+ \frac{1}{3!}x(x^2-1^2)[\Delta^3 f(a-\omega) - \tfrac{1}{2}\Delta^4 f(a-2\omega)] \\ &+ \frac{1}{4!}x^2(x^2-1^2)\Delta^4 f(a-2\omega) + \cdots. \end{aligned} \tag{54}$$

Then we replace the differences of even order within the brackets by differences of odd order by using the identities,

$$\left.\begin{aligned} \Delta^2 f(a-\omega) &= \Delta f(a) - \Delta f(a-\omega), \\ \Delta^4 f(a-2\omega) &= \Delta^3 f(a-\omega) - \Delta^3 f(a-2\omega), \quad \text{etc.} \end{aligned}\right\} \tag{55}$$

We thus obtain the '*Newton–Stirling*' *formula*

$$\begin{aligned} f(a + x\omega) = f(a) &+ x\cdot\tfrac{1}{2}[\Delta f(a) + \Delta f(a-\omega)] + \frac{1}{2!}x^2\Delta^2 f(a-\omega) \\ &+ \frac{1}{3!}x(x^2-1^2)\cdot\tfrac{1}{2}[\Delta^3 f(a-\omega) + \Delta^3 f(a-2\omega)] + \frac{1}{4!}x^2(x^2-1^2)\Delta^4 f(a-2\omega) \\ &+ \frac{1}{5!}x(x^2-1^2)(x^2-2^2)\cdot\tfrac{1}{2}[\Delta^5 f(a-2\omega) + \Delta^5 f(a-3\omega)] \\ &+ \frac{1}{6!}x^2(x^2-1^2)(x^2-2^2)\Delta^6 f(a-3\omega) + \cdots. \end{aligned} \tag{56}$$

It will be observed that in this formula the mean differences, $\tfrac{1}{2}[\Delta f(a) + \Delta f(a-\omega)]$, $\tfrac{1}{2}[\Delta^3 f(a-\omega) + \Delta^3 f(a-2\omega)]$, etc., are completely symmetrical with respect to increasing and decreasing arguments.

An alternative transformation of the 'Newton–Gauss' interpolation formula (53) is to substitute for $\tfrac{1}{2}f(a)$, $\tfrac{1}{2}\Delta^2 f(a-\omega)$, $\tfrac{1}{2}\Delta^4 f(a-2\omega)$, etc., their values obtained from the identities,

$$\left.\begin{aligned} f(a) &= f(a+\omega) - \Delta f(a), \\ \Delta^2 f(a-\omega) &= \Delta^2 f(a) - \Delta^3 f(a-\omega), \\ \Delta^4 f(a-2\omega) &= \Delta^4 f(a-\omega) - \Delta^5 f(a-2\omega), \quad \text{etc.} \end{aligned}\right\} \tag{57}$$

Equation (53) then becomes

$$f(a + x\omega) = \tfrac{1}{2}[f(a) + f(a+\omega)] + (x - \tfrac{1}{2})\Delta f(a)$$
$$+ \frac{1}{2!}x(x-1)\cdot\tfrac{1}{2}[\Delta^2 f(a-\omega) + \Delta^2 f(a)] + \frac{1}{3!}x(x-1)(x-\tfrac{1}{2})\Delta^3 f(a-\omega)$$
$$+ \frac{1}{4!}(x+1)x(x-1)(x-2)\cdot\tfrac{1}{2}[\Delta^4 f(a-2\omega) + \Delta^4 f(a-\omega)] + \cdots. \quad (58)$$

This is the '*Newton–Bessel*' *formula*. It will be observed that it is symmetrical with respect to the argument $(a + \tfrac{1}{2}\omega)$, one of its principal advantages.

An alternative useful form of formula (58) is obtained by writing

$$x - \tfrac{1}{2} = y, \quad (59)$$

when it takes the form

$$f(a + \tfrac{1}{2}\omega + y\omega) = \tfrac{1}{2}[f(a) + f(a+\omega)] + y\Delta f(a)$$
$$+ \frac{1}{2!}(y^2 - \tfrac{1}{4})\cdot\tfrac{1}{2}[\Delta^2 f(a-\omega) + \Delta^2 f(a)] + \frac{1}{3!}y(y^2-\tfrac{1}{4})\Delta^3 f(a-\omega)$$
$$+ \frac{1}{4!}(y^2-\tfrac{1}{4})(y^2-\tfrac{9}{4})\cdot\tfrac{1}{2}[\Delta^4 f(a-2\omega) + \Delta^4 f(a-\omega)] + \cdots. \quad (60)$$

(v) *Propositions* 3–6

After Propositions 1 and 2 of his *Methodus differentialis* (of which we have given an account in §§(i) and (ii)), Newton proceeds to enunciate his interpolation formulae involving central differences with no more explanation than that they can be 'easily inferred'!

Proposition 3. Newton formulates the problem in the context of finding a polynomial curve which will pass through the extremities, $B_2, \ldots, B_8$, (in the adjoining diagram) of equidistant ordinates at unit distances apart. He distinguishes two cases: when the number of ordinates is odd and when the number of ordinates is even. We shall presently find that the two cases correspond exactly to using the 'Newton–Stirling' or the 'Newton–Bessel' interpolation formula. (It is a strange irony that in giving an account of Newton's published work of 1711, we have to hyphenate his name with Gauss, Stirling, and Bessel!)

Case 1. We first consider the case of an odd number of ordinates. In the adjoining difference table (taken from *Methodus differentialis*) distances are measured from the central ordinate $A_5(=k)$. The quantities labelled $l, m, \ldots,$ and r are:

$$\left.\begin{array}{ll} l = \tfrac{1}{2}(b_4 + b_5), & m = c_4, \\ n = \tfrac{1}{2}(d_3 + d_4), & o = e_3, \\ p = \tfrac{1}{2}(f_2 + f_3), & q = g_2, \\ r = \tfrac{1}{2}(h_1 + h_2), & i_1. \end{array}\right\} \quad (61)$$

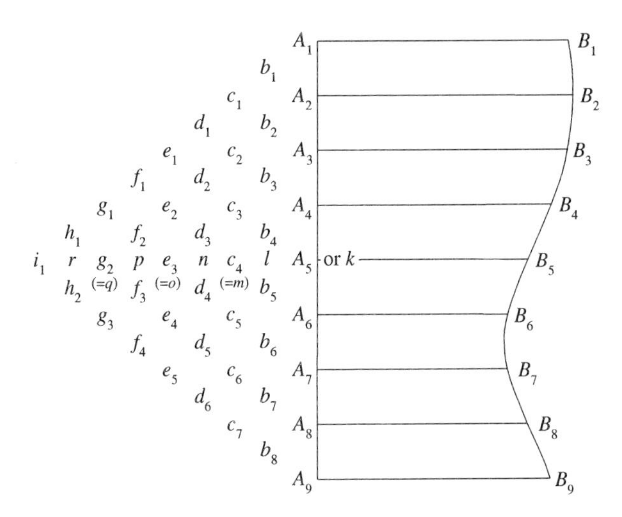

Newton determines the values of l, m, n, etc., by requiring that the ordinate at x be represented by the sextic polynomial,

$$A + Bx + Cx^2 + Dx^3 + Ex^4 + Fx^5 + Gx^6. \tag{62}$$

In following Newton's evaluation of these constants it is convenient to use the divided differences,

$$l' = l, \qquad m' = \tfrac{1}{2}m, \qquad n' = \frac{1}{3!}n, \qquad o' = \frac{1}{4!}o \qquad p' = \frac{1}{5!}p, \qquad \text{and} \qquad q' = \frac{1}{6!}q. \tag{63}$$

In the notation of the difference table on p. 492, following equation (44),

$$\begin{aligned} l' &= \tfrac{1}{2}[\Delta f(0) + \Delta f(-1)] = \tfrac{1}{2}\{f(+1) - f(0) + [f(0) - f(-1)]\} \\ &= \tfrac{1}{2}[f(+1) - f(-1)] = B + D + F, \end{aligned} \tag{64}$$

$$m' = \tfrac{1}{2}\Delta^2 f(-1) = \tfrac{1}{2}[f(+1) - 2f(0) + f(-1)] = C + E + G. \tag{65}$$

$$\begin{aligned} n' &= \tfrac{1}{12}[\Delta^3 f(-1) + \Delta^3 f(-2)] \\ &= \tfrac{1}{12}[C_0^3 f(+2) - C_1^3 f(+1) + C_2^3 f(0) - C_3^3 f(-1) \\ &\quad + C_0^3 f(+1) - C_1^3 f(0) + C_2^3 f(-1) - C_3^3 f(-2)] \\ &= \tfrac{1}{12}\{f(+2) - f(-2) - 2[f(+1) - f(-1)]\} = D + 5F, \end{aligned} \tag{66}$$

$$\begin{aligned} o' &= \tfrac{1}{24}\Delta^4 f(-2) = \tfrac{1}{24}[C_0^4 f(+2) - C_1^4 f(+1) + C_2^4 f(0) - C_3^4 f(-1) + C_4^4 f(-2)] \\ &= \tfrac{1}{24}\{f(+2) + f(-2) - 4[f(+1) + f(-1)] + 6f(0)\} \\ &= \tfrac{1}{12}[4C + 16E + 64G - 4(C + E + G)] = E + 5G \end{aligned} \tag{67}$$

$$p' = \tfrac{1}{240}[\Delta^5 f(-2) + \Delta^5 f(-3)] = F; \qquad q' = \tfrac{1}{720}\Delta^6 f(-3) = G, \tag{68}$$

where, in evaluating the various differences, we have made use of equation (37). The equations determining the coefficients A, B, C, etc., are therefore,

$$\left.\begin{aligned} A &= k, \\ B + D + F &= l' = l, \\ C + E + G &= m' = 2m, \\ D + 5F &= n' = 6n, \\ E + 5G &= o' = 24o, \\ F &= p' = 120p, \\ G &= q' = 720q, \end{aligned}\right\} \tag{69}$$

and the solutions of these equations are:

$$\left.\begin{array}{llll} A = k, & B = l' - n' + 4p', & C = m' - o' + 4q', & \\ D = n' - 5p', & E = o' - 5q' & F = p', \quad \text{and} & G = q'. \end{array}\right\} \tag{70}$$

The required expansion (62), in terms of the differences, is therefore

$$\begin{aligned} k &+ (l' - n' + 4p')x + (m' - o' + 4q')x^2 + (n' - 5p')x^3 + (o' - 5q')x^4 + p'x^5 + q'x^6 \\ &= k + l'x + m'x^2 + n'x(x^2 - 1) + o'x^2(x^2 - 1) + p'x(x^2 - 1)(x^2 - 4) \\ &\quad + q'x^2(x^2 - 1)(x^2 - 4). \end{aligned} \tag{71}$$

Reverting to the ordinary differences l, m, etc., we have Newton's expression (as he writes):

$$k + xl + \tfrac{1}{2}x^2 m + \tfrac{1}{6}(x^3 - x)n + \tfrac{1}{24}(x^4 - x^2)o + \tfrac{1}{120}(x^5 - 5x^3 + 4x)p + \cdots. \tag{72}$$

By comparison with equation (56), equation (72) is, in fact, the 'Newton–Stirling' formula!

Case 2. When the number of ordinates is even the origin of the abscissa is shifted to the centre of A_4 and A_5 as indicated in the adjoining scheme. This corresponds to the transformation (59). Now letting

$$\left.\begin{array}{llll} k = \tfrac{1}{2}[f(0) + f(1)], & l = b_4, & m = \tfrac{1}{2}(c_3 + c_4), & n = d_3, \\ & o = \tfrac{1}{2}(e_2 + e_3), \quad \text{and} & f_2 = p, & \end{array}\right\} \tag{73}$$

Newton finds:

$$k + xl + \frac{1}{2!\cdot 4}(4x^2 - 1)m + \frac{1}{3!\cdot 4}x(4x^2 - 1)n + \frac{1}{4!\cdot 16}(4x^2 - 1)(4x^2 - 9)o + \cdots. \tag{74}$$

By comparison with equation (60), it is clear that in this instance he has established the 'Newton–Bessel' formula!

Propositions 4–6. In Proposition 4, the formulae of Proposition 3 are generalized for the case of unequal distances between the ordinates by a procedure strictly along the lines of Propositions 1 and 2, and he obtains central-difference formulae applicable to the two cases when the number of ordinates is odd and even.

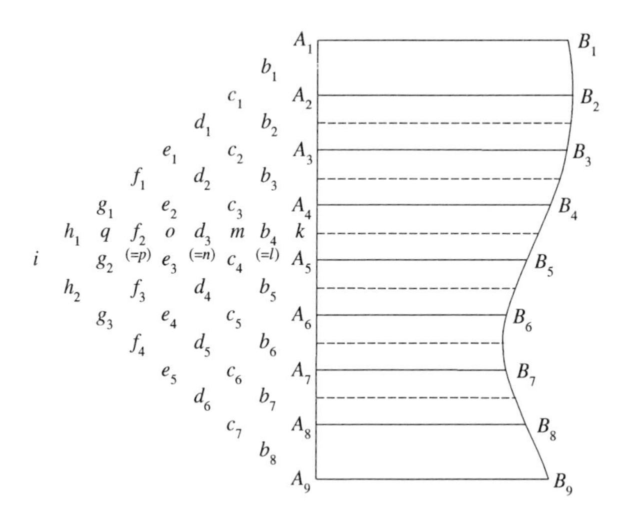

Proposition 5 merely points out that the methods described can be used to determine the values of the interpolated curves at intermediate points, and in Proposition 6 it is pointed out that the approximate areas of any curve can be found by a simple integration of the interpolated polynomial.

This completes our partial account of Newton's *Methodus differentialis.*

(*b*) *Lemma V*

After the long preamble on Newton's *Methodus differentialis*, we return to Lemma V of Book III of the *Principia*, his first published statement on the calculus of finite differences. The Lemma, as we shall find out, is no more than an algorithm for computing.

Lemma V

> *To find a curved line of the parabolic kind which shall pass through any given number of points.*

The problem is to find a polynomial curve which will pass through the extremities, A, B, C, D, E, and F of the ordinates, AH, BI, CK, DL, ME, and NF, depicted in the adjoining diagram; *and* to find the value of the ordinate RS at an intermediate point, R on the curve.

Newton's illustrative diagram accompanying this lemma has been slightly altered to bring it in conformity with that on p. 496 illustrating Proposition 3 of his *Methodus differentialis.* As in that diagram the differences are set out in triangular form.

Newton considers two cases: when the ordinates are equidistant and when they are not.

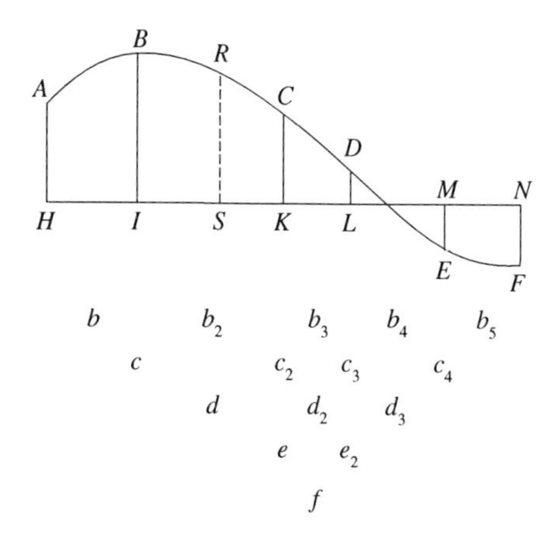

Case 1. When the ordinates are equidistant, Newton's definitions are:

$$AH - BI = b, \quad BI - CK = b_2, \quad CK - DL = b_3, \quad DL - EM = b_4, \quad EM - FN = b_5, \tag{75}$$

$$\left.\begin{aligned} b - b_2 = c, \quad b_2 - b_3 = c_2, \quad b_3 - b_4 = c_3, \quad b_4 - b_5 = c_4;\\ c - c_2 = d, \quad c_2 - c_3 = d_2, \quad c_3 - c_4 = d_3;\\ d - d_2 = e, \quad d_2 - d_3 = e_2;\\ e - e_2 = f;\end{aligned}\right\} \tag{76}$$

and

$$AH = a, \qquad -HS = p, \qquad -\tfrac{1}{2}pIS = q, \qquad \tfrac{1}{3}qSK = r, \qquad \tfrac{1}{4}rSL = s, \qquad \tfrac{1}{5}sSM = t. \tag{77}$$

Defining the further quantities q, r, s, and t by the equations,

$$\left.\begin{aligned} p &= -HS,\\ \tfrac{1}{2}p(-IS) &= \tfrac{1}{2}HS.IS = q,\\ \tfrac{1}{3}qSK &= \tfrac{1}{6}HS.IS.SK = r,\\ \tfrac{1}{4}rSL &= \tfrac{1}{24}HS.IS.SK.SL = s,\\ \tfrac{1}{5}sSM &= \tfrac{1}{120}HS.IS.SK.SL.SM = t,\end{aligned}\right\} \tag{78}$$

Newton simply states that

$$RS = a + bp + cq + dr + es + ft. \tag{79}$$

It is manifest that equation (79) is no more than an expression of the *forward interpolation* formula (cf. equation (13)):

$$\begin{aligned} f(x) = f(a) &+ (x - x_1)\Delta f(a) + \tfrac{1}{2}(x - x_1)(x - x_2)\Delta^2 f(a)\\ &+ \frac{1}{3!}(x - x_1)(x - x_2)(x - x_3)\Delta^3 f(a) + \frac{1}{4!}(x - x_1)(x - x_2)(x - x_3)(x - x_4)\Delta^4 f(a)\\ &+ \frac{1}{5!}(x - x_1)(x - x_2)(x - x_3)(x - x_4)(x - x_5)\Delta^5 f(a). \end{aligned} \tag{80}$$

Case 2. When the ordinates are at unequal distances, Newton defines *divided differences.* His definitions are:

$$\left.\begin{array}{llll} b = \dfrac{AH - BI}{HI}, & b_2 = \dfrac{BI - CK}{IK}, & b_3 = \dfrac{CK - DL}{KL}, & \text{etc.,} \\ c = \dfrac{b - b_2}{HK}, & c_2 = \dfrac{b_2 - b_3}{IL}, & c_3 = \dfrac{b_3 - b_4}{KM}, & \text{etc.,} \\ d = \dfrac{c - c_2}{HL}, & d_2 = \dfrac{c_2 - c_3}{IM}, & \text{etc.} & \end{array}\right\} \tag{81}$$

Defining the further quantities (cf. equations (78)),

$$AH = a; \quad -HS = p; \quad -pIS = q; \quad qSK = r; \quad rSL = s; \quad sSM = t \tag{82}$$

he now states,

$$RS = a + bp + cq + dr + es + ft + \cdots. \tag{83}$$

Again, it is manifest that equation (83) is the expression of the forward interpolation formula for divided differences as given by equation (18):

$$\begin{aligned} f(x) = f(x_1) + (x - x_1)[x_1x_2] + (x - x_1)(x - x_2)[x_1x_2x_3] \\ + (x - x_1)(x - x_2)(x - x_3)[x_1x_2x_3x_4] + \cdots. \end{aligned} \tag{84}$$

The corollary which follows is a statement of Proposition 6 of *Methodus differentialis.* We shall quote it in full.

> COR. Hence the areas of all curves may be nearly found; for if some number of points of the curve to be squared are found, and a parabola be supposed to be drawn through those points, the area of this parabola will be nearly the same with the area of the curvilinear figure proposed to be squared: but the parabola can be always squared geometrically by methods generally known.

135. Lemmas VI–XI

These are the lemmas on which Newton founds his solution of 'the most difficult (*longe difficillium*) problem' of determining 'from three given observations, the orbit of a comet moving in a given parabola' in Proposition XLI (see §136 below).

(*a*) *Lemma VI*

In Lemma VI, Newton explains how the algorithm of Lemma V can be applied in the context of the *longe difficillium* problem.

Lemma VI

Certain observed places of a comet being given, to find the place of the same at any intermediate given time.

Let *HI*, *IK*, *KL*, *LM* (in the preceding fig.) [page 499] represent the times between the observations; *HA*, *IB*, *KC*, *LD*, *ME*, five observed longitudes of the comet; and *HS* the given time between the first observation and the longitude required. Then if a regular curve *ABCDE* is supposed to be drawn through the points *A*, *B*, *C*, *D*, *E*, and the ordinate *RS* is found out by the preceding lemma, *RS* will be the longitude required.

By the same method, from five observed latitudes, we may find the latitude at a given time.

If the differences of the observed longitudes are small, let us say 4 or 5 degrees, then three or four observations will be sufficient to find a new longitude and latitude; but if the differences are greater, as of 10 or 20 degrees, five observations ought to be used.

(*b*)

Lemma VII

Through a given point P to draw a right line BC, whose parts PB, PC, cut off by two right lines AB, AC, given in position, may be one to the other in a given ratio.

From the given point *P* suppose any right line *PD* to be drawn to either of the right lines given, as *AB*; and produce the same towards *AC*, the other given right line, as far as *E*, so as *PE* may be to *PD* in the given ratio. Let *EC* be parallel to *AD*. Draw *CPB*, and *PC* will be to *PB* as *PE* to *PD*. Q.E.F.

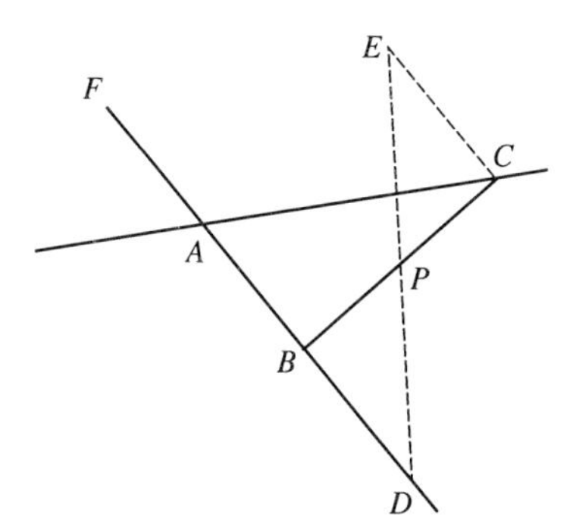

On first reading, one is astonished that Newton should have wanted to include this very elementary lemma among so many others that are barely explained; but the construction is central to Lemma VIII and his resolution of the *longe difficillimum* problem.

(c)

Lemma VIII

Let ABC be a parabola, having its focus in S. By the chord AC bisected in I cut off the segment ABCI, whose diameter is $I\mu$ and vertex μ. In $I\mu$ produced take μO equal to one-half of $I\mu$. Join OS, and produce it to ξ, so that $S\xi$ may be equal to 2SO. Now, supposing a comet to revolve in the arc CBA, draw ξB, cutting AC in E: I say, the point E will cut off from the chord AC the segment AE, nearly proportional to the time.

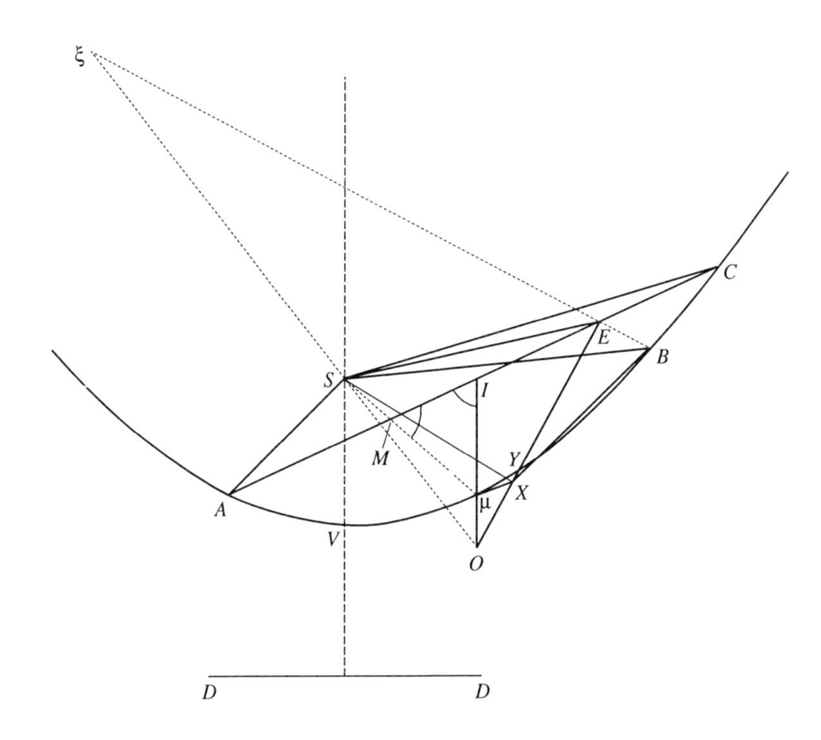

Description of the figure

$ACB\mu VA$ is a parabola with focus at S and vertex at V;
DD is the directrix;
AC is a chord;
A, B, and C are points on the parabola, later to be identified as the three given positions of a comet;
I is the mid-point of AC;
$I\mu$ is a diameter of the parabola passing through I and parallel to SV;
μX is the tangent at μ;

$I\mu$ is extended to the point O so that

$$\mu O = \tfrac{1}{2}I\mu = \tfrac{1}{3}IO; \tag{1}$$

SO is extended to ξ so that

$$SO = \tfrac{1}{3}O\xi; \tag{2}$$

ξB intersects AC at E;
EO intersects the tangent at μ at X;
μX being parallel to AC, by (1),

$$OX = \tfrac{1}{3}OE. \tag{3}$$

By (2) and (3) (and Lemma VII!)

$$SO:O\xi = XO:OE; \quad \text{and} \quad \therefore \quad SX \text{ is } \parallel \text{ to } \xi B; \tag{4}$$

and by (1)

$$\mu X = \tfrac{1}{3}IE. \tag{5}$$

The first part of the lemma states:

$$\text{area of } AEX\mu A\text{: area of } AECBY\mu A = AE:AC. \tag{6}$$

The proof of equation (6) depends on a theorem in conics*, that the area of a half-segment such as $AI\mu A$ is given by

$$\text{area of } AI\mu A = \tfrac{2}{3}(AI\,.\,I\mu)\sin \angle AI\mu. \tag{7}$$

Proof of equation (6):

* I have not seen this theorem (which Newton could casually pass over as *patent ex conics*!) stated in any of the current textbooks on conic sections that I have consulted. But I did find it in George Salmon's *A treatise on conic sections* (Longmans, Green, and Co., 1879, sixth edition), Chapter XIX, p. 372. For its historical interest, I reproduce it below.

To find the area of the segment of a parabola cut off by any right line.
Draw the diameter bisecting it, then the parallelogram PR' is equal to PM', since they are the complements of parallelograms about the diagonal; but since TM is bisected at V', the parallelogram PN' is half PR'; if, therefore, we take a number of points P, P', P'', & c., it follows that the sum of all the parallelograms PM' is double the sum of all the parallelograms PN', and therefore ultimately that the space $V'PM$ is double $V'PN$; hence the area of the parabolic segment $V'PM$ is to that of the parallelogram $V'NPM$ in the ratio 2:3.

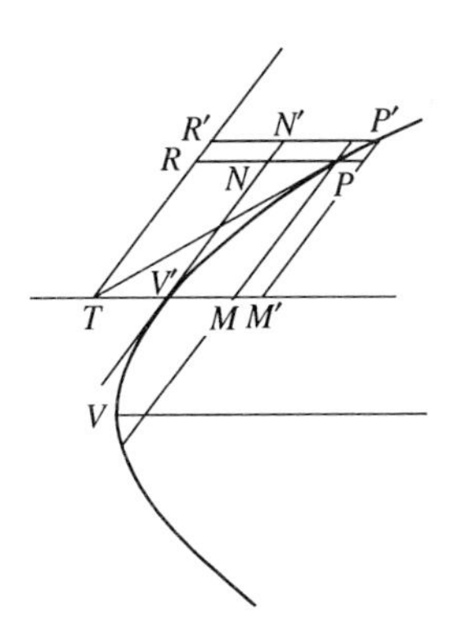

First we observe that

$$\text{area of } AEX\mu A = \text{area of } AI\mu A + \text{area of } I\mu XEI. \tag{8}$$

By equation (7)

$$\text{area of } AI\mu A = \tfrac{2}{3}(I\mu . AI)\sin \angle AI\mu; \tag{9}$$

and

$$\text{area of } I\mu XEI = \tfrac{1}{2}I\mu(IE + \mu X)\sin \angle AI\mu. \tag{10}$$

But by equation (5),

$$\mu X = \tfrac{1}{3}IE; \qquad \therefore \quad IE + \mu X = \tfrac{4}{3}IE. \tag{11}$$

Therefore,

$$\text{area of } I\mu XEI = \tfrac{2}{3}(I\mu . IE)\sin \angle AI\mu. \tag{12}$$

Now combining equations (9) and (12), we obtain,

$$\text{area of } AEX\mu A = \tfrac{2}{3}(I\mu . AE)\sin \angle AI\mu. \tag{13}$$

Again, by equation (7)

$$\text{area of } ACBY\mu A = \tfrac{2}{3}(I\mu . AC)\sin \angle AI\mu. \tag{14}$$

From a comparison of equations (13) and (14) we conclude:

$$\text{area of } AEX\mu A\text{: area } ACBY\mu A = AE : AC. \tag{15}$$

Manifestly,

$$\text{area of } \triangle ASE\text{: area of } \triangle ASC = AE : AC. \tag{16}$$

By the addition of equations (15) and (16), we obtain:

$$\text{area of } ASEX\mu A\text{: area of } ASCBY\mu A = AE : AC. \tag{17}$$

Turning to the proof of the lemma, we first observe that since SX and ξB are parallel (by equation (4))

$$\text{area of } \triangle SEB = \text{area of } \triangle EXB. \tag{18}$$

Therefore, successively,

$$\begin{aligned} \text{area of } ASEX\mu A &= \text{area of } (\, ASEX\mu A + \triangle EXB - \triangle SEB) \\ &= \text{area of } (ASEBX\mu A - \triangle SEB) \\ &= \text{area of } ASBX\mu A; \end{aligned} \tag{19}$$

and finally by equation (17)

$$\text{area of } ASBX\mu A\text{: area of } ASCBY\mu A = AE : AC. \tag{20}$$

But

$$\text{area of } ASBX\mu A \text{ is very nearly} = \text{area of } ASBY\mu A. \tag{21}$$

Therefore, as Newton concludes

But the area $ASBY\mu A$ is nearly equal to the area $ASBX\mu A$; and this area $ASBY\mu A$ is to the area $ASCY\mu A$ as the time of description of the arc AB is to the time of description of the whole arc AC; and, therefore, AE is to AC nearly in the proportion of the times. Q.E.D.

COR. When the point B falls upon the vertex μ of the parabola, AE is to AC accurately in the proportion of the times.

The brevity of Newton's demonstration of this lemma in less than 15 lines of prose is to be contrasted with three pages of ours.

Scholium

If we join $\mu\xi$ cutting AC in δ, and in it take ξn in proportion to [μB as $27MI$ to $16\,M\mu$]*, and draw Bn, this Bn will cut the chord AC, in the ratio of the times, more accurately than before; but the point n is to be taken beyond or on this side the point ξ, according as the point B is more or less distant from the principal vertex of the parabola than the point μ.

The Scholium is concerned with improving the error in the approximation of Lemma VIII, that the intersection of $B\xi$ (in the diagram on p. 502) with chord AC at E divides it in the ratio of the times of the description of the sectors $ASB\mu A$ and $SCBS$ by the comet. The 'extreme' case to consider is clearly the one in which $B\xi$ intersects AC at its mid-point I and, when on Lemma VIII, the sectors $ASB\mu A$ and $SCBS$ will be described in equal times. The purpose of the Scholium is to provide a division of the chord (at K say) which will correspond to equal times.

The adjoining illustration specializes the figure of Lemma VIII (p. 502) for the case when ξI extends to meet the arc AC at B, the place of the second observation. The line BX is drawn parallel to the axis of the parabola VS and meeting the tangent at μ at X. The axis VS is prolonged to intersect $\mu\xi$ and $B\xi$ at q and p, respectively. (We shall not be concerned, at present, with the rest of the lettering in the diagram.)

Since $O\mu I$ and $VSqp$ are parallel, it follows that

$$\left.\begin{aligned} &Sq : O\mu = \xi S : \xi O = \tfrac{2}{3} \quad \text{(from equation (2))} \\ \text{and} \quad &Sq = \tfrac{2}{3}O\mu = \tfrac{1}{3}I\mu \qquad \text{and} \qquad O\mu = \tfrac{1}{2}I\mu \quad \text{(by equation (1)).} \end{aligned}\right\} \tag{22}$$

* Corrected to '$I\delta$ as $3MI$ to $4M\mu$', to allow for a minor oversight, by E. T. Whiteside in *Mathematical papers of Sir Isaac Newton*, (Cambridge University Press, 1974) pp. 506–507.

The solution to the problem posed in the Scholium was found independently. Perhaps, I may narrate the sequence of events. Though no clear formulation of the problem is enunciated in the text, the intended meaning was clear. And it was not difficult to find a solution along the lines that would have been natural for Newton and in conformity with his style while writing the first edition of *Principia*. I attribute the circumlocution and the interspersed oversights in the unpublished manuscript to the state of harrassment that he must have been in, while writing the Scholium, under the pressure of revising for the second edition of the *Principia*, in 1713, some 25 years later.

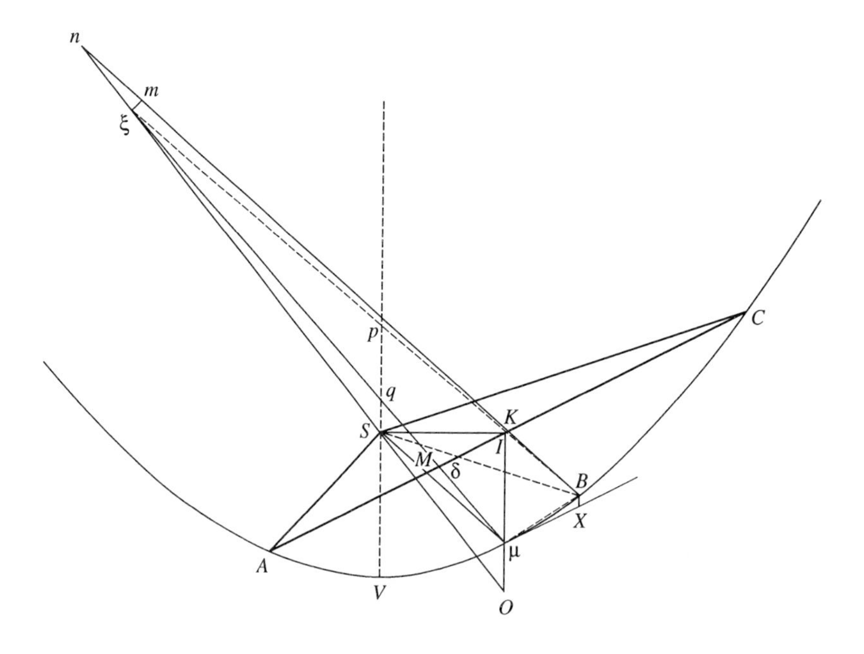

And since

$$\xi S : SO = 2 \qquad \text{and} \qquad I\mu : \mu O = 2, \tag{23}$$

we infer that

$$S\mu \text{ is parallel to } BI\xi. \tag{24}$$

We can now disregard the construction that led to the inference (24) and consider only the final disposition of the lines relevant to our purposes.

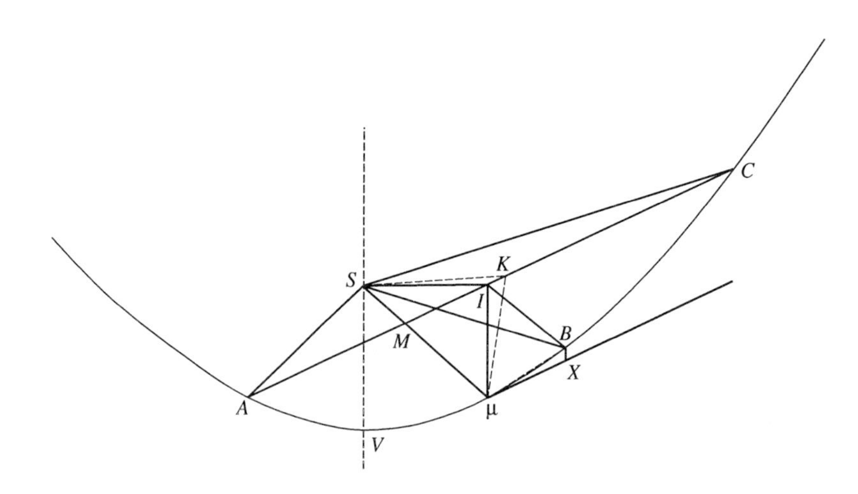

To recall:

I is the mid-point of AC;
$I\mu$ is parallel to SV and μX is the tangent at μ;
IB is drawn parallel to $S\mu$;
B is the second place of observation;
$I\mu M$ is an isosceles triangle; and
$SK\mu$ a triangle adjoining to $SI\mu$ (to be specified presently).

First, we shall exhibit the origin of the approximation in Lemma VIII, in this case.

$$\begin{aligned}\text{The area of } ASB\underset{\leftrightarrow}{\mu}A &= \text{sector } ASM\mu A + \triangle S\underset{\leftrightarrow}{B\mu}\\ &= \text{sector } ASM\mu A + \triangle S\mu I(\because S\mu \| IB)\\ &= \text{sector } AMI\mu A + (\triangle ASM + \triangle SMI)\\ &= \text{segment } AMI\mu + \triangle ASI\\ &= \text{segment } I\underset{\sim}{\mu B}CI + \triangle CSI\\ &= \text{Area } I\underset{\sim}{\mu B}CSI. \qquad (25)\end{aligned}$$

[The arrow below $B\mu$ signifies that $B\mu$ is a *right line*.]

[The wavy line below μB signifies that μB is an *arc* of the parabola]

The occurrence of $\underset{\leftrightarrow}{B\mu}$ on the left-hand side of the equation instead of $\underset{\sim}{\mu B}$ as on the right-hand side is the extent to which the required equality of the sectors $ASB\mu A$ and $SBCS$ fails to be satisfied. Newton achieves the required equality by displacing the point I to K such that

$$\text{segment } \mu B\mu = \triangle SK\mu - \triangle SI\mu. \qquad (26)$$

It is precisely, when I noticed later, this very same requirement in Newton's manuscript ('*Pone sectorem SBμ aequlem triangulo SKμ verissimé*') that I came to be convinced that, left to himself, Newton would most certainly have determined the position of K directly as below.

Since BX and MI, to second order in the small angle $B\mu X$ (or, as Newton expressed '*etiem chorda Bμ quamproximé parallela*' CA, $B\mu = MI$) (by equation (33) below)

$$4S\mu . BX = MI^2, \qquad (27)$$

or

$$BX = \frac{MI^2}{4S\mu}. \qquad (28)$$

The area of the segment $\mu B\mu$ (being the complement of the segment $B\mu XB$) is

$$\text{segment } \mu B\mu = \tfrac{1}{3}BX . MI \sin\beta = \frac{MI^3}{12S\mu}\sin\beta, \qquad (29)$$

where $\beta = \angle IM\mu = \angle \mu IM$. On the other hand,

$$\triangle SK\mu - \triangle SI\mu = \tfrac{1}{2}(IK . S\mu)\sin\beta. \qquad (30)$$

Hence the required placement of K, to compensate for the correction of Lemma VIII in this case, is given by

$$\frac{IK}{MI} = \frac{1}{6}\left(\frac{MI}{S\mu}\right)^2, \tag{31}$$

a ratio, which, under normally occurring conditions, is likely to be negligible. Q.E.D.(!)

(*d*)

Lemma IX

The right lines $I\mu$ and μM, and the length $AI^2/4S\mu$, are equal among themselves.

In the figure on p. 502 for illustrating Lemma VIII, $S\mu$ intersects AC at M. Since, by a well-known (!) theorem on conics, the tangent μX at μ bisects the $\angle S\mu O$, the triangle $I\mu M$ is isosceles and therefore,

$$I\mu = \mu M. \tag{32}$$

And by the equation of the parabola referred to the diameter $I\mu$ and the tangent at μ,

$$AI^2 = 4(a \operatorname{cosec}^2 \omega)I\mu = 4S\mu . I\mu, \qquad (33)^*$$

where, as Newton states, '$4S\mu$ is the latus rectum of the parabola referred to the vertex μ'.

(*e*) *Lemma X and 'Lambert's theorem'*

Lemma X is, for celestial mechanics, the most significant among the present group of lemmas. In it is derived one of the basic theorems of the subject, however, misnamed 'Lambert's' theorem (e.g., E. T. Whittaker, *A treatise on analytical dynamics of particles and rigid bodies*, Cambridge University Press, 1937, 4th edition, p. 31; and, J. M. A. Danby *Fundamentals of celestial mechanics*, Willmann-Bell, Inc., Richmond, VA., 2nd edition, p. 217). But Lagrange knew, and more recently (70 years ago!) A. N. Kriloff provided an elegant demonstration of the equivalence (Monthly Notices of the Royal Astronomical Society, Vol. 84, pp. 392–395 (1926)). Kriloff's demonstration will be included in the account which follows.

* The proof of this equation follows from the brief description on p. 99 of the geometry of a parabola referred to a diameter and the tangent at any point. Thus, representing a point $P = (x, y)$ on the parabola $y^2 = 4ax$ parametrically by $(a\mu^2, 2a\mu)$ then

$$SP^2 = (x_P - x_S)^2 + (y_P - y_S)^2 = (a\mu^2 - a)^2 + 4a^2\mu^2 = a^2(1 + \mu^2)^2.$$

Therefore,

$$SP = a(1 + \mu^2) = a(1 + \cot^2 \omega) = a \operatorname{cosec}^2 \omega = b.$$

Lemma X

Produce $S\mu$ to N and P, so that μN may be one-third of μI, and SP may be to SN as SN to $S\mu$; and in the time that a comet would describe the arc $A\mu C$, if it was supposed to move always forwards with the velocity which it has in a height equal to SP, it would describe a length equal to the chord AC.

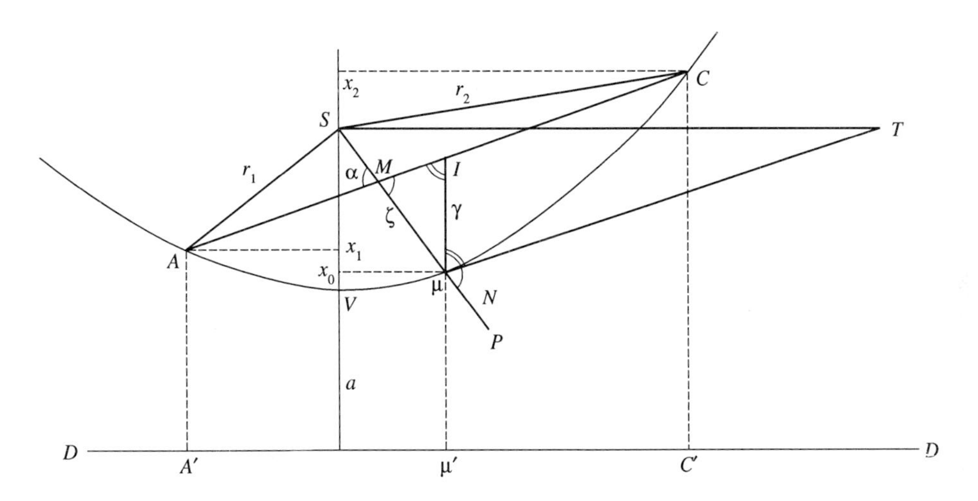

The figure coincides, in part, with that illustrating Lemma X (or, as Newton would say 'the same things supposed'). The extra additions are:

(1) A', μ', and C' are the projections of A, B, and C on the directrix, DD;
(2) x_0, x_1, and x_2 are the projections of μ, A, and C on the axis SV of the parabola;

$$S\mu = \zeta, \qquad SA = r_1, \qquad SC = r_2, \qquad I\mu = \gamma, \qquad AC = 2AI = c,$$

and (34)

$$SV = a = \tfrac{1}{4}\text{ latus rectum};$$

(3) $S\mu$ is extended to N and P such that

$$\mu N = \tfrac{1}{3}I\mu = \tfrac{1}{3}M\mu \quad \text{(by Lemma IX)}, \tag{35}$$

and

$$SP = SN^2/S\mu; \tag{36}$$

$$\alpha = \angle MI\mu = \angle IM\mu \quad (\because \triangle \mu IM \text{ is isosceles}); \tag{37}$$

(4) μT is tangent to the parabola at μ.

If the comet continues to move along the tangent μT at μ, with its instantaneous velocity at μ, and arrives at T at the same time the comet, describing its orbit along the parabola, arrives at C, then by the area theorem,

$$\text{area of sector } SC\mu AS = \text{area of } \triangle S\mu T. \tag{38}$$

Making use of equation (7), we have

$$\begin{aligned}\text{area of sector } SC\mu AS &= \text{area of } \triangle SCA + \text{area of segment } AC\mu A\\ &= \tfrac{1}{2}(AC.SM)\sin\alpha + \tfrac{2}{3}(AC.I\mu)\sin\alpha\\ &= \tfrac{1}{2}AC(SM + \tfrac{4}{3}I\mu)\sin\alpha. \end{aligned} \tag{39}$$

But by construction (cf. equation (35))

$$\tfrac{4}{3}I\mu = \tfrac{4}{3}M\mu = M\mu + \tfrac{1}{3}M\mu = M\mu + \mu N = MN. \tag{40}$$

Hence,

$$\text{area of sector } SC\mu AS = \tfrac{1}{2}(AC.SN)\sin\alpha. \tag{41}$$

On the other hand,

$$\text{area of } \triangle ST\mu = \tfrac{1}{2}(S\mu.\mu T)\sin\alpha. \tag{42}$$

Therefore, by the assumed equality of the two areas, $SC\mu AS$ and $\triangle S\mu T$,

$$AC.SN = S\mu.\mu T \qquad \text{or} \qquad \frac{\mu T}{AC} = \frac{SN}{S\mu}. \tag{43}$$

Now by Corollary VI of Proposition XVI of Book I, the velocities, V_μ and V_P at μ and P of particles describing parabolic orbits are 'inversely as the square root of the ratio of the distances of the body from the focus'; therefore

$$\frac{V_\mu}{V_P} = \left(\frac{SP}{S\mu}\right)^{1/2}. \tag{44}$$

But by construction (equation (36))

$$SP = \frac{SN^2}{S\mu}. \tag{45}$$

Hence,

$$\frac{V_\mu}{V_P} = \frac{SN}{S\mu} = \frac{\mu T}{AC} \quad \text{(by equation (43)).} \tag{46}$$

Therefore if τ is the time for describing the arc $A\mu C$,

$$\frac{\mu T}{V_\mu} = \frac{AC}{V_P} = \tau. \tag{47}$$

Q.E.D.

Newton adds the corollary,

> COR. Therefore a comet, with that velocity which it hath in the height $S\mu + \frac{2}{3}I\mu$, would in the same time nearly describe the chord AC.

'Lambert's theorem'

We shall now show, following Kriloff, the equivalence of Newton's relation (47) to the misnamed Lambert's theorem (equation (66) below).

Lemma IX in our present notation (equations (34) and (35)) states:

$$\gamma = I\mu = M\mu = \frac{AI^2}{4S\mu} = \frac{c^2}{16\zeta}. \tag{48}$$

Also,

$$SN = S\mu + \mu N = \zeta + \tfrac{1}{3}I\mu = \zeta + \frac{c^2}{48\zeta}. \tag{49}$$

From the definitions (34) and (36) it now follows:

$$SP = \frac{SN^2}{S\mu} = \frac{1}{\zeta}\left(\zeta + \frac{c^2}{48\zeta}\right)^2. \tag{50}$$

Now, by equation (47) defining τ,

$$\tau = \frac{AC}{V_P} = \frac{c}{V_P}, \tag{51}$$

and by Corollary VI of Proposition XVI, Book I, expressed in standard astronomical units,

$$V_P = \frac{\kappa\sqrt{2}}{\sqrt{SP}}, \tag{52}$$

where, by the relations given in the footnote on p. 480

$$\kappa = \frac{2\pi}{\text{sidereal year }[1 + (m_\oplus + m_☾)/M_\odot]^{1/2}} = 0.017202098$$

$$\text{or } 0.01720212 \text{ (as Newton gives).} \tag{53}$$

By combining equations (50), (51), and (52), we obtain

$$\kappa\tau\sqrt{2} = c\sqrt{SP} = \frac{c}{\sqrt{\zeta}}\left(\zeta + \frac{c^2}{48\zeta}\right). \tag{54}$$

Now the distances $S\mu$ $(=\zeta)$, SA $(=r_1)$, and $SC = (r_2)$ of the points μ, A, and C from the focus S, are, by the definition of a parabola, related to their respective abscissae, x_0, x_1, and x_2, by

$$x_0 = \zeta + 2a, \qquad x_1 = r_1 + 2a, \qquad \text{and} \qquad x_2 = r_2 + 2a, \tag{55}$$

where $2a$ denotes the semi latus rectum. Eliminating a from the relations (55), we obtain

$$\tfrac{1}{2}(r_1 + r_2) - \zeta = \tfrac{1}{2}(x_1 + x_2) - x_0. \tag{56}$$

Since I is the mid-point of AC,

$$\tfrac{1}{2}(x_1 + x_2) - x_0 = I\mu = \gamma = \frac{c^2}{16\zeta} \quad \text{(by equation (48)).} \tag{57}$$

Equation (56) thus gives,

$$\tfrac{1}{2}(r_1 + r_2) - \zeta = \frac{c^2}{16\zeta}: \tag{58}$$

in other words, we have the following quadratic equations for ζ:

$$\zeta^2 - \tfrac{1}{2}(r_1 + r_2)\zeta + \frac{c^2}{16} = 0. \tag{59}$$

For solving equation (59), it is convenient to set

$$r_1 + r_2 = \tfrac{1}{2}(g + f) \qquad \text{and} \qquad c = \tfrac{1}{2}(g - f). \tag{60}$$

With these substitutions, equation (59) becomes,

$$\zeta^2 - \tfrac{1}{4}(g + f)\zeta + \tfrac{1}{64}(g - f)^2 = 0; \tag{61}$$

on simplification, equation (61) reduces to,

$$[\zeta - \tfrac{1}{8}(g + f)]^2 = \tfrac{1}{16}gf; \tag{62}$$

and yields,

$$\zeta = \tfrac{1}{8}(g + f) + \tfrac{1}{4}\sqrt{(gf)} = \tfrac{1}{8}(\sqrt{g} + \sqrt{f})^2. \tag{63}$$

Returning to equation (54), we first find

$$\begin{aligned} \zeta + \frac{c^2}{48\zeta} &= \tfrac{1}{8}(\sqrt{g} + \sqrt{f})^2 + \frac{1}{24}\frac{(g - f)^2}{(\sqrt{g} + \sqrt{f})^2} \\ &= \tfrac{1}{24}[3(\sqrt{g} + \sqrt{f})^2 + (\sqrt{g} - \sqrt{f})^2] \\ &= \tfrac{1}{6}[g + f + \sqrt{(gf)}]. \end{aligned} \tag{64}$$

Now, equation (54) together with equation (59) gives

$$\begin{aligned} \kappa\tau\sqrt{2} &= \tfrac{1}{2}(g - f)\frac{2\sqrt{2}}{\sqrt{g} + \sqrt{f}}\cdot\frac{1}{6}[g + f + \sqrt{(gf)}] \\ &= \frac{\sqrt{2}}{6}(\sqrt{g} - \sqrt{f})[g + f + \sqrt{(gf)}] = \frac{\sqrt{2}}{6}(g^{3/2} - f^{3/2}). \end{aligned} \tag{65}$$

Reverting to the variables r_1, r_2, and c, equation (55) gives

$$6\kappa\tau = (r_1 + r_2 + c)^{3/2} - (r_1 + r_2 - c)^{3/2}, \tag{66}$$

which is 'Lambert's' theorem. And to quote Kriloff,

> A lot of different proofs of the formula [66] can be found in modern treatises on theoretical astronomy, but Newton's remains unsurpassed in its wonderful dynamical insight.

(*f*)

Lemma XI

If a comet void of all motion was let fall from the height SN, or $S\mu + \frac{1}{3}I\mu$, *towards the Sun, and was still impelled to the Sun by the same force uniformly continued by which it was impelled at first, the same, in one-half of that time in which it might describe the arc AC in its own orbit, would in descending describe a space equal to the length* $I\mu$.

'The things above being supposed' a comet, during a time τ, that it takes to describe the arc $A\mu C$ would describe a chord of length AC with a uniform velocity V_P that it will instantaneously have at the distance $SP = SN^2/S\mu$.

It was stated in Proposition XL (see p. 480) that 'the velocity of every comet will always be to the velocity of any planet, supposed to be revolved at the same distance in a circle about the Sun, nearly as the square root of double the distance of the planet from the centre of the Sun at the distance of the comet from the Sun's centre'. Or, in the present context, the velocity along a parabolic arc, at the distance SP from the focus, is $\sqrt{2}$ times the velocity in a circular orbit of radius SP. Therefore, during the interval of time τ, that the comet describes the arc $A\mu C$ along the parabola and a chord of length AC at a distance SP from the focus, it will describe a circular arc of radius SP and length $AC/\sqrt{2}$. Therefore, if v_0 is the velocity of circular motion in a circle of radius SP,

$$\frac{1}{\sqrt{2}} AC = v_0 \tau. \tag{67}$$

The centripetal acceleration, w, towards the centre of a circle of radius SP is

$$w = \frac{v_0^2}{SP}. \tag{68}$$

By Galileo's theorem, the height to which a body will ascend under the constant acceleration w in the time τ is

$$\tfrac{1}{2} w \tau^2 = \frac{(v_0 \tau)^2}{2SP} = \frac{AC^2}{4SP}; \tag{69}$$

and in *half the time*, the height will be

$$h = \frac{AI^2}{4SP} \quad (\text{since } AI = \tfrac{1}{2}AC). \tag{70}$$

For the acceleration along a circular orbit of radius SN (instead of SP) the corresponding height H will be

$$H = h\frac{SP^2}{SN^2} = \frac{1}{4} AI^2 \frac{SP}{SN^2} = \frac{AI^2}{4S\mu} = I\mu = M\mu, \tag{71}$$

by equation (35) and Lemma IX. Q.E.D.

(Again one must marvel at Newton's ability to explain all this in less than 16 lines of prose. Or is it that his succinctness is a measure of his increasing impatience at this stage of writing the *Principia*?)

136. Propositions XLI and XLII

In these last two propositions of the *Principia* Newton addresses himself to the problem of determining the orbit of a comet about the Sun; and to much else. Barely five pages, out of some forty that are devoted to Propositions XLI and XLII, are concerned with the algorithm of his method; the remaining pages are devoted to 'graphical operations' and 'arithmetical calculus'; and to comments on the physics of comets, much in the style of the *Queries* in his *Opticks*. In the account which follows, we shall concentrate almost exclusively on the five pages describing his method of determining the parabolic orbit of a comet which, to quote Kriloff, is

> one of the most admirable achievements of Newton ... and [is] as perfect as anything else he ever wrote.

As the last taste of sweets, the sweetest last,
W. Shakespeare (Richard II)

Proposition XLI. Problem XXI

From three given observations to determine the orbit of a comet moving in a parabola.

As depicted in the adjoining diagram, the problem is to determine the orbital elements of a comet from three sightings, 1, 2, and 3 from the Earth in its course around the Sun.

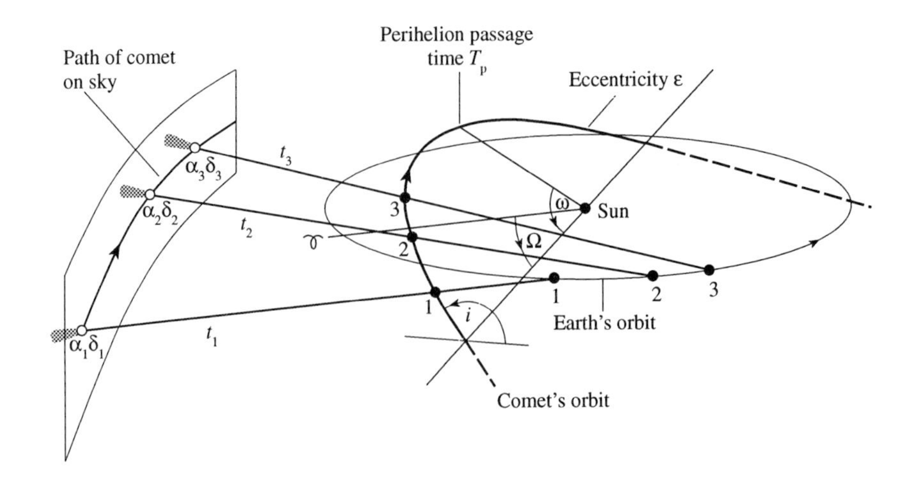

Newton begins with the statement

> This being a problem of very great difficulty, I tried many methods of resolving it; and several of those problems, the composition whereof I have given in the first Book,* tended to this purpose. But afterwards I [succeeded in arriving at] the following solution, which is somewhat more simple.

(a) *Recapitulation*

Newton derives his method of solution from Lemmas VI–XI considered in §135. It is convenient to collect in one place the constructions in those lemmas and the relations that are at the base of his method. All the relations included in the adjoining diagram, except the last, are taken directly from §135.

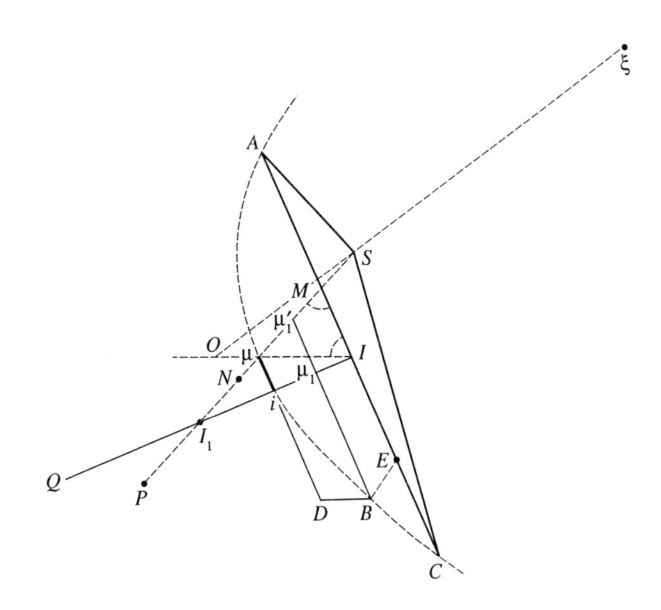

(i) $I\mu = M\mu = AI^2/4S\mu$;
(ii) $\mu O = \frac{1}{2}I\mu$; $IO = \frac{3}{2}I\mu$; $O\xi = 3OS$;
(iii) $N\mu = \frac{1}{3}I\mu = \frac{1}{3}M\mu$; $SN = S\mu + \frac{1}{3}I\mu$;
(iv) $SP = SN^2/S\mu$; $V_P = \kappa\sqrt{2}/\sqrt{SP}$;
(v) $\tau = AC/V_P =$ time to describe the arc $A\mu C$;
(vi) $BD = \mu_1\mu = (\mu_1 B)^2/4S\mu = I\mu(\mu D/AI)^2$.

* The reference here is to the problems and lemmas in Sections IV and V of Book I which have been omitted in Part I (see p. 127).

The references are in §135:

(i) Lemma IX;
(ii) equations (1) and (2);
(iii) equation (35);
(iv) equations (36) and (52);
(v) equation (47).

It remains to explain and establish relation (vi). The line $B\mu_1\mu_1'$ is drawn parallel to AC; it intersects the sides of the isosceles triangle $M\mu I$ at μ_1 and μ_1'. Clearly, $\mu\mu_1 = \mu\mu_1'$. The line BD is drawn parallel to $I\mu$. The tangent μD at μ, being parallel to AC, is also parallel to $B\mu_1\mu_1'$; and therefore,

$$\mu_1 B = \mu D \qquad \text{and} \qquad BD = \mu_1\mu. \tag{1}$$

By construction, $\mu_1 B$ and $\mu_1\mu$ are the oblique coordinates with respect to the diameter $I\mu$ and the tangent μD at μ. Therefore (cf. equation (28), §135)

$$\mu_1\mu = \tfrac{1}{4}(B\mu_1)^2/S\mu. \tag{2}$$

A comparison of equation (2) and relation (i) shows that

$$\mu_1\mu = I\mu\left(\frac{\mu D}{AI}\right)^2 \quad (=BD) \tag{3}$$

Q.E.D.

(*b*) *The formulation of the problem*

What the observations provide are three directions in the ecliptic plane of the Earth's orbit in which the comet is sighted at three different (well-chosen, by criteria that will be specified in §(d) below) instants of time T_1, T_2, and T_3; and the intervals of time, τ_1 and τ_2, between the first and the second, and the second and the third sightings, respectively. In other words, if A, B, and C are the positions of the comet at the three instants, we know, in the first instance, only the directions, T_1A, T_2B, and T_3C, in which they are sighted. Besides, (as will be shown in §(d) below), we can also deduce the perpendicular distance of B from the chord AC ($=T_2V$ in the illustrations on p. 524).

The determination of the true parabolic orbit of the comet in the solar system requires the resolution of two intertwined problems: the solution of the problem in the orbital plane of the comet (to which Lemmas VI–XI refer) and the transformation of the solution to the ecliptic plane in which the observations are made. We shall first consider the first of the two problems, for therein lies the essence of Newton's method.

(c) Newton's method of solution in the orbital plane

The problem, as it presents itself in the orbital plane, is that the three directions, T_1A, T_2B, and T_3C in which the comet is sighted, are delineated; and we are also given an approximate estimate of the distance BE from the position B, at the second sighting, and the point E on the chord AC, joining the positions at the first and the third sightings, which divides it in the ratio,

$$\frac{AE}{AC} = \frac{\tau_1}{\tau_1 + \tau_2}, \tag{4}$$

in accordance with Lemma VIII (see the upper figure (a) in the adjoining diagram). It should be noted that the definition of E as dividing AC in the ratio (4) is only an approximate one; but it can be a very good approximation if the time of the second sighting is well chosen as explained in the corollary to Lemma VIII and in §(d) below.

We may distinguish two distinct steps in Newton's procedure. They correspond to requiring of the true parabolic orbit that it satisfy two conditions which we may describe as *kinematical* and *dynamical*.

We 'take at pleasure the point B, for the place of the comet'* at the second sighting. The *strict* kinematic requirements are that the points B, E, and ξ (as defined in Lemma VIII, equation (2)) are collinear; and further that the triangle $M\mu I$ is isosceles (as required by Lemma IX).

First, we try to correct the approximate (or guessed!) position of E. Through E we draw the chord AC, by the algorithm of Lemma VII, which will be divided at E in the ratio (4). Next, we determine the vertex μ of the parabola passing through A, B, and C by the following procedure.

At I, the mid-point of AC, erect the perpendicular IQ. Take some point I_1 on IQ 'at pleasure'; and through the mid-point i of II_1, draw a line parallel to AC and intersecting I_1S at μ. Join $S\mu$, and let M be the point of intersection of $S\mu$ with AC. The requirement is that $\triangle M\mu I$ is isosceles. If it is not, the construction should be repeated till the equality of $I\mu$ and $M\mu$ is satisfied. Having determined μ in this fashion, extend $I\mu$ to O so that $IO = \frac{3}{2}I\mu$. Join SO and extend SO to ξ so that $\xi O = 3SO$. The intersection E' of $B\xi$ with

* Not quite 'at pleasure'! We place it tentatively at the estimated distance X which the comet will describe during the time $\tau_1 + \tau_2 = \tau$ with a velocity (in a parabolic orbit) at the mean distance from S of the Earth from the Sun (the semimajor axis of the Earth's orbit), namely (cf. equations (52) and (53), §135; also the footnote on p. 480)

$$X = \frac{2\pi}{\text{sidereal year}}(a\sqrt{2})\tau = 0.01720212(a\sqrt{2})\tau.$$

Newton explains later (see p. 528) how the placement of B need not be 'at random, but nearly true'.

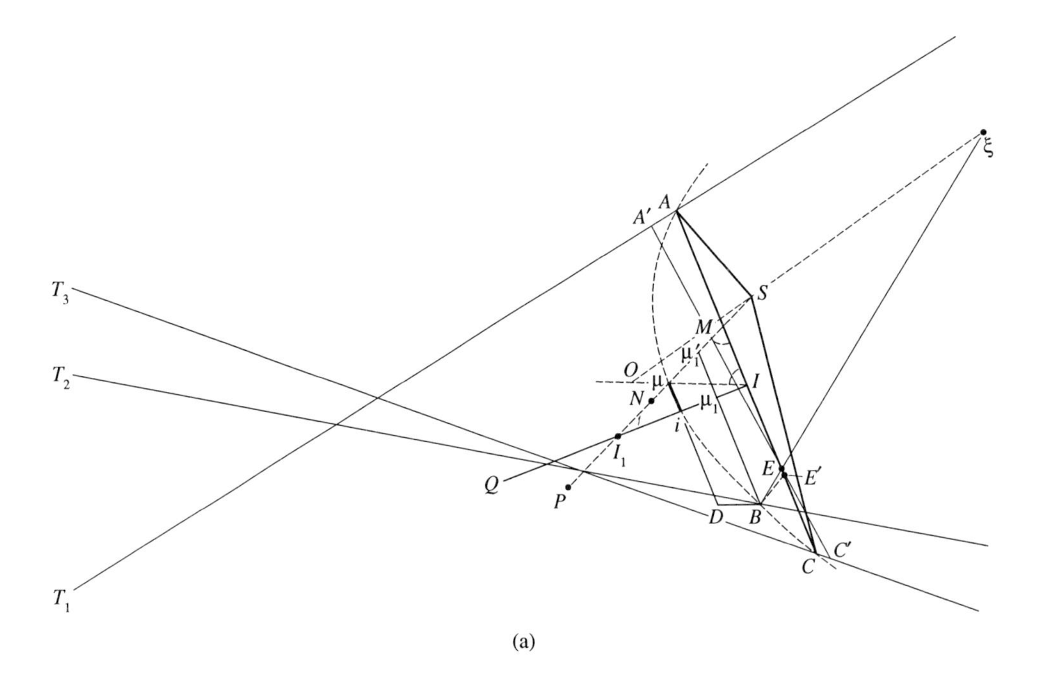
ξ
A
A′
S
T3
M
μ1′
O
μ
I
T2
N
μ1
i
I1
E
E′
Q
P
D
B
C′
C
T1

(a)

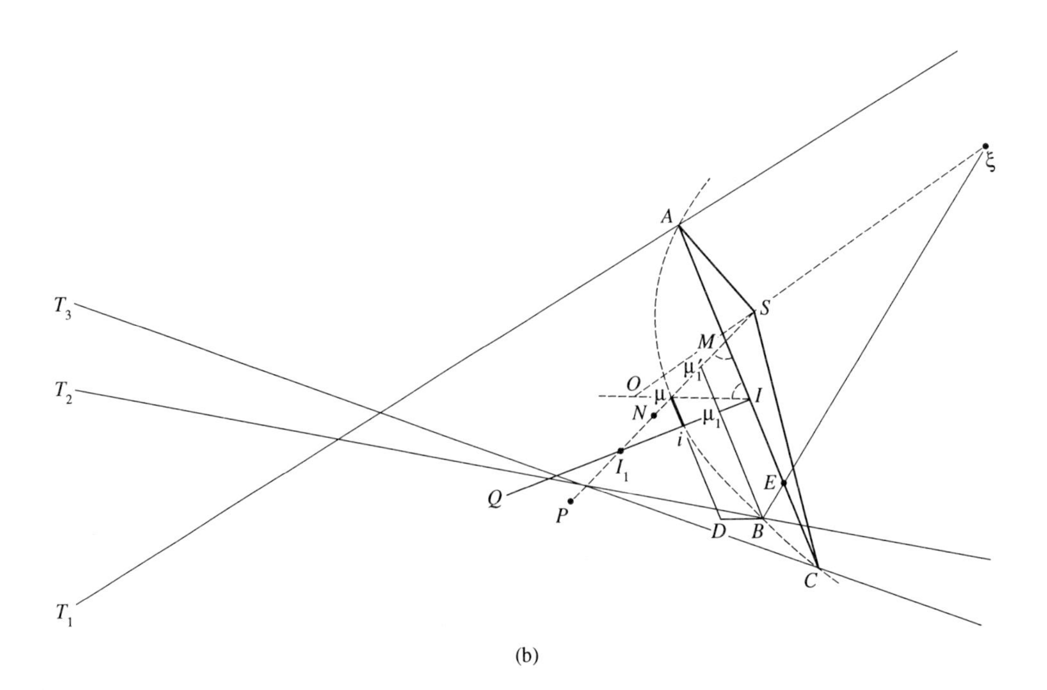
ξ
A
S
T3
M
μ1′
O
μ
I
T2
N
μ1
i
I1
E
Q
P
D
B
C
T1

(b)

AC is the new corrected position of E. Through this new corrected position E' draw once again a chord $A'C'$ which is divided at E' in the same ratio (4), that is,

$$\frac{A'E'}{A'C'} = \frac{\tau_1}{\tau_1 + \tau_2}, \tag{5}$$

and repeat the same construction. In this fashion we obtain the iterated diagram (the lower figure (b)) which, hopefully, will satisfy the kinematic requirements. If not, one must iterate once more. At the second and later stages of iteration, a useful check is provided by the construction described in establishing equation (3), namely by drawing the lines BD and $B\mu_1\mu_1'$ parallel respectively to $I\mu$ and AC and verifying if equation (3) is satisfied.

Another check, mentioned by Kriloff, that the kinematical requirements are satisfied, is provided by the relation (equivalent to Lemma XI)

$$I\mu = M\mu = \left(\frac{\kappa\tau}{2\sqrt{2}}\right)^2 \frac{1}{SN^2}, \tag{6}$$

which follows from equation (54) §135, namely,

$$\kappa\tau\sqrt{2} = c\sqrt{SP} = 2AI\sqrt{SP}. \tag{7}$$

We find,

$$I\mu = M\mu = \tfrac{1}{4}AI^2\frac{SP}{SN^2} = \frac{AI^2}{4S\mu}, \tag{8}$$

which is Lemma IX. We may also use equation (6) to correct an approximate position of the vertex μ to a corrected position μ' by letting

$$I\mu' = M\mu' = \left(\frac{\kappa\tau}{2\sqrt{2}}\right)^2 \frac{1}{(S\mu + \frac{1}{3}I\mu)^2}, \tag{9}$$

in an iterative procedure.

After satisfying the kinematical requirements in this fashion, we turn to the dynamical requirement provided by equations (49), (51), and (52) of §135. We require that

$$\tau_1 + \tau_2 = \frac{AC}{V_P} = \frac{AC.\sqrt{SP}}{\kappa\sqrt{2}}, \tag{10}$$

or,

$$SP = \frac{(S\mu + \frac{1}{3}I\mu)^2}{S\mu} = 2\left[\frac{\kappa(\tau_1 + \tau_2)}{AC}\right]^2. \tag{11}$$

For a point B 'chosen at pleasure' we should not expect that equation (11) will be satisfied. We *must* find a position B, by trial and error, such that both the kinematical and the dynamical requirements are satisfied. Once this has been accomplished, the parabola passing through the definitive positions A and C, and having its focus at S, can be readily constructed.

Q.E.I.

(d) Newton's formulation of the solution

As has increasingly become his wont in these later mathematical sections of the *Principia*, Newton is concise almost to a fault; and to an extent that his theoretical insights are often obscured; we shall encounter an example presently. Thus, even with the account of his method in the simpler context of the orbital plane of the comet in §(c), additional explanations are needed beyond paraphrasing his text and transcribing his prose into formulae.

I.

> Select three observations distant one from another by intervals of time nearly equal; but let that interval of time in which the comet moves more slowly be somewhat greater than the other; namely, so that the difference of the times may be to the sum of the times as the sum of the times is to about 600 days; or that the point E may fall nearly upon M and may err therefrom rather towards I than towards A. If such direct observations are not at hand, a new place of the comet must be found, by Lemma VI.

The reason for the stipulation,

$$\frac{\tau_1 - \tau_2}{\tau_1 + \tau_2} \simeq \frac{\tau_1 + \tau_2}{600\text{ days}}, \tag{12}$$

derives from the corollary to Lemma VIII that the division of the chord AC at E in the ratio,

$$AE:EC = \tau_1:\tau_2, \tag{13}$$

becomes increasingly accurate as the line drawn through B, parallel to AC, comes closer to the tangent at the vertex μ; and a necessary condition for this to happen is

$$|\tau_1 - \tau_2| \leqslant \tau_1 + \tau_2; \tag{14}$$

and further that in this limit,

$$(\tau_1 - \tau_2) \rightarrow \text{proportionality with } (\tau_1 + \tau_2)^2, \tag{15}$$

a proportionality that is asserted but by no means obvious!

The proportionality (15) follows from equation (3) of §(a), namely,

$$\mu_1\mu = I\mu\left(\frac{B\mu_1}{AI}\right)^2, \tag{16}$$

when B approaches the vertex μ and the line $B\mu_1\mu_1'$ tends to coincidence with the tangent at μ.

Proof:
From the adjoining diagram appropriate to this limit, when BE also becomes parallel to μM (as is apparent from figure (b) in the following paragraph II), it is manifest that

$$\mu_1\mu \sin\beta = \mu\mu_0, \tag{17}$$

where $\mu\mu_0$, bisecting $\mu_1\mu_1'$ is perpendicular to it, and

$$\beta = \angle \mu IM = \angle \mu MI. \tag{18}$$

It is also clear that in the limit considered,

$$D\mu = B\mu_1 \simeq B\mu_1' \simeq B\mu_0 = EM. \tag{19}$$

Now if ρ denotes the radius of curvature of the parabola at μ, then

$$B\mu_0 \to \rho\delta\theta \qquad \text{and} \qquad \mu\mu_0 \to \rho(1 - \cos\delta\theta) \simeq \tfrac{1}{2}\rho(\delta\theta)^2, \tag{20}$$

where $\delta\theta$ is the angle subtended by $B\mu_0$ at the centre of curvature. Hence,

$$\frac{\mu_1\mu}{(B\mu_1)^2} \to \frac{1}{2\rho}\operatorname{cosec}\beta, \tag{21}$$

and equation (16) gives

$$\rho = \frac{AI^2}{2I\mu}\operatorname{cosec}\beta. \tag{22}$$

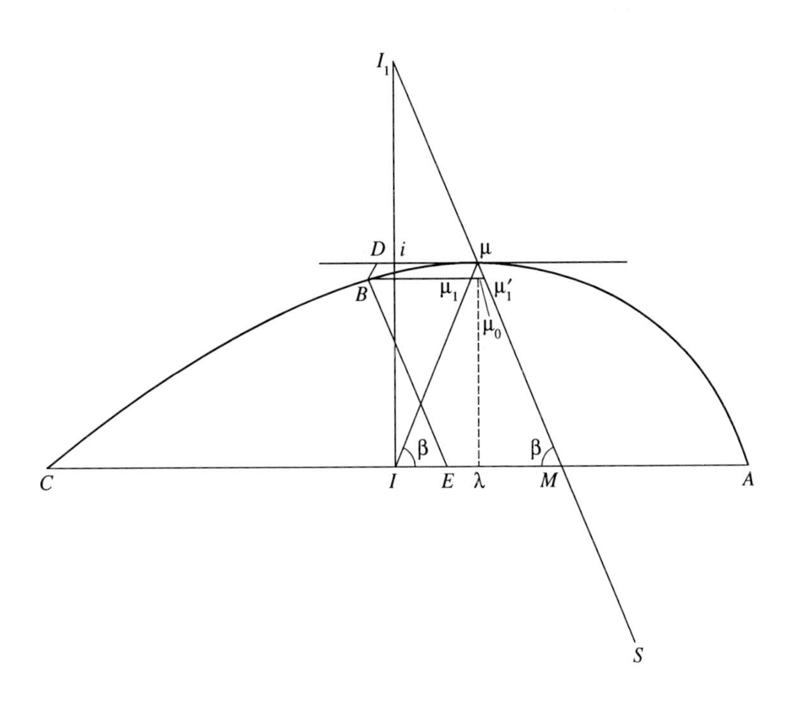

Therefore, in the limit $B \to \mu$,

$$EM = \rho\delta\theta = \left(\frac{AI^2}{2I\mu}\operatorname{cosec}\beta\right)\delta\theta; \tag{23}$$

or, alternatively,

$$[(AI + EM) - (AI - EM)] = \frac{\operatorname{cosec}\beta}{4I\mu}[(AI + EM) + (AI - EM)]^2\delta\theta, \tag{24}$$

establishing the required proportionality. Q.E.D.

One wonders why Newton did not include equation (16), of which he was certainly aware by the proportionality (15) that he asserts as a corollary to Lemma IX. Also, we may note that in the examples he considers

$$\tau_1 - \tau_2 \sim 0{\cdot}1(\tau_1 + \tau_2). \tag{25}$$

II. Consider next the following text:*

> Let S represent the Sun; T_1, T_2, T_3 three places of the Earth in the Earth's orbit; T_1A, T_2B, T_3C three observed longitudes of the comet; τ_1 the time between the first observation and the second; τ_2 the time between the second and the third; X the length which in the whole time $\tau_1 + \tau_2$, the comet might describe with that velocity which it has in the mean distance of the Earth from the Sun, which length is to be found by Cor. III, Prop. XL, Book III.
>
> [Draw] T_2V a perpendicular upon the chord T_1T_3. In the mean observed longitude T_2B, take at pleasure the point B, for the place of the comet in the plane of the ecliptic; and from thence, towards the Sun S, draw the line BE, which may be to the perpendicular T_2V as the product of SB and ST_2^2 is to the cube of the hypotenuse of the right-angled triangle whose sides are SB and the tangent of the latitude of the comet in the second observation to the radius T_2B.
>
> And through the point E (by Lem. VII) draw the right line AEC, whose parts AE and EC, terminating in the right lines T_1A and T_3C, may be one to the other as the times τ_1 and τ_2: then A and C will be nearly the places of the comet in the plane of the ecliptic in the first and third observations, if B was its place rightly assumed in the second.

The text includes a definition and two constructions. The definition is of X: as already defined in the footnote on p. 519, it is the distance a comet would describe in the total

* For convenience, the text has been divided into three paragraphs.

time $\tau_1 + \tau_2$ with the velocity it would have in a parabolic orbit at a distance from the focus equal to the semimajor axis, a, of the Earth's orbit:

$$\left.\begin{aligned} X &= \frac{2\pi}{365.26\text{ days}}(a\sqrt{2})\tau = \kappa(a\sqrt{2})\tau \\ \text{where} \qquad & \\ \kappa &= 0.01720212 \quad \text{(Newton's value).} \end{aligned}\right\} \tag{26}$$

Before describing the two constructions, it is important to distinguish the quantities that refer to the plane of the orbit of the comet and to the plane of the ecliptic. We shall distinguish the latter by a subscript 'e'. Thus, E_e is the projection on the plane of the ecliptic of the point B on the plane of the orbit.

In the accompanying diagrams, Newton's original* (above) and a simplified version (below) (illustrating the relative dispositions of the Earth, the Sun and the comet, as projected on the ecliptic plane, at the three sightings), T_2V is drawn perpendicular to the chord T_1T_3 and towards the Sun at S. Next, from the position, B_e of the comet, along the line of sight T_2B_e, taken 'at pleasure', the line $(BE)_e$ is drawn towards S. The length $(BE)_e$ is the projection of the length BE (in the orbital plane of the comet) on to the ecliptic plane. This length BE can be considered as the 'height' through which the comet falls towards the Sun during the time τ_1. The length T_2V being the corresponding height of fall at the distance ST_2 of the Earth, it follows from the inverse-square law of attraction that

$$BE : T_2V = ST_2^2 : SB^2. \tag{27}$$

In order to obtain $(BE)_e$, we must apply the factor $(SB)_e/SB$. Therefore,

$$(BE)_e = T_2V\frac{(ST_2^2)^2(SB)_e}{(SB)^3}. \tag{28}$$

where

$$SB = [(SB)_e^2 + (T_2B)_e^2 \tan^2 \beta_2]^{1/2} \tag{29}$$

and β_2 is the latitude of the comet at the second sighting (compare figure on p. 514).

By Lemma VIII, the line $(BE)_e$ will cut the line $(AC)_e$ in approximately the ratio,

$$(AE)_e : (EC)_e = \tau_1 : \tau_2; \tag{30}$$

and this is approximate, because Lemma VIII applies ('nearly', even then) when the line $(BE)_e$ is directed towards ξ, three times as far as S. But the restriction (14), on the choice of the second observation (as described in paragraph I above), makes this a good enough approximation needing only the 'cancelling' of the initial placement of $(AC)_e$ by a corrected placement as described in the paragraph below.

* The illustration is from *Sir Isaac Newton's Principia*, reprinted for Sir William Thomson and Hugh Blackburn (James Maclehose, Publisher to the University, Glasgow, 1871) p. 492; the one in Cajori's edition has several misprints.

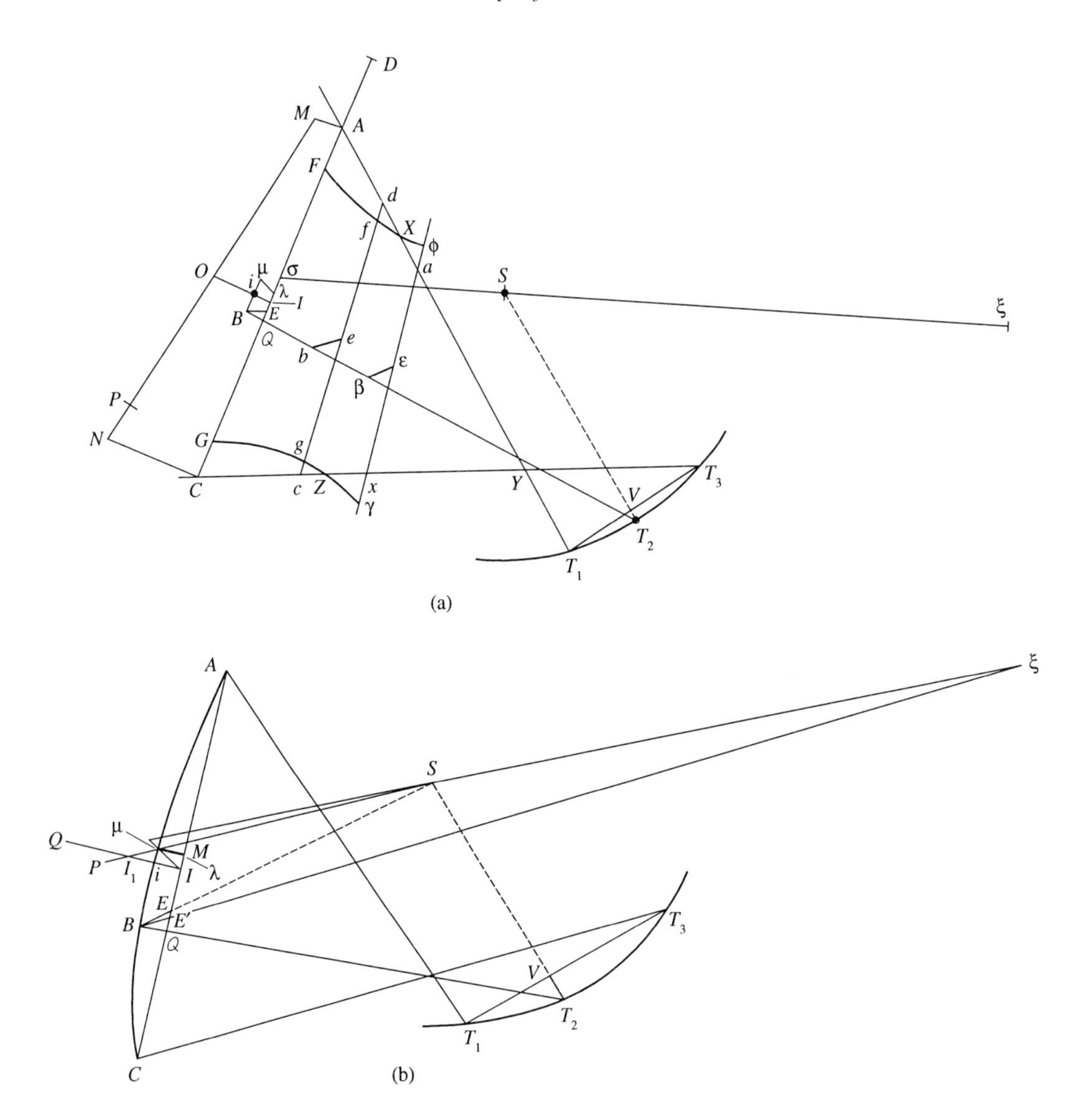

III.

Upon AC, bisected in I, erect the perpendicular Ii. Through B imagine the line Bi drawn parallel to AC. Imagine the line Si drawn, cutting AC in λ, and complete the parallelogram $iI\lambda\mu$. Take $I\sigma$ equal to $3I\lambda$; and through the Sun S imagine the line $\sigma\xi$ drawn equal to $3S\sigma + 3i\lambda$. Then, cancelling the letters A, E, C, I, from the point B towards the point ξ, imagine the new line BE drawn, which may be to the former BE as the square of the ratio of the distance BS to the quantity $S\mu + \frac{1}{3}i\lambda$. And through the point E draw again the right line AEC by the same rule as before; that is, so that its parts AE and EC may be one to the other as the times τ_1 and τ_2 between the observations. Thus A and C will be the places of the comet more accurately.

In this paragraph, Newton describes, ostensibly, in the plane of the ecliptic, the construction described in §(c) in the context of satisfying what we have called the *kinematical* requirements. But the construction as described and explained in this paragraph is *precisely* as in §(c) in the orbital plane of the comet allowing for S being 'far away'. It might appear that in this instance Homer has nodded: for while on projection from the orbital plane to the ecliptic plane, paralled lines remain parallel, orthogonality and equality of lines in different directions are not maintained. But under the conditions Newton considers this problem, namely, when $\tau_1 - \tau_2 \ll \tau_1 + \tau_2$, B is close to the vertex μ, and the distance $S\mu \gg I\mu$, the distinction is indifferent. Thus, Newton ignores, justifiably in the context, the difference in the directions SI and Si. His reasons for insisting on the conditions he states—'If such observations are not at hand, a new place of the comet is to be computed by Lemma VI'—are now apparent.

IV.

Upon AC, bisected in I, erect the perpendiculars AM, CN, IO, of which AM and CN may be the tangents of the latitudes in the first and third observations, to the radii T_1A and T_3C. Join MN, cutting IO in O. Draw the rectangular parallelogram $iI\lambda\mu$, as before. In IA produced take ID equal to $S\mu + \frac{2}{3}i\lambda$. Then in MN, towards N, take MP, which may be to the above-found length X as the square root of the ratio of the mean distance of the Earth from the Sun (or of the semidiameter of the Earth's orbit) to the distance OD. If the point P fall upon the point N; A, B, and C will be three places of the comet, through which its orbit is to be described in the plane of the ecliptic. But if the point P falls not upon the point N, in the right line AC take CG equal to NP, so that the points G and P may lie on the same side of the line NC.

In this paragraph, Newton turns to what we have called the dynamical requirements (or, as one would say at the present time, compliance with 'Lambert's' theorem).

In the orbital plane, the requirement, as given in equation (54) §135, is

$$\kappa\tau\sqrt{2} = AC.\sqrt{SP}. \tag{31}$$

But by definition (26),

$$X = \kappa(a\sqrt{2})\tau. \tag{32}$$

Hence,

$$X = aAC.\sqrt{SP}, \tag{33}$$

or, alternatively,

$$aAC : X = \sqrt{a} : \sqrt{(aSP)}, \tag{34}$$

where it may be recalled that

$$SP = \frac{1}{S\mu}(S\mu + \tfrac{1}{3}I\mu)^2 = S\mu\left(1 + \frac{I\mu}{3S\mu}\right)^2. \tag{35}$$

We have now to express the requirement (34) in terms of the corresponding quantities defined in the ecliptic plane. First, we note that the distances SA and SC in the orbital plane translate into distances $(SA)_e$ and $(SC)_e$ in the ecliptic plane, by (cf. equation (29))

$$\left.\begin{aligned} SA^2 &= [(SA)_e^2 + (AT_1)_e^2 \tan^2 \beta_1] \\ &\text{and} \\ SC^2 &= [(SC)_e^2 + (CT_3)_e^2 \tan^2 \beta_3] \end{aligned}\right\} \tag{36}$$

where β_1 and β_3 are the latitudes of the comet at the first and the third sightings. Therefore, the projected positions, on the ecliptic plane of the points A and C, on the orbital plane are at M and N at distances

$$AM = (T_1A)_e \tan \beta_1 \qquad \text{and} \qquad CN = T_3C \tan \beta_3 \qquad \text{normal to } (AC)_e. \tag{37}$$

Since mid-points of a line project on to mid-points of the projected line, the mid-point I of AC on the orbital plane will project on the mid-point O of MN on the ecliptic plane at a distance,

$$(SI)^2 = (SO)_e^2 + IO^2. \tag{38}$$

We now require the distance $(SP)_e$ on the ecliptic plane that corresponds to the distance SP on the orbital plane, since it is in terms of SP that the dynamical requirement is expressed. If we neglect the difference in the elevation of I and μ when S is sufficiently 'far way' (as in the context considered)

$$S\mu^2 \simeq (S\mu)_e^2 + IO^2. \tag{39}$$

We now find, successively,

$$\begin{aligned} SP = S\mu\left(1 + \frac{I\mu}{3S\mu}\right)^2 &\simeq \left[1 + \frac{IO^2}{(S\mu)_e^2}\right]^{1/2} (S\mu)_e\left(1 + \frac{2I\mu}{3S\mu}\right) \\ &= \left[1 + \frac{IO^2}{(S\mu_e)^2}\right]^{1/2}\left[(S\mu)_e + \frac{2}{3}\frac{I\mu}{S\mu}(S\mu)_e\right] \\ &= \left[1 + \frac{IO^2}{(S\mu_e)^2}\right]^{1/2}[(S\mu)_e + \tfrac{2}{3}(I\mu)_e] \\ &\simeq \{[(S\mu)_e + \tfrac{2}{3}(I\mu)_e]^2 + IO^2\}^{1/2} = OD. \end{aligned} \tag{40}$$

as Newton defines. (In the reductions, we have neglected terms such as $(I\mu)_e/(S\mu)_e$.)

The dynamical requirement (34) now translates into the requirement that MP, evaluated in accordance with the relation

$$aMP:X = \sqrt{a}:\sqrt{(aOD)}, \qquad (41)$$

agrees with MN into which the chord AC, in the orbital plane, is projected on to the ecliptic plane. If MN and MP do not agree, for B_e chosen 'at pleasure' the location of B_e along the line of sight should be changed. And the equality of MN and MP must be achieved by a process described in the following paragraph.

V.

> By the same method as the points E, A, C, G were found from the assumed point B, from other points b and β assumed at pleasure, find out the new points e, a, c, g; and ε, α, κ, γ. Then through G, g, and γ draw the circumference of a circle $Gg\gamma$, cutting the right line T_3C in Z: and Z will be one place of the comet in the plane of the ecliptic. And in AC, ac, $a\kappa$, taking AF, af, $a\phi$, equal respectively to CG, cg, $\kappa\gamma$; through the points F, f, and ϕ, draw the circumference of a circle $Ff\phi$, cutting the right line AT_1 in X; and the point X will be another place of the comet in the plane of the ecliptic. And at the points X and Z, erecting the tangents of the latitudes of the comet to the radii T_1X and T_3Z, two places of the comet in its own orbit will be determined. Lastly, if (by Prop. XIX, Book I) to the focus S a parabola is described passing through those two places, this parabola will be the orbit of the comet. Q.E.I.

The method outlined is one of 'trial and error'. And as Newton describes the process, it can be made to converge rapidly by graphical interpolation.

And finally, Newton provides the *raison d'etre* for the algorithm that he has described, in terms of Lemmas VI–XI:

VI.

> The demonstration of this construction follows from the preceding lemmas, because the right line AC is cut in E in the proportion of the times, by Lemma VII, as it ought to be by Lemma VIII; and BE, by Lemma XI, is a portion of the right line BS or $B\xi$ in the plane of the ecliptic, intercepted between the arc ABC and the chord AEC; and MP (by Cor., Lem. X) is the length of the chord of that arc, which the comet should describe in its proper orbit between the first and third observations, and therefore is equal to MN, providing B is a true place of the comet in the plane of the ecliptic.

It should be particularly noted that the final process of trial and error is directly based on Lemma X which, as we have shown, is entirely equivalent to 'Lambert's' theorem, a fact not generally recognized.*

And finally, Newton explains how the initial placement of the point B along the line of sight T_2B need not be 'at random, but nearly true' (See illustration 16) on p. 524).

> If the angle $A\mathcal{Q}T_2$, at which the projection of the orbit in the plane of the ecliptic cuts the right line T_2B, is roughly known, at that angle with T_3B, draw the line AC [such that
>
> $$AC.S\mathcal{Q}^{1/2} = \tfrac{4}{3}T_1T_3.ST_2^{1/2}]\dagger \tag{42}$$
>
> and drawing the line SEB so as its part EB may be drawn equal to T_2V‡, the point B [or B_e in our notation] will be determined which we are to use for the first time.

Equation (42) asserts no more than equation (10), namely that the comet describes the chord AC uniformly with the velocity V_P in the same time that the Earth describes the arc T_1T_3 with the uniform circular velocity $V_\oplus$:

$$AC:V_P = (\text{arc } T_1T_3):V_\oplus, \tag{43}$$

And allowing for the factor $\sqrt{2}$ between the velocities along a parabola and a circle at equal distances, we require,

$$AC.SP^{1/2} = (T_1T_3.ST_2^{1/2})\sqrt{2}. \tag{44}$$

Newton approximates this relation by replacing SP by $S\mathcal{Q}$ and $\sqrt{2}\,(=1.414)$ by $\tfrac{4}{3}$ so that

$$AC.S\mathcal{Q}^{1/2} \simeq \tfrac{4}{3}T_1T_3.ST_2^{1/2}. \tag{45}$$

* For example, as A. N. Kriloff has pointed out, a blatant mistake by Julius Bauschinger occurs in his *Die bahnbestimmung der himmelskorper* (Verlag von Wilhelm Engelmann, 1906, p. 388). As Kriloff writes:

> ...it is impossible to agree with Bauschinger's words:
>
> '*Newton used the property that the chord is subdivided by the mean radius vector proportionally to the time, hence he replaces, as observed by Lambert, the sectors by the corresponding triangles; what his method is wanting in order to render it perfect is Euler–Lambert's theorem, instead of which he was satisfied with an approximate relation between the chord, the radii vectors, and the time.*'
>
> The whole sentence printed in italics is wrong; not only did Newton not use the subdivision of the chord by the mean radius vector but he invented the wonderfully exact construction of Lemma VIII. As to the relation between the chord, the radii vectors, and the time, the relation of Lemma X is just the so-called Euler–Lambert theorem, which is simply an analytical expression of Newton's Lemma X. I wish to correct this statement in a standard treatise of great excellence, because it may induce the student not to pay due attention to Newton's method, and to overlook its real meaning and perfection.

† There is a misprint at this point: $SQ^{1/2}$ ($= S\mathcal{Q}^{1/2}$ in our present notation) and $ST_2^{1/2}$ should be interchanged.

‡ Clearly not T_2V but (in our present notation) in accordance with the earlier

$$T_2V(ST_2)^2(SB)_e[(SB)_e^2 + (T_2B)_e^2\tan^2\beta_2]^{-3/2}$$

prescription (cf. equation (2)). From these frequent misprints in these latter sections of the *Principia* it seems to me that neither Cotes nor Halley have shown much perspicacity in their reading of the *Principia*.

This completes our account of Newton's solution to the problem enunciated in Proposition XLI. The remaining 25 pages of this proposition are devoted to an account of an example—*calculus arithmeticus*—and of his inferences on the physics of the comets. We shall not attempt a summary; but the following brief extracts may provide a flavour.

From these observations, by constructions of figures and calculations, I deduced the longitudes and latitudes of the comet:

All this I determined by scale and compass, and the chords of angles, taken from the table of natural sines, in a pretty large figure, in which, to wit, the radius of the Earth's orbit (consisting of 100000 parts) was equal to $16\frac{1}{3}$ inches of an English foot.

Lastly, in order to discover whether the comet did truly move in the orbit so determined, I investigated its places in this orbit partly by arithmetical operations, and partly by scale and compass, to the times of some of the observations, as may be seen in the following table:

	The Comet's						
	Distance from Sun	Longitude computed	Latitude computed	Longitude observed	Latitude observed	Difference longitude	Difference latitude
12 Dec.	2792	ℑ $6°32'$	$8°18\frac{1}{2}$	ℑ $6°31\frac{1}{2}$	$8°26$	$+1$	$-7\frac{1}{2}$
29 Dec.	8403	𝔛 $13.13\frac{2}{3}$	28.00	𝔛 $13.11\frac{3}{4}$	$28.10\frac{1}{12}$	$+2$	$-10\frac{1}{12}$
5 Feb.	16669	♉ 17.00	$15.29\frac{2}{3}$	♉ $16.59\frac{7}{8}$	$15.27\frac{2}{5}$	$+0$	$+2\frac{1}{4}$
5 Mar.	21737	$29.19\frac{3}{4}$	12. 4	$29.20\frac{6}{7}$	$12.\ 3\frac{1}{2}$	-1	$+\frac{1}{2}$

The observations of this comet from the beginning to the end agree as perfectly with the motion of the comet in the orbit just now described as the motions of the planets do with the theories from whence they are calculated, and by this agreement plainly evince that it was one and the same comet that appeared all that time, and also that the orbit of that comet is here rightly defined.

Now if one reflects upon the orbit described, and duly considers the other appearances of this comet, he will be easily satisfied that the bodies of comets are solid, compact, fixed, and durable, like the bodies of the planets; for if they were nothing else but the vapours or exhalations of the Earth, of the Sun, and other planets, this comet, in its passage by the neighbourhood of the Sun, would have been immediately dissipated; for the heat of the Sun is as the density of its rays, that is, inversely as the square of the distance of the places from the Sun.

(*e*)

Proposition XLII. Problem XXII

To correct a comet's orbit found as above.

This proposition is concerned with a last correction to the cometary orbit in the ecliptic plane derived by the algorithm of Proposition XLI. Newton's procedure consists of three steps or 'operations', as he calls them, consistently with his increasing tendency to giving imperial instructions rather than elucidating explanations.

Operation I

> Assume that position of the plane of the orbit which was determined according to the preceding proposition; and select three places of the comet, deduced from very accurate observations, and at great distances one from the other ... it will be convenient that in one of those times the comet be in its perigee, or at least not far from it.

Let 1, 2, and 3 denote the successive positions of the comet. With the Sun as the focus, draw on the orbital plane, that has been derived, a parabola 'by arithmetical operations, according to Proposition XXI, Book I'.*

The parabola having been constructed, determine the areas A_{12} and A_{23} of the sectors included between the radius vectors joining the Sun to the three locations, 1, 2, and 3:

$$A_{12} = \frac{1}{2}\int_1^2 r^2\, d\varphi \quad \text{and} \quad A_{23} = \frac{1}{2}\int_2^3 r^2\, d\varphi. \tag{46}$$

Now if T_1, T_2, and T_3 are the observed recorded times of the three sightings, then, if the derived orbit is consistent with the law of areas, we should find that

$$C = \frac{A_{12}}{A_{23}} \quad \text{and} \quad D = \frac{T_2 - T_1}{T_3 - T_2} \quad \text{are equal.} \tag{47}$$

Similarly, if we evaluate (graphically, as Newton undoubtedly did) the time required for the comet to describe the full arc AC by the formula,

$$S = \int_{T_1}^{T_3} \frac{ds}{v}, \tag{48}$$

* This proposition of Book I, along with the others of Section IV and V were not included in Part I of this book. For, according to Newton's own testimony 'the composition, whereof... given in the first Book, tended' to the '*longe difficillimum*' problem; and were later abandoned in favour of the simpler solution presented in Proposition XLI. This proposition is the only place in Books I or III that Newton returns to those propositions. But what he needs in the present context is the simple construction described in the Scholium at the end of Section IV. This construction is presented in the appendix following this subsection.

where ds is the infinitesimal arc length measured along the orbit and v is the instantaneous velocity,

$$v = \frac{\kappa\sqrt{2}}{\sqrt{r}}, \tag{49}$$

as theoretically determined, then we should find that

$$S \text{ and the recorded time } T_{13} (= T_3 - T_1) \text{ are equal.} \tag{50}$$

In other words, ideally, we should find that

$$C - D \quad \text{and} \quad S - T_{13} \quad \text{are both zero.} \tag{51}$$

But due to the inherent uncertainties in the graphical and arithmetical operations, the required equalities (51) will mostly not obtain and

$$C - D \quad \text{and} \quad S - T_{13} \quad \text{will differ from zero;} \tag{52}$$

but, presumably by not much.

Newton now seeks to achieve the desired accuracy by effecting simultaneously small changes in the longitude and in the inclination of the orbital plane. As we shall presently see, the method Newton adopts is, in effect, a first-order Taylor expansion of a function $F(x; y)$, for increments Δx and Δy in the two variables x and y: thus,

$$F(x + \Delta x; y + \Delta y) \simeq F(x; y) + \Delta x \frac{\partial F}{\partial x} + \Delta y \frac{\partial F}{\partial x}. \tag{53}$$

More precisely, the 'operations' II and III, that Newton uses towards this end are the following.

Operations II and III

First, effect a small change $P(\sim 20\text{–}30')$ in the longitude K, keeping the inclination I unchanged; redetermine the locations of the comet at the times of the three observations on the altered orbital plane, draw the parabola through these points, and determine, as before, the ratio of the corresponding areas, $a_{12}/a_{23} = c$ and of the times, $t_{12}/t_{23} = d$ in the new orbit. The change effected in $C - D$ will be small compared to $|C - D|$:

$$(C - D) - (c - d) \ll |C - D|, \tag{54}$$

and similarly

$$(S - T_{13}) - (S - t_{13}) \ll |S - T_{13}|. \tag{55}$$

Next, repeat the same procedure, after effecting a small change Q (again, 20–30′) in the inclination I, keeping the longitude K unchanged; and determine the resulting changes in the corresponding ratios of the areas, $\alpha_{12}/\alpha_{23} = \gamma$ and of the times, $\tau_{12}/\tau_{23} = \delta$. Again, we should find

$$(C - D) - (\gamma - \delta) \ll |C - D| \tag{56}$$

and

$$(S - T_{13}) - (S - \tau_{13}) \ll |S - T_{13}|. \tag{57}$$

Now, considering $(C - D)[K; I]$ and $(S - T_{13})[K; I]$ as functions of the two variables, K and I, we can express equations (54) and (56) in the forms

$$\left.\begin{aligned} (C - D)[K + P; I] &= C - D + [(c - d) - (C - D)], \\ \text{and} \qquad & \\ (C - D)[K; I + Q] &= C - D + [(\gamma - \delta) - (C - D)]. \end{aligned}\right\} \tag{58}$$

For simultaneous increases of mP in K and nQ in I, where m and n are at present unspecified 'numbers', we should find,

$$(C - D)[K + mP; I + nQ] = (C - D) + m[(c - d) - (C - D)] + n[(\gamma - \delta) - (C - D)], \tag{59}$$

and similarly,

$$(S - T_{13})[K + mP; I + nQ] = (S - T_{13}) + m(T_{13} - t_{13}) + n(T_{13} - \tau_{13}). \tag{60}$$

Therefore, if the parabola drawn on the orbital plane, for the latitude $K + mP$ and inclination $I + nQ$, should satisfy the required equalities, then m and n should be so determined that, simultaneously,

$$\left.\begin{aligned} (C - D) &= m[(C - D) - (c - d)] + n[(C - D) - (\gamma - \delta)] \\ \text{and} \qquad & \\ (S - T_{13}) &= m(t_{13} - T_{13}) + n(\tau_{13} - T_{13}). \end{aligned}\right\} \tag{61}$$

For m *and* n determined in this fashion, the longitude and the inclination of the 'true' orbital plane are

$$K + mP \quad \text{and} \quad I + nQ. \tag{62}$$

Besides, if we now express the coordinates of a point on the original parabola in terms of a parameter $\bar{P}$ in the manner,

$$(x, y) = (A\bar{P}^2, 2A\bar{P}) \tag{63}$$

where $2A$ denotes the transverse diameter (or, more commonly, the semilatus rectum); and if p and π, and $2a$ and 2α are the corresponding parameters and transverse diameters of the two other parabolae for $(K + P; I)$ and $(K; I + Q)$, the corrected parameters and semidiameters for the true orbit will be given by

$$\bar{P} + m(p - \bar{P}) + n(\pi - \bar{P}) \tag{64}$$

and

$$\frac{1}{A} + m\left(\frac{1}{a} - \frac{1}{A}\right) + n\left(\frac{1}{\alpha} - \frac{1}{A}\right). \tag{65}$$

Q.E.I.

This completes our account of Newton's mathematical basis for the computation of cometary orbits. In the remaining 13 pages devoted to this proposition, Newton considers the results of his own and of Halley's computations of many cometary orbits with special attention to the possible occurrence of periodic orbits and to matters arising therefrom. Again, we shall not attempt a summary. But the following extract may suggest their tenor.

> But the periodic times of the revolutions of comets, and the transverse diameters of their orbits, cannot be accurately enough determined but by comparing comets together which appear at different times. If, after equal intervals of time, several comets are found to have described the same orbit, we may thence conclude that they are all but one and the same comet revolved in the same orbit; and then from the times of their revolutions the transverse diameters of their orbits will be given, and from those diameters the elliptic orbits themselves will be determined.
>
> To this purpose the orbits of many comets ought to be computed, supposing those orbits to be parabolic; for such orbits will always nearly agree with the phenomena.

(*f*) *Appendix*

Scholium at the end of Section IV Book I: Required to draw a conic with focus at S and passing through given points B, C, D.

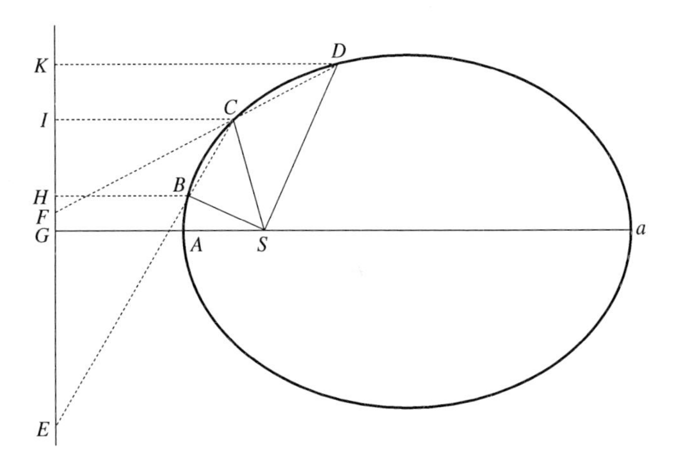

Construction:

Join CB and DC and extend them to E and F, respectively, such that

$$\frac{EB}{EC} = \frac{SB}{SC} \quad \text{and} \quad \frac{FC}{FD} = \frac{SC}{SD}. \tag{66}$$

Join EF and prolong it at both ends; and drop

$$SG \quad \text{and} \quad BH \quad \text{perpendiculars to} \quad EF. \tag{67}$$

Prolong GS indefinitely; and mark off points A and a on GS such that

$$\frac{GA}{AS} = \frac{Ga}{aS} = \frac{HB}{BS}. \tag{68}$$

Then

$$\left.\begin{array}{l} A \text{ is the vertex, } Aa, \text{ is the principal axis, and a conic,} \\ \text{with focus at } S, \text{ will pass through } B, C, \text{ and } D. \end{array}\right\} \tag{69}$$

And 'I say, the thing is done'.

Proof:

By the similarity of the $\triangle$s ICE and HBE,

$$\frac{IC}{HB} = \frac{EC}{EB} = \frac{SC}{SB}, \quad \text{by construction (equation (66)),} \tag{70}$$

or

$$\frac{IC}{SC} = \frac{HB}{SB} = \frac{GA}{AS}, \quad \text{by construction (equation (68)).} \tag{71}$$

And, by the similarity of the $\triangle$s KDF and ICF,

$$\frac{KD}{IC} = \frac{DF}{FC} = \frac{SD}{SC}, \quad \text{by construction (equation (66))} \tag{72}$$

or

$$\frac{KD}{DS} = \frac{IC}{SC} = \frac{GA}{AS}, \quad \text{by equation (71).} \tag{73}$$

Hence

$$\frac{HB}{SB} = \frac{IC}{SC} = \frac{KD}{SD} = \frac{GA}{AS}. \tag{74}$$

Therefore the curve passing through B, C, and D is a conic whose focus is at S and directrix is EFK. Q.E.F.

As Newton points out, we shall have an ellipse, a parabola, or a hyperbola according as

$$GA \text{ is } >, =, \qquad \text{or} \qquad < AS,$$

'the point a in the first case falling on the same side of the line GF as the point A; in the second, going off to an infinite distance; in the third falling on the other side of GF'.

And Newton concludes this Scholium with a handsome compliment to M. de la Hire

> That excellent geometer M. de la Hire has solved this problem much after the same way, in his *Conics*, Prop. XXV, Book VIII.

137. The general Scholium

In the penultimate paragraph of the general Scholium which concludes Book III of the *Principia*, we have Newton's testament on gravity written with a sentiment and a style that are unsurpassed.

> Hitherto we have explained the phenomena of the heavens and of our sea by the power of gravity, but have not yet assigned the cause of this power. This is certain, that it must proceed from a cause that penetrates to the very centres of the Sun and planets, without suffering the least diminution of its force; that operates not according to the quantity of the surfaces of the particles upon which it acts (as mechanical causes used to do), but according to the quantity of the solid matter which they contain, and propagates its virtue on all sides to immense distances, decreasing always as the inverse square of the distances. Gravitation towards the Sun is made up out of the gravitations towards the several particles of which the body of the Sun is composed; and in receding from the Sun decreases accurately as the inverse square of the distances as far as the orbit of Saturn, as evidently appears from the quiescence of the aphelion of the planets; nay, and even to the remotest aphelion of the comets, if those aphelions are also quiescent. But hitherto I have not been able to discover the cause of those properties of gravity from phenomena, and I frame [feign] no hypotheses; for whatever is not deduced from the phenomena is to be called an hypothesis; and hypotheses, whether metaphysical or physical, whether of occult qualities or mechanical, have no place in experimental philosophy. In this philosophy particular propositions are inferred from the phenomena, and afterwards rendered general by induction. Thus it was that the impenetrability, the mobility, and the impulsive force of bodies, and the laws of motion and of gravitation, were discovered. And to us it is enough that gravity does really exist, and act according to the laws which we have explained, and abundantly serves to account for all the motions of the celestial bodies, and of our sea.

There is only one additional statement one can make, at this date, three hundred years later: the quest for the 'cause of gravity' still continues.

Miscellania

As stated in the Preface, my primary purpose in writing this book was to present to the 'common reader' an account of those parts of the *Principia* that are directly related to Newton's sequential development of his mathematical theory of gravity: the basic concepts; the Laws of Motion; the propositions, the lemmas, and the Scholia; the rules of reasoning; the universal law of gravitation and how 'it abundantly serves to account for all the motions of the celestial bodies and of our sea'. For this purpose, it sufficed to restrict myself to Books I and III, omitting only Sections IV and V (on theorems in the analytic geometry of conic sections) and Section X (on motions on surfaces and on the oscillations of 'pendulous' bodies). Even so, to leave out Book II (with the sole exception of Proposition XXIV on the principle of equivalence) is to ignore parts of the *Principia* that are as illustrative of Newton's perception at the best (if not the very best) of Books I and III. On this account, in this *Miscellania*, we shall consider some propositions from Book II that provide some indication of Newton's range of perception.

We shall first consider Lemma III and Propositions XV–XVIII of Section IV on *the circular motion of bodies in resisting mediums* in which we find Newton formulating and solving a fundamental problem on the effect of air-drag on the orbit of an artificial satellite as it descends towards the Earth—a problem that was to become central to modern-day celestial mechanics.

Next, we consider Proposition XXXIV. In the Scholium that follows this proposition, Newton formulates, almost casually, a problem in the *calculus of variations* whose solution, as stated, requires the derivation of the relevant Euler–Lagrange equations.

Leaving the *Principia*, we shall follow Proposition XXXIV by Newton's solution of the problem of minimum descent under gravity along a smooth frictionless curve—a problem which was posed by Johann Bermoulli on New Years' Day 1697, as a challenge to 'the sharpest mathematicians in the whole world'. The reason for considering the problem of least resistance and of least time of descent, in juxtaposition, is that the two problems require for their solution very similar applications of the calculus of variations.

And finally, we shall turn to Propositions XLIV–XL in which the velocity of sound in elastic and inelastic media is derived—a problem as sophisticated in physics as the problem in the *calculus of variations*, in the Scholium to Proposition XXXIV, is sophisticated in mathematics.

❖25❖

The effect of air-drag on the descent of bodies

Connoisseurs of independent discovery may like to know that Newton's result was independently re-derived by Morduchow and Volpe in 1973.

D.G. King-Hele and D.M.C. Walker
(*Vistas in Astronomy*, Vol. 30, p. 271, 1987)

138. Newton's problem and its solution

In Propositions XV and XVI, Newton formulates a problem on the effect of air-drag on the descent of bodies under centripetal attraction. As a prelude to Newton's method of solution, we shall state the problem simply and provide the solution as one might today.

We consider a body, under the centripetal attraction,

$$-\gamma r^{-n} \quad (\gamma \text{ a constant and } n \text{ unspecified}), \tag{1}$$

that is descending in an atmosphere with density varying as

$$\rho = \rho_0 r^{-1} \quad (\rho_0 = \text{constant}), \tag{2}$$

and experiencing a drag proportional to the ambient density and to the square of the velocity in the direction its motion:

$$\begin{aligned} \text{drag} &= -\tfrac{1}{2}C\rho v^2 \quad (C \text{ a constant}) \\ &= -\tfrac{1}{2}\rho_0 C\frac{v^2}{r} \quad (\text{for the atmospheric law (2)}) \\ &= -D\frac{v^2}{r} \quad (D = \tfrac{1}{2}\rho_0 C), \end{aligned} \tag{3}$$

where D is a constant for the problem.

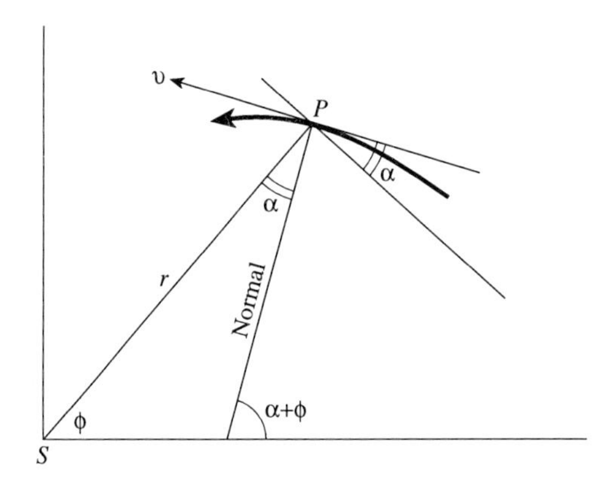

In the accompanying diagram:

P denotes the position of the body at some instant of time t;
r is its distance from the centre, S;
v is its velocity in the direction of its motion;
α is the angle which the normal at P makes with the radius vector;
ϕ is the azimuthal angle; and
$\alpha + \phi$ is the angle which the inward normal at P makes with the horizontal.

The radius of curvature, κ, by its definition is given by

$$\frac{1}{\kappa} = \frac{d}{ds}(\alpha + \phi) \tag{4}$$

where s measures the arc-length along the orbit.

With the foregoing definitions, the radial and the transverse components of the velocity are:

$$\frac{\mathrm{d}r}{\mathrm{d}t} = -v \sin \alpha \qquad \text{and} \qquad r\frac{\mathrm{d}\phi}{\mathrm{d}t} = +v \cos \alpha, \tag{5}$$

and the tangential and the normal accelerations are given by

$$v\frac{\mathrm{d}v}{\mathrm{d}s} = \frac{\mathrm{d}v}{\mathrm{d}t} = \frac{\gamma}{r^n}\sin \alpha - \frac{D}{r}v^2, \tag{6}$$

and

$$\frac{v^2}{\kappa} = v^2\frac{\mathrm{d}(\phi + \alpha)}{\mathrm{d}s} = v\frac{\mathrm{d}}{\mathrm{d}t}(\phi + \alpha) = \frac{\gamma}{r^n}\cos \alpha. \tag{7}$$

Newton considers the case when

$$\alpha = \text{constant}, \tag{8}$$

and the orbits are *equi-angular spirals.** Then equation (7) and the second of equations (5) give

$$v\,\frac{\mathrm{d}\phi}{\mathrm{d}t} = \frac{\gamma}{r^n}\cos\alpha = \frac{v^2}{r}\cos\alpha. \tag{9}$$

It follows

$$v = \frac{\sqrt{\gamma}}{r^{(n-1)/2}}. \tag{10}$$

Differentiating equation (10) with respect to t and making use of the first of equations (5), we find

$$\left.\begin{aligned}\frac{dv}{dt} &= -\tfrac{1}{2}(n-1)\sqrt{\gamma}\,\frac{\dot{r}}{r^{(n+1)/2}} = \tfrac{1}{2}(n-1)v\sqrt{\gamma}\,\frac{\sin\alpha}{r^{(n+1)/2}}\\ &= \tfrac{1}{2}(n-1)\gamma\,\frac{\sin\alpha}{r^n}.\end{aligned}\right\} \tag{11}$$

On the other hand, by inserting for v its solution (10) in equation (6), we have

$$\frac{\mathrm{d}v}{\mathrm{d}t} = \frac{\gamma}{r^n}\sin\alpha - \frac{D}{r}v^2 = \frac{\gamma}{r^n}(\sin\alpha - D). \tag{12}$$

From a comparison of equations (11) and (12) we now obtain

$$\sin\alpha - D = \tfrac{1}{2}(n-1)\sin\alpha \tag{13}$$

or

$$D = \tfrac{1}{2}(3-n)\sin\alpha. \tag{14}$$

For the particular case of the inverse-square law of attraction, $n = 2$,

$$D = \tfrac{1}{2}\sin\alpha. \tag{15}$$

Newton obtains the foregoing solutions in Propositions XV and XVI and considers their implications in great detail in their context. We now turn to the consideration of these propositions.

139. Lemma III and Proposition XV

Since Newton starts, from the outset, with the assumption that the orbits are equi-angular spirals, he begins with a geometrical lemma relating to such orbits.

Lemma III

Let PQR be a spiral cutting all the radii SP, SQ, SR, etc., in equal angles. Draw the right line PT touching the spiral in any point P, and cutting the radius SQ in T; draw PO, QO perpendicular to the spiral, and meeting in O, and join SO: I

* It is more conventional to define the complement of α, the inclination of the inward normal to the direction of motion.

say, that if the points P and Q approach and coincide, the angle PSO will become a right angle, and the ultimate ratio of the rectangle $TQ \cdot 2PS$ to PQ^2 will be the ratio of equality.

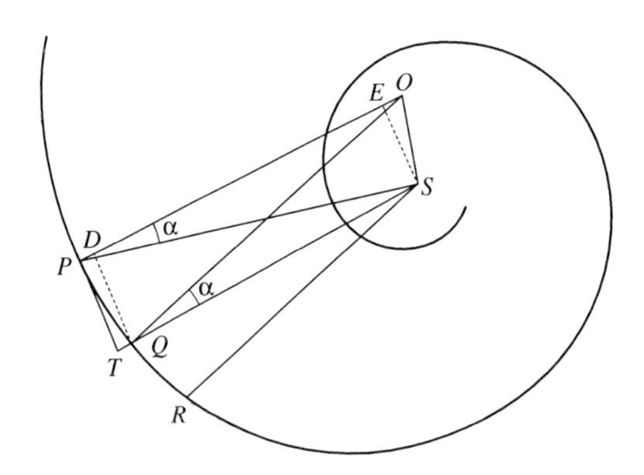

Proof:

It is manifest that

$$\alpha = \angle OPS = 90^\circ - \angle SPT = 90^\circ - \angle SQR = \angle OQS. \qquad (1)^*$$

Hence, a circle passing through the points S, O, and P will pass through Q as well. As P tends to coincidence with Q,

> this circle will touch the spiral in the place of coincidence PQ, and will therefore cut the right line OP perpendicularly. Therefore, OP will become a diameter of this circle, and the angle OSP, being in a semicircle, becomes a right one.

Draw QD and SE perpendicular to OP (and therefore parallel to TP). Then in the limit, $P \to Q$.

$$\cos\alpha = \frac{PD}{TQ} = \frac{PS}{PO}. \qquad (2)$$

At the same time,

$$\frac{PD}{PQ} = \frac{PQ}{2PO}, \qquad (3)$$

the second equality following from the fact that $\angle PQD = \angle QPT$ is the angle which the tangent at P, of a circle of radius PO, makes at P. From equations (2) and (3) it follows:

$$\frac{PD}{TQ}\cdot\frac{PQ}{PD} = \frac{PS}{PO}\cdot\frac{2PO}{PQ} \quad \text{i.e.,} \quad \frac{PQ}{TQ} = 2\frac{PS}{PQ}, \qquad (4)$$

or

$$PQ^2 = 2TQ.PS. \qquad \text{Q.E.D.}$$

* The 'equal angles' of Newton are the complements of α.

The same otherwise

$TQ \cos \alpha =$ Distance, in the direction of the inward normal, travelled in a time dt, by the action of the centripetal acceleration, v^2/κ

$$= \frac{1}{2}\frac{v^2}{\kappa}(dt)^2 = \frac{(ds)^2}{2\kappa} = \frac{PQ^2}{2\kappa} = \frac{PQ^2}{2OP}. \tag{5}$$

Hence

$$TQ = \frac{PQ^2}{2OP \cos \alpha} = \frac{PQ^2}{2PS}, \tag{6}$$

or

$$PQ^2 = 2TQ.PS. \tag{7}$$

(Newton uses arguments similar to the foregoing in his proof of Proposition XV).

Proposition XV. Theorem XII

If the density of a medium in each place thereof be inversely as the distance of the places from an immovable centre, and the centripetal force be as the square of the density: I say, that a body may revolve in a spiral which cuts all the radii drawn from that centre in a given angle.

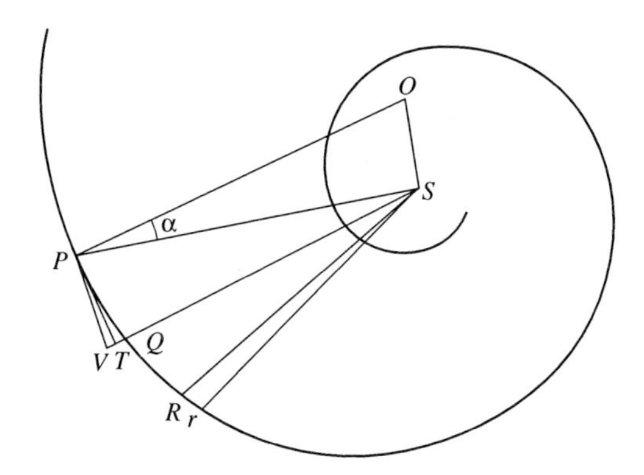

In the adjoining diagram (which is virtually the same as the one for Lemma III with the deletions only of the dotted lines ES and DQ), it is now assumed that the arcs PQ and QR are described in two successive intervals, time dt apart, under the simultaneous action of a retarding force,

$$\mathbb{R} = \tfrac{1}{2}\rho C v^2, \tag{8}$$

as in §138, equation (3) and an inverse-square law of centripetal attraction,

$$-\gamma r^{-2} \quad (\gamma = \text{a constant}). \tag{9}$$

Also the right line SQT is extended to V such that

$$SV = SP; \tag{10}$$

and an arc Rr extending QR is marked off such that

$$\text{area of } \triangle QSr = \text{area of } \triangle PSQ, \tag{11}$$

that is,

$$QS.Qr\cos\alpha = PS.PQ\cos\alpha \tag{12}$$

(on the assumption that the orbit is an equi-angular spiral); and therefore,

$$\frac{Qr}{PQ} = \frac{SP}{SQ}. \tag{13}$$

In the absence of the drag, the body will describe arcs $(PQ)_0$ and $(PR)_0$ in time intervals at $\mathrm{d}t$ and $2\,\mathrm{d}t$ which, in the limit of the coincidence of P and Q, must be in the ratio 1:2, that is,

$$(PR)_0 = 2(PQ)_0. \tag{14}$$

However, with the drag D operating, the body will experience decrements $\Delta(PQ)$ and $\Delta(PR)$ in the times $\mathrm{d}t$ and $2\,\mathrm{d}t$ of amounts,

$$\Delta(PQ) = \tfrac{1}{2}\mathbb{R}(\mathrm{d}t)^2 \qquad \text{and} \qquad \Delta(PR) = \tfrac{1}{2}\mathbb{R}(2\,\mathrm{d}t)^2 = 2\mathbb{R}(\mathrm{d}t)^2. \tag{15}$$

Also, in the limit,

$$Pr \to 2PQ. \tag{16}$$

In the same limit,

$$\begin{aligned} Rr = Pr - PR &= 2PQ - PR \\ &= 2[(PQ)_0 - \Delta(PQ)] - [(PR)_0 - \Delta(PR)] \\ &= \Delta(PR) - 2\Delta(PQ) = \mathbb{R}\,(\mathrm{d}t)^2, \end{aligned} \tag{17}$$

by equation (15). By the same equation,

$$\Delta(PQ) = \tfrac{1}{2}\mathbb{R}(\mathrm{d}t)^2 = \tfrac{1}{2}Rr. \tag{18}$$

On the other hand, TQ being the decrement in SQ by the action of the centripetal attraction $-\gamma r^{-2}$,

$$TQ = \tfrac{1}{2}\gamma(SP)^{-2}(\mathrm{d}t)^2, \tag{19}$$

or

$$\gamma(\mathrm{d}t)^2 = 2TQ.(SP)^2 = SP.PQ^2, \tag{20}$$

by Lemma II (equation (7)).

A comparison of equations (18) and (20) now yields the relation:

$$\gamma Rr = \gamma\mathbb{R}(\mathrm{d}t)^2 = \mathbb{R}SP.PQ^2. \tag{21}$$

And by equation (20)

$$v_P = \frac{\mathrm{d}s}{\mathrm{d}t} = \sqrt{\gamma}\,\frac{PQ}{PQ\sqrt{SP}} = \frac{\sqrt{\gamma}}{\sqrt{SP}}. \tag{22}$$

Similarly,

$$v_Q = \frac{\sqrt{\gamma}}{\sqrt{SQ}}. \tag{23}$$

Since, by assumption, PQ and QR are described in equal times,

$$\frac{QR}{PQ} = \frac{\sqrt{SP}}{\sqrt{SQ}} = \frac{\sqrt{(SP.SQ)}}{SQ}. \tag{24}$$

This last equation, together with equation (13), gives

$$\frac{Rr}{PQ} = \frac{Qr - QR}{PQ} = \frac{1}{SQ}[SP - \sqrt{(SP.SQ)}]$$

$$= \frac{1}{SQ}\{SP - [SP.(SV - VQ)]^{1/2}\} \tag{25}$$

or since $SV = SP$ by construction,

$$\frac{Rr}{PQ} = \frac{1}{SQ}\{SP - [SP.(SP - VQ)]^{1/2}\}. \tag{26}$$

Hence

$$\frac{Rr}{PQ} = \frac{SP}{SQ}\left\{1 - \left(1 - \frac{VQ}{SP}\right)^{1/2}\right\} = \frac{1}{2}\frac{VQ}{SQ}. \tag{27}$$

Therefore, by equation (21)

$$\mathbb{R} = \gamma\frac{Rr}{SP.PQ^2} = \frac{1}{2}\gamma\frac{VQ}{SP.SQ.PQ}. \tag{28}$$

From the similarity of the $\triangle$s PVQ and SPV,

$$\frac{VQ}{PQ} = \frac{PV}{SP} = \cos(90^\circ - \alpha) = \sin\alpha. \tag{29}$$

Therefore

$$\mathbb{R} = \frac{1}{2}\gamma\frac{\sin\alpha}{SP.SQ} = \tfrac{1}{2}\sin\alpha\frac{v_P v_Q}{\sqrt{(SP.SQ)}}, \tag{30}$$

by making use of equations (22) and (23). Now passing to the limit, $P \to Q$, we obtain

$$\mathbb{R} = \tfrac{1}{2}\rho C v^2 = \frac{1}{2}\frac{v^2}{r}\sin\alpha. \tag{31}$$

In other words

$$\rho C = \frac{1}{r}\sin\alpha. \tag{32}$$

Therefore the density must vary inversely as r if the orbit is to be an equi-angular spiral. As Newton concludes:

> Therefore in a medium whose density is inversely as SP the distance from the centre, a body will revolve in this spiral. Q.E.D.

The analysis leading to this conclusion is a veritable *tour de force*.

Newton follows this proposition with a number of corollaries,

> COR. I. The velocity in any place P, is always the same wherewith a body in a non-resisting medium with the same centripetal force would revolve in a circle, at the same distance SP from the centre.

This follows from equations (21) and (22) according to which we may write,

$$v = \frac{\sqrt{\gamma}}{\sqrt{r}}. \tag{33}$$

> COR. II. The density of the medium, if the distance SP be given, is as OS/OP, but if that distance is not given, as $OS/OP.SP$. And thence a spiral may be fitted to any density of the medium.

This is a restatement of equations (31) and (32) since $OS/OP = \sin\alpha$.

Corollary III

By virtue of equations (27), (20), and (21),

$$\tfrac{1}{2}Rr\colon TQ = \frac{1}{4}\frac{VQ.PQ}{SQ} : \frac{1}{2}\frac{PQ^2}{SP} = \frac{1}{2}\frac{VQ}{PQ}\cdot\frac{SP}{SQ}$$

$$\rightarrow \frac{1}{2}\frac{VQ}{PQ} = \frac{1}{2}\frac{OS}{OP} = \tfrac{1}{2}\sin\alpha, \tag{34}$$

or, replacing Rr and TQ by their equivalents (18) and (19), we obtain

$$\frac{\mathbb{R}}{\gamma(SP)^{-2}} = \tfrac{1}{2}\sin\alpha. \tag{35}$$

> The spiral therefore being given, there is given the proportion of the resistance to the centripetal force; and, conversely, from that proportion given the spiral is given.
>
> COR. IV. Therefore the body cannot revolve in this spiral, except where the force of resistance is less than half the centripetal force.

This follows directly from equation (35) according to which

$$\mathbb{R} \leqslant \tfrac{1}{2}\gamma SP^{-2} \tag{36}$$

the equality sign holding only when $\alpha = \pi/2$. And when $\alpha = \pi/2$, the spiral will coincide with the radial direction PS; and the body will descend to the centre with the velocity (33) which is a factor $\sqrt{2}$ less than the corresponding parabolic velocity.

Corollary V

It is manifest from the illustration on p. 540 that

$$ds = d(PS)\operatorname{cosec} \alpha: \tag{37}$$

and therefore,

$$s = PS \operatorname{cosec} \alpha. \tag{38}$$

And since the velocity at any point on the spiral orbit at a radial distance r from the centre is the same as on the radial orbit at the same distance, it follows that

> the time of the descent in the spiral will be to the time of the descent in the right line SP in the same given ratio, and therefore given.

Corollary VI

From the polar equation of the equi-angular spiral,

$$\log(r/r_0) = \phi \tan \alpha \quad (r_0 \text{ is a constant}), \tag{39}$$

it follows that after each revolution the radial distance increases by a constant ratio; thus,

$$\log \frac{r(\phi + 2\pi)}{r(\phi)} = 2\pi \tan \alpha; \tag{40}$$

and after n revolutions by the ratio,

$$\exp(2\pi n \tan \alpha). \tag{41}$$

Alternatively, the number of revolutions the spiral describes between R_1 and R_0 is,

$$\frac{1}{2\pi}(\cot \alpha)\log(R_1/R_0), \tag{42}$$

and the time to make these number of revolutions will be as cosec $\alpha \times$ the time for direct descent from R_1 to R_0.

Corollary VII

From equation (40) it follows that, in the accompanying illustration of the equi-angular spiral, the radial distances,

$$AS,\ BS,\ CS, \text{ etc.} \tag{43}$$

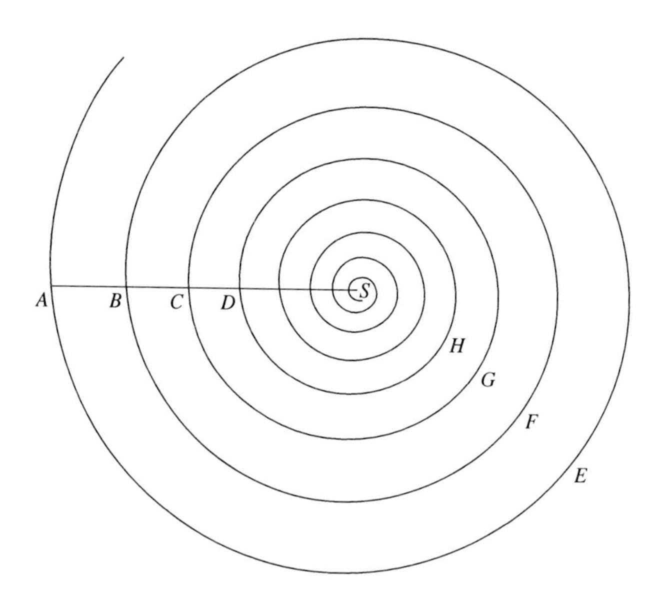

form a geometric series with the common ratio,

$$q = e^{-2\pi \tan \alpha}; \tag{44}$$

and the perimeters of the successive revolutions,

$$AEB,\ BFC,\ CGD, \text{ etc.}, \tag{45}$$

form a similar geometric series with the same common ratio. On the other hand, the velocities,

$$v_A,\ v_B,\ v_C,\ v_D, \text{ etc.}, \tag{46}$$

at A, B, C, etc., from a geometric series with the common ratio $q^{-1/2}$. Consequently, the times for successive revolutions will form a geometric series with the common ratio $q^{3/2}$. Therefore, the total time for descent from A to S along the spiral orbit will be proportional to the sum 'going on *ad infinitum*'

$$AS^{3/2}[1 + q^{3/2} + q^3 + q^{9/2} + \cdots] = \frac{AS^{3/2}}{1 - q^{3/2}}. \tag{47}$$

Therefore,

the time for the first revolution (AEB): the time for the descent to S

$$= \frac{1}{1 - (BS/AS)^{3/2}} = \frac{AS^{3/2}}{AS^{3/2} - BS^{3/2}}. \tag{48}$$

If the spiral is closely wound ($AB \ll AS$), this ratio tends to

$$1: \ \{1 - [(AS - AB)/AS]^{3/2}\} \simeq \frac{2}{3}\frac{AS}{AB}. \tag{49}$$

Corollaries VIII and IX

In these two corollaries, Newton makes some qualitative remarks on the general case when the variation of density is different from the inverse radius and the motions become eccentric.

140. Propositions XVI–XVIII

Proposition XVI. Theorem XIII

> *If the density of the medium in each of the places be inversely as the distance of the places from the immovable centre, and the centripetal force be inversely as any power of the same distance: I say, that the body may revolve in a spiral intersecting all the radii drawn from that centre in a given angle.*

In this proposition, the analysis of Proposition XV is generalized to the case when the centripetal attraction is as the inverse nth power of r; that is as

$$-\gamma r^{-n}, \tag{1}$$

as in §138. The analysis for this case follows so very closely that of the earlier proposition, that each equation of §139 is replaced by an equation to which it reduces when $n = 2$.* Thus, the following primed equations replace the unprimed equations of §139:

$$2TQ.SP^n = PQ^2.SP^{n-1}; \tag{20$'$}$$

$$\mathbb{R} = \gamma \frac{Rr}{PQ^2.SP^{n-1}}; \tag{21$'$}$$

$$v_P = \frac{\sqrt{\gamma}}{\sqrt{SP^{n-1}}}; \tag{22$'$}$$

$$Rr = \tfrac{1}{2}(3-n)\frac{VQ.PQ}{SQ}; \tag{27$'$}$$

$$\mathbb{R} = \tfrac{1}{2}(3-n)\gamma\frac{VQ}{SP^{n-1}.SQ.PQ} \to \tfrac{1}{2}(3-n)\gamma\frac{\sin\alpha}{SP^n}; \tag{28$'$, 30$'$}$$

$$\mathbb{R} = \tfrac{1}{2}(3-n)\frac{v^2}{r}\sin\alpha; \tag{31$'$}$$

and

$$\frac{\mathbb{R}}{\gamma SP^{-n}} = \tfrac{1}{2}(3-n)\sin\alpha. \tag{35$'$}$$

The corollaries which Newton adjoins follow directly from equation (35$'$).

* We have replaced Newton's 'n' by '$n + 1$' to be in accord with the notation of §138. In particular, $n = 2$ (instead of $n = 1$) corresponds to the inverse-square law of attraction.

Corollary I

The resistance is to the centripetal force as $\frac{1}{2}(3-n)OS$ to OP.

Corollary II

If the centripetal force be inversely as SP^3, $3-n=0$; and therefore the resistance and density of the medium will be nothing, as in Prop IX, Book I.

Corollary III

If the centripetal force be inversely as any power of the radius SP, whose index is greater than the number 3, the positive resistance will be changed into a negative.

In the Scholium and in Propositions XVII and XVIII which follow, Newton encapsulates, in broad terms, the procedures adopted in Propositions XV and XVI. We quote him in full,

Scholium

This proposition and the former, which relate to mediums of unequal density, are to be understood as applying only to the motion of bodies that are so small, that the greater density of the medium on one side of the body above that on the other is not to be considered. I suppose also the resistance, other things being equal, to be proportional to its density. Hence, in mediums whose force of resistance is not as the density, the density must be so much augmented or diminished, that either the excess of the resistance may be taken away, or the defect supplied.

Proposition XVII. Problem IV

To find the centripetal force and the resisting force of the medium, by which a body, the law of the velocity being given, shall revolve in a given spiral.

Let that spiral be PQR. From the velocity, with which the body goes over the very small arc PQ, the time will be given; and from the altitude TQ, which is as the centripetal force, and the square of the time, that force will be given. Then from the difference RSr of the areas PSQ and QSR described in equal intervals of time, the retardation of the body will be given; and from the retardation will be found the resisting force and density of the medium.

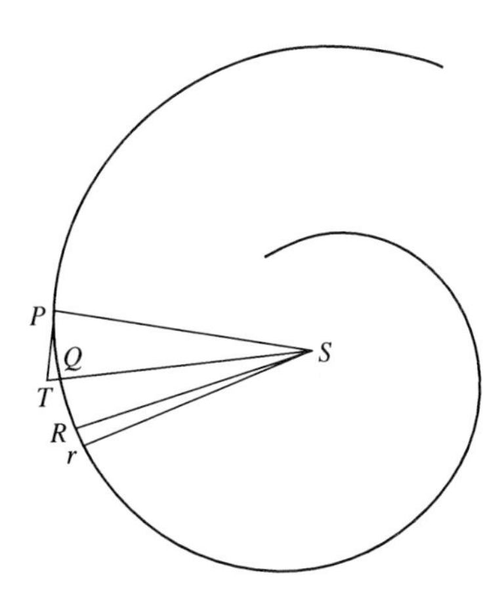

Proposition XVIII. Problem V

The law of centripetal force being given, to find the density of the medium in each of the places thereof, by which a body may describe a given spiral.

From the centripetal force the velocity in each place must be found; then from the retardation of the velocity the density of the medium is found, as in the foregoing proposition.

But I have explained the method of managing these problems in the tenth proposition and second lemma of this Book; and will no longer detain the reader in these complicated investigations.

141. An alternative method of solution of Newton's problem

Newton's principal concern in the solution of the problem considered in Propositions XV and XVI is to relate the drag ($\mathbb{R}$) with the centripetal attraction ($-\gamma r^{-n}$), as he does in Corollary IV of Proposition XV and in all three corollaries of Proposition XVI. The most direct route to achieve this end is to consider the motion along the spiral as one-dimensional motion (as along a string) with the arc-length, s, as the independent variable. But to complete the solution, as Newton recognizes in Corollaries V–VII of Proposition XV, one must express the coordinates r and ϕ in terms of the time, t, as the independent variable. While Newton accomplishes this transformation indirectly, it may be useful to present the solution starting directly with the relevant equations of motion.

The components of the drag, $\frac{1}{2}\rho C v^2$, in the direction of motion, in the radial, r, and in the transverse, ϕ, direction are, respectively,

$$\tfrac{1}{2}\rho C \frac{\dot{r}}{(\dot{r}^2 + r^2\dot{\phi}^2)^{1/2}}(\dot{r}^2 + r^2\dot{\phi}^2) \qquad \text{and} \qquad \tfrac{1}{2}\rho C \frac{r\dot{\phi}}{(\dot{r}^2 + r^2\dot{\phi}^2)^{1/2}}(\dot{r}^2 + r^2\dot{\phi}^2). \tag{1}$$

For a variation of density, $\rho = \rho_0 r^{-1}$,

$$\left.\begin{aligned} &\text{the drag in the } r\text{-direction} = D\frac{\dot{r}}{r}(\dot{r}^2 + r^2\dot{\phi}^2)^{1/2} \\ \text{and} \\ &\text{the drag in the } \phi\text{-direction} = D\dot{\phi}(\dot{r}^2 + r^2\dot{\phi}^2)^{1/2}, \end{aligned}\right\} \tag{2}$$

where

$$D = \tfrac{1}{2}\rho_0 C. \tag{3}$$

The relevant equations of motion are, therefore,

$$\ddot{r} - r(\dot{\phi})^2 = -\frac{\gamma}{r^n} + D\frac{\dot{r}}{r}(\dot{r}^2 + r^2\dot{\phi}^2)^{1/2}, \tag{4}$$

$$\frac{1}{r}(r^2\dot{\phi})^{\cdot} = D\dot{\phi}(\dot{r}^2 + r^2\dot{\phi}^2)^{1/2}. \tag{5}$$

We shall now show that equations (4) and (5) allow an equi-angular spiral as an exact solution when

$$r = \mathrm{e}^{\phi \tan \alpha}. \tag{6}$$

For this solution

$$\left.\begin{aligned} &\dot{r} = r\dot{\phi}(\tan \alpha); \qquad \dot{\phi} = \frac{\dot{r}}{r}\cot \alpha, \\ \text{and} \\ &(\dot{r}^2 + r^2\dot{\phi}^2)^{1/2} = r\dot{\phi} \sec \alpha. \end{aligned}\right\} \tag{7}$$

Making these substitutions, we may readily verify that equations (4) and (5) become:

$$\ddot{r} - \frac{\dot{r}^2}{r}\cot^2 \alpha = -\frac{\gamma}{r^n} + D\frac{\dot{r}^2}{r}\operatorname{cosec} \alpha, \tag{8}$$

and

$$\ddot{r} + \frac{\dot{r}^2}{r} = D\frac{\dot{r}^2}{r}\operatorname{cosec} \alpha. \tag{9}$$

By subtracting equation (8) from equation (9) we obtain:

$$\dot{r}^2 \operatorname{cosec}^2 \alpha = \frac{\gamma}{r^{n-1}}, \tag{10}$$

or

$$\dot{r} = -\sqrt{\gamma}\,\frac{\sin \alpha}{r^{(n-1)/2}}, \tag{11}$$

with the sign and the constant of integration chosen appropriately for an orbit descending with zero velocity from infinity. Equation (11) directly integrates to give

$$r^{(n+1)/2} = -\frac{\sqrt{\gamma}}{2}(n+1)(t+t_0)\sin \alpha, \tag{12}$$

where t_0 is a constant of integration.

Equations (6) and (12) provide the complete solution to the problem under the stated premises. Parenthetically, we may note that for an inverse-square law of attraction, $n = 2$, equation (12) gives

$$r^{3/2} = -\tfrac{3}{2}\gamma^{1/2}(t + t_0)\sin\alpha, \tag{13}$$

which is equivalent to Newton's solution, §139, equation (47).

Returning to equation (11), differentiating with respect to t, and substituting $\dot{r}$ for the same equation, we obtain

$$\ddot{r} = +\frac{\sqrt{\gamma}}{2}(n-1)\frac{\dot{r}}{r^{(n+1)/2}}\sin\alpha = -\tfrac{1}{2}\gamma\sin^2\alpha\,\frac{n-1}{r^n}. \tag{14}$$

On the other hand, from equation (9) it follows that

$$\ddot{r} = (D\operatorname{cosec}\alpha - 1)\frac{\dot{r}^2}{r} = (D\operatorname{cosec}\alpha - 1)\gamma\sin^2\alpha\,\frac{1}{r^n}, \tag{15}$$

where we have substituted for $\dot{r}^2$ from equation (10). From a comparison of equations (14) and (15), we obtain

$$(D\operatorname{cosec}\alpha - 1) = -\tfrac{1}{2}(n-1), \tag{16}$$

or

$$D\operatorname{cosec}\alpha = \tfrac{1}{2}(3-n), \tag{17}$$

which is Newton's result. Q.E.D. (!)

❖26❖

The solid of least resistance

142. Introduction

In the Scholium to Proposition XXXIV, Newton formulates his problem of the solid of revolution which experiences the least resistance in moving through a 'rare medium' with a constant velocity in the direction of its axis. With hardly a hint of an explanation, Newton states an integral (which, as we shall presently verify, is a first integral of the appropriate Euler–Lagrange equations) and concludes with the assertion:

> ... the solid described by the revolution of this figure about its axis ... will be less than any other circular solid whatsoever, described of the same length and breadth.

Newton's solution to his problem of a solid of revolution of least resistance is the earliest modern problem—albeit a special one—treated by the method of the calculus of variations. (The problem of the brachistochrone, he was to treat in 1697, is the second (see Chapter (28).) The calculus of variations was uncharted territory when Newton's *Principia* was published: it was to be discovered by Euler in 1744.

Newton's contemporaries (including Leibniz) were baffled by his assertions in the Scholium. Newton's letter in response to some queries by David Gregory (who included it in a collection *Newtoni methodus fluzionum*) provided some information which was later published as an appendix by Andrew Motte in his 1729 English translation of the *Principia.* Additional explicatory details, included in a second letter of Newton's to Gregory, were rescued by J. C. Adams in an appendix to his *A catalogue of the Portsmouth collection of books and papers written by or belonging to Sir Isaac Newton* (Cambridge, 1888). With the details of Newton's calculations thus made available, a reconstruction of his method of solution has been made by O. Bolza,* A. R. Forsyth,† and E. T. Whiteside.‡ The presentation in §144 is in part based on these accounts.

* *Bibliotheca Mathematica* Ser. 3, 13, 146–149, 1913.
† In *Isaac Newton (1642–1727); a memorial volume*, edited by W. J. Greenstreet, G. Bell & Sons Ltd., London, 1927, pp. 75–86.
‡ E. T. Whiteside. *The mathematical papers of Isaac Newton*, Vol. VI, Cambridge University Press, 1974, pp. 456–480.

143.

Proposition XXXIV. Theorem XXVIII

If in a rare medium, consisting of equal particles freely disposed at equal distances from each other, a globe and a cylinder described on equal diameters move with equal velocities in the direction of the axis of the cylinder, the resistance of the globe will be but half as great as that of the cylinder.

Newton begins with the statement:

> For since the action of the medium upon the body is the same (by Cor. v of the Laws) whether the body move in a quiescent medium, or whether the particles of the medium impinge with the same velocity upon the quiescent body, let us consider the body as if it were quiescent, and see with what force it would be impelled by the moving medium.

Consider then a parallel stream of particles, with velocity v and of unit net flux normal to itself, incident on a plane surface under the circumstance envisaged in the illustration below.

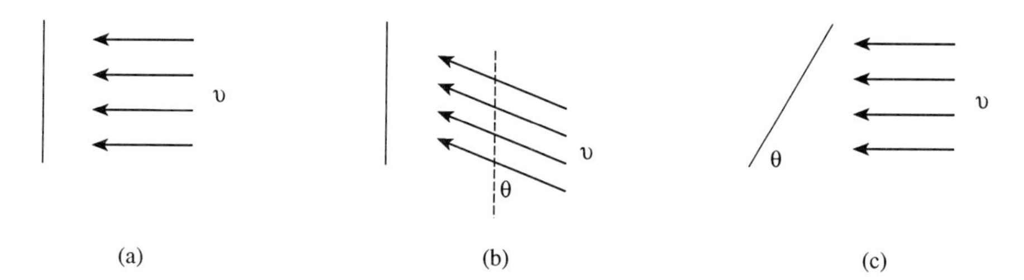

In case (a) the stream is incident normally to the plane surface. The pressure exerted on the surface is assumed to be of the form,

$$p = Cv^2, \tag{1}$$

where C is a constant.

In cases (b) and (c) the pressure exerted will be

$$Cv^2 \sin^2 \theta; \tag{2}$$

since the stream is incident at an angle θ to the plane, the flux normal to the plane is reduced by a factor $\sin \theta$; and since the direction of the velocity of the incident particles is also inclined to the plane by the angle θ the momentum of the incident particles is reduced by the same factor $\sin \theta$.

Consider*, first, a right circular cylinder of diameter $2R$ moving uniformly with a velocity v along the axis of the cylinder in a 'rare medium'. By equation (1), it is manifest that the

* I am grateful to Mr Aaron Grant for correcting an earlier version of this proposition.

pressure P_C experienced by the cylinder is given by the area πR^2, presented by the face of the cylinder to the oncoming stream of particles, times Cv^2, i.e., by

$$P_C = \pi(Cv^2)R^2. \tag{3}$$

Newton deduces the pressure P_G exerted on a globe of the same radius R to the same oncoming stream of particles from the pressure exerted on the cylinder and on the globe disposed as in the accompanying illustration.

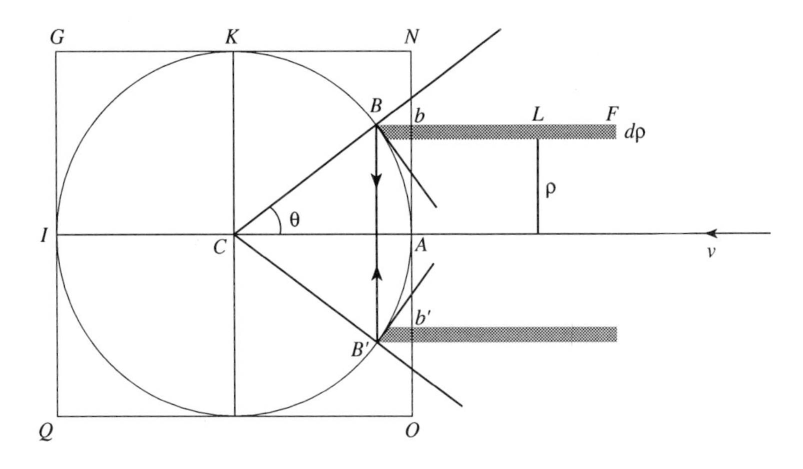

Towards this end, consider an annular circular stream of particles impinging on the face of the cylinder in a ring of area $2\pi\rho\, d\rho$. The surface area of the ring projected on the cap BAB' of the globe is clearly,

$$2\pi\rho\, d\rho \times \sec\theta \qquad \theta = \angle(AC, CB). \tag{4}$$

The pressure exerted on the globe, by the annular stream of particles considered is by equation (2)

$$(Cv^2)(2\pi\rho\, d\rho \sec\theta)\cos^2\theta \tag{5}$$

—$\cos^2\theta$ not $\sin^2\theta$ since the angle θ defined here is the complement of the angle θ in equation (2).

Of the resulting pressure acting on this annular strip on the globe, the component in the directions $\overrightarrow{BB'}$ and $\overleftarrow{B'B}$ will cancel by symmetry; and only the component in the direction $\overrightarrow{FB}$ will contribute to the total pressure P_G exerted on the globe. Therefore the pressure dP_G exerted on the globe in the direction $\overrightarrow{FB}$ is

$$(Cv^2)(2\pi\rho\, d\rho\cos\theta)\cos\theta. \tag{6}$$

Since

$$\rho = R\sin\theta \qquad \text{and} \qquad d\rho = R\cos\theta\, d\theta, \tag{7}$$

the total pressure resisting the globe is therefore

$$P = (Cv^2)2\pi R^2 \int_0^{\pi/2} \cos^3\theta\sin\theta\, d\theta \tag{8}$$

or

$$P_{\mathrm{G}} = \tfrac{1}{2}\pi(Cv^2)R^2. \tag{9}$$

Therefore, by equation (3) and (10),

$$P_{\mathrm{G}}/P_{\mathrm{C}} = \tfrac{1}{2}. \tag{10}$$

This is the result stated by Newton as follows:

> And therefore if the particles of the medium are at rest, and the cylinder and globe move with equal velocities, the resistance of the globe will be half the resistance of the cylinder. Q.E.D.

144. The Scholium, I: the frustum of minimum resistance

The Scholium consists of two parts. In the first part, the problem is posed of a frustum (i.e., a truncated circular cone) of minimum resistance for a given height and base. The problem more explicitly formulated is the following. In the adjoining diagram, *CEBHC*

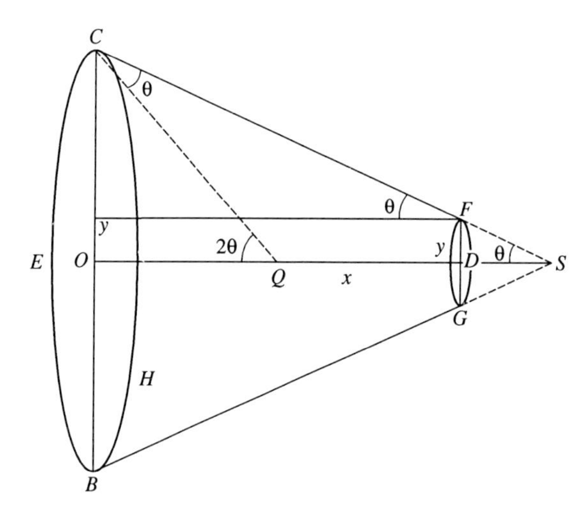

is the circular base and *OD* is the height of the frustum, both of which are given. It is required to find the radius *DF* of the upper side (or, equivalently, the vertex of the cone) of the frustum for minimum resistance. Newton simply states (without a hint of an explanation) that the solution is obtained by the following construction.

Let Q be the mid-point of OD. Join QC; and extend OQ to S such that

$$QS = QC. \tag{1}$$

'Then S will be the vertex of the cone whose frustum is sought.' From the illustration, it is clear that the semiangle, θ, of the cone is given by

$$\cot 2\theta = \frac{OD}{2OC}. \tag{2}$$

Though Newton does not provide in the Scholium any clue to the construction he describes, he did provide sufficient information to David Gregory in his letter to him, which, as we have stated in §142, is included as an appendix in Motte's 1729 English translation of the *Principia*. However, the problem is so simple (as it must have been to Newton!) that one can readily construct the proof.

Proof:

For the frustum depicted, the net resistance is clearly

$$\left.\begin{aligned}\mathbb{R} &= (Cv^2)\pi[(CO^2 - FD^2)\sin^2\theta + FD^2]\\ &= (Cv^2)\pi(CO^2\sin^2\theta + FD^2\cos^2\theta)\\ &= (Cv^2)\pi CO^2(\sin^2\theta + y^2\cos^2\theta),\end{aligned}\right\} \tag{3}$$

where

$$y = \frac{FD}{CO}. \tag{4}$$

Now letting

$$y = 1 - x\tan\theta, \tag{5}$$

we find

$$\mathbb{R} = (Cv^2)\pi CO^2(1 + x^2\sin^2\theta - x\sin 2\theta). \tag{6}$$

The minimum of $\mathbb{R}$ clearly occurs when

$$\cot 2\theta = \tfrac{1}{2}x = \frac{OD}{2OC}, \tag{7}$$

in agreement with Newton's result (2).

An important corollary of equation (7) is that in the limit of an infinitely thin frustum,

$$\theta \to \pi/4 \quad (x \to 0). \tag{8}$$

145. The Scholium, II: the solid of least resistance

Before presenting Newton's solution as reconstructed from his letters to David Gregory, it is useful to present the solution as one might today (though at one point we shall have to appeal to a result of Newton's!). We shall then be in a better position to appreciate that Newton's manner of solving the problem is not really different from ours.

Let $CNGT$ represent the curve which, by rotation about its axis OA, generates the solid of revolution of least resistance. Let nN, the tangent to the curve at N, make an angle θ with oN parallel to OA, the axis. 'The same things supposed', the resistance $\mathbb{R}$ experienced by the solid is given by

$$\mathbb{R} = (Cv^2)\int(2\pi y\sin\theta)(\sin^2\theta)\,\mathrm{d}s, \tag{1}$$

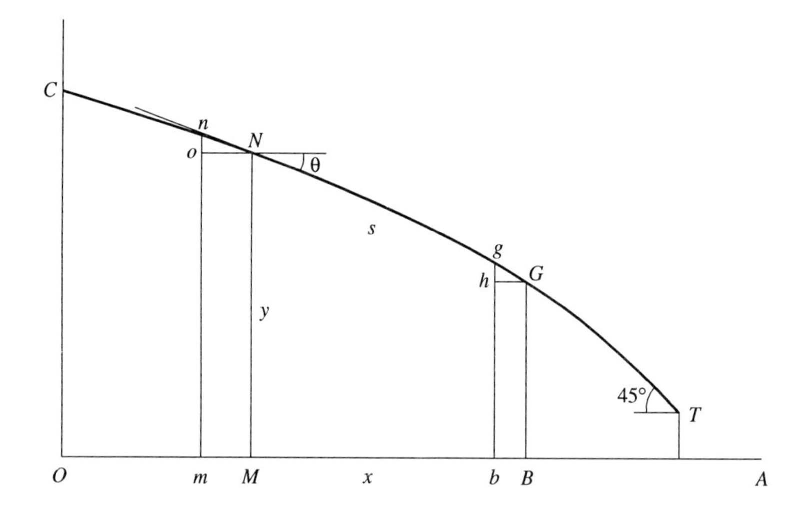

where s is the arc-length measured along the curve. We shall assume, consistently with our premises, that the coordinates, x and y, of any point along the curve are functions of an independent variable t, and let

$$\dot{x} = \frac{\mathrm{d}x}{\mathrm{d}t}, \qquad \dot{y} = \frac{\mathrm{d}y}{\mathrm{d}t} \qquad \text{so that} \qquad \dot{s} = \frac{\mathrm{d}s}{\mathrm{d}t} = (\dot{x}^2 + \dot{y}^2)^{1/2}, \tag{2}$$

and

$$\sin\theta = \frac{\dot{y}}{\sqrt{(\dot{x}^2 + \dot{y}^2)}} \qquad \text{and} \qquad \cos\theta = \frac{\dot{x}}{\sqrt{(\dot{x}^2 + \dot{y}^2)}}. \tag{3}$$

Now, $\mathbb{R}$ takes the form,

$$\mathbb{R} = 2\pi(Cv^2)\int \frac{y\dot{y}^3}{\dot{x}^2 + \dot{y}^2}\,\mathrm{d}t = 2\pi(Cv^2)\int F\,\mathrm{d}t \quad \text{(say)}, \tag{4}$$

where

$$F = \frac{y\dot{y}^3}{(\dot{x}^2 + \dot{y}^2)} = F(\dot{x}, y, \dot{y}). \tag{5}$$

To obtain a solution to the problem, the integral,

$$\int F\,\mathrm{d}t, \tag{6}$$

must be made a minimum.

As we now know, a necessary (but by no means a sufficient*) condition for the integral (6) to be an extremal is that F must satisfy the Euler–Lagrange equations:

$$\frac{\partial F}{\partial x} = \frac{\mathrm{d}}{\mathrm{d}s}\left(\frac{\partial F}{\partial \dot{x}}\right); \tag{7}$$

* Some of these other conditions are considered in §146 below.

and

$$\frac{\partial F}{\partial y} = \frac{\mathrm{d}}{\mathrm{d}s}\left(\frac{\partial F}{\partial \dot{y}}\right). \tag{8}$$

The special feature of the problem on hand is

$$F \text{ is } \textit{independent} \text{ of } x. \tag{9}$$

By virtue of this independence of F on x, it follows from equation (7) that

$$\frac{\partial F}{\partial \dot{x}} = \text{an arbitrary constant} = -K \quad \text{(say);} \tag{10}$$

or, by equation (5),

$$\frac{y\dot{y}^3\dot{x}}{(\dot{x}^2 + \dot{y}^2)^2} = K. \tag{11}$$

Substituting for $\dot{x}$ and $\dot{y}$ from equations (3), and letting

$$q = \cot\theta \tag{12}$$

we find that equation (11) becomes

$$K = y\,\frac{q}{(1 + q^2)^2}, \tag{13}$$

or, alternatively,

$$y = K\left(q^3 + 2q + \frac{1}{q}\right). \tag{14}$$

The integration can be completed by noting

$$\frac{\mathrm{d}x}{\mathrm{d}q} = \frac{\mathrm{d}x}{\mathrm{d}y}\frac{\mathrm{d}y}{\mathrm{d}q} = \cot\theta\,\frac{\mathrm{d}y}{\mathrm{d}q} = q\,\frac{\mathrm{d}y}{\mathrm{d}q}, \tag{15}$$

or

$$\frac{\mathrm{d}x}{\mathrm{d}q} = qK\left(3q^2 + 2 - \frac{1}{q^2}\right) = K\left(3q^3 + 2q - \frac{1}{q}\right). \tag{16}$$

Therefore

$$x = K(\tfrac{3}{4}q^4 + q^2 - \log q) + \text{constant}. \tag{17}$$

To determine the constant of integration in equation (17), we must make use of a result of Newton's, in his solution of the frustum of minimum resistance, derived expressly for its application in this context. The solid of revolution of least resistance, we are presently considering, must terminate where the infinitesimal terminal frustum is one of minimum resistance; for, otherwise, by a slight distortion of the bounding edge, we can reduce the net resistance of the solid. Therefore, by equation (8) of the last section, at the bounding edge

$$\theta = \pi/4 \qquad \text{and} \qquad q = 1. \tag{18}$$

If

$$x = x_1 \qquad \text{and} \qquad y = y_1 \qquad \text{at the bounding edge,} \tag{19}$$

then, by equations (14) and (15),

$$y_1 = 4K \qquad \text{and} \qquad x_1 = \tfrac{7}{4}K; \tag{20}$$

and the solid of revolution of least resistance is parametrically described by

$$\left.\begin{aligned} x_1 - x &= K(\tfrac{7}{4} - \tfrac{3}{4}q^4 - q^2 + \log q) \\ \text{and} \qquad y_1 - y &= K\left(4 - q^3 - 2q - \frac{1}{q}\right). \end{aligned}\right\} \tag{21}$$

From equations (15) and (16) it follows that

$$\text{both} \quad \frac{dx}{dq} \quad \text{and} \quad \frac{dy}{dq} \quad \text{vanish} \tag{22}$$

when

$$3q^4 + 2q^2 - 1 = 0 \qquad \text{or} \qquad q^2 = \tfrac{1}{3}. \tag{23}$$

The solution curve (21) has therefore a *cusp* at

$$\cot\theta = 1/\sqrt{3} \qquad \text{or} \qquad \theta = 60°. \tag{24}$$

The behaviour of the derived solution is illustrated below.

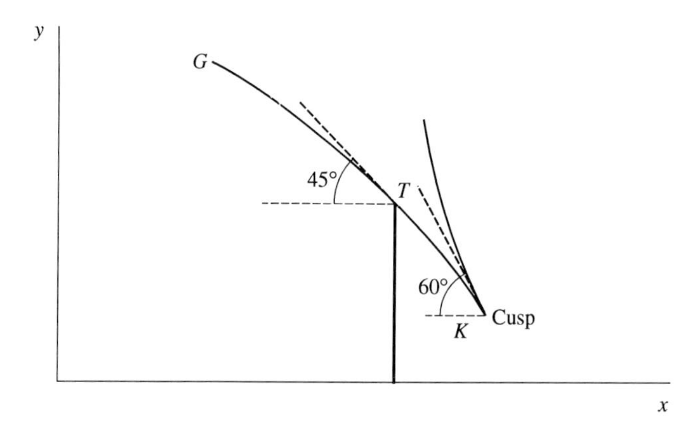

Newton's manner of solution

Newton readily enough establishes that the solution to the problem requires the minimization of the integral (cf. equations (5) and (6)),

$$\int F\,ds, \tag{25}$$

where

$$F = \frac{y\dot{y}^3}{\dot{x}^2 + \dot{y}^2} \tag{26}$$

or, as Newton writes in his flux notation,

$$F = \frac{y(\mathrm{d}y)^3}{(\mathrm{d}x)^2 + (\mathrm{d}y)^2} \tag{27}$$

with the obvious meanings

$$\mathrm{d}x = AD = Ee \qquad \text{and} \qquad \mathrm{d}y = AB \tag{28}$$

in the adjoining diagram. With the further definitions,

$$\alpha d = eH = \delta x, \qquad \alpha D = aA = \delta y \qquad \text{and} \qquad CD = e'e = \varepsilon, \tag{29}$$

Newton's procedure (as explained in his letter to Gregory*) is to compare the contributions to the integrand by the slice $BDdHEB$ and the slice $BCdHEB$ from the slightly displaced curve BCd obtained by joining B and d by some curve.

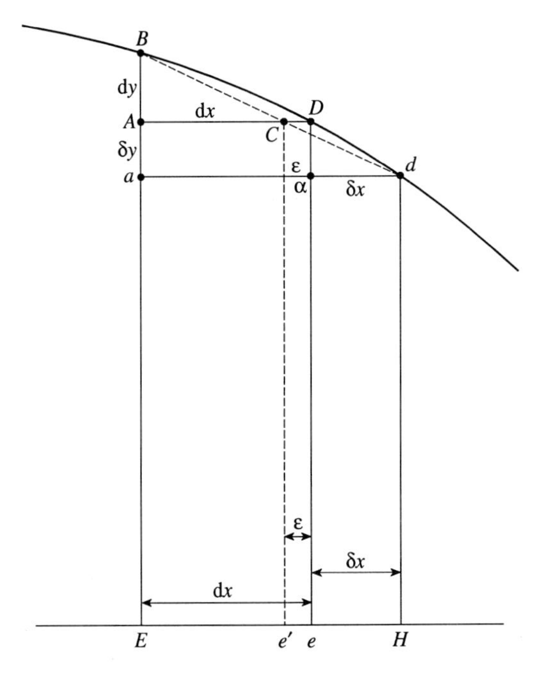

The contributions to the integrand derived from the arc BDd (as the sum of the trapezia $DeEB$ and $dHeD$) and from the displaced arc BCd (as the sum of the trapezia $Ce'EB$ and $dHe'C$) are respectively,

$$(\mathrm{d}y)^3\left[\frac{y}{(\mathrm{d}x)^2 + (\mathrm{d}y)^2} + \frac{y - \delta y}{(\mathrm{d}x + \delta x)^2 + (\mathrm{d}y)^2}\right] \tag{30}$$

* Transcribed on p. 658 in Cajori's edition of the *Principia*.

and

$$(\mathrm{d}y)^3\left[\frac{y}{(\mathrm{d}x-\varepsilon)^2+(\mathrm{d}y)^2}+\frac{y-\delta y}{(\mathrm{d}x+\delta x+\varepsilon)^2+(\mathrm{d}y)^2}\right]. \tag{31}$$

Newton correctly concludes that the condition for a local minimum of the integral (25) is that the two foregoing expressions must be equal, that is,

$$y(\mathrm{d}y)^3\left[\frac{1}{(\mathrm{d}x)^3+(\mathrm{d}y)^2}-\frac{1}{(\mathrm{d}x-\varepsilon)^2+(\mathrm{d}y)^2}\right]$$
$$=(y-\delta y)(\mathrm{d}y)^3\left[\frac{1}{(\mathrm{d}x+\delta x+\varepsilon)^2+(\mathrm{d}y)^2}-\frac{1}{(\mathrm{d}x+\delta x)^2+(\mathrm{d}y)^2}\right]. \tag{32}$$

Now passing to the limit $\varepsilon = 0$ we obtain

$$y(\mathrm{d}y)^3\frac{\partial}{\partial(\mathrm{d}x)}\frac{1}{(\mathrm{d}x)^2+(\mathrm{d}y)^2}=(y-\delta y)(\mathrm{d}y)^3\frac{\partial}{\partial(\mathrm{d}x)}\frac{1}{(\mathrm{d}x+\delta x)^2+(\mathrm{d}y)^2}. \tag{33}$$

This equality must obtain for arbitrary δx and δy. Therefore,

$$y(\mathrm{d}y)^3\frac{\partial}{\partial(\mathrm{d}x)}\frac{1}{(\mathrm{d}x)^2+(\mathrm{d}y)^2}=\text{constant}, \tag{34}$$

or, in modern terminology,

$$y\dot{y}^3\frac{\partial}{\partial\dot{x}}\frac{1}{\dot{x}^2+\dot{y}^2}=\text{constant}, \tag{35}$$

that is,

$$\frac{y\dot{y}^3\dot{x}}{(\dot{x}^2+\dot{y}^2)^2}=\text{constant}. \tag{36}$$

Voilà!

Setting, as before,

$$q=\dot{x}/\dot{y}=\cot\theta, \tag{37}$$

we recover equation (13):

$$y\frac{q}{(1+q^2)^2}=y_1\frac{q_1}{(1+q_1^2)^2}\quad(=\text{constant}=K), \tag{38}$$

where q_1 is the value of q at some fixed point y_1 on the extremal curve.

Returning to the diagram on p. 560 we can write the left-hand side of equation (38), alternatively in the form

$$y\cos\theta\sin^3\theta, \tag{39}$$

where

$$y=MN,\qquad \cos\theta=mM/nN,\qquad\text{and}\qquad \sin\theta=on/nN. \tag{40}$$

Substituting the equivalents (40) in (39), we have

$$MN \frac{mM}{(nN)^4} (on)^3. \tag{41}$$

For the right-hand side of equation (38), we have the analogous expression

$$BG \frac{bB}{(gG)^4} (hg)^3. \tag{42}$$

Therefore, equation (38) translates into

$$MN \frac{mM}{(nN)^2} (on)^3 = BG \frac{bB}{(gG)^4} (hg)^3. \tag{43}$$

Without loss of generality, we can assume (as does Newton, as we shall see presently)

$$on = hg, \tag{44}$$

when equation (43) becomes

$$MN \frac{mM}{(nN)^4} = BG \frac{bB}{(gG)^4}. \tag{45}$$

Newton writes this equation in the form

$$\overline{Gg}^4 : \overline{Nn}^4 :: BG \times bB : MN \times mM, \tag{46}$$

as quoted by Motte in his 1729 translation of the *Principia* (p. 321).

Returning to equation (38) (or, more strictly, equation (45)) Newton determines the constant of integration by terminating the solid of revolution at $q_1 = 1$ in accordance with the result (equation (8), §144) he had established earlier in his treatment of the frustum of minimum resistance. Accordingly, the required solution is

$$y \cos\theta \sin^3\theta = \tfrac{1}{4} y_1. \tag{47}$$

In the adjoining figure (which is substantially the same as that on p. 560)

$$y_1 = BG \tag{48}$$

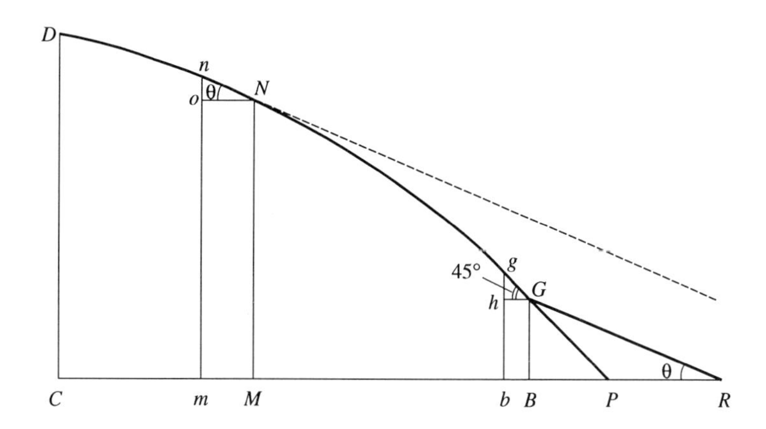

and the line GR is drawn parallel to the tangent nN at N. Manifestly,

$$\cos\theta = \frac{BR}{GR} \qquad \text{and} \qquad \sin\theta = \frac{BG}{GR}. \tag{49}$$

With these substitutions, equation (47) becomes

$$MN = \frac{(GR)^4}{4BR(BG)^2}. \tag{50}$$

This is the result which Newton states in the Scholium.

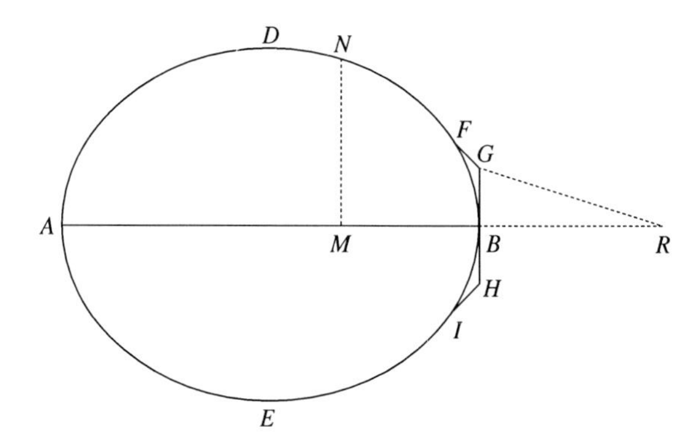

> If the figure $DNFG$ be such a curve, that if, from any point thereof, as N, the perpendicular NM be let fall on the axis AB, and from the given point G there be drawn the right line GR parallel to a right line touching the figure in N, and cutting the axis produced in R, MN becomes to GR as GR^3 to $4BR.GB^2$, the solid described by the revolution of this figure about its axis AB, moving in the before-mentioned rare medium from A towards B, will be less resisted than any other circular solid whatsoever, described of the same length and breadth.

This is the only concrete result (without so much as a hint as to its origin) that Newton states in the Scholium. No surprise that in Leibniz's personal copy of the *Principia* we find the marginal comment

investigandum est isoclinis [*isolaba*?] *facillimè progrediens.*

It is clear that the equation

$$y\frac{q}{(1+q^2)^2} = \tfrac{1}{4}y_1 \tag{51}$$

is the same as the solution given in equation (21). Newton completes the integration of this equation to obtain the solution for x (in the papers rescued by J. C. Adams) in essentially the same manner as we have.

As we have stated earlier, Newton's solution to the problem of a solid of revolution of least resistance, anticipating a special case of the Euler–Lagrange equations, is the most sophisticated mathematical problem treated in the *Principia.* And yet, Section VII of Book II (which includes the Scholium to Proposition XXXIV) has been brusquely brushed aside by an eminent historian with the comment:

> Not much of anything in Section VII is satisfactory (!)

146. Amplifications

While Newton's solution to the problem of the solid of revolution of least resistance described in §145 is satisfactory so far as it goes, a modern treatment will require further criteria to be satisfied before one can be certain that a solution of the Euler–Lagrange equations (as Newton's is) provides a true extremal or only a local extremal. A modern treatment of the problem has been given by A. R. Forsyth in an investigation to which we have already referred in the introductory Section §142. Forsyth considers a large class of problems of which Newton's is a special case. In this section, we shall restrict ourselves to Newton's problem.

I. Newton's kernel-function,

$$F = \frac{y\dot{y}^3}{\dot{x}^2 + \dot{y}^2} \tag{1}$$

satisfies the identity,

$$F = \dot{x}\frac{\partial F}{\partial \dot{x}} + \dot{y}\frac{\partial F}{\partial \dot{y}}. \tag{2}$$

Proof:

$$\frac{\partial F}{\partial \dot{x}} = -2\frac{y\dot{y}^3\dot{x}}{(\dot{x}^2 + \dot{y}^2)^2}, \tag{3}$$

and

$$\frac{\partial F}{\partial \dot{y}} = +\frac{y}{(\dot{x}^2 + \dot{y}^2)^2}(3\dot{x}^2\dot{y}^2 + \dot{y}^4). \tag{4}$$

Therefore,

$$\dot{x}\frac{\partial F}{\partial \dot{x}} + \dot{y}\frac{\partial F}{\partial \dot{y}} = \frac{y\dot{y}^3}{(\dot{x}^2 + \dot{y}^2)} = F. \tag{5}$$

Q.E.D.

II. The Euler–Lagrange equations

$$\frac{\partial F}{\partial x} = \frac{\mathrm{d}}{\mathrm{d}t}\left(\frac{\partial F}{\partial \dot{x}}\right); \qquad \frac{\partial F}{\partial y} = \frac{\mathrm{d}}{\mathrm{d}t}\left(\frac{\partial F}{\partial \dot{y}}\right), \tag{6}$$

are equivalent to a single equation by virtue of the identity (2).

Proof:

$$0 = \dot{x}\left[\frac{\partial F}{\partial x} - \frac{\mathrm{d}}{\mathrm{d}t}\left(\frac{\partial F}{\partial \dot{x}}\right)\right] = \dot{x}\,\frac{\partial F}{\partial x} - \frac{\mathrm{d}}{\mathrm{d}t}\left(\dot{x}\,\frac{\partial F}{\partial \dot{x}}\right) + \ddot{x}\,\frac{\partial F}{\partial \dot{x}}$$
$$= \dot{x}\,\frac{\partial F}{\partial x} - \frac{\mathrm{d}}{\mathrm{d}t}\left(F - \dot{y}\,\frac{\partial F}{\partial \dot{y}}\right) + \ddot{x}\,\frac{\partial F}{\partial \dot{x}}$$
$$= \dot{x}\,\frac{\partial F}{\partial x} + \ddot{y}\,\frac{\partial F}{\partial \dot{y}} + \ddot{x}\,\frac{\partial F}{\partial \dot{x}} - \frac{\mathrm{d}F}{\mathrm{d}t} + \dot{y}\,\frac{\mathrm{d}}{\mathrm{d}t}\left(\frac{\partial F}{\partial \dot{y}}\right)$$
$$= \dot{x}\,\frac{\partial F}{\partial x} + \ddot{y}\,\frac{\partial F}{\partial \dot{y}} + \ddot{x}\,\frac{\partial F}{\partial \dot{x}} - \left(\dot{x}\,\frac{\partial F}{\partial x} + \dot{y}\,\frac{\partial F}{\partial y} + \ddot{x}\,\frac{\partial F}{\partial \dot{x}} + \ddot{y}\,\frac{\partial F}{\partial \dot{y}}\right) + \dot{y}\,\frac{\mathrm{d}}{\mathrm{d}t}\left(\frac{\partial F}{\partial \dot{y}}\right)$$
$$= \dot{y}\left[\frac{\mathrm{d}}{\mathrm{d}t}\left(\frac{\partial F}{\partial \dot{y}}\right) - \frac{\partial F}{\partial y}\right]. \tag{7}$$

Q.E.D.

III. The Legendre test: We first verify that as a consequence of the identity (2),

$$\frac{\partial F}{\partial \dot{x}} = \frac{\partial F}{\partial \dot{x}} + \dot{x}\,\frac{\partial^2 F}{\partial \dot{x}^2} + \dot{y}\,\frac{\partial^2 F}{\partial \dot{x}\,\partial \dot{y}}, \tag{8}$$

or

$$\frac{\dot{y}}{\dot{x}}\,\frac{\partial^2 F}{\partial \dot{x}\,\partial \dot{y}} = -\frac{\partial^2 F}{\partial \dot{x}^2}. \tag{9}$$

Similarly,

$$\frac{\dot{x}}{\dot{y}}\,\frac{\partial^2 F}{\partial \dot{x}\,\partial \dot{y}} = -\frac{\partial^2 F}{\partial \dot{y}^2}. \tag{10}$$

Therefore,

$$\frac{1}{\dot{y}^2}\,\frac{\partial^2 F}{\partial \dot{x}^2} = -\frac{1}{\dot{x}\dot{y}}\,\frac{\partial^2 F}{\partial \dot{x}\,\partial \dot{y}} = \frac{1}{\dot{x}^2}\,\frac{\partial^2 F}{\partial \dot{y}^2} = P \quad \text{(say)}. \tag{11}$$

On evaluating P we find:

$$P = -\frac{1}{\dot{x}\dot{y}}\,\frac{\partial}{\partial \dot{y}}\left(-2\,\frac{y\dot{y}^3\dot{x}}{(\dot{x}^2+\dot{y}^2)^2}\right) = \frac{2y\dot{y}}{(\dot{x}^2+\dot{y}^2)^3}\,(3\dot{x}^2 - \dot{y}^2), \tag{12}$$

or, letting (as in §145, equation (3))

$$\sin\theta = \frac{\dot{y}}{(\dot{x}^2+\dot{y}^2)^{1/2}} \quad \text{and} \quad \cos\theta = \frac{\dot{x}}{(\dot{x}^2+\dot{y}^2)^{1/2}}, \tag{13}$$

we have the alternative form,

$$P = 2\,\frac{y}{(\dot{x}^2+\dot{y}^2)^{3/2}}\,\sin\theta\,(3\cos^2\theta - \sin^2\theta). \tag{14}$$

Legendre's test requires that the quantity P shall not change its sign anywhere along the range of the curve, and that the sign must be positive for a minimum.

We observe that Newton's solution satisfies the test so long as

$$\cot \theta > 1/\sqrt{3}; \tag{15}$$

that is, if the solution curve terminates before the cusp that occurs at $\theta = 60°$ (cf. §145, equation (24)). Actually, as we have seen, Newton terminates the solution at $\theta = 45°$.

IV. The Weierstrass test: We shall formulate the Weierstrass test for the special case of Newton's problem.

For Newton's kernel-function,

$$F(\dot{x}, \dot{y}) = \frac{y\dot{y}^3}{\dot{x}^2 + \dot{y}^2}, \tag{16}$$

the critical Weierstrass function, W, is

$$W = F(\lambda, \mu) - \lambda \frac{\partial F}{\partial \dot{x}} - \mu \frac{\partial F}{\partial \dot{y}}, \tag{17}$$

where λ and μ are the direction cosines of any direction through any chosen point (x, y) along the solution curve so that $\lambda^2 + \mu^2 = 1$. Making use of the derivatives of F given in equations (3) and (4), we find:

$$W = y\left\{\mu^3 + \frac{2\lambda}{(\dot{x}^2 + \dot{y}^2)^2}\,\dot{y}^3\dot{x} - \frac{\mu}{(\dot{x}^2 + \dot{y}^2)^2}(3\dot{x}^2\dot{y}^2 + \dot{y}^4)\right\}, \tag{18}$$

or, writing $\dot{x}$ and $\dot{y}$ in terms of θ as in equations (13), we have

$$W = y\{\mu^3 + 2\lambda \sin^3\theta \cos\theta - \mu(\sin^2\theta + 2\sin^2\theta\cos^2\theta)\}. \tag{19}$$

The Weierstrass test requires that for the existence of an extremal, the function W maintains the same sign at all points on the solution curve for all directions (λ, μ) which do not coincide with the tangent at that point.

Let χ denote the angle between the arbitrary direction (λ, μ) and the tangent to the curve at that point, so that

$$\lambda = \cos(\theta + \chi) \qquad \text{and} \qquad \mu = \sin(\theta + \chi). \tag{20}$$

The function W now takes the form

$$W = y\{\sin^3(\theta + \chi) + 2\cos(\theta + \chi)\sin^3\theta\cos\theta - \sin(\theta + \chi)(\sin^2\theta + 2\sin^2\theta\cos^2\theta)\}; \tag{21}$$

or, as we may verify.*

$$W = y\sin^2\chi\sin(3\theta + \chi). \tag{22}$$

* Perhaps, a few steps of this verification may be noted:

$$\begin{aligned}
&\sin^3(\theta + \chi) - \sin(\theta + \chi)\sin^2\theta - 2\sin^2\theta\cos\theta\,[\sin(\theta + \chi)\cos\theta - \cos(\theta + \chi)\sin\theta] \\
&= \sin^3(\theta + \chi) - \sin(\theta + \chi)\sin^2\theta - 2\sin^2\theta\cos\theta\sin\chi \\
&- \sin^3\theta\cos^3\chi + \cos^3\theta\sin^3\chi + 3\sin^2\theta\cos\theta\sin\chi\cos^2\chi \\
&\quad - \sin^3\theta\cos\chi - 3\sin^2\theta\cos\theta\sin\chi + 3\sin\theta\cos^2\theta\sin^2\chi\cos\chi \\
&= \sin^2\chi[-\sin^3\theta\cos\chi + \cos^3\theta\sin\chi - 3\sin^2\theta\cos\theta\sin\chi + 3\sin\theta\cos^2\theta\cos\chi] \\
&= \sin^2\chi\,[-4\sin^3\theta\cos\chi + 3\sin\theta\cos\chi + 4\cos^3\theta\sin\chi - 3\cos\theta\sin\chi] \\
&= \sin^2\chi\sin(3\theta + \chi).
\end{aligned}$$

Manifestly, W changes sign as χ ranges from 0 to 2π. Hence *Newton's problem admits no solution.* What this conclusion means is that it must be possible to construct solids of revolution, S_{mm} (say), which offer less resistance than Newton's solid, S_m. The meridional curves of S_{mm} would not, however, have continuously varying tangents, as is the case with S_m, but would consist of a series of arcs which change directions discontinuously at their points of intersection. If we restrict ourselves to curves with continuously varying tangents, then Newton's solid provides the unique solution.

❖27❖

The problem of the brachistochrone

I do not like to be dunned and teezed by foreigners about Mathematical things.

I. Newton to J. Flamsteed
(in a letter dated 16 January, 1699

147. Introduction

The problem of the brachistochrone is the following: let A and B be given points, A higher than B; and suppose they are joined by a smooth wire in the vertical plane through AB. A particle starts from rest at A and slides down the wire to B under the action of gravity. The problem is to find the curve which minimizes the time of descent.

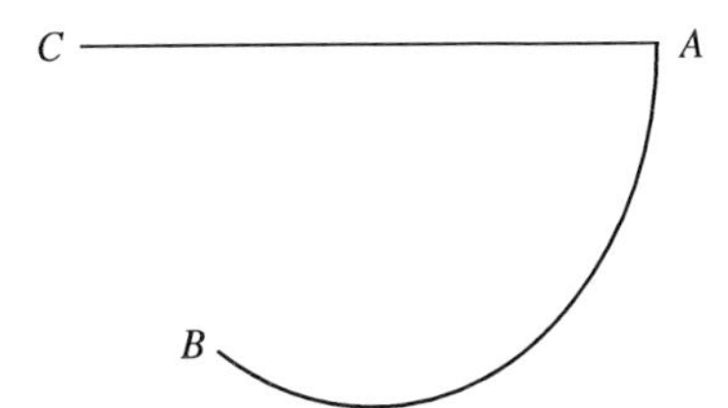

As we have stated already (p. 538), this problem* was issued by Johann Bernoulli on New Year's Day 1697, as a challenge to 'the sharpest mathematicians in the whole world'. Bernoulli's principal object(?) in issuing the challenge was to demonstrate to the 'whole world' the superiority of the German–Swiss mathematicians. Having had no response from Holland and France(?) Bernoulli had it forwarded to Charles Montagu, then the President of the Royal Society.† The rest of the story is best told as recalled some thirty years later by Catherine Barton (Conduit), Newton's niece.

> On January 29, 1697, 'in the midst of the hurry of the great recoinage' [Newton] did not come home [in Jermyn street] till 4 [in the afternoon] from the Tower

* Actually one of two.
† See the translated extract from a letter of Bernoulli to Basange de Beauval on p. 573.

> there to find awaiting him a printed paper [a folio half-sheet] bearing a twin mathematical challenge from a young Groningen professor, Johann Bernoulli, addressed generally 'to the sharpest mathematicians flourishing throughout the world'. Newton did not hestitate straightaway to attack their problem, indeed, (to continue Catherine's story) he 'did not sleep till he had solved it which was by 4 in the morning'.

[It may be noted parenthetically that for the Newton of 1686, who had solved effortlessly the far more difficult problem of the solid of revolution of least resistance (see Chapter 26), the present problem should have been 'child's play' (to quote Littlewood once again). Apparently, ten years later and with pressure of recoinage as the Master of the Mint, it took Newton most of the night till '4 in the morning'. Clearly, even then he was not 'merely a chip off the old block; but the old block itself'.]

Newton's response was sent to Montagu in a letter dated 30 January 1697 (the day after he had received the communication from Montagu). It was read out to the assembled Royal Society on 24 February 1697. It was subsequently published, anonymously, in the *Philosophical Transactions* for January 1696/7 (Vol. 17, No. 224).

The solution as Newton responded in his highest imperial style (of the later propositions of Book III) is given below:

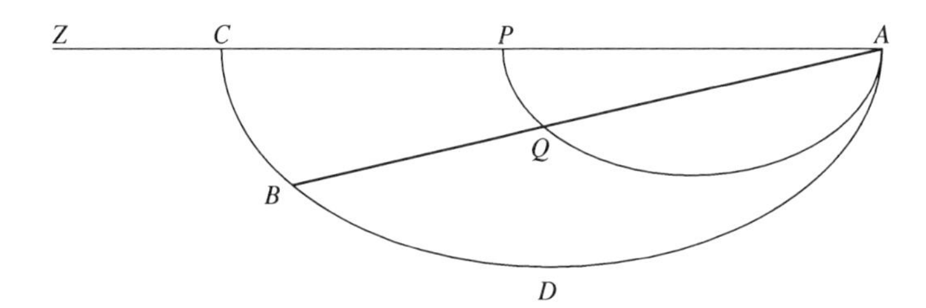

> From the given point A draw the unbounded straight line $APCZ$ parallel to the horizontal and upon this same line describe both any cycloid AQP whatever, meeting the straight line AB (drawn and, if need be, extended) in the point Q, and then another cycloid ADC whose base and height shall be to the previous one's base and height [as $AC:AP$] respectively as AB to AQ. This most recent cycloid will then pass through the point B and be the curve in which a heavy body shall, under the force of its own weight, most swiftly reach the point B from the point A. As was to be found.

On seeing this solution to his problem given as instructions to an inquisitive child, Bernoulli could immediately recognize the author as Newton. As Bernoulli wrote later in a letter to Basange de Beauval,*

* I am indebted to Professor Raghavan Narasimhan for this translation from the French original.

Thus it is, Dear Sir, that my problem remained unsolved after it had been examined by several people in Holland. It was then sent to England where I had great hopes that it would have a happier end, since in England there are some excellent geometers skilled in using our methods or similar methods. In fact, the January issue of the *Philosophical Transactions*, which you were kind enough to send me, shows that I was not mistaken, since it includes a construction of the curve of quickest descent that deals with the problem perfectly. Although it's author, in excessive modesty, does not reveal his name, we can be certain beyond any doubt that the author is the celebrated Mr. Newton: for, even if we had no information other than this sample, we should have recognized him by his style even as the lion by its paw . . . It is only to be wished that Mr. Newton had published the solution and the method by which he found the desired curve. . . .

148. Newton's anonymous solution

It is somewhat surprising that Bernoulli should have complained that Newton had not disclosed his method of solution. For it should have been obvious to anyone familiar with Newton's Proposition L of Book I,* that 'to cause a pendulous body to oscillate in a given cycloid' is virtually the 'cause' for a particle to slide down a cycloid, under the action of gravity consistently with the dynamical requirements. Newton must have instantly recognized this fact.†

In the adjoining diagram, the cycloid ACB is traced by a point P, on the circumference of the circle IPJ, of radius $OP = a$, rolling uniformly along the horizontal line, $AIBX$. Let $t = \angle MOP$ in radians. Then,

$$AI = \text{arc } PI = at. \tag{1}$$

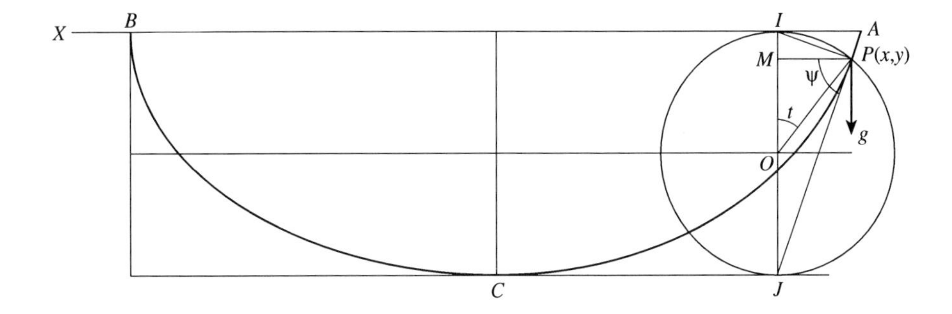

* Not, however, included in this book.

† Leibniz, apparently, did not realize this connection with Proposition L of Book I when he 'incautiously remarked, that only those who were "masters of the calculus" could succeed with the problem'. Not that calculus is not needed to 'succeed with the problem' entirely, but only that the connection of the problem to the cycloid should be obvious to anyone endowed with sufficient insight.

The rectangular coordinates (x, y) of P, as one can directly read out, are:

$$\left.\begin{aligned} x &= at - a\sin t = a(t - \sin t) \\ \text{and} \qquad y &= a - a\cos t = a(1 - \cos t). \end{aligned}\right\} \tag{2}$$

An immediate consequence of this parametric representation of the cycloid is that *all cycloids are similar*. Newton's construction of the solution is based on this elementary fact. But it remains to show that a particle will slide down the cycloid under the action of gravity 'without the services of an archangel'. (We are not, at present, addressing ourselves to the brachistochrone property of the cycloid.)

The instantaneous centre of the rolling circle is at I. Therefore, P is moving at right angles to PI, that is, towards J, the opposite end of the diameter through I. Thus, PJ is the tangent to the curve at P and PI is the normal. The component of the acceleration due to gravity, acting on P, in the direction of its motion is

$$g \sin \angle MPJ = g \sin \psi, \tag{3}$$

while the component along the normal PI is supported by the wire.

Q.E.D.

Further consequences of equation (3) may be noted. Since

$$g \sin \psi = g\,\frac{PJ}{IJ} = \frac{g}{2a}\,PJ = \frac{g}{4a} \times \text{arc } CP, \tag{4}$$

the motion of P along the arc ACB will be simple harmonic, with period $2\pi\sqrt{(4a/g)}$. The period is thus independent of the starting point.

If, instead of sliding down the cycloid, P is attached by a string $A'C$ to A', at the cusp where the two equal branches, $A'A$ and $A'B$, of the cycloid join, and constrained to move between the cycloidal arcs $A'A$ and $A'B$, the same motion, that is described in the previous paragraph, will result. This is the cycloidal pendulum of Huygens. It has the isochronous property of the period of the pendulum being independent of amplitude.

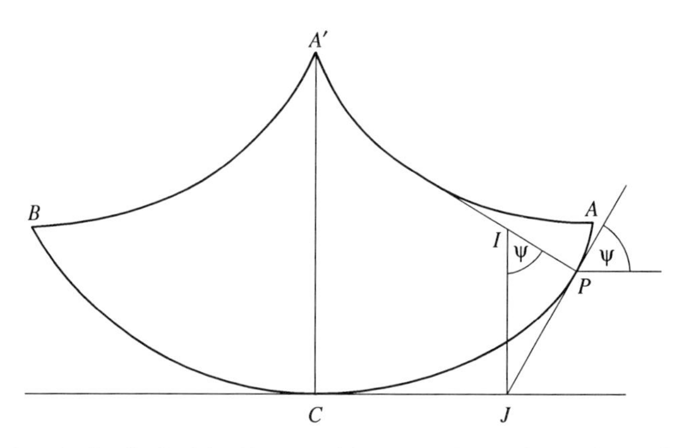

All of the foregoing is included in Propositions L–LII of Book I of the *Principia*.

149. The solution derived from its extremal property

While it is highly probable that Newton first 'saw in his mind's eye' the solution to Bernoulli's challenge in the manner we have described in §148, he must have realized, at the same time, that the solution does not address itself to its brachistochrone property; and that for that end he must apply the same method by which he had obtained earlier the solution to his problem of the solid of revolution of least resistance. Newton described his application of this method to the present problem in letters to Fatio de Duillier and David Gregory, now available. But first, we shall derive the solution as one might today.

Consider a smooth curve joining A and B, A higher than B, down which a particle, starting from rest at A, slides under the influence of gravity acting in the vertical y-direction. The problem is to find the curve down which the particle will slide to B in the shortest time. Assuming that the x- and the y-coordinates of the particle are smooth functions of the arc-length s, measured along the curve, let

$$x = x(s), \qquad y = y(s), \qquad \text{and} \qquad \mathrm{d}s = \sqrt{(\mathrm{d}x^2 + \mathrm{d}y^2)}. \tag{1}$$

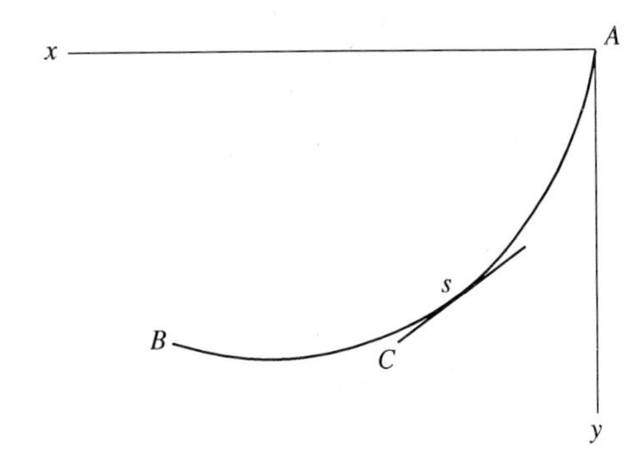

We further define,

$$\dot{x} = \frac{\mathrm{d}x}{\mathrm{d}t} \qquad \text{and} \qquad \dot{y} = \frac{\mathrm{d}y}{\mathrm{d}t}, \qquad \text{so that} \quad \mathrm{d}s = \mathrm{d}t\sqrt{(\dot{x}^2 + \dot{y}^2)}. \tag{2}$$

In Proposition XL of Book I, it has been shown that the velocity acquired by a particle descending a height y from rest is $\sqrt{(2gy)}$ and independent of any lateral displacement. Therefore, the time required for descent, from A to B, is

$$t = \int_A^B \frac{\mathrm{d}s}{\sqrt{(2gy)}}. \tag{3}$$

On the other hand, by the definitions (1) and (2)

$$\mathrm{d}s = \mathrm{d}t\sqrt{(\dot{x}^2 + \dot{y}^2)} = \dot{y}\,\mathrm{d}t\sqrt{(1 + \dot{x}^2/\dot{y}^2)} = \mathrm{d}y\sqrt{[1 + (\mathrm{d}x/\mathrm{d}y)^2]}. \tag{4}$$

We can therefore write,

$$t = \int_A^B \left(\frac{1 + x'^2}{2gy}\right)^{1/2} \mathrm{d}y, \tag{5}$$

where

$$x' = \frac{\mathrm{d}x}{\mathrm{d}y}. \tag{6}$$

Consider now the net change, δt, in the time of descent that results from an arbitrary displacement $\delta x(y)$ at each point $x(y)$ except at the end points, A and B, where it vanishes. Since,

$$\delta x' = \delta \frac{\mathrm{d}x}{\mathrm{d}y} = \frac{\mathrm{d}}{\mathrm{d}y}\,\delta x = (\delta x)', \tag{7}$$

we obtain from equation (5)

$$\delta t = \int_A^B \frac{x'\delta x'}{[2gy(1+x'^2)]^{1/2}}\,\mathrm{d}y = \int_A^B \frac{x'}{[2gy(1+x'^2)]^{1/2}}\frac{\mathrm{d}}{\mathrm{d}y}(\delta x)\,\mathrm{d}y. \tag{8}$$

By an integration by parts, we obtain

$$\delta t = \frac{x'\delta x}{[2gy(1+x'^2)]^{1/2}}\bigg|_B^A - \int_B^A \delta x \left\{\frac{\mathrm{d}}{\mathrm{d}y}\frac{x'}{[2gy(1+x'^2)]^{1/2}}\right\}\mathrm{d}y. \tag{9}$$

The integrated part vanishes since $\delta x = 0$ at A and B, and we are left with

$$\delta t = -\int_B^A \delta x \left\{\frac{\mathrm{d}}{\mathrm{d}y}\frac{x'}{[2gy(1+x'^2)]^{1/2}}\right\}\mathrm{d}y. \tag{10}$$

If the chosen curve is indeed the curve of quickest descent, then δt must vanish for all increments $\delta x(y)$. Therefore, the condition for the chosen curve to have an extremal character is

$$\frac{\mathrm{d}}{\mathrm{d}y}\frac{x'}{[2gy(1+x'^2)]^{1/2}} = O, \tag{11}$$

or,

$$x' = C[2gy(1+x'^2)]^{1/2}, \tag{12}$$

where C is a constant. Equation (12) can be reduced to the form,

$$(a-y)\left(\frac{\mathrm{d}x}{\mathrm{d}y}\right)^2 = y \qquad \text{where} \quad a = 1/(2gC^2); \tag{13}$$

or, alternatively,

$$\frac{\mathrm{d}x}{\mathrm{d}y} = \left(\frac{y}{a-y}\right)^{1/2}. \tag{14}$$

The solution of this equation is readily verified to be

$$x = \tfrac{1}{2}a(t - \sin t) \qquad \text{and} \qquad y = \tfrac{1}{2}a(1-\cos t). \tag{15}$$

Q.E.F.

Newton's manner of solution

Newton starts with the integral (3), which has to be minimized, and writes it in the form,

$$\int_0^t F(t)\,\mathrm{d}t, \tag{16}$$

where

$$F = \left(\frac{\dot{x}^2 + \dot{y}^2}{2gy}\right)^{1/2}. \tag{17}$$

In Newton's flux notation (suppressing the factor $(2g)^{-1/2}$), we write

$$F = \left[\frac{1}{y}(\mathrm{d}x^2 + \mathrm{d}y^2)\right]^{1/2}. \tag{18}$$

In the adjoining diagram,

$$eD = y, \qquad Ee = B\alpha = \mathrm{d}x \qquad \text{and} \qquad AH = D\alpha = \mathrm{d}y. \tag{19}$$

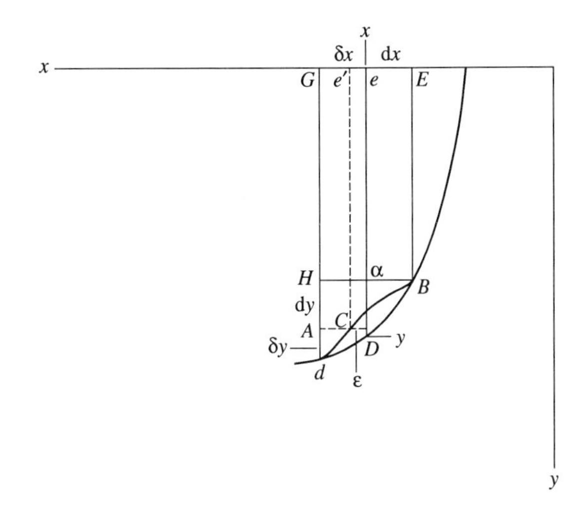

With the further definitions,

$$eG = DA = \delta x, \qquad Ad = \delta y, \qquad \text{and} \qquad CD = e'e = \varepsilon, \tag{20}$$

Newton's procedure is to compare the contributions to the integrand by the slide *BDdGE* and the slice *BCdGE* from the slightly displaced arc *BCd* joining *B* and *d*.

The contributions to the integrand from the arc *BDd* (as the sum of the trapezia *DeEB* and *dGeD*) and the arc *BCd* (as the sum of the trapezia *Ce′EB* and *dGe′C*) arc proportional, respectively, to

$$\frac{1}{\sqrt{y}}\,[(\mathrm{d}x)^2 + (\mathrm{d}y)^2]^{1/2} + \frac{1}{\sqrt{(y + \delta y)}}\,[(\mathrm{d}x + \delta x)^2 + (\mathrm{d}y)^2]^{1/2}, \tag{21}$$

and

$$\frac{1}{\sqrt{y}}[(\mathrm{d}x+\varepsilon)^2+(\mathrm{d}y)^2]^{1/2}+\frac{1}{\sqrt{(y+\delta y)}}[(\mathrm{d}x+\delta x-\varepsilon)^2+(\mathrm{d}y)^2]^{1/2}. \tag{22}$$

The condition for a local minimum of the integral of F, defined in equation (18), is

$$\frac{1}{\sqrt{y}}\{[(\mathrm{d}x+\varepsilon)^2+(\mathrm{d}y)^2]^{1/2}-[(\mathrm{d}x)^2+(\mathrm{d}y)^2]^{1/2}\}$$
$$=\frac{1}{\sqrt{(y+\delta y)}}\{[(\mathrm{d}x+\delta x)^2+(\mathrm{d}y)^2]^{1/2}-[(\mathrm{d}x+\delta x-\varepsilon)^2+(\mathrm{d}y)^2]^{1/2}\}. \tag{23}$$

Passing to the limit $\varepsilon \to 0$, we obtain,

$$\frac{1}{\sqrt{y}}\frac{\partial}{\partial(\mathrm{d}x)}[(\mathrm{d}x)^2+(\mathrm{d}y)^2]^{1/2}=\frac{1}{\sqrt{(y+\delta y)}}\frac{\partial}{\partial(\mathrm{d}x)}[(\mathrm{d}x+\delta x)^2+(\mathrm{d}y)^2]^{1/2}. \tag{24}$$

This equality must obtain for arbitrary increments δx and δy. Therefore

$$\frac{1}{\sqrt{y}}\frac{\partial}{\partial(\mathrm{d}x)}[(\mathrm{d}x)^2+(\mathrm{d}y)^2]^{1/2}=\text{constant}, \tag{25}$$

or, in current terminology,

$$\frac{1}{\sqrt{(2gy)}}\frac{\partial}{\partial\dot{x}}[(\dot{x})^2+(\dot{y})^2]^{1/2}=\text{constant}, \tag{26}$$

restoring the factor $1/\sqrt{(2g)}$. Effecting the differentiation, we recover equation (12):

$$\frac{\dot{x}}{\sqrt{\{2gy[(\dot{x})^2+(\dot{y})^2]\}}}=\text{constant}. \tag{27}$$

The solution can now be completed as before.

Q.E.D.

·28·

The velocity of sound and of long waves in canals

Newton was the first person who dealt with [this problem] ... his theory, although imperfect, is a monument [to] his genius.

The way in which he arrives at this [formula] is one of the most remarkable traits of his genius.

Marquis de Laplace

150. Introduction

One can search in vain to find an adequate account of Newton's investigations on the velocity of sound in Propositions XLVII–XL of Book II. The best that I have been able to find (which is not saying very much in the context) is that of Rayleigh's:*

> The first theoretical investigation of the velocity of sound was made by Newton, who assumed that the relation between pressure and density was that formulated in Boyle's law. If we assume $p = \kappa\rho$, we see that the velocity of sound is expressed by $\sqrt{\kappa}$, or $\sqrt{p} \div \sqrt{\rho}$, in which the dimensions of p (=force ÷ area) are $[M]\ [L]^{-1}\ [T]^{-2}$, and those of ρ (=mass ÷ volume) are $[M]\ [L]^{-3}$. Newton expressed the result in terms of the 'height of the homogeneous atmosphere', defined by the equation,
>
> $$g\rho h = p,$$
>
> where p and ρ refer to the pressure and the density at the earth's surface. The velocity of sound is thus $\sqrt{(gh)}$, or the velocity which would be acquired by a body falling freely under the action of gravity through half the height of the homogeneous atmosphere. [The velocity, 279·945 metres per second, derived on this assumption falls] short of the result of direct observations by about a sixth part.

* *The theory of sound*, Vols. I & II (first edition, MacMillan Company, 1877); reprinted by Dover Publications, New York, 1945: §246 (pp. 18–19) Vol. II. In contrast to Rayleigh's account, that by the editor of the Dover reprint (pp. xix–xx) is written in manifest ignorance of Newton's propositions and Laplace's account in *Mecanique celeste*, Livre XII., pp. 119, *et seq.*

> Newton's investigation established that the velocity of sound should be independent of the amplitude of the vibration, and also of the pitch, but the discrepancy between his calculated value (published in 1687) and the experimental value was not explained until Laplace (1816) pointed out that the use of Boyle's law involved the assumption that in the condensations and rarefactions accompanying sound, the temperature remains constant, in contradiction to the known fact that, when air is suddenly compressed, its temperature rises....
>
> Laplace considered instead that the condensations and rarefactions, concerned in the propagation of sound, take place with such rapidity that the heat and cold produced have not time to pass away, and that therefore the relation between volume and pressure is sensibly the same as if the air were confined in an absolutely non-conducting vessel.

In other words, the changes in pressure, δp, and density, $\delta \rho$, take place (as one would now say) adiabatically, when $\delta p/p = \gamma\, \delta\rho/\rho$, where γ denotes the ratio of the specific heats, with the result that Newton's value for the velocity of sound should be multiplied by a factor $\sqrt{\gamma}$.

Where, in the foregoing, does one sense cause for Laplace's exhilaration?

151. Propositions XLI–XLIII: Newton's conception of wave propagation

In Propositions XLI–XLIII, Newton develops, in part, his conception of wave propagation as a basis for his explanation of (what was till then totally obscure) the transmission of sound through air. But some other essential ingredients of his conception are explained only later, in the context of analytical needs, in Propositions XLVII–XL. We shall, therefore, not attempt a detailed analysis of these preliminary propositions but provide instead a general account, following a lucid presentation by Sir George B. Airy.*

The transmission of sound in air is based on the notion of propagation of waves in which the movements of the different elements of air are of limited extents while the relative movements of neighbouring elements are propagated over long distances without much attenuation (see the adjoining figure).

Now, if the points along the lines (β), (γ), ..., (ζ) represent the positions of elements of air at successive equal intervals of time, T, $T + \frac{1}{4}\tau$, ..., $T + \tau$, it is manifest that we have states of condensations and states of rarefactions travelling progressively in one direction. This is the conception of a wave, depending on relatively slight oscillatory motions of elements of air travelling over relatively long distances in one direction. Newton considers waves of this kind as at the base of the transmission of sound.

* *On sound and atmospheric vibrations with the mathematical elements of music*, MacMillan, London and Cambridge, 1868, pp. 22–27.

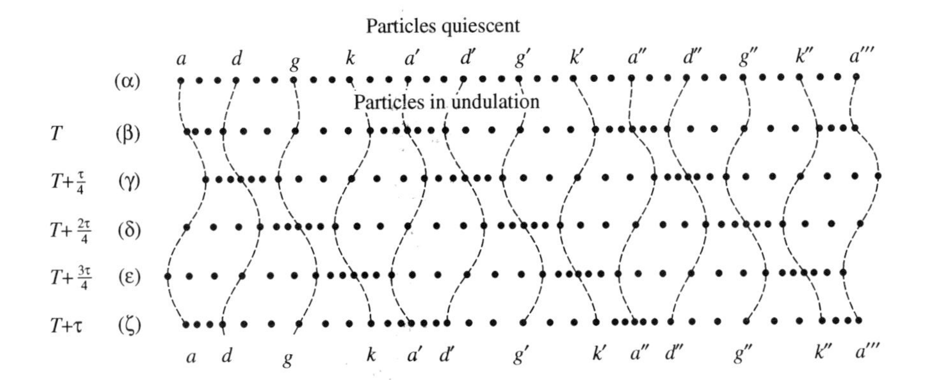

1. In the topmost line (α), the elements of air, represented by points a, b, c, d, etc., are placed at equal distances in air that is still (quiescent).
2. In the next line (β), the elements are assumed (arbitrarily) to be distributed, as shown, condensed as at a, a' etc., and expanded as at g, g', etc.
3. In the third line (γ), the points of condensation occur at about d, d', etc., while points of rarefaction occur at k, k', etc.
4. In the fourth line (δ), the states of condensation and of rarefaction have travelled a little further along in the same direction; and so on in the successive lines (ε) and (ζ).

The basic idea in this conception of a wave is that of a *state of displacement* travelling continuously in one direction, while the motions of the individual elements are small and oscillatory. And the major step that Newton took beyond formulating this conception is to demonstrate (as he does in Propositions XLVII and XLIX) how the condensations and rarefactions produced by such oscillatory motions can be maintained consistently with the elasticity properties of air. Simple as these notions are, it is perhaps not possible for us to realize the novelty and the boldness of the conception and the difficulty in formulating analytically the underlying ideas.

152. Propositions XLIV–XLVI: the propagation of long waves in canals

In Proposition XLIV, Newton considers an extremely simple example of oscillation of liquid columns in a U-shaped canal; and by a stroke of intuition extends it to solve the problem of long waves in canals. These propositions are explained in so exemplary a fashion that one cannot do better than quote him verbatim. (The crucial arguments are underlined.)

Proposition XLIV. Theorem XXXV

If water ascend and descend alternately in the erected legs KL, MN of a canal or pipe; and a pendulum be constructed whose length between the point of suspension and the centre of oscillation is equal to half the length of the water in the canal: I say, that the water will ascend and descend in the same times in which the pendulum oscillates.

I measure the length of the water along the axes of the canal and its legs, and make it equal to the sum of those axes; and take no notice of the resistance of the water arising from its attrition by the sides of the canal. Let, therefore, *AB*, *CD* represent the mean height of the water in both legs; and when the water in the leg *KL* ascends to the height *EF*, the water will descend in the leg *MN* to the height *GH*. Let *P* be a pendulous body, *VP* the thread, *V* the point of

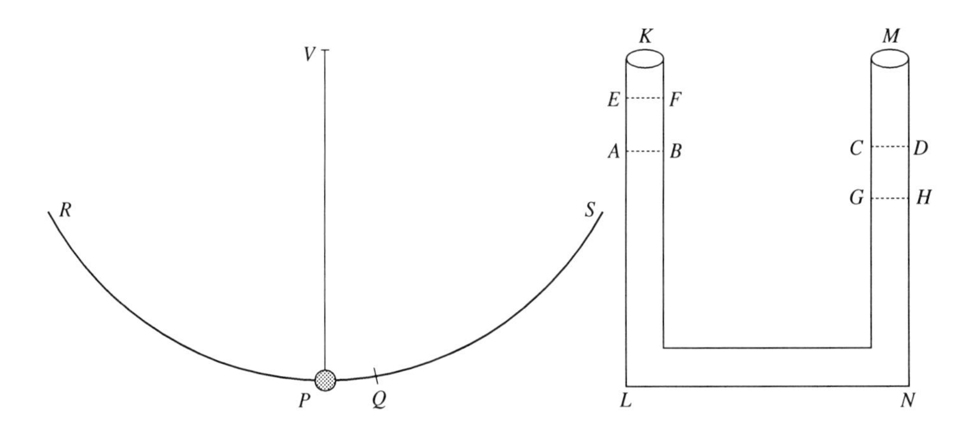

suspension, *RPQS* the cycloid which the pendulum describes, *P* its lowest point, *PQ* an arc equal to the height *AE*. The force with which the motion of the water is accelerated and retarded alternately is the excess of the weight of the water in one leg above the weight in the other; and, therefore, when the water in the leg *KL* ascends to *EF*, and in the other leg descends to *GH*, that force is double the weight of the water *EABF*, and therefore is to the weight of the whole water as *AE* or *PQ* to *VP* or *PR*. The force also with which the body *P* is accelerated or retarded in any place, as *Q*, of a cycloid, is (by Cor., Prop. LI, Book I) to its whole weight as its distance *PQ* from the lowest place *P* to the length *PR* of the cycloid. Therefore the motive forces of the water and pendulum, describing the equal spaces *AE*, *PQ*, are as the weights to be moved; and therefore if the water and pendulum are quiescent at first, those forces will move them in equal times, and will cause them to go and return together with a reciprocal motion. Q.E.D.

COR. I. Therefore the reciprocations of the water in ascending and descending are all performed in equal times, whether the motion be more or less intense or remiss.

COR. II. If the length of the whole water in the canal be of $6\frac{1}{9}$ feet of French measure, the water will descend in one second of time, and will ascend in another second, and so on by turns *in infinitum*; for a pendulum of $3\frac{1}{18}$ such feet in length will oscillate in one second of time.

COR. III. But if the length of the water be increased or diminished, the time of the reciprocation will be increased or diminished as the square root of the length.

Proposition XLV. Theorem XXXVI

The velocity of waves varies as the square root of the breadths.

This follows from the construction of the following proposition.

Proposition XLVI. Problem X

To find the velocity of waves.

Let a pendulum be constructed, whose length between the point of suspension and the centre of oscillation is equal to the breadth of the waves, and in the time that the pendulum will perform one single oscillation the waves will advance forwards nearly a space equal to their breadth.

That which I call the breadth of waves is the transverse measure lying between the deepest part of the hollows, or the tops of the ridges. Let *ABCDEF* represent the surface of stagnant water ascending and descending in successive

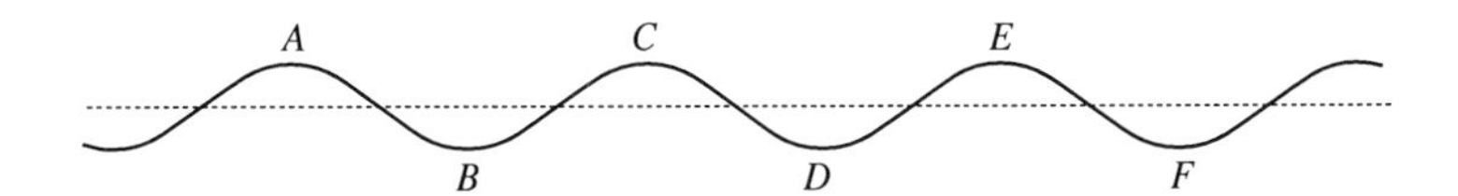

waves; and let *A*, *C*, *E*, etc., be the tops of the waves; and let *B*, *D*, *F*, etc., be the intermediate hollows. Because the motion of the waves is carried on by the successive ascent and descent of the water, so that the parts thereof, as *A*, *C*, *E*, etc., which are highest at one time, become lowest immediately after; and because the motive force, by which the highest parts descend and the lowest ascend, is the weight of the elevated water, that alternate ascent and descent will be analogous to the reciprocal motion of the water in the canal, and will observe the same laws as to the times of ascent and descent; and therefore (by Prop. XLIV) if the distances between the highest places of the waves *A*, *C*, *E* and the lowest *B*, *D*, *F* be equal to twice the length of any pendulum, the highest parts *A*, *C*, *E* will become the lowest in the time of one oscillation, and in the time of another oscillation will ascend again. Therefore with the passage of each wave, the time of two oscillations will occur; that is, the wave will describe its breadth in the time that pendulum will oscillate twice; but a pendulum of four times that length, and therefore equal to the breadth of the waves, will just oscillate once in that time. Q.E.I.

COR. I. Therefore, waves, whose breadth is equal to $3\frac{1}{18}$ French feet, will advance through a space equal to their breadth in one second of time; and therefore in one minute will go over a space of $183\frac{1}{3}$ feet; and in an hour a space of 11000 feet, nearly.

COR. II. And the velocity of greater or less waves will be augmented or diminished as the square root of their breadth.

These things are true upon the supposition that the parts of water ascend or descend in a straight line; but, in truth, that ascent and descent is rather performed in a circle; and therefore I give the time defined by this proposition as only approximate.

The theory of long waves in canals

It is clear that in Proposition XLVI, Newton, in effect, considers the propagation of long waves in canals, that is, of waves of infinitesimal height (or, amplitude) travelling along a straight canal with a horizontal bed and vertical sides.

Let the x-axis be taken along the length of the canal and the y-axis in the vertical upward direction. We shall suppose that the motion takes place in the two dimensions x and y. Let the ordinate of the free surface at x and at time t be

$$y_0 + \eta \quad \text{where } y_0 \text{ specifies the undisturbed surface.} \tag{1}$$

We shall suppose that the pressure at any point (x, y) is sensibly the same as the static pressure at that depth, namely,

$$p - p_0 = g\rho(y_0 + \eta - y), \tag{2}$$

where p_0 denotes the uniform pressure above the free surface ($y \geqslant y_0$) and g denotes the value of gravity. Under these assumptions, the equation of hydrostatic equilibrium is

$$\frac{\partial p}{\partial x} = g\rho \frac{\partial \eta}{\partial x}. \tag{3}$$

It follows that the horizontal velocity u is a function of x and y only since the horizontal component of the acceleration is the same in a plane perpendicular to the x-direction.

The equation governing the horizontal motion is

$$\frac{\partial u}{\partial t} + u\frac{\partial u}{\partial x} = -\frac{1}{\rho}\frac{\partial p}{\partial x}. \tag{4}$$

If the velocities, as we shall suppose, are sufficiently small, we can neglect the non-linear inertial term, and we can write

$$\frac{\partial u}{\partial t} = -\frac{1}{\rho}\frac{\partial p}{\partial x} = -g\frac{\partial \eta}{\partial x}. \tag{5}$$

It is convenient to introduce the Lagrangian displacement, $\xi(x, t)$, defined by

$$u(x, t) = \frac{\partial}{\partial t}\,\xi(x, t). \tag{6}$$

The equation of continuity can be obtained by ascertaining the volume of the fluid which has, up to the time t, entered the space bounded by the planes x and $x + \delta x$; thus,

$$-\delta x\,\frac{\partial}{\partial x}\,(\xi h b) = \eta b \delta x \tag{7}$$

where h and b denote the height and the breadth of the canal. Therefore,

$$\eta = -h\,\frac{\partial \xi}{\partial x}, \tag{8}$$

or,

$$\frac{\partial \eta}{\partial t} = -h\,\frac{\partial u}{\partial x} \quad \text{by equation (6).} \tag{9}$$

Equations (5) and (9) together give,

$$\frac{\partial^2 \eta}{\partial t^2} = gh\,\frac{\partial^2 \eta}{\partial x^2}. \tag{10}$$

Defining the velocity,

$$c = \sqrt{(gh)}, \tag{11}$$

equation (10) takes the standard form,

$$\frac{\partial^2 \eta}{\partial t^2} = c^2\,\frac{\partial^2 \eta}{\partial x^2}. \tag{12}$$

As is well known, the general solution of this equation is

$$\eta = f(x - ct) + k(x + ct), \tag{13}$$

where f and k are arbitrary functions of the arguments specified. In particular, a solution, representing a progressive wave of frequency ω and wavelength λ is given by

$$\eta = \eta_0 \cos\left(\omega t - \frac{2\pi}{\lambda}\,x\right), \tag{14}$$

where

$$\omega^2 = \frac{4\pi^2}{\lambda^2}\,c^2 = \frac{4\pi^2}{\lambda^2}\,gh \quad \text{by equation (11).} \tag{15}$$

Therefore,

$$\text{period of oscillation } (=2\pi/\omega) \times \text{velocity } (=\sqrt{(hg)}) = \text{wavelength } (=\lambda). \tag{16}$$

This is the relation that Newton states at the conclusion of Proposition XLVI (without any calculation!). His concluding remark: ‘the time defined by the proposition is only approximate’, refers to the strictly infinitesimal amplitudes to which Proposition XLVI applies in contrast to Proposition XLV which is *exact* by the comparison with a cycloidal pendulum.

153. A standard treatment of the velocity of sound

In Proposition XLVII–L of Book II, Newton proceeds to describe and develop his method of determining the velocity of sound. But unlike in Propositions XLIV–XLVI, his explanations are not easy to follow since he attempts to explain, at the same time, his underlying conception of wave propagation *and* the underpinnings of the analytical framework. (Mozart might have had similar difficulties had he not on hand, ready made, an adequate musical notation.) We shall, therefore, preface Newton's treatment of the problem by a more conventional treatment closest to his.

We shall adopt, what we now call, a Lagrangian description, Newton having in effect discovered it long before it became to be so called. Let $\xi(x, t)$ denote then the 'Lagrangian displacement'! Then, by definition, the substratum, originally confined in the interval $(x, x + \delta x)$ at time t will, at a later time $t + \delta t$, find itself in the interval

$$x + \xi \qquad \text{and} \qquad x + \xi + \left(1 + \frac{\partial \xi}{\partial x}\right)\delta x. \tag{1}$$

Therefore, by the equation of continuity,

$$\rho\left(1 + \frac{\partial \xi}{\partial x}\right) = \rho_0 \tag{2}$$

where ρ_0 denotes the undisturbed density and ρ is the density at time t. Corresponding to the *expansion*, $(1 + \partial\xi/\partial x)$, we have the density *condensation*,

$$S = \frac{\rho - \rho_0}{\rho_0} = -\frac{\partial\xi/\partial x}{1 + \partial\xi/\partial x} \simeq -\frac{\partial \xi}{\partial x}, \tag{3}$$

to the first order. Also, by definition, the velocity is

$$u = \frac{\partial \xi}{\partial t}. \tag{4}$$

Now, the equation of motion, ignoring the inertial and similar non-linear terms, is

$$\rho_0 \frac{\partial u}{\partial t} = -\frac{\partial p}{\partial x} = -\left(\frac{\partial p}{\partial \rho}\right)_0 \frac{\partial \rho}{\partial x} = -\left(\rho\frac{\partial p}{\partial \rho}\right)_0 \frac{\partial S}{\partial x}. \tag{5}$$

Therefore,

$$\frac{\partial u}{\partial t} = -\frac{\kappa}{\rho_0}\frac{\partial S}{\partial x} = +\frac{\kappa}{\rho_0}\frac{\partial^2 \xi}{\partial x^2} \quad \text{(by equation (3))}, \tag{6}$$

where

$$\kappa = \left(\rho\frac{\partial p}{\partial \rho}\right)_0 \tag{7}$$

is the *elasticity coefficient*. (In the foregoing, the subscripts 0 distinguish the equilibrium values.) Combining equations (4) and (6) we obtain,

$$\frac{\partial^2 \xi}{\partial t^2} = c^2 \frac{\partial^2 \xi}{\partial x^2}, \tag{8}$$

where

$$c^2 = \frac{\kappa}{\rho_0} = \left(\frac{\partial p}{\partial \rho}\right)_0 \tag{9}$$

has the dimensions of a square of the velocity.

We observe that equation (8) is of the same form as the equation we encountered in the theory of long waves in canals (equation (12), §152). The general solution is accordingly of the same form (equation (13), §152); in particular, the solution for a progressive sound wave of frequency ω and wavelength λ is given by

$$\xi = \xi_0 \cos\left(\omega t - \frac{2\pi}{\lambda} x\right) \quad (\xi_0 = \text{a constant}). \tag{10}$$

Its phase velocity (or more simply, the velocity of propagation) is

$$c = \sqrt{\frac{\kappa}{\rho_0}} = \omega . \frac{\lambda}{2\pi}, \tag{11}$$

which is the same as the relation (16) of §152.

154. Propositions XLVII–L

For the same reasons explained at the beginning of §153, we shall refrain from analysing these propositions in detail; but we shall present the same arguments in a different order and in a (superficially) different language; and relate them at various points to what Newton states explicitly.

The basic underlying notion is that of a 'pulse' by which Newton means an interval a wavelength long in a wave train in which the amplitude is (effectively) assumed to vary like, say,

$$\cos\left(\omega t - \frac{2\pi}{\lambda} x\right) \quad (0 \leqslant x \leqslant x + \lambda); \tag{1}$$

for example, the segments between a and a' in the different lines $(\beta), \ldots, (\zeta)$, in the illustration on p. 581 in §151. We observe that the pulse is no more than the Lagrangian displacement defined in equation (10), §153. But Newton now requires to prove that the posited assumption,

$$\xi = \xi_0 \cos\left(\omega t - \frac{2\pi}{\lambda} x\right) \quad (\xi_0 = \text{a constant}), \tag{2}$$

follows consistently from the known elasticity properties of air, as he states in Proposition XLVII:

Proposition XLVII. Theorem XXXVII

If pulses are propagated through a fluid, the several particles of the fluid, going and returning with the shortest reciprocal motion, are always accelerated or retarded according to the law of the oscillating pendulum.

The proof goes as follows.

From equation (2) it follows that the condensation, S, is given by

$$S = -\frac{\partial \xi}{\partial x} = -\xi_0 \frac{2\pi}{\lambda} \sin\left(\omega t - \frac{2\pi}{\lambda} x\right). \tag{3}$$

The resulting elasticity force per unit mass is

$$-\frac{\kappa}{\rho_0}\frac{\partial S}{\partial x} = -\xi_0 \frac{\kappa}{\rho_0}\frac{4\pi^2}{\lambda^2} \cos\left(\omega t - \frac{2\pi}{\lambda} x\right) = -\frac{\kappa}{\rho_0}\frac{4\pi^2}{\lambda^2}\xi, \tag{4}$$

where

$$\kappa = \left(\rho \frac{\partial p}{\partial \rho}\right)_0 = \text{the elasticity coefficient.} \tag{5}$$

The restoring force being thus proportional to the distance, the particles of the fluid are 'always accelerated or retarded according to the law of the oscillating pendulum'.

Q.E.D. (!)

It also follows from equations (3) and (4) that

$$\frac{\partial^2 \xi}{\partial t^2} = -\omega^2 \xi \tag{6}$$

and

$$\omega^2 = \frac{\kappa}{\rho_0}\frac{4\pi^2}{\lambda^2}. \tag{7}$$

In terms of the velocity,

$$c = \sqrt{\frac{\kappa}{\rho_0}} = \sqrt{\left(\frac{\partial p}{\partial \rho}\right)_0}, \tag{8}$$

we can rewrite equation (7) in the form,

$$\omega = c.\frac{2\pi}{\lambda}, \tag{9}$$

or, in words,

$$\text{period of oscillation } (=2\pi/\omega) \times \text{velocity of propagation } (=c[=\sqrt{(\kappa/\rho_0)}]) = \text{wavelength } (=\lambda). \tag{10}$$

This is the first time that this relation between the period, the wavelength, and the velocity of propagation was established. That this same relation obtains for light as well is noted in the opening sentence of the concluding Scholium:

> The last Propositions [XLVIII–L which state the same relation in different forms] respect the motions of light and sounds.

All of the foregoing relations are included in the demonstration of Proposition XLVIII; but Newton selects equation (8) for its statement:

Proposition XLVIII. Theorem XXXVIII

> *The velocities of pulses propagated in an elastic fluid are in a ratio compounded of the square root of the ratio of the elastic force directly, and the square root of the ratio of the density inversely; supposing the elastic force of the fluid to be proportional to its condensation.*

In the last phrase, Newton reiterates the assumption,

$$\text{elastic force} = -\frac{\kappa}{\rho_0}\frac{\partial S}{\partial x}, \tag{11}$$

that is at the base of the derivation of equation (4).

Newton appears to be dissatisfied with his demonstration of Proposition XLVII; for he writes:

> This proposition will be made clearer from the constructions of the following [proposition].

Proposition XLIX. Problem XI

> *The density and elastic force of a medium being given, to find the velocity of the pulses.*

This proposition refers to the relation (8) and essentially the same arguments, as in Proposition XLVII, are advanced, though with some variations. Thus, the principal result (namely, equation (8)) is now expressed in the statement:

> Therefore the time $(=\lambda/c)$ in which a pulse runs over the space BC $(=\lambda)$ is to the time of one oscillation $(=1/\omega)$ composed of the going and returning of the pendulum as V is to A $(=2\pi)$,

that is, $(\lambda/c).\omega = 2\pi$, which is a repetition of what was stated earlier in the demonstration of Proposition XLVIII:

> The pulses advance a space equal to their breadths $(=\lambda)$ in the times $(=\lambda/c)$ of going once and returning once $(=2\pi/\omega)$; that is, they go over spaces proportional to the times,

that is, $\lambda/c = 2\pi/\omega$. And the entire Proposition L is towards the same end.

Proposition L. Problem XII

To find the distances of the pulses.

Let the number of the vibrations of the body, by whose tremor the pulses are produced, be found to any given time. By that number divide the space which a pulse can go over in the same time, and the part found will be the breadth of one pulse. Q.E.I.

For, if N denotes the number of vibrations in a time T and L is the distance the pulse travels during this time, then

$$N = T(\omega/2\pi) \qquad \text{and} \qquad L = Tc \tag{12}$$

and

$$\frac{N}{L} = \frac{\omega}{2\pi c} = \frac{1}{\lambda}. \tag{13}$$

Q.E.I.

From these variations on the same theme, it would appear that Newton was experiencing the difficulty (as we have said earlier) that Mozart might have, had he had no musical notation in which to express himself. Perhaps Laplace had the same reaction when he wrote:

> The way at which he [Newton] arrives at this [formula] is one of the most remarkable traits of his genius.

After studying these propositions on the velocity of sound in air, the most important fact that emerges, beyond the sentiment expressed by Laplace, is that all the propositions are couched in terms of the coefficient of elasticity (as the factor κ/ρ_0 multiplying $-\partial S/\partial x$—explicitly, for example, in the enunciation of Proposition XLVIII) and *unspecified*, otherwise. Occasionally, it appears that for purposes of clarification Newton specializes the coefficient of elasticity to p/ρ $(=gh=$the height of the homogeneous atmosphere$)$ that follows from Boyle's law. Clearly, to replace p/ρ by $\gamma p/\rho$ is obvious, once one had become aware of the relevant experiments of Gay-Lussac (and of Desormes and Clément (1819) and of Gay-Lussac and Welter*). This is stated, not to detract from Laplace's sensitiveness in recognizing for the first time the need for the required replacement, but rather to emphasize that all that Laplace had to do was to replace 1 by $\sqrt{\gamma}$, taking Newton's theory intact. Indeed, that is what Laplace himself says in his first announcement of his discovery in 1816.

> The speed of sound is equal to the product of the speed given by the Newtonian formula by the square root of the ratio of the specific heat of air at constant pressure to its specific heat at constant volume.

* Be it noted that all of the authors quoted were professional friends of Laplace.

155. Scholium

After having been in the realm of abstract thoughts, contemplating on the workings of nature, Newton turns to confronting his formula,

$$c = \sqrt{\left(\frac{\partial p}{\partial \rho}\right)_0} \tag{1}$$

for the velocity of sound, with what one measures for air. As we have seen in §150, if one assumes that the changes in pressure and density take place in accordance with Boyle's law—Newton had in fact no other choice—then the velocity of sound in air at a given temperature, T, is given by

$$c = \sqrt{\frac{p}{\rho}} = \sqrt{(gh)}, \tag{2}$$

where h is the height of the homogeneous atmosphere at the same temperature. If, following Rayleigh's calculation in §150, we consider air at 0° Celsius temperature and barometric pressure 760 mm of mercury, then

$$p = 760 \times 13{\cdot}5953\ (=\text{density of Hg}) \times 980{\cdot}939\ (=g), \qquad \rho = 0{\cdot}001293, \tag{3}$$

and

$$c = 279{\cdot}95 \quad \text{meters per second} \tag{4}$$

which 'falls short of direct observations by about a sixth part'. Let us now see how Newton does his first comparison with observations.

> Since they [the sounds] arise from tremulous bodies, they can be nothing else but pulses of the air propagated through it (by Prop. XLIII); and this is confirmed by the tremors which sounds, if they be loud and deep, excite in the bodies near them, as we experience in the sounds of drums; for quick and short tremors are less easily excited. But it is well known that any sounds, falling upon strings in unison with the sonorous bodies, excite tremors in those strings. This is also confirmed from the velocity of sounds; for since the specific gravities of rain water and quicksilver are to one another as about 1 to $13\frac{2}{3}$, and when the mercury in the barometer is at the height of 30 inches of our measure, the specific gravities of the air and of rain water are to one another as about 1 to 870, therefore the specific gravities of air and quicksilver are to each other as 1 to 11890. Therefore when the height of the quicksilver is at 30 inches, a height of uniform air, whose weight would be sufficient to compress our air to the density we find it to be of, must be equal to 356700 inches, or 29725 feet of our measure; and this is that very height of the medium, which I have called A in the construction of the foregoing proposition. A circle whose radius is 29725 feet is 186768 feet in circumference. And since a pendulum $39\frac{1}{5}$ inches in length completes one oscillation, composed of its going and return, in two seconds of

> time, as is commonly known, it follows that a pendulum 29725 feet, or 356700 inches in length will perform a like oscillation in $190\frac{3}{4}$ seconds. Therefore in that time a sound will go right onwards 186768 feet, and therefore in one second 979 feet.

When Newton arrived at the value 968 feet per second at the time (*c.* 1686) of writing the first edition of the *Principia*—a value he later revised to 979 feet per second in his second 1713 edition of the *Principia*—the then extant determinations of the velocity of sound were disparate: 1474 feet per second by Mersenne and 600 feet per second by Roberval. Newton, therefore, set about measuring the velocity of sound himself by adjusting a pendulum to swing synchronously with the return of an echo in the portico of length 208 feet in Nevile's court (below the Wren Library) in Trinity College. He found that the echo was faster in return than the period ($28\frac{3}{4}$ min.) of a pendulum of 8 inches and longer than the period ($19\frac{1}{6}$ min.) of a pendulum of 5 inches. From these measurements, Newton concluded that the velocity of sound was in the range 920 and 1085 feet per second* bracketing his theoretical value. Newton was satisfied, thinking that all was well.

But the situation became greatly altered in 1713 when he came to revising the *Principia* for its second edition; and it remained the same in 1726 when the third edition was published. Newton's final assessment in the third edition is:

> Now by experiments it actually appears that sounds do really advance in one second of time about 1142 feet of English measure, or 1070 feet of French measure.
>
> The velocity of sounds being known, the intervals of the pulses are known also. For M. Sauveur, by some experiments that he made, found that an open pipe about five Paris feet in length gives a sound of the same tone with a viol string that vibrates a hundred times in one second. Therefore there are near 100 pulses in a space of 1070 Paris feet, which a sound runs over in a second of time; and therefore one pulse fills up a space of about $10\frac{7}{10}$ Paris feet, that is, about twice the length of the pipe. From this it is probable that the breadths of the pulses, in all sounds made in open pipes, are equal to twice the length of the pipes.

Newton must have been baffled, not to say disappointed. Search as he might, he could find no flaw in his theoretical framework—neither could Euler, Lagrange, and Laplace; nor, indeed, anyone down to the present. Newton correctly surmised that the source of the discrepancy must lie in an under-estimate of κ, the coefficient of elasticity. And he speculated on the effect of the admixture of water vapour and of the 'crassitude' of solid

* The details mentioned can be found in the first edition of the *Principia* (1686, p. 371).

particles; but he could reach no certain quantitative conclusion. He could not of course have foreseen the advances in the science of heat that were to come some hundred years later.

But undaunted he ends this section in a note of optimism; and one may add, the optimism was justified!

> Moreover, from the Corollary of Prop. XLVII* appears the reason why the sounds immediately cease with the motion of the sonorous body, and why they are heard no longer when we are at a great distance from the sonorous bodies than when we are very near them. And besides, from the foregoing principles, it plainly appears how it comes to pass that sounds are so mightily increased in speaking trumpets; for all reciprocal motion tends to be increased by the generating cause at each return. And in tubes hindering the dilatation of the sounds, the motion decays more slowly, and recurs more forcibly; and therefore is the more increased by the new motion impressed at each return. And these are the principal phenomena of sounds.

* The Corollary simply states that 'the number of pulses propagated is the same as the number of vibrations of the tremulous body'.

Epilogue

Ben Johnson said of Shakespeare in the First Folio:

He was not of an age, but for all time!

It could be equally said of Newton:

He was not of an age, but for all time!